Nonlinear Systems and Applications

An International Conference

Academic Press Rapid Manuscript Reproduction

Proceedings of an International Conference on Nonlinear Systems and Applications
Held at the University of Texas at Arlington, July 19–23, 1976

Nonlinear Systems and Applications

An International Conference

edited by

V. Lakshmikantham

Department of Mathematics
University of Texas at Arlington
Arlington, Texas

Academic Press, Inc.

NEW YORK SAN FRANCISCO LONDON 1977

A Subsidiary of Harcourt Brace Jovanovich, Publishers

ACADEMIC PRESS, INC.
111 Fifth Avenue, New York, New York 10003

United Kingdom Edition published by
ACADEMIC PRESS, INC. (LONDON) LTD.
24/28 Oval Road, London NW1

Library of Congress Cataloging in Publication Data

Main entry under title:

Nonlinear systems and applications.

Held at the University of Texas, Arlington, July 19-23, 1976.
1. Differential equations, Nonlinear–Congresses. 2. Nonlinear theories–Congresses. 3. Stability–Congresses. I. Lakshmikantham, V.
QA372.N66 516′.7 77-7190
ISBN 0–12–434150–0

PRINTED IN THE UNITED STATES OF AMERICA

Contents

*Indicates the author who presented the paper at the conference.

Contributed Papers

*Indicates the author who presented the paper at the conference.

*Indicates the author who presented the paper at the conference.

*Indicates the author who presented the paper at the conference.

*Indicates the author who presented the paper at the conference.

List of Contributors

AMES, W. F.,* Department of Mathematics, Georgia Institute of Technology, Atlanta, Georgia 30332.

ANDERSON, D. H., Department of Mathematics, Southern Methodist University, Dallas, Texas.

ANTOSIEWICZ, H. A.,* Department of Mathematics, University of Southern California, Los Angeles, California 90007.

ARONSON, D. G., School of Mathematics, University of Minnesota, Minneapolis, Minnesota.

BABU, A. J. G., Department of Mathematics, University of Alabama, University, Alabama.

BAJAJ, P. N., Department of Mathematics, Wichita State University, Wichita, Kansas.

BANKS, H. T.,* Division of Applied Mathematics, Brown University, Providence, Rhode Island 02912.

BARRETT, T. W., Department of Physiology and Biophysics, University of Tennessee Center for the Health Sciences, Memphis, Tennessee.

BAXTER, J. E., Department of Biochemistry, University of Tennessee Center for the Health Sciences, Memphis, Tennessee.

BELLENI-MORANTE, A., Istituto di Matematica Applicata Universita di Firenze, Firenze, Italy.

BERKOWITZ, D. P., Department of Mathematics, Pomona College, Claremont, California.

BERNFELD, S. R., Department of Mathematics, University of Texas at Arlington, Arlington, Texas.

BOWMAN, T. T., Center for Applied Mathematics, University of Florida, Gainesville, Florida.

BROCKETT, R. W.,* Department of Electrical Engineering, Harvard University, Cambridge, Massachusetts 02138

BURTON, T. A., Department of Mathematics, Southern Illinois University, Carbondale, Illinois.

BUSONI, G., Istituto di Matematica Applicata, Universita di Firenze, Firenze, Italy.

CALEF, D. F., Department of Mathematics, Pomona College, Claremont, California.

CAPASSO, V., Istituto di Analisi Matematica, Universita di Bari, Bari, Italy.

CESARI, L.,* Department of Mathematics, University of Michigan, Ann Arbor, Michigan 48104.

CHANCE, J. E., Department of Mathematics, Pan American University, Edinburg, Texas.

CHANDRA, J.,* U.S. Army Research Office, Box CM, Duke Station, Durham, North Carolina 27706.

CHOUDHURY, A. K., Department of Mathematics, Howard University, Washington, D.C.

CHUKWU, E. N., Department of Mathematics, Cleveland State University, Cleveland, Ohio.

*Indicates contributor of an invited address or research report.

COOKE, K. L.,* Department of Mathematics, Pomona College, Claremont, California 91711.

COUGHLIN, P., Department of Economics, Middlebury College, Middlebury, Vermont.

CUSHING, J. M., Department of Mathematics, University of Arizona, Tucson, Arizona.

DANIEL, P. L., Department of Medical Computer Science, University of Texas Health Science Center, Dallas, Texas.

DAVIS, P. WM., Department of Mathematics, Worcester Polytechnic Institute, Worcester, Massachusetts.

DIETZ, K., WHO, World Health Organization, 1211-Geneva 27, Switzerland.

DRAGAN, V., Institute Mat. Academy RPR, Str. Mihail Eminescu 47, Bucuresti 3, Rumania.

DRANE, J. W., Department of Statistics, Southern Methodist University, Dallas, Texas.

EISENFELD, J., Department of Mathematics, University of Texas at Arlington, Arlington, Texas, and Department of Medical Computer Science, University of Texas Health Science Center, Dallas, Texas.

ELDERKIN, R. H., Department of Mathematics, Pomona College, Claremont, California.

FARRIS, F. A., Department of Mathematics, Pomona College, Claremont, California.

FITZGIBBON, W. E., Department of Mathematics, University of Houston, Houston, Texas.

FLEISHMAN, B. A.,* Department of Mathematical Sciences, Rensselaer Polytechnic Institute, Troy, New York 12181.

FRANCIS, A. A., Department of Economics, University of the West Indies, Kingston, Jamaica.

GILBERT, R. P.,* Department of Mathematics, University of Delaware, Newark, Delaware 17911.

GOEL, N. S.,* School of Advanced Technology, State University of New York, Binghamton, New York 13901.

GROSSMAN, Z., CERN, European Organization for Nuclear Research, 1211-Geneva 23, Switzerland.

GUMOWSKI, I., CERN, European Organization for Nuclear Research, 1211-Geneva 23, Switzerland.

GUNN, C. F. Department of Mathematics, Pomona College, Claremont, California.

GUPTA, C. P., Department of Mathematics, Northern Illinois University, DeKalb, Illinois.

HADDOCK, J. R., Department of Mathematics, Memphis State University, Memphis, Tennessee.

HALANAY, A.,* Institute Mat. Academy RPR, Str. Mihail Eminescu 47, Bucuresti 3, Rumania.

HALLAM, T. G.,* Department of Mathematics, Florida State University, Tallahassee, Florida, 32306.

HARRIS, W. A., JR.,* Department of Mathematics, University of Southern California, Los Angeles, California 90007.

HERRON, I. H., Department of Mathematics, Howard University, Washington, D.C.

HICKERNELL, F. J., Department of Mathematics, Pomona College, Claremont, California.

HSIA, W. S., Department of Mathematics, University of Alabama, Montgomery, Alabama.

JONES, G. D., Department of Mathematics, Murray State University, Murray, Kentucky.

JUDY, M. M., Department of Biophysics, University of Texas Health Science Center, Dallas, Texas.

KALABA, R.,* Department of Economics, University of Southern California, Los Angeles, California 90007.

KANNAN, R.,* Department of Mathematical Sciences, University of Missouri at St. Louis, 8001 Natural Bridge Rd., St. Louis, Missouri 63121.

KAPLAN, J. L.,* Department of Mathematics, Boston University, Boston, Massachusetts 02215.

KASS, S. N., Department of Mathematics, Pomona College, Claremont, California.

LADDE, G. S.,* Department of Mathematics, State University of New York, Potsdam, New York 13676.

LAKSHMIKANTHAM, V.,* Department of Mathematics, University of Texas at Arlington, Arlington, Texas 76019.

*Indicates contributor of an invited address or research report.

LEELA, S.,* Department of Mathematics, State University of New York at Genesee, Genesee, New York 14454.

LEVEL, E. V., Department of Mathematics, Pomona College, Claremont, California.

LIGHTBOURNE, J. H., III, Department of Mathematics, Pan American University, Edinburg, Texas.

MCCALLA, C., Department of Mathematics, Howard University, Washington, D.C.

MAHAR, T. J., Courant Institute of Mathematical Sciences, New York, New York.

MANSFIELD, F. I., Department of Mathematics, Pomona College, Claremont, California.

MAROTTO, F. R., Department of Mathematics, Boston University, Boston, Massachusetts.

MERRILL, S. J., Division of Mathematical Sciences, The University of Iowa, Iowa City, Iowa.

MISHELEVICH, D. J., Department of Medical Computer Science, University of Texas Health Science Center, Dallas, Texas.

MITCHELL, R. W., Department of Mathematics, University of Texas at Arlington, Arlington, Texas.

MIZE, C. E., Department of Pediatrics, University of Texas Health Science Center, Dallas, Texas.

NUSSBAUM, R. D.,* Department of Mathematics, Rutgers University, New Brunswick, New Jersey 08903.

OǦUZTÖRELI, M. N.,* Department of Mathematics, The University of Alberta, Edmonton, Alberta, Canada T6G 2G1.

PACE, D. A.,* Department of Mathematics, University of Texas at Arlington, Arlington, Texas.

PLANT, R. E., Department of Mathematics, University of California, Davis, California.

POPOV, V. M.,* Department of Mathematics, University of Florida, Gainesville, Florida 32611.

RANKIN, S. M., III, Department of Mathematics, Murray State University, Murray, Kentucky.

REICH, S.,* Department of Mathematics, University of Chicago, 5734 University Avenue, Chicago, Illinois 60637.

REISCH, J. S., Department of Medical Computer Science, University of Texas Health Science Center, Dallas, Texas.

SAFFER, S. I., Department of Medical Computer Science, University of Texas Health Science Center, Dallas, Texas.

SENGER, R. L., Department of Medical Computer Science, University of Texas Health Science Center, Dallas, Texas.

SILJAK, D. D.,* Department of Electrical Engineering, University of Santa Clara, Santa Clara, California 95053.

SINGH, K. L., Department of Mathematics, Texas A & M University, College Station, Texas.

SPINGARN, K., Hughes Aircraft Company, Los Angeles, California.

STEIN, R. B., Department of Mathematics, University of Alberta, Edmonton, Alberta, Canada.

STERNBERG, R. L.,* Department of the Navy, Office of Naval Research, 495 Summer Street, Boston, Massachusetts 02210.

TARANTO, R. G., Department of Mathematics, Pomona College, Claremont, California.

THAMES, H. D., JR., Department of Biomathematics, University of Texas System Cancer Center, M.D. Anderson Hospital and Tumor Institute, Houston, Texas.

TSOKOS, C. P.,* Department of Mathematics, University of Southern Florida, Tampa, Florida 33620.

TSOKOS, J. O., Department of Chemistry, University of South Florida, Tampa, Florida.

WEBB, G. F.,* Department of Mathematics, Vanderbilt University, Memphis, Tennessee 37203.

WRIGHT, A. K., Department of Biochemistry, University of Tennessee Center for the Health Sciences, Memphis, Tennessee.

YOUNG, E. C., Department of Mathematics, Florida State University, Tallahassee, Florida.

ZAGUSTIN, E., Department of Civil Engineering, California State University, Long Beach, California.

*Indicates contributor of an invited address or research report.

Preface

An International Conference on Nonlinear Systems and Applications was held at the University of Texas, Arlington, July 19–23, 1976. The conference was sponsored by the U.S. Army Research Office, Durham, North Carolina; the University of Texas Health Science Center at Dallas; and the University of Texas at Arlington. It is a pleasure to acknowledge the financial support received from the sponsoring agencies, which made the conference possible.

The aim of the conference was to feature recent advances in nonlinear systems and applications and to stimulate discussions for new directions and for future research. The integration and cooperation among scientists across the traditional disciplinary barriers proved extremely rewarding and stimulating.

The present volume consists of the proceedings of this conference. It includes papers that were delivered as invited addresses and research reports as well as contributed papers. The three different but related areas of nonlinear analysis, namely, theory, methods, and applications, emerged as the theme of the conference, highlighted by the interaction of mathematics and the life sciences.

The theoretical contributions included abstract evolution equations and nonlinear semigroups; controllability, reachability, and degeneracy of dynamical systems and delay-differential systems; oscillation theory and boundary value problems for nonlinear differential equations; stability theory for differential and difference equations; and qualitative behavior of nonlinear equations with time lags.

There is a group of papers concerned with the various methods used in solving equations. These include the application of the maximum principle for nonlinear diffusion equations, differential inequalities and singular perturbation, Lyapunov functions and stability, analytic methods for solving elliptic equations, dynamic programming, monotone methods for obtaining approximate solutions of nonlinear boundary value problems, and approximation techniques for delay systems.

Most of the applications were in the area of the life sciences. They included such topics as neuromuscular systems, oscillation of cooperatively coupled enzymes, epidemic models, population dynamics and immune response models, identification problems, and compartmental analysis. There were also some applications in economics, including optimal harvesting and labor market systems.

I wish to express my special thanks to my colleagues and friends Bill Beeman, Steve Bernfeld, Jagdish Chandra (U.S. Army), Dorothy Chestnut, Jerome Eisenfeld, S. Leela (SUNY), Mike Lord, David Mishelevich, A. R. Mitchell, R. W. Mitchell, and Bennie Williams for assisting me in planning and organizing the conference; to my secretaries Mrs. Gloria Brown, Sandra Weber, and Virginia Becknell for their assistance during the conference; and to Mrs. Sandra Weber for excellent typing of the proceedings.

Contributed Papers

MONOTONICALLY CONVERGENT UPPER AND LOWER BOUNDS FOR CLASSES OF CONFLICTING POPULATIONS

W. F. Ames
Georgia Institute of Technology

1. *Introduction*

There exist many examples of populations consisting of elements which influence one another through cooperation and competition. Important examples include chemical reactions, businesses, societies, biological species etc. Many of the existing mathematical models for these interactions are discussed in Goel, Maitra and Montroll [1]. Among the models that will be of concern here are those of Volterra - Lotka - Verhulst

$$\frac{dN_i}{dt} = k_i N_i [1 - \frac{1}{2}(1 + sgn\ k_i) N_i \theta_i^{-1}] + \beta_i^{-1} \cdot N_i \sum_{j=1}^{n} a_{ij} N_j \ , \qquad (1)$$

where $N_i(t)$ is a population "size", θ_i is a saturation level and k_i, β_i and a_{ij} are population parameters; the generalized Gompertz model

$$\frac{dN_i}{dt} = k_i N_i \ G(N_i/\theta_i) + H(N_i)\{V_i(t) + \beta_i^{-1} \sum_{j=1}^{n} a_{ij} N_j\} \qquad (2)$$

where G is an arbitrary saturation inducing function such that $G(s) \to 0$ as $s \to 1$; the generalized war equations of Richardson -Waltman

$$\frac{dN_i}{dt} = N_i \ K_i(N_1, \ldots, N_n), \qquad (3)$$

where, in each of these cases, $i = 1,2,\ldots,n$. All of these models, like many of those which arise in the study of invariant solutions of partial differential equations (see e.g. Ames and Ginsberg [2]) have the common feature that all are special cases of the system

$$\frac{dy_i}{dt} = g_i(y_i) f_i(t, y_1, y_2, \ldots, y_n), \qquad (4)$$

$i = 1,2,\ldots,n,$ together with appropriate initial data.

Efficient computational methods for such systems should not only preserve the positivity of the solutions but should provide iteratively improvable upper and lower bounds. One manner in which this can be accomplished for a subset of system (4) is to construct antitone functional operators from (4) which are oscillatory contraction mappings. A preliminary example is presented in Section 2 followed by the general result in Section 3. A detailed chemical kinetics example is presented in Section 4 followed by some remarks on an application of the method of lines to parabolic partial differential equations.

2. *Introductory Example*

A simple birth and death (predator-prey) process is described by the dimensionless Volterra equations

$$x'(t) = ax(1 - y), \quad y'(t) = -cy(1 - x), \tag{5}$$

where $a, c > 0$ and $x(0)$, $y(0)$ are prescribed. Here it is expected that $x(t) \geq 0$, $y(t) \geq 0$. Of the several iterative forms that can be adopted for (5) the sequences

$$\begin{gathered} x_0(t) = 0; \quad y_n' = -cy_n(1 - x_n), \quad x_{n+1}' = ax_{n+1}(1 - y_n) \\ y_n(0) = y(0), \quad x_n(0) = x(0), \quad n \geq 0 \end{gathered} \tag{6}$$

are particularly attractive, as will be seen.

The integrated forms of (6) are easiest to analyze. For $n = 0$, since $x_0(t) = 0$,

$$y_0(t) = y(0)\, exp(-ct) \tag{7}$$

and

$$x_1(t) = x(0) exp\{at - ay(0)[1 - exp(-ct)]/c\} \tag{8}$$

can be obtained explicitly. The iteration for $n \geq 1$ is

$$y_n(t) = y(0)exp[-c(t - \int_0^t x_n(\tau)\,d\tau)] \tag{9a}$$

$$x_{n+1}(t) = x(0)exp[a(t - \int_0^t y_n(\tau)d\tau)] \tag{9b}$$

which maintains the positivity of x and y. The iteration can be summarized as

$$\begin{aligned} x_0(t) &= 0 \\ y_n(t) &= T_1 x_n \\ x_{n+1}(t) &= T_2 y_n. \end{aligned} \tag{10}$$

It is not difficult to show that this iteration is antitone by assuming that $\eta_{n-1}(t) < \eta_n(t)$, $\eta = x$ or y, and verifying that $\eta_n(t) > \eta_{n+1}(t)$ by induction. Thus the oscillating behavior,

$$0 \leq \eta_0 < \eta_2 < \eta_4 < \ldots < \eta_{2n} < \ldots < \eta < \ldots < \eta_{2n+1} < \ldots < \eta_5 < \eta_3 < \eta_1 , \tag{11}$$

which provides convergent upper and lower bounds for x and y, is obtained. The actual computation of (6) can be done in either differential or integral form.

The study of convergence and the development of error bounds is easily accomplished. In particular from (9b) and with elementary inequalities, on $0 < t < T$,

$$0 \leq x_{2n+1} - x_{2n} < L_1 \int_0^t (y_{2n-1} - y_{2n})d\tau, \tag{12}$$

with

$$L_1 = x(0)\ a\ exp[aT].$$

In a similar fashion, from (9a),

$$0 \le y_{2n-1} - y_{2n} < L_2 \int_0^t (x_{2n-1} - x_{2n-1})d\tau \tag{13}$$

with

$$L_2 = cy(0) \exp[c \int_0^T x_1 d\tau].$$

Combining the results of (12) and (13) and converting the two fold integral to a single integral there results

$$0 \le x_{2n+1} - x_{2n} < L_1 L_2 \int_0^t \int_0^t (x_{2n-1} - x_{2n-2})d\tau d\tau$$

$$= L_1 L_2 \int_0^t (t - \tau)(x_{2n-1} - x_{2n-2})d\tau$$

$$< M_1 T \int_0^t (x_{2n-1} - x_{2n-2})d\tau$$

$$< K_1 T (M_1 T)^n / n!$$

where $x_1 < K_1$ on $0 < t \le T$.

A similar result holds for the sequence y_n, whereupon

$$0 \le \eta_{2n+1} - \eta_{2n} < KT(MT)^n/n! \tag{14}$$

$$0 < t \le T,$$

for $M = Max\{M_1, M_2\}$, $K = Max\{K_1, K_2\}$. Uniform convergence to solutions of the original differential equations follows as a result of (14).

3. *Remarks on the Theory*

Collatz [3] contains some results which guarantee monotone or alternating convergence if the pertinent operators are monotonically decomposable. However, the construction of the decomposition is not discussed except in certain special cases. Nevertheless the result is useful and is briefly discussed here followed by a theorem of Ames and Ginsberg [2] which handle equations of concern here under rather general conditions.

Given the equation $u = Tu$ with the domain $T = D \subseteq R$, range $T = W \subseteq R^*$ where R and R^* are linear partially ordered metric spaces. Suppose that

$$T = T_1 + T_2 \tag{15}$$

is representable by the sum of an isotone operator T_1 and an antitone operator T_2, where both are continuous and have the same domain D. Starting with two elements v_0, $w_0 \in D$, selected so that the entire interval $[v_0, w_0] \in D$, define the iteration

$$\begin{aligned} v_{n+1} &= T_1 v_n + T_2 w_n \\ w_{n+1} &= T_1 w_n + T_2 v_n, \qquad n = 0,1,2,\ldots . \end{aligned} \tag{16}$$

If $v_0 \le v_1 \le w_1 \le w_0$ then under the assumption that $v_n \le w_n$ an inductive proof verifies that

$$v_n \le v_{n+1} \le w_{n+1} \le w_n, \quad M_n = [v_n, w_n] \in D.$$

Thus the elements v_n, w_n appear as

$$v_0 \le v_1 \le v_2 \le \cdots \le v_n \le \cdots \le w_n \le \cdots \le w_2 \le w_1 \le w_0 .$$

The important special case T_1 = Null operator and $T_2 = T$ is the antitone case in which alternating behavior is obtained.

Theorems providing for the construction of the operator decomposition are not available in general but there are special cases. In particular one due to Ames and Ginsberg [2] applies to the initial value problem

$$\frac{dy_i}{dx} = g_i(y_i) f_i(x, y_1, y_2, \ldots, y_n) \tag{17}$$

$$y_i(x_0) = \beta_i > 0, \quad i = 1,2,\ldots,n$$

The result is stated here for positive solutions where the vectors y and f represent their corresponding components. Q will be the Banach space with norm $||z(x)|| = \max_i |z_i(x)|$ containing positive solutions of the initial value problem (17). Further the quantities $G_i(u) = \int \frac{du}{g_i(u)}$, $i = 1,2,\ldots,n$ and the results will be stated for $x \in I = [x_0, x_0 + h]$, $h > 0$.

Suppose

1) G_i^{-1} is a positive continuous strictly antitone operator;
2) $f(x,y)$ is such that each component is positive and bounded on I and f_i is a strictly monotone increasing function with respect to all components of y.
3) Each f_i satisfies a Lipschitz condition on a closed region about the first iterate $y^{(1)}$, i.e.

$$f_i(x,\xi) - f_i(x,\eta) < K_i \; ||\xi - \eta||, \quad \xi > \eta.$$

Then

a) There exists a sequence of iterates $\{y_i^{(p)}\}$ satisfying

$$y_i^{(0)} = \beta_i$$

$$\frac{dy_i^{(p)}}{dx} = g_i(y_i^{(p)}) f_i(x, y_1^{(p-1)}, y_2^{(p-1)}, \ldots, y_n^{(p-1)}),$$

$$y_i^{(p)}(x_0) = \beta_i, \quad p = 1,2,\ldots; \quad i = 1,2,\ldots,n.$$

where

$$y_i^{(p)}(x) = G_i^{-1}\left[\int_{x_0}^{x} f_i(u, y^{(p-1)}(u))\,du + G_i(\beta_i)\right].$$

b) The sequence $\{y_i^{(p)}\}$ maintains the order relations for all $x \in I$ and all $i = 1,2,\ldots,n$

$$y_i^{(1)}(x) < y_i^{(3)}(x) < \ldots < y_i^{(2p+1)}(x) < \ldots <$$
$$y_i^{(2p)}(x) < \ldots < y_i^{(4)}(x) < y_i^{(2)}(x).$$

c) The sequence $\{y_i^{(p)}\}$ has a limit $y_i(x)$ which satisfies the initial value problem (17).

Various modifications of the assumptions can be treated.

These are summarized in the table below where subscripts have been dropped.

VARIATIONS OF GENERAL THEOREM FOR STRICTLY POSITIVE $f(x,y)$ AND y

	$f(x,y)$ strictly monotone increasing	$f(x,y)$ strictly monotone decreasing
G^{-1} strictly antitone	oscillatory convergence -- even iterates monotonically decreasing from above and odd iterates monotonically increasing from below	one-sided convergence from above or below
G^{-1} strictly monotone increasing	one-sided convergence from above or below	oscillatory convergence even iterates monotonically decreasing from above and odd iterates monotonically increasing from below

VARIATIONS OF GENERAL THEOREM FOR STRICTLY NEGATIVE $f(x,y)$ AND y

	$f(x,y)$ strictly monotone increasing	$f(x,y)$ strictly monotone decreasing
G^{-1} strictly antitone	oscillatory convergence -- even iterates monotonically increasing from below and odd iterates monotonically decreasing from above	one-sided convergence from above or below
G^{-1} strictly monotone increasing	one-sided convergence from above or below	oscillatory convergence even iterates monotonically increasing from below and odd iterates monotonically decreasing from above

4. *Chemical Reaction Example*

The chemical reaction (Benson[4])

$$A_1 + A_2 \xrightarrow{k_1} A_3$$

$$\begin{aligned} A_2 + A_3 &\xrightarrow{k_2} A_4 \\ A_5 + A_2 &\xrightarrow{k_3} A_6 \\ A_6 + A_2 &\xrightarrow{k_4} A_7 , \end{aligned} \tag{18}$$

taking place at constant volume, is modeled, using the "law of mass action", by

$$\frac{d}{dt}\begin{bmatrix} x_1 \\ x_2 \\ x_3 \\ x_4 \\ x_5 \\ x_6 \\ x_7 \end{bmatrix} = \begin{bmatrix} -k_1 & 0 & 0 & 0 \\ -k_1 & -k_2 & -k_3 & -k_4 \\ k_1 & -k_2 & 0 & 0 \\ 0 & k_2 & 0 & 0 \\ 0 & 0 & -k_3 & 0 \\ 0 & 0 & k_3 & -k_4 \\ 0 & 0 & 0 & k_4 \end{bmatrix} \begin{bmatrix} x_1 \\ x_3 \\ x_5 \\ x_6 \end{bmatrix} x_2 \tag{19}$$

In (19) $x_i \geq 0$ are the concentrations of A_i, $k_i > 0$ $i = 1,2,\ldots,7$, and $x_1(0)$, $x_2(0)$, $x_5(0)$, $x_3(0) = x_4(0) = x_6(0) = x_7(0) = 0$ are prescribed. While A_4 is the desired product the presence of the A_5 leads to undesirable byproducts. In the situation $\alpha = k_2/k_1 < 1$ and $\beta = k_4/k_3 < 1$.

By elementary transformations on the rows alone the canonical form (Ames [5])

$$\frac{d}{dt}\begin{bmatrix} x_1 \\ x_1 + x_3 \\ x_5 \\ x_5 + x_6 \end{bmatrix} = \begin{bmatrix} -k_1 & 0 & 0 & 0 \\ 0 & -k_2 & 0 & 0 \\ 0 & 0 & -k_3 & 0 \\ 0 & 0 & 0 & -k_4 \end{bmatrix} \begin{bmatrix} x_1 \\ x_3 \\ x_5 \\ x_6 \end{bmatrix} x_2 \tag{20}$$

is obtained, together with the algebraic relations

$$x_2(t) = 2x_1 + x_3 + 2x_5 + x_6 + [x_2(0) - 2x_1(0) - 2x_5(0)] \tag{21}$$

$$x_7(t) = x_5(0) - x_6 - x_5,\ x_4(t) = x_1(0) + x_3(0) - x_3(t) - x_1(t) . \tag{22}$$

Further reduction of (20) is readily accomplished. For example division of the second equation by the first, followed by an integration, yields

$$x_3(t) = \frac{1}{1-\alpha}\{(x_1(0)/x_1(t))^{1-\alpha} - 1\}x_1(t). \tag{23}$$

In a similar fashion

$$x_6(t) = \frac{1}{1-\beta}\{(x_5(0)/x_5(t))^{1-\beta} - 1\}x_5(t). \tag{24}$$

Finally, eliminating x_2 from the first and third equations results in

$$x_5(t)/x_5(0) = [x_1(t)/x_1(0)]^{\gamma}, \quad \gamma = {}^{k_3}/k_1 . \tag{25}$$

As a consequence of these preliminaries only one, say $x_1(t)$, need be computed and the remaining components follow directly from equations (21) to (25).

To compute x_1 an algorithm will be developed from the integrated form of the first equation of (20),

$$x_1(t) = x_1(0)\, exp[-k_1 \int_0^t x_2(\tau)d\tau]. \tag{26}$$

Beginning with the obvious upper bound $x_2^{(0)}(t) = x_2(0)$, for $x_2(t)$, compute recursively

$$x_1^{(n+1)} = x_1(0)\, exp[-k_1 \int_0^t x_2^{(n)}(\tau)d\tau],$$

$x_5^{(n+1)}$ from (25),

$x_6^{(n+1)}$ from (24), (27)

$x_3^{(n+1)}$ from (23),

$x_2^{(n+1)}$ from (21) and $x_4^{(n+1)}$, $x_7^{(n+1)}$ from (22).

From (19) it is obvious that x_2 is monotone decreasing in t. To discover the properties of the foregoing algorithm, for

$t > 0$, suppose $x_2^{(n)}(t) < x_2^{(n+1)}(t)$. Then, from (26), it follows that $x_1^{(n+1)} > x_1^{(n+2)}$ and, from (25), $x_5^{(n+1)} > x_5^{(n+2)}$. Since x_3 and x_6 are governed by similar expressions only the properties of the iterative structure of $x_3^{(n)}$ will be examined. Clearly, from (23),

$$x_3^{(n+1)} - x_3^{(n+2)} = \frac{x_1(0)}{1-\alpha}\left[\left(\frac{x_1^{(n+1)}}{x_1(0)}\right)^{\alpha} - \frac{x_1^{(n+1)}}{x_1(0)}\right] - \left[\left(\frac{x_1^{(n+2)}}{x_1(0)}\right)^{\alpha} - \frac{x_1^{(n+2)}}{x_1(0)}\right] \tag{28}$$

is positive when the function $f(u) = u^{\alpha} - u$, $0 < u < 1$, $0 < \alpha < 1$, is monotonically increasing. This occurs for $f'(u) > 0$, i.e. for $u < \alpha^{1/(1-\alpha)}$. Since x_1 is not produced in the system there will occur a time t_0 and a value of n, say n_0, such that

$$\frac{x_1^{(n_0)}(t_0)}{x_1(0)} < \alpha^{1/(1-\alpha)} ,$$

whereupon (28) becomes positive. Similarly, there exists t_1 and n_1 such that

$$\frac{x_5^{(n_1)}(t_1)}{x_5(0)} < \beta^{1/(1-\beta)} .$$

Thus for all $t > t^* = max(t_0, t_1)$ and all $n > max(n_0, n_1)$ there follows $x_3^{(n+1)} > x_3^{(n+2)}$ and $x_6^{(n+1)} > x_6^{(n+2)}$, whereupon, from (21), $x_2^{(n+1)} > x_2^{(n+2)}$. Thus, ultimately, the algorithm becomes oscillatory and provides two sided bounds.

If, on the other hand, it is assumed that $x_2^{(n)} > x_2^{(n+1)}$ then an analogous argument demonstrates that, ultimately $x_2^{(n+1)} < x_2^{(n+2)}$ for t and n sufficiently large.

5. *Bounds for Parabolic Equations*

In this section a suggestion is given for the development of bounds for nonlinear parabolic equations of the form

$$u_t = [F(u)]_{xx} \tag{29}$$

by the use of lines (see Liskovets [6] for details). With $v = F(u)$ equation (29) becomes

$$[F^{-1}(v)]_t = v_{xx} \, . \tag{30}$$

With $v_i(t) = v(i\Delta x, t)$ choose a discretization of v_{xx}, say $g(v_{i-1}, v_i, v_{i+1})$, whereupon a method of lines

$$[F^{-1}(v_i)]_t = g(v_{i-1}, v_i, v_{i+1}) \tag{31}$$

is generated. Integrating (31) the iteration

$$v_i^{(n)} = F\left[\int_0^t g(v_{i-1}^{(n-1)}(\tau),\ v_i^{(n-1)}(\tau),\ v_{i+1}^{(n-1)}(\tau))d\tau + F^{-1}(v_i(0))\right] \tag{32}$$

is obtained, with $v_i^{(0)} = v_i(0)$. The properties of this iteration depend, as before, on F and g. Thus it appears that the use of the method of lines is useful in generating bounds for parabolic equations of certain types.

REFERENCES

[1] Goel, N. S., Maitra, S. C., and Montroll, E. W., *Nonlinear Models of Interacting Populations*, Academic Press, New York, 1971.

[2] Ames, W. F., and Ginsberg, M., *Bilateral algorithms and their applications*, Computational Mechanics (Lecture Notes in Mathematics #461, J. T. Oden (Ed.)), Springer-Verlag, New York, 1975, 1-32.

[3] Collatz, L., *Functional Analysis and Numerical Mathematics*, Academic Press, New York, 1966, 350-57.

[4] Benson, S., *Foundations of Chemical Kinetics*, McGraw Hill, New York, 1960.

[5] Ames, W. F., *Nonlinear Ordinary Differential Equations in Transport Processes*, Academic Press, New York, 1968, 87-94.

[6] Liskovets, O. A., *The Method of Lines* (English Translation), J. Diff. Eqts. 1(1965), 1308.

A NEW VIEW ON CONSTRAINED PROBLEMS

H. A. Antosiewicz
University of Southern California

1. Constrained problems for ordinary differential equations and their solution in the abstract setting of bifurcation theory continue to attract wide attention. This is a brief overview of a unified approach to some of the most frequently used techniques such as the methods of Cesari, Hale, Mawhin, and Cronin (see, e.g., [1], [4], [7], [10], [12]).

All details which are omitted here may be found in the references. A full exposition will appear elsewhere.

2. There is a wealth of qualitative results that have one feature in common: they can be reduced to considerations of an equation of the form

$$(1) \qquad Lu = Fu$$

where F is a continuous nonlinear mapping of a normed (function) space U into a Banach (function) space W, and L is a linear mapping with domain in U and range in W, which has no continuous inverse.

Generally, U is the space of continuous mappings into R^n that satisfy the constraints of the given problem, such as the space of continuous ω-periodic mappings or the space of continuous mappings that are bounded in R_+, W coincides with U, and L is the mapping induced by the linear differential operator involved, and F is the Niemitzky operator generated by the nonlinearities.

It is well known that if the nullspace $N(L)$ of L has finite dimension, there is a continuous projection P onto it, and if L is closed and has closed range $R(L)$, then L admits a continuous right (pseudo-) inverse K such that $KL = I - P$. Moreover, if $R(L)$ is closed and $\mathrm{codim}\, R(L) = \dim N(L)$, there is a continuous projection Q whose nullspace is $R(L)$, and for any isomorphism J of the range of P onto the range of Q, $L + JP$ is bijective and its inverse $(L + JP)^{-1} = K(I - Q) + J^{-1}Q$ is continuous. Thus, if these assumptions hold, the equation (1) is equivalent to the equation

(2) $$u = Pu + J^{-1}QFu + K(I - Q)Fu,$$

and this equation can be solved in a variety of ways, for example using degree theory (see, e.g., [7],[12]) or the classical fixed point theorems of Banach and Schauder (see, e.g., [4],[10],[11]). In the latter case, two steps are involved in essence: first, the solution of the equation

(3) $$u_w = w + K(I - Q)Fu_w$$

for u_w in terms of a parameter $w \in N(L)$, which is finite dimensional; and second, the determination of the parameter $w \in N(L)$ such that

(4) $$QFu_w = 0.$$

Often, especially when (1) arises from a constrained problem for an ordinary differential equation, the mapping $K(I - Q)$ can be shown to be compact, and then much more can be done. In particular, when the underlying space U is a Hilbert space, or even only pre-Hilbert, elementary spectral theory can be invoked at once to yield a very simple approach to a variety of different results.

3. As illustration, we indicate here briefly some well-known considerations for the existence of an ω-periodic solution to the differential equation

(5) $$x' = f(t,x)$$

when f, for simplicity's sake, is assumed to be continuous in $R \times R^n$ and $f(t + \omega, x) = f(t,x)$ for all $(t,x) \in R \times R^n$.

Let U be the space of continuous ω-periodic mappings of R into R^n, with the usual supremum norm, and let V be the pre-Hilbert space induced by the norm $u \mapsto ||u||_2$ where

$$||u||_2 = \left(\int_0^\omega ||u(t)||^2 dt \right)^{1/2} .$$

Define $Lu : t \to u'(t)$ for each continuously differentiable $u \in U$, denote by F the Niemitzky operator in U generated by f, and let P,Q both be the projection P_0 in U that assigns to each $u \in U$ the constant mapping equal to the mean value of u. Then L is a closed linear mapping whose range $R(L)$ is the set of all $u \in U$ with $P_0 u = 0$, and hence L admits a continuous right inverse K which is defined for each $v \in R(L)$ by

$$(6) \qquad Kv : t \to \int_0^t v(s)ds - \frac{1}{\omega} \int_0^\omega \int_0^t v(s)ds\, dt.$$

Clearly, the differential equation (5) has an ω-periodic solution if and only if the equation

$$Lu = Fu$$

has a solution in U or, equivalently, if and only if the equation

$$(7) \qquad u = P_0 u + P_0 Fu + K(I - P_0)Fu$$

that corresponds to (2) with $J = I$ has a solution.

There are many different techniques by which (7) can be shown to have a solution; most require smallness assumptions on f in (5) and hence F in (7), which frequently are too stringent in applications.

Such smallness requirements become largely unnecessary on observing that K, as given by (6), may be viewed as a compact mapping of V into U. Hence, since the identity mapping of U into V is continuous. $K(I - P_0)$ is a compact linear operator in V. Moreover, if $u_k(t) = exp\, \mu_k t$, $\mu_k = 2\pi i k/\omega$ for each $k \geq 1$, then

$$Ku_k = \frac{1}{\mu_k} u_k .$$

Thus, if P_m for some integer $m \geq 1$ denotes the projection in V such that for each $u \in V$

$$P_m u : t \mapsto \sum_{-m}^{m} A_j u_j(t)$$

where the constants A_j are the usual Fourier coefficients, then $P_m P_0 = P_0 P_m = P_0$, $P_m K(I - P_0) = K(P_m - P_0)$ and

$$||K(I - P_m)u||_2 \leq \frac{1}{m + 1}||u||_2. \tag{8}$$

This suggest at once that (7) can be solved very conveniently by a projection method, setting $u = v + w$ where $v = P_m u$ and $w = (I - P_m)u$. Indeed (7) then becomes the familiar system

$$\begin{aligned} v &= P_0 v + P_0 F(v + w) + K(P_m - P_0)F(v + w) \\ w &= K(I - P_m)F(v + w) \end{aligned} \tag{9}$$

in which the first equation is finite dimensional. More important, in the second equation, any bound on the norm of the right-hand side, or its Lipschitz constant, can be made arbitrarily small whenever $m \geq 1$ is chosen large enough because, in view of (8), they carry the factor $1/(m + 1)$.

This is the crux of the entire approach. It applies, of course, more generally by virtue of the classical results of elementary spectral theory.

Analogous arguments are used in the methods pioneered by Cesari [4] and Hale [3]; by contrast these methods, however, are based more strongly on considerations of the given mapping L rather than on utilizing fully the properties of its pseudo inverse K.

ACKNOWLEDGMENT

This work was done with partial support from the U.S. Army Research Office (Durham).

REFERENCES

[1] H. A. Antosiewicz, Boundary value problems for nonlinear ordinary differential equations, *Pacific J. Math.* 17(1966), 191-197.

[2] H. A. Antosiewicz, Un analogue du principe du point fixe de Banach, *Ann. Mat. Pura Appl.* 74(1966), 61-64.

[3] S. Bancroft, J. K. Hale, and D. Sweet, Alternative problems for nonlinear functional equations, *J. Differential Equations* 4(1968), 40-56.

[4] L. Cesari, Functional analysis and periodic solutions of nonlinear differential equations, *Contrib. Differential Equations* 1(1963), 149-187.

[5] L. Cesari, Functional analysis and Galerkin's method, *Michigan Math. J.* 11(1964), 385-414.

[6] L. Cesari, Functional analysis and differential equations, in "*SIAM Studies in Applied Mathematics*", *Vol.* 5, pp. 143-155, 1969.

[7] J. Cronin, Equations with bounded nonlinearities, *J. Differential Equations* 14(1963), 581-596.

[8] J. K. Hale, *Oscillations in Nonlinear Systems*, McGraw-Hill, New York, 1963.

[9] J. K. Hale, *Ordinary Differential Equations*, Wiley-Interscience, New York, 1969.

[10] J. K. Hale, *Applications of Alternative Problems*, Lecture Notes 71-1, Brown University, Providence, R.I., 1971.

[11] A. C. Lazer, On Schauder's fixed point theorem and forced second-order nonlinear oscillations, *J. Math. Anal. Appl.* 21(1968), 421-425.

[12] J. Mawhin, Equivalence theorems for nonlinear operator equations and coincidence degree theory for some mappings in locally convex topological vector spaces, *J. Differential Equations*, 12(1972), 610-636.

[13] N. Rouche et J. Mawhin, *Equations differentielles ordinaires, I, II*, Masson et Cie, Paris, 1973.

[14] S. A. Williams, A connection between the Cesari and Leray-Schauder methods, *Michigan Math. J.* 15(1968), 441-448.

DELAY SYSTEMS IN BIOLOGICAL MODELS: APPROXIMATION TECHNIQUES*

H. T. Banks†
Brown University

Introduction

Delay systems are playing an increasingly important role in the modeling of a number of biological systems. We shall, in the first part of our discussions, give a brief indication of several areas where delay systems models appear to be very useful, and perhaps even essential in some cases.

Because of an increased use of delay system models in biological applications, recent literature [21] has evidenced an interest by some investigators in approximation techniques. We shall, in the second part of our presentation, outline some approximation ideas that we feel will prove useful in the study of delay systems such as those currently under investigation in connection with several of the biomodels mentioned in the first part of our lecture as well as in the study of more traditional delay systems models of dynamical systems arising in mechanical, physical, chemical, aerodynamical, etc., modeling projects.

I. *Delays in Biomodels*

A. Protein Synthesis Models.

A number of dynamical models for protein synthesis are based on the Jacob-Monod hypothesis as described in Figure 1 ([17], [19]).

Here DNA synthesis of messenger RNA takes place in the nucleus of the cell. At the ribosomes (in the cytoplasm) mRNA combines with transfer RNA and translation occurs (i.e., the message

* Invited lecture, International Conference on Nonlinear Systems and Applications, July 19-23, 1976, Arlington, Texas.

† This research was supported in part by the National Science Foundation under grant NSF GP 28931x3.

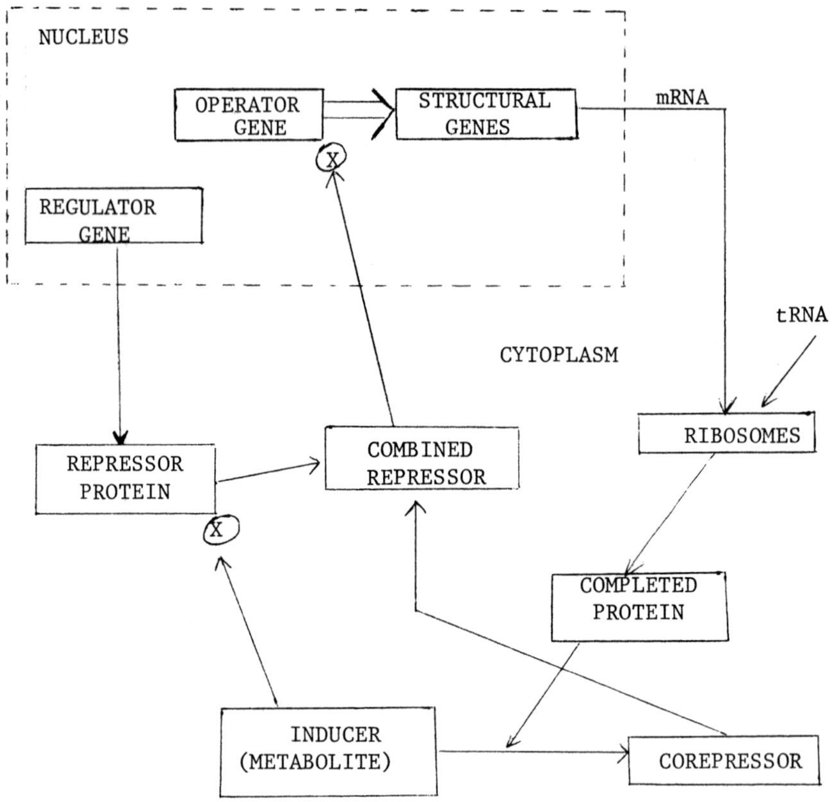
NUCLEUS
OPERATOR GENE
STRUCTURAL GENES
mRNA
X
REGULATOR GENE
tRNA
CYTOPLASM
RIBOSOMES
COMBINED REPRESSOR
REPRESSOR PROTEIN
X
COMPLETED PROTEIN
INDUCER (METABOLITE)
COREPRESSOR

FIGURE 1

in DNA which has been transcribed into the mRNA is "translated" into a specific sequence of amino acids resulting in a completed polypeptide chain). The ribosomes are then released and the protein (polypeptide chain) thus produced acts as an enzyme to convert a metabolite (inducer) to a corepressor molecule. Before this conversion the inducer molecule acts as a "deactivator" for a regulator-gene-produced repressor protein by combining with it and "blocking" its function in aiding to repress action of the operator gene. Upon conversion of the inducer to the corepressor, one has increased amounts of the repressor protein to combine with the corepressor to form combined repressor molecules which block action of the operator gene in the nucleus. The model thus represents an example of <u>end-product inhibition</u> [28], [29] since the product (of the reaction catalyzed by the completed polypeptide chain) combines with the repressor protein and this complex blocks synthesis of mRNA by combining with the operator gene to turn it off.

The DNA synthesis of mRNA is enzyme regulated. Furthermore, translation at the ribosomes is also enzyme catalyzed and thus a number of mathematical models (e.g., [11], [12], [14], [27], [28], [29]) for protein synthesis are based on a sequence of enzyme-substrate reactions linked in some type of chain or cascade. A schematic for one type of such sequence is depicted in Figure 2.

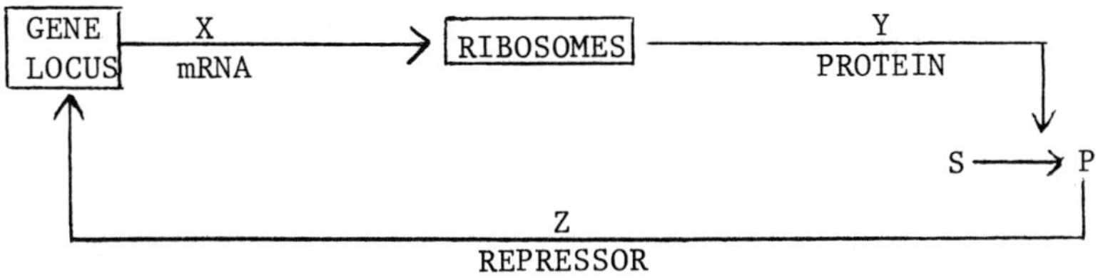

FIGURE 2

Mathematically, one writes equations

$$\dot{X} = v_1 - v_2$$

$$\dot{Y} = v_3 - v_4$$

$$\dot{Z} = v_5 - v_6$$

where X, Y, Z represent concentrations of mRNA, protein, and repressor respectively and the velocity terms $v_i = v_i(X,Y,Z)$ are concentration dependent, their form depending upon the kinetic assumptions made. Under certain kinetic velocity approximation hypotheses, one obtains the model

$$
\begin{aligned}
\dot{X}(t) &= \frac{a}{K + kZ(t)} - bX(t) \\
\dot{Y}(t) &= cX(t - r) - dY(t) \qquad (1) \\
\dot{Z}(t) &= eY(t) - gZ(t)
\end{aligned}
$$

which has been proposed in the literature (a number of other models currently under investigation are in fact modifications of this model). Here r is a generalized "transport" delay incorporating the time for transport of mRNA from the nucleus to the ribosomes plus translation time - i.e., time to complete the polypeptide chain.

B. Respiratory-Circulatory Models.

Respiratory models [7], [13], [22], [26] usually entail study of the principal gases N_2, O_2, H_2O, CO_2 and often encompass in actuality combined models for (i) the respiratory system (gas dynamics) with compartments such as alveolar space, dead space, airways, and atmosphere which are connected to a "gas pump" and (ii) the circulatory system (fluid dynamics) with compartments such as pulmonary vein, left heart, aorta, arteries, tissue, veins, right heart, and pulmonary artery. The respiratory components and circulatory components are connected via a diffusion compartment representing the gas exchange taking place during blood flow from the pulmonary artery to the pulmonary vein as it perfuses the alveoli.

In such compartmental models, usually one is concerned mainly with mass balance equations and diffusive mechanisms. A schematic of a simplified version of such a compartmental model is given in Figure 3.

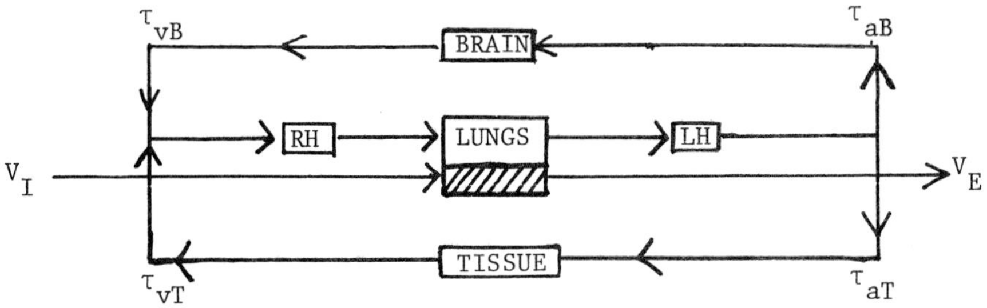

FIGURE 3

Based on accepted or hypothesized biomechanical and biochemical mechanisms, one writes mass balance equations for the alveolar partial pressures $P_{A(j)}$, brain concentrations $C_{B(j)}$, and tissue concentrations $C_{T(j)}$ (for the various species $j = O_2, CO_2, N_2, H_2O$) in terms of variables such as the arterial concentrations $C_{a(j)}$, venous concentrations $C_{v(j)}$, inspired and expired volumes V_I, V_E, and control parameters consisting of blood flow rate Q and respiratory minute volume RMV (controlled itself via sensors for $C_{a(j)}$ in the carotid body). In addition, equilibrium equations of the form $C_{a(j)} = f(P_{A(j)})$ must be written.

We shall not write specific equations (we refer those interested in details to examples such as those found in [13], [22]) since to do so correctly would entail a rather lengthy and digressive discussion at this point. However, if done carefully such an endeavor will result in a model involving quite complicated delay systems since transport delays τ (see Figure 3) are involved in relating the arterial concentrations $C_{aB(j)}(t)$, $C_{aT(j)}(t)$ at the brain and tissue entrance with the arterial concentrations $C_{a(j)}(t)$ at the lung exit and in relating the mixed venous concentrations $C_{v(j)}$ entering the lungs with the venous concentrations $C_{vB(j)}$, $C_{vT(j)}$ at the brain and tissue exits. That is,

$$C_{aB(j)}(t) = C_{a(j)}(t - \tau_{aB})$$

$$C_{aT(j)}(t) = C_{a(j)}(t - \tau_{aT})$$

$$C_{v(j)}(t) = g(C_{vB(j)}(t - \tau_{vB}), C_{vT(j)}(t - \tau_{vT})).$$

In general the transport delays $\tau_{aB}, \tau_{aT}, \tau_{vB}, \tau_{vT}$ are functions of the blood flow rate Q, which in turn may be a function of $C_{a(O_2)}$ and $C_{a(CO_2)}$. This leads to variable delays which must be defined via a set of additional equations involving the blood flow. However, in some models it is a reasonable approximation to assume constant delays (the resulting model then involves differential difference equations with constant time delays). For example, in surgical models (anesthesia studies) one often assumes that the patient is endotracheally intubated and on controlled (constant) alveolar ventilation with $C_{a(O_2)}$, $C_{a(CO_2)}$ (and, also Q) held roughly constant.

C. Ecological Models.

Many of the theoretical models for ecological systems under investigation during recent years are based on extensions of various reasonably well-known basic population models to allow treatment of communities of "populations" ([6], [15], [20], [32]). Of importance in many such ecomodels is the fact that several trophic levels may be involved. For example, (see [6]), a typical four level, ten species community might be schematically represented as follows:

Level 4: Carnivores

Level 3: Herbivores

Level 2: Vegetation

Level 1: Space, nutrients (soil)

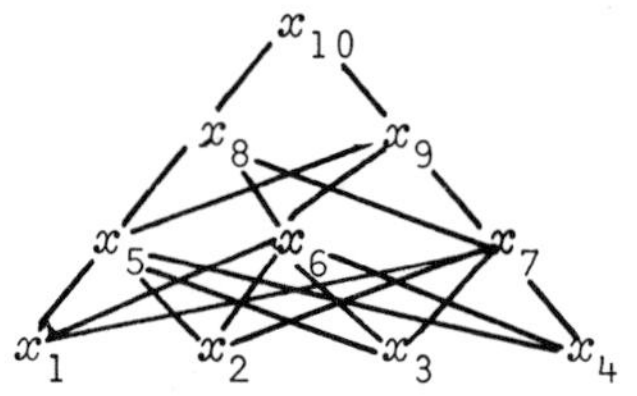

Here the x_i represent numbers (or biomass) of various species and the lines indicate a utilization as nutrient sources by species of a higher tropic level of species at the next level. Typical equations that one might write to describe these interdependences have the form

(2) $$\dot{x}_i(t) = \alpha_i \frac{x_i(t)}{k_i}\left\{k_i - \sum_j \beta_{ij}x_j(t)\right\},$$

where k_i is the carrying capacity (possibly itself a function of "population"densities) of the ith species. However, a more realistic model (see [15], [33]) would include the assumption that changes at one trophic level don't affect the next level immediately but only after some time delay (in place of $x_j(t)$ in (2) use $x_j(t - r)$, or even better $\int_{-r}^{0} x_j(t + \theta)d\theta$). Typical model equations would then have the form

$$\dot{x}_i(t) = \alpha_i \frac{x_i(t)}{k_i}\left\{k_i - \sum_j \beta_{ij}x_j(t - r_{ij})\right\}$$

or

$$\dot{x}_i(t) = \alpha_i \frac{x_i(t)}{k_i}\left\{k_i - \sum_j \int_{-r_{ij}}^{0} \beta_{ij}(\theta)x_j(t + \theta)d\theta\right\}.$$

D. Population and Epidemic Models.

As we have observed earlier many ecological models have as their basis traditional population and/or epidemic models. Since the importance of time delays in such modeling is by now quite well-known (indeed several presentations elsewhere in these proceedings deal with such models) we shall simply list for the benefit of those interested only a few basic and widely-available papers. In addition to the discussions in [8], [9], [16], [31] one might also consult the references on this topic found in [1] and [30].

II. *Approximation of Delay Systems*

Let us turn now to our discussion of approximation ideas for delay systems. One idea which has received widespread attention in the engineering and mathematical literature for some years (e. g., see [10], [23]) and is now being discussed in connection with biological applications [21] is based on truncated Taylor series expansions. A heuristic development is as follows.

Consider the simple delay equation

$$\dot{x}(t) = A_0 x(t) + A_1 x(t - r) \tag{3}$$

and expand in a Taylor series

$$x(t - r) \simeq x(t) - r\dot{x}(t) + \frac{r^2}{2!}\ddot{x}(t) - \frac{r^3}{3!}\dddot{x}(t) + \ldots . \tag{4}$$

We then truncate (4) and substitute into (3), obtaining the higher order ordinary differential equation

$$\dot{x}(t) = A_0 x(t) + A_1\{x(t) - r\dot{x}(t) + \ldots + \frac{(-r)^N}{N!} x^{(N)}(t)\} . \tag{5}$$

This is clearly an intuitively appealing approach, but several potential difficulties are immediately apparent: (i) In general, such power series expansions are valid only when the delay r is sufficiently small. Can we expect such an idea to be useful in modeling projects where the delay r is substantial relative to our time scale? (ii) For general and frequently encountered delay systems, the solutions x need not be smooth so that indeed the expansion in (4) is likely not to be valid even for r small.

It is desirable, of course, to have a method of approximation that is valid for <u>all</u> r and for x having only the smoothness expected of solutions. We shall outline (heuristically, but to be made rigorous in our subsequent discussions) a slightly different approach involving a finite sequence of first term Taylor expansions which also has intuitive appeal.

Solutions x of (3) are usually only absolutely continuous (or in $W_2^{(1)}$, C^1, etc., but almost never C^∞, let alone analytic), so we use the approximation

$$x(t) - x(t - \delta) = \int_{t-\delta}^{t} \dot{x}(s)ds \simeq \dot{x}(t - \delta)\delta. \tag{6}$$

Secondly, the delay r may not be small, but for N a large positive integer, $\frac{r}{N}$ certainly will be. Hence, we use $\delta = \frac{r}{N}$ in (6) and obtain what is equivalent to a first term Taylor expansion (at $t - \frac{r}{N}$) truncation

$$x(t) \simeq x(t - \frac{r}{N}) + \dot{x}(t - \frac{r}{N})\frac{r}{N} . \tag{7}$$

To implement this idea in approximating (3), we partition the interval $[-r,0]$ and use a finite sequence of approximations as in (7) at the points $t - \frac{r}{N}$, $t - \frac{2r}{N}, \ldots,$ $t - (N - 1)\frac{r}{N}$, $t - r$. That is, consider again (3) and make the following identifications of variables and approximations:

$$\begin{aligned}
Y_1^N(t) &= x(t - \tfrac{r}{N}) \\
Y_2^N(t) &= Y_1^N(t - \tfrac{r}{N}) = x(t - \tfrac{2r}{N}) \\
&\vdots \\
Y_N^N(t) &= Y_{N-1}^N(t - \tfrac{r}{N}) = x(t - \tfrac{Nr}{N}) = x(t - r)
\end{aligned} \tag{8}$$

and

$$\begin{aligned}
x(t) &\simeq x(t - \tfrac{r}{N}) + \dot{x}(t - \tfrac{r}{N})\tfrac{r}{N} \\
Y_1^N(t) &\simeq Y_1^N(t - \tfrac{r}{N}) + \dot{Y}_1^N(t - \tfrac{r}{N})\tfrac{r}{N} \\
&\vdots \\
Y_{N-1}^N(t) &\simeq Y_{N-1}^N(t - \tfrac{r}{N}) + \dot{Y}_{N-1}^N(t - \tfrac{r}{N})\tfrac{r}{N} .
\end{aligned} \tag{9}$$

Combining (8) and (9), we obtain the system

$$\begin{aligned}
\dot{Y}_1^N(t) &= \tfrac{N}{r}\{x(t) - Y_1^N(t)\} \\
\dot{Y}_2^N(t) &= \tfrac{N}{r}\{Y_1^N(t) - Y_2^N(t)\} \\
&\vdots \\
\dot{Y}_N^N(t) &= \tfrac{N}{r}\{Y_{N-1}^N(t) - Y_N^N(t)\} .
\end{aligned} \tag{10}$$

The approximation to (3) is then obtained by adjoining to (10) the equation

(11) $$\dot{x}(t) = A_0 x(t) + A_1 Y_N^N(t).$$

This results in an $N + 1$ vector ordinary differential equation approximation to the delay system (3). As we shall see presently, this can be viewed as a special case of the general scheme which we proceed now to sketch briefly.

Consider the n-vector nonlinear functional differential equation and initial data

(12) $$\begin{aligned} \dot{x}(t) &= L(x_t) + f(x(t),x_t,t), \quad t > 0 \\ (x(0),x_0) &= (\eta,\phi), \end{aligned}$$

where x_t denotes the function $\theta \to x(t+\theta)$, $\theta \in [-r,0]$. We consider this in the space $Z \equiv R^n \times L_2^n(-r,0) = R^n \times L_2([-r,0], R^n)$ and although our comments are valid for L a general linear operator (so as to include all such delay systems arising in applications), we shall here, for ease in exposition, take $L(\phi) = A_0\phi(0) + A_1\phi(-r)$ where A_0, A_1 are $n \times n$ matrices.

We shall treat the nonlinear system (12) as a perturbation of the linear system

(13) $$\begin{aligned} \dot{x}(t) &= L(x_t) \\ (x(0),x_0) &= (\eta,\phi). \end{aligned}$$

For $z = (\eta,\phi) \in Z$, we define the linear operators $S(t): Z \to Z$, $t \geq 0$, by

(14) $$S(t)(\eta,\phi) = (x(t;\eta,\phi),x_t(\eta,\phi))$$

where x is the solution of (13) corresponding to (η,ϕ). That is, $S(t)$ is the solution operator for the <u>linear</u> system (13) in the space Z. Under standard hypotheses on L (see [3]) one finds that $\{S(t)\}_{t \geq 0}$ is a C_0-semigroup of linear operators with an easily computed infinitesimal generator $\mathcal{A}$.

Next, we let π_1, π_2 be the coordinate projections in Z, i.e., $\pi_1(\eta,\phi) = \eta$, $\pi_2(\eta,\phi) = \phi$, and for $f : R^n \times L_2^n(-r,0) \times R^1 \to R^n$, we define $F : Z \times R^1 \to Z$ by

(15) $$F(z,t) \equiv (f(\pi_1 z, \pi_2 z, t), 0).$$

We then consider the implicit equation for a function $z : [0,t_1] \to Z$, $t_1 > 0$, given by

(16) $$z(t;\phi) = S(t)(\phi(0),\phi) + \int_0^t S(t - \sigma)F(z(\sigma),\sigma)d\sigma.$$

Observe that formally this is nothing more than the integrated form of $\dot{z}(t) = \mathcal{A}z(t) + F(z(t),t)$.

We make the following assumptions:

(H_0) The function $f = f(y,\psi,t)$ has the form $f(y,\psi,t) = N_1(y,\psi) + N_2(y,\psi)v(t)$ where $v \in L_2$ and $N_i : R^n \times L_2^n(-r,0) \to R^n$ satisfies (H_1) - (H_3) below;

(H_1) [continuity] The mappings $(y,\psi) \to N_i(y,\psi)$ are continuous on $R^n \times L_2^n(-r,0)$;

(H_2) [local lipschitz] Given a bounded subset $\mathcal{B}$ of $R^n \times L_2^n(-r,0)$, there exist constants $K_i = K_i(\mathcal{B})$ such that for (x,ϕ), $(y,\psi) \in \mathcal{B}$ one has

$$|N_i(x,\phi) - N_i(y,\psi)| \leq K_i\{|x - y| + |\phi - \psi|\};$$

(H_3) [growth] (a) $N_i(0,0) = 0$, and
(b) There exist constants $\hat{K}_i$ such that as $|(y,\psi)|_Z \to \infty$, the N_i satisfy $|N_i(y,\psi)| \leq \hat{K}_i\{|y| + |\psi|\}$.

We then can establish the equivalence results stated in the following theorem.

THEOREM 1: Suppose f satisfies (H_0) - (H_3) and $\{S(t)\}$ is as defined earlier. Then for $\phi \in L_2^n(-r,0)$ the equation (16) defines a unique continuous function $z : [0,t_1] \to Z$. Further, for $\phi \in W_2^{(1)}([-r,0],R^n)$ we have

$$z(t;\phi) = (x(t;\phi), x_t(\phi)), \quad t \in [0,t_1],$$

where z is given by (16) and x is the solution of the nonlinear system (12) with $\eta = \phi(0)$.

Given the equivalence of (12) and (16) stated in Theorem 1, we may approximate (12) by approximating (16). To do this, we first approximate solutions of the linear system (13) via approximations of the solution semigroup defined in (14).

Let $\{S^N(t)\}$ be a sequence of C_0-semigroups in Z satisfying $|S^N(t)| \leq Me^{\beta t}$ for all N (for a fixed M,β) such that

$$(17) \qquad S^N(t)z \to S(t)z, \quad \text{for} \quad z \in Z, \quad N \to \infty,$$

uniformly in t on $[0,t_1]$.

Further suppose we have a sequence $\{Z^N\}$ of finite dimensional subspaces of Z, projections $P^N : Z \to Z^N$ of Z onto Z^N satisfying $P^N(\phi(0),\phi) \to (\phi(0),\phi)$ on all initial data $(\phi(0),\phi)$ of interest in our problem.

We then define the approximating equations

$$(18) \qquad z^N(t) = S^N(t)P^N(\phi(0),\phi) + \int_0^t S^N(t-\sigma)F(z^N(\sigma),\sigma)d\sigma.$$

One can show that this defines (implicitly) a unique continuous function $z^N : [0,t_1] \to Z$. Further, convergence of the $\{z^N\}$ can, with a little argument, be established.

THEOREM 2: Suppose f satisfies (H_0) - (H_3) and $\{S^N\}$ satisfies (17). Then for each $\phi \in W_2^{(1)}([-r,0],R^n)$, we have $z^N(t) \to z(t)$, uniformly in t on compact intervals, where z^N, z are defined by (18) and (16) respectively.

Let us turn to a specific example that falls within the framework of the general theory just outlined. These are what we shall term "averaging" approximations (see [5]). We partition the interval $[-r,0]$ by $\{t_j^N\}_{j=0}^N$ where $t_j^N = \frac{-jr}{N}$ and let χ_j^N, $j = 2,3,\ldots,N$ and χ_1^N be the characteristic functions of $[t_j^N, t_{j-1}^N)$, $j = 2,\ldots,N$ and $[t_1^N,t_0^N] = [\frac{-r}{N},0]$ respectively. Define $S^N(t) = e^{\mathcal{A}^N t}$ where the operators $\mathcal{A}^N : Z \to Z$ are given by

$$\mathcal{A}^N(\eta,\phi) = (A_0\eta + A_1\phi_N^N, \sum_{j=1}^N \tfrac{N}{r}\{\phi_{j-1}^N - \phi_j^N\}\chi_j^N),$$

with

$$\phi_0^N \equiv \eta, \quad \phi_j^N \equiv \frac{N}{r}\int_{t_j^N}^{t_{j-1}^N} \phi(s)ds, \quad j = 1,2,\ldots,N.$$

One can argue [5] that $\{\mathcal{A}^N\}$ converges to $\mathcal{A}$ (the infinitesimal generator of $\{S(t)\}$) in a sense so that the Trotter-Kato approximation theorem in C_0-semigroup theory may be used to insure that (17) obtains. Taking

$$Z^N \equiv \{(\eta,\phi) \mid \eta \in R^n, \quad \phi = \sum_{j=1}^{N} \nu_j^N \chi_j^N, \quad \nu_j^N \in R^n\}$$

and $P^N : Z \to Z^N$ given by

$$P^N(\eta,\phi) \equiv (\eta, \sum_{j=1}^{N} \phi_j^N \phi_j^N$$

(this definition of P^N corresponds to a classical least squares approximation of (η,ϕ) in Z-norm), one verifies easily that Z^N, P^N are as required earlier and that Theorem 2 thus is valid for these "averaging" approximations.

In this case, one finds further that (18) is actually equivalent to an ordinary differential equation in the finite dimensional space Z^N and can be written in terms of "Fourier" coefficients relative to an appropriately chosen basis. That is, letting $z^N(t) = \sum_{j=0}^{N} w_j^N(t)e_j^N$, where $e_0^N \equiv (1,0)$, $e_j^N \equiv (0,\chi_j^N)$, $j = 1,2,\ldots,N$, one finds that (18) is equivalent to

(19)
$$\dot{w}_0^N(t) = A_0 w_0^N(t) + A_1 w_N^N(t) + f(w_0^N(t), \sum_1^N w_j^N(t)e_j^N, t)$$

$$\dot{w}_j^N(t) = \frac{N}{r}\{w_{j-1}^N(t) - w_j^N(t)\}, \quad j = 1,2,\ldots,N.$$

If we identify the "components" w_j^N with the Y_j^N of our earlier heuristic discussions (see (10), (11)) we see immediately that the averaging approximation case of our general approximation theory does indeed lead to the same equations as the first term

Taylor expansion truncation ideas.

Details for the averaging approximations can be found in [5]. We would like to point out that this approximation is but one special case of the general theoretical framework sketched here and partially developed earlier in [3]. In subsequent efforts by this author and other investigators, it has been shown that certain spline and other polygonal types of approximations can also be used in the context of this framework to generate approximation techniques (see also [2]). Details of both the theoretical and numerical aspects of these findings will be reported in forthcoming publications.

While the conditions (H_0) - (H_3) on f can probably be relaxed somewhat without affecting the veracity of Theorem 2, we observe that these conditions are already sufficiently general so as to include a number of the nonlinearities that arise in applications. Included are f of the form stated in (H_0) with N_i given, for example, by

$$N_i(y,\psi) = h_i(y) + \int_{-r}^{0} k_i(\theta)\psi(\theta)d\theta$$

or

$$N_i(y,\psi) = h_i(y) + \int_{-r}^{0} k_i(\theta)\ \sin\ \psi(\theta)d\theta$$

with h_i anyone of the following:

(i) $$h_i(y) = \begin{cases} 0 & y \leq 0 \\ \dfrac{y^p}{1 + y^p} & y > 0, \quad p \text{ a positive integer;} \end{cases}$$

(ii) $$h_i(y) = \begin{cases} 1 & y \leq 0 \\ \dfrac{1}{1 + y^p} & y > 0; \end{cases}$$

(iii) $h_i(y) = y \sin y$ or $y \cos y$;

(iv) $h_i(y) = y$.

Nonlinearities such as those in (i) arise from biochemical kinetic velocity approximation terms, while (ii), of course, corresponds to the nonlinearities in the protein synthesis models we have mentioned briefly earlier. The terms in (iii) arise in certain particle accelerator models currently under investigation while (iv) (under (H_0)) yields in (12) the type of bilinear control systems often found in the literature.

Approximation schemes such as those discussed here should prove useful in several areas of investigation of delay systems. In particular, our own experiences suggest immediately several important uses:

(1) Numerical (approximate) solutions of functional differential equations: Examples reveal that one obtains very good numerical approximations to solutions of both linear and nonlinear equations (we shall comment further on this momentarily).

(2) Stability analysis: It can be shown that certain types of stability (or instability) are preserved under the approximations. It is often much easier to investigate the approximating (finite-dimensional) system than the original (infinite-dimensional) delay system.

(3) Optimal control: Taking $v = u$ in the f of (H_0), one can pose the problem of minimizing $\Phi(u) \equiv J(x(u),u)$ over some closed convex $\mathcal{U} \subset L_2(0,t_1)$ subject to (12) (the original delay system). Approximate problems of minimizing $\Phi^N(u) \equiv J(\pi_1 z^N(u),u)$ over $\mathcal{U}$ subject to (18) (or in case of the averaging approximations $\Phi^N(u) \equiv J(w_0^N(u),u)$ subject to (19)) can then be formulated. If $\bar{u},\bar{u}^N$ denote the respective solutions of these problems, one can argue that for a generous class of payoff functions Φ, $\Phi^N(\bar{u}^N) \to \Phi(\bar{u})$ and $\bar{u}^N \to \bar{u}$ (in a precisely specified sense; e.g., see [2], [3], [5]).

We have had some experience using the averaging approximation scheme for linear system optimal control problems. The numerical results one obtains here are quite good as is attested to by the details for a substantial number of examples reported in [4]. Our experience to date with nonlinear systems involves only approximate numerical solution to the delay equations themselves (numerical work on optimal control problems with nonlinear systems is in progress and results will hopefully be announced soon). Here again, the results are quite satisfactory. During a recent visit to the University of Graz, we investigated, in cooperation with F. Kappel and H. Fröschl of that University, use of the averaging approximation scheme (among others) to obtain numerical solutions for several of the protein model examples (e.g., see system (1)) discussed in Part I of this presentation. The preliminary results were very good, in each case yielding, for example, approximate solutions at $N = 8$ with a relative error in each coordinate of less than 1.5%.

In closing, we remark that other investigators are also currently using abstract semigroup theory to develop approximation schemes for functional differential equations. Specifically, Reddien and Webb [24], and Sasai and Ishigaski [25] have applied the theory of nonlinear contraction semigroups to nonlinear delay systems where the right side (i.e., f in (12)) satisfies a global Lipschitz condition while Kappel and Schappacher [18] have extended certain ideas from nonlinear semigroups to develop a local semigroup theory that can be used to treat systems of nonlinear nonautonomous delay systems with the right side satisfying a local Lipschitz condition analogous to (H_2) in our discussions here.

REFERENCES

[1] H. T. Banks, *Modeling and Control in the Biomedical Sciences*, Springer Lecture Notes in Biomath., Vol. 6, 1975.

[2] H. T. Banks, *Approximation methods for optimal control problems with delay-differential systems*, Séminaires IRIA: analyse et contrôle de systemes, May, 1976.

[3] H. T. Banks, and J. A. Burns, *An abstract framework for approximate solutions to optimal control problems governed by hereditary systems*, in Proceedings, International Conference on Differential Equations, (University of Southern California, September, 1974), H. A. Antosiewicz, editor, Academic Press, New York, 1975, 10-25.

[4] H. T. Banks, J. A. Burns, E. M. Cliff, and P. R. Thrift, *Numerical solutions of hereditary control problems via an approximation technique*, Brown University LCDS Technical Report 75-6, Providence, R. I., October, 1975.

[5] H. T. Banks, and J. A. Burns, *Hereditary control problems: numerical methods based on averaging approximations*, to appear (preprint).

[6] H. Barclay, and P. van den Driessche, *Time lags in ecological systems*, J. Theor. Biol. 51(1975), 347-56.

[7] J. Comroe, *Physiology of Respiration*, Year Book Med. Publ., Chicago, 1965.

[8] K. L. Cooke, *Functional-differential equations: some models and perturbation problems*, in Differential Equations and Dynamical Systems, J. Hale and J. P. LaSalle, editors, Academic Press, New York, 1967, 167-83.

[9] K. Cooke, and J. Yorke, *Equations modeling population growth, economic growth and gonorrhea epidemiology*, in Ordinary Differential Equations, L. Weiss, editor, Academic Press, New York, 1972, 35-53.

[10] L. E. El'sgol'ts, *Introduction to the Theory of Differential Equations with Deviating Arguments*, Holden-Day, San Francisco, 1966.

[11] B. Goodwin, *Oscillatory behavior in enzymatic control processes*, Adv. Enzyme Reg. 3(1965), 425-49.

[12] J. S. Griffith, *Mathematics of cellular control processes, I; II*, J. Theor. Biol. 20(1968), 202-08; 209-16.

[13] F. Grodins, J. Buell, and A. Bart, *Mathematical analysis and digital simulation of the respiratory control system*, J. Applied Physiol. 22(1967), 260-76.

[14] Z. Grossman, and I. Gumowski, *Self-sustained oscillations in the Jacob-Monod mode of gene regulation*, 7th IFIP Conference on Optimization Techniques, September, 1975, Nice, France.

[15] C. Holling, *Resilience and stability of ecological systems*, Ann. Rev. Ecology Systematics 4(1973), 1-23.

[16] G. E. Hutchinson, *Circular causal systems in ecology*, Ann. N. Y. Acad. Sci. 50(1948), 221-46.

[17] F. Jacob, and J. Monod, *On the regulation of gene activity*, Cold Spring Harbor Symp. Quant. Biol. 26(1961), 193-211; 389-401.

[18] F. Kappel, and W. Schappacher, *Autonomous nonlinear functional differential equations*, to appear (preprint).

[19] J. W. Kimball, *Cell Biology*, Addison-Wesley, Reading, Massachusetts, 1968.

[20] R. May, *Time-delay versus stability in population models with two and three trophic levels*, Ecology 54(1973), 315-25.

[21] A. Mazanov, and K. P. Tognetti, *Taylor series expansion of delay differential equations - a warning*, J. Theor. Biol. 46(1974), 271-82.

[22] H. T. Milhorn, *The Application of Control Theory to Physiological Systems*, Saunders, Philadelphia, 1966, Chapter 15.

[23] N. Minorsky, *Self-excited oscillations in dynamical systems possessing retarded actions*, J. Applied Mechanics 9(1942), A65-A71.

[24] G. W. Reddien, and G. F. Webb, *Numerical approximation of nonlinear functional differential equations with L_2 initial functions*, to appear (preprint).

[25] H. Sasai, and H. Ishigaski, *Convergence of approximate solutions to functional differential systems by projection methods*, to appear (preprint).

[26] E. Selkurt, *Physiology*, Little, Brown & Co., Boston, 1966.

[27] J. M. Smith, *Mathematical Ideas in Biology*, Cambridge University Press, 1968.

[28] J. Tiwari, A. Fraser, and R. Beckman, *Genetical feedback repression*, J. Theor. Biol. 45(1974), 311-26.

[29] C. F. Walter, *Kinetic and thermodynamic aspects of biological and biochemical control mechanisms*, in Biochemical Regulatory Mechanisms in Eukaryotic Cells, E. Kun & S. Grisolia, editors, Wiley, New York, 1972, 355-489.

[30] P. Waltman, *Deterministic Threshold Models in the Theory of Epidemics*, Springer Lecture Notes in Biomath., Vol. 1, 1974.

[31] P. J. Wangersky, and W. J. Cunningham, *Time lag in prey-predator population models*, Ecology 38(1957), 136-39.

[32] F. Williams, *Dynamics of microbial populations*, in Systems Analysis and Simulation in Ecology, B. Patten, editor, Academic Press, New York, 1971, Chapter 3.

[33] D. Winter, K. Banse, and G. Anderson, *The dynamics of phytoplankton blooms in Puget Sound*, Mar. Biol. 29(1975), 139-76.

CONVERGENCE OF VOLTERRA SERIES ON INFINITE INTERVALS AND BILINEAR APPROXIMATIONS

R. W. Brockett*
Harvard University

1. Introduction

The results of [1-4] each in a somewhat different way demonstrate that the so called bilinear systems, i.e. those of the form

$$\dot{x}(t) = Ax(t) + u(t)Bx(t); \; y(t) = cx(t); \; x(0) = x_0 \qquad (\dagger)$$

can, subject to a very mild hypothesis, be used to approximate the more general "linear-analytic" systems

$$\dot{x}(t) = f[x(t)] + u(t)g[x(t)]; \; y(t) = h[x(t)]; \; x(0) = x_0 \qquad (*)$$

to an arbitrary degree of accuracy as measured in the L_∞ norm on $[0,T]$. There are, however a number of very interesting properties of nonlinear systems such as stability, subharmonic response, etc. which relate to the infinite interval $[0,\infty)$. In this paper we establish some basic results on the nature of this approximation on $[0,\infty)$. Our methods are based on Volterra series developments and in the process we establish some new results on the convergence of Volterra series on $[0,\infty)$.

We say that (*) is a complex linear-analytic system in $\mathcal{D} \subset \mathbb{C}^n$ if f, g, and h are analytic at every point in $\mathcal{D}$. We say that (*) is real linear-analytic if f, g and h are real analytic. Clearly bilinear systems are examples of complex linear-analytic systems in $\mathbb{C}^n$. Given a linear analytic system in $\mathcal{D}$ and an interval I in the time axis we associate an extended real number $r(I)$ defined to be the largest real number such that the Volterra series converges in the L_∞ norm on I for all measurable $u(\;)$ such that

* This work was supported in part by the U.S. Office of Naval Research under the Joint Services Electronics Program Contract N00014-75-C-0648 & the Army Research Office under Grant DAAG29-76-G-0139.

$$\sup_{0 \leq t \leq T} |u(t)| \leq r(I)$$

we call $r(I)$ the radius of convergence of the Volterra series. We remind the reader (see [2]) that the Volterra series associated with (*) is the series solution

$$y(t) = w_0(t) + \int_0^t w_1(t,\sigma)u(\sigma)d\sigma + \int_0^t\int_0^{\sigma_1} w_2(t,\sigma_1,\sigma_2)u(\sigma_1)u(\sigma_2)d\sigma_1 d\sigma_2 + \ldots$$

and such a power series development is known to exist on any finite interval $[0,T)$ if f, g, h are analytic and the free solution passing through x_0 exists on $[0,T+\varepsilon)$ for some $\varepsilon > 0$. Explicit formulas for the w_i in terms of the Taylor series of f, g and h and certain constructions from multilinear algebra are given in [2]. The Volterra series for bilinear systems is very easy to display as was pointed out some time ago [5].

The classical Peano-Baker series development shows that for bilinear systems the radius of convergence is infinite for finite time intervals.

2. *The Radius of Convergence* $r([0,\infty))$

We now establish two basic results on the convergence of Volterra series on the interval $[0,\infty)$. The second of these gives an explicit formula for the radius of convergence for time invariant bilinear systems.

THEOREM 1. Consider the complex linear analytic system

$$\dot{x}(t) = f[x(t)] + u(t)g[x(t)]; \quad y(t) = h[x(t)]$$

with $f(0) = 0 = x(0)$. If the Jacobian of f evaluated at $x = 0$ has eigenvalues with negative real parts then the radius of convergence of the Volterra series $r([0,T))$ does not go to zero as T goes to infinity.

Proof: Suppose it does. Then for each positive integer n there exists u and T_n such that with $|u(t)| \leq 1/n$ on $[0,T_n)$ the Volterra series diverges for some $t_n \in [0,T_n)$. But by the Weirstrass theorem this means that for some complex α, $|\alpha| = 1$ and for $\tilde{u} = \alpha u$ the partial sums for $y(t_n)$ approach ∞. However, a routine application of the Gronwall–Bellman inequality regarding (*) as a complex equation shows that there exists an $\varepsilon > 0$ such that for all $|u(\cdot)| < \varepsilon$ the solution of (*) are bounded on $[0,\infty)$ which contradicts the above statement.

Notice that even if we have a real linear-analytic system the convergence of the Volterra series is considerably clarified by passing to its complexification precisely because the radius of convergence of the taylor series at z_0 is the distance to the nearest singularity if we work over $\mathbb{C}$. The next result provides a further example by relating the radius of convergence of the Volterra series for bilinear systems to the very well known circle criterion for asymptotic stability.

THEOREM 2. Let u and y be related by the time invariant, real, bilinear system

$$\dot{x}(t) = Ax(t) + u(t)Bx(t); \quad y(t) = cx(t); \quad x(\ ) = x_0$$

with the eigenvalues of A having negative real parts. Suppose that this realization is minimal in the sense of [6]. Then the Volterra series

$$y(t) = ce^{At}x(0) + \int_0^t ce^{A(t-\sigma)}Be^{A\sigma}x(0)u(\sigma)d$$

$$+ \int_0^t \int_0^{\sigma_1} ce^{A(t-\sigma_1)}Be^{A(\sigma_1-\sigma_2)}Be^{A\sigma_2}x(0)u(\sigma_1)u(\sigma_2)d\sigma_1 d\sigma_2 + \ldots$$

converges on $[0,\infty)$ in the L_∞ norm for all measurable $u(\ )$ provided $|u(t)| \leq \alpha$ and

$$\sup_{-\infty<\omega<\infty} ||B_1(Ii\omega - A)^{-1}B_2|| \leq \beta$$

with $B_2B_1 = B$, $\alpha\beta < 1$, and $||\cdot||$ indicating the Euclidean norm.

Proof: One recognizes, via Parseval's relation, β as the norm on the operator

$$L(y)(t) = \int_0^t B_1 e^{A(t-\sigma)} B_2 y(\sigma) d\sigma$$

induced by the $L_2[0,\infty)$ topology on $y(\cdot)$. Thus if $|u(\cdot)| < \alpha$ then the L_2 norm of the nth term of Volterra series is bounded from above by $(\alpha\beta)^n$. Hence we have convergence in the $L_2[0,\infty)$ topology. But because y satisfies a differential equation with a bounded right-hand side this means that all derivatives of y are in $L_2[0,\infty)$ as well. Thus by standard arguments we may conclude that the series converges in the $L_\infty[0,\infty)$ topology.

On the other hand, if the supremum in the theorem statement is β, and if we notice that the supremum is necessarily attained for some $\omega < \infty$ because $B_1(I i\overset{\omega}{\infty} - A)^{-1}B_2$ vanishes at $\omega = \infty$, then we see that for some complex vector x_0 and some ω_0

$$B_1(Ii\omega_0 - A)^{-1}B_2 x_0 = \lambda x_0$$

with $|\lambda| = \beta$. If we let $u(\cdot)$ be the constant λ^{-1}

$$\dot{x}(t) = (A + \lambda^{-1}B)x(t) = (A + \lambda^{-1}B_2B_1)x(t)$$

But the eigenvalues of $A + \lambda^{-1}B_2B_1$ are easily seen to be the zeros of

$$det\ (\lambda^{-1}B_1(Is - A)^{-1}B_2 + I)$$

and hence one or more lie on the imaginary axis. Thus we see that if $|u(t)|$ is permitted to exceed $1/\beta$ x can grow exponentially fast as a function of t. By the minimality hypothesis this implies that y grows exponentially fast and hence convergence on $[0,\infty)$ is impossible.

There are two points about this result which are noteworthy.

(a) As a criterion for asymptotic stability the circle criterion (see [7] for a discussion) is only a sufficient condition and not generally a necessary one. However theorem 2 shows that as a criterion for Volterra series convergence, the circle criterion defines exactly the radius of convergence. The stability results are, themselves an immediate corollary of the convergence result so it appears that theorem 2 is in a mathematical sense a more fundamental result than the circle criterion.

(b) In the case of bilinear systems the radius of convergence of the Volterra series is independent of the initial state $x(0)$. For more general classes of linear analytic systems this is not the case.

COROLLARY: A necessary and sufficient condition for a time invariant bilinear system to have a Volterra series with an infinite radius of convergence is that the Volterra series be finite with each kernel being realizable by an asymptotically stable system.

Proof: The stability condition together with finiteness of the Volterra series clearly implies an infinite radius of convergence. From the theorem we see that we must have $B_1(Is - A)^{-1}B_2 = 0$. This implies that the McMillan degree (see [7]) of the triple (A, B_2, B_1) is zero. Thus from the well known work of Kalman [8] there is a change of basis such that A is upper triangular and B_2B_1 is strictly upper triangular. From the results of [2] this implies that the Volterra series is finite. Clearly stability of the individual is necessary as well.

3. *Some Nonlinear Phenomena*

There are a few qualitative properties of nonlinear systems which are especially noteworthy because they (i) are observed to occur and (ii) cannot be accounted for on the basis of linear models. We mention a few of these here:

(i) stable free oscillations

(ii) stable subharmonic responses

(iii) jump phenomena

Each of these represents an aspect of periodic behavior. Experimentally one can never set the initial conditions exactly so as to correspond to a periodic trajectory and thus they are observed as a form of limiting behavior reflecting what happens for large values of time.

The concepts of stability which we have in mind require a

discussion. In each case we want to work with input-output models and want to capture an idea which is related to "local" structural stability but with an important difference due to the way in which we control the nature of the perturbations using the input-output model.

Suppose that we model an oscillator as

$$\dot{x}(t) = f[x(t)] + u(t)g[x(t)]; \; y(t) = h[x(t)]; \; x(0) = x_0$$

together with the Volterra series

$$y(t) = w_0(t) + \int_0^t w(t,\sigma)u(\sigma)d\sigma$$

$$+ \int_0^t \int_0^{\sigma_1} w(t,\sigma_1\sigma_2)u(\sigma_1)u(\sigma_2)d\sigma_1 d\sigma_2 + \ldots$$

with $w_0(t)$ periodic. We will say that the oscillator is stable at $w_0(t)$ if the solution with $u(t) \equiv 0$ is periodic and orbitally stable and the first term of the Volterra series has McMillan degree equal to $\dim x$ and is weakly stable. That is to say, the kernel $w_1(t_1\sigma)$ has a realization

$$\dot{x} = A(t)x + b(t)u(t); \quad y(t) = c(t)x(t)$$

with A, b and c periodic and all but one of the Floquet multipliers of $\Phi_A(t,\sigma)$ being inside the unit disk, the remaining one being $+1$.

Let $u(\cdot)$ be periodic with least period T and suppose that

$$\dot{x}(t) = f[x(t)] + u(t)g[x(t)]; \; y(t) = h[x(t)]; \; x_0 = x(0)$$

has a solution which is periodic with least period T/n for some integer $n \geq 2$. We say that this system exhibits <u>stable frequency division</u> if the Volterra expansion for

$$\dot{x}(t) = f[x(t)] + u(t)g[x(t)] + v(t)g[x(t)]; \; y(t) = h[x(t)]$$

$$x = x(0)$$

is

$$y(t) = w_0(t) + \int_0^t w_1(t,\sigma)v(\sigma)d\sigma$$

$$+ \int_0^t \int_0^\sigma w_2(t,\sigma_1,\sigma_2)u(\sigma_1)u(\sigma_2)d\sigma_2 d\sigma_2 + \ldots$$

with $w_1(t,\sigma)$ having McMillan degree equal to $dim\ x$ and being asymptotically stable.

A system of this form will be said to exhibit jump phenomena if for some periodic $u(\cdot)$ there exists two or more stable periodic responses.

THEOREM 3. Bilinear systems do not exhibit stable free oscillations or stable subharmonic response. If for a given periodic function $u(\cdot)$ with period T there is a periodic response $x(\cdot)$ with period T then there is necessarily an infinity of periodic responses.

Proof: The Volterra series for a bilinear system is

$$y(t) = ce^{At}x_0 + \int_0^t ce^{A(t-\sigma)}Be^{A\sigma}u(\sigma)d\sigma$$

$$+ \iint ce^{A(t-\sigma_1)}Be^{A\sigma_1-\sigma_2}Be^{A\sigma_2}u(\sigma_1)u(\sigma_2)d\sigma_1 d\sigma_2 + \ldots$$

If for $u = 0$ there is an oscillation then clearly A has an eigenvalue on $Re\ s = 0$. This means that small changes in x_0 produce changes in both the amplitude and the phase of the periodic solution and hence that we do not have a stable oscillation.

If $u(T)$ is periodic of period T and x is periodic of period T then we see that

$$x(T) = \Phi_{A+uB}(T,0)x(0)$$

Thus we see that the initial condition $\beta x(0)$ also gives a periodic response for all β.

It is more appropriate to consider in this context a slightly more general class of models, namely those of the form

$$\dot{x}(t) = Ax(t) + u(t)Bx(t) + u(t)b; \; y(t) = cx(t); \; x(0) = x_0$$

In this case theorem 3 remains true if we modify the last sentence to say "more than one periodic response" in place of "a periodic response".

REFERENCES

[1] A. J. Krener, *Linearization and Bilinearization of Control Systems*, Proc. 1974 Allerton Conference on Circuit and System Theory, Univ. of Illinois, Urbana, Illinois 1974.

[2] R. W. Brockett, *Volterra Series and Geometric Control Theory*, Automatica, 12(1976), pp. 167-176.

[3] H. J. Sussmann, *Semigroup representations, bilinear approximation of input-output maps, and generalized inputs*, Algebraic System Theory, Springer Lecture Notes, 1976.

[4] M. Fliess, *Series de Volterra et series formells non commutatives*, C. R. Acad. Sci. Paris, t. 280(1975), p. 965-967.

[5] C. Bruni, G. DiPillo and G. Koch, *On the mathematical models of bilinear systems*, Ricerche di Automatica, 2(1972), p.11-26.

[6] R. W. Brockett, *On the algebraic structure of bilinear systems*, in Variable Structure Systems, Academic Press, New York, 1972, pp. 153-168, (R. Mohler and A. Ruberti eds.).

[7] R. W. Brockett, *Finite Dimensional Linear Systems*, J. Wiley, New York, 1970.

[8] R. E. Kalman, *The mathematical description of linear dynamical systems*, SIAM J. on Control, Vol. 1, 1963, pp. 152-192.

DIFFERENTIAL INEQUALITIES FOR LIAPUNOV FUNCTIONS

T. A. Burton
Southern Illinois University

We consider a system of ordinary differential equations

(1) $$x' = f(t,x)$$

where $f : [0,\infty) \times R^n \to R^n$ is continuous, and we suppose that there is a function $V : [0,\infty) \times R^n \to [0,\infty)$ which is continuous, locally Lipschitz in x, and satisfies $V'_{(1)}(t,x) \leq 0$. The object of the paper is to study differential inequalities relating V' and (1) in order to conclude that solutions of (1) are continuable, bounded, or approach certain limit sets. We notice that when V is autonomous and differentiable then $V'_{(1)} = |grad\ V||f| \cos\theta$ with θ being the angle between $grad\ V$ and f. Thus, it seems crucial that the differential inequality should include at least

(2) $$V' \leq \Gamma(t,|x|,|f|,V).$$

Consequences of (2) are investigated in some detail when

$$\Gamma = -\lambda(t) r(V) h(\mu(x)) |\mu'_{(1)}(x)|^{\alpha}$$

for appropriate functions λ, r, h, and μ with $\alpha \geq 1$. The following is a typical consequence of our results.

THEOREM. Let λ be a positive continuous scalar function, $\alpha \geq 1$, and let M be a positive constant. If $V'_{(1)}(t,x) \leq -\lambda(t)|f(t,x)|^{\alpha}$ for $t \geq 0$ and $|x| \geq M$, then each solution of (1) can be continued for all future time.

Full details will appear in a paper by the same title in "Journal of Nonlinear Analysis: Theory, Methods and Applications". Numerous examples are given.

NONLINEAR OSCILLATIONS

by

Lamberto Cesari
University of Michigan

1. *Introduction.*

In this lecture we wish to report recent advances on the alternative or bifurcation method. As in previous work, the results are essentially in the large, that is, not of the perturbation variety.

We shall report on new existence theorems for periodic nonlinear Liénard systems under qualitative assumptions only. In technical terms, certain networks of nonlinear electric circuits will admit of forced oscillations of the same period of the external forces. In dealing with this problem, auxiliary and bifurcation equations were discussed as a system, a priori bounds were established, and topological tools used such as the Borsuk-Ulam theorem.

Also, we attempt to show the connection of the alternative or bifurcation method with the finite element theory. The methods of approximation of the finite element theory give rise to projection operators in suitable function spaces, which have important properties. In particular, they make it possible to treat the auxiliary equation by the Banach fixed point theorem for contraction mappings. Then, as usual, in the alternative method, existence statements, methods of successive approximations, and error bounds estimates follow naturally.

Finally, we mention here recent developments in our previous work concerning boundary value problems for hyperbolic systems of first order quasi linear partial differential equations. Also, we touch on the resonance problem with monochromatic laser beams which has motivated this work on boundary value problems for hyperbolic systems.

2. *The Alternative, or Bifurcation Process.*

We present here, for the sake of simplicity, only the basic framework of the alternative or bifurcation method, as described in more details in recent expositions [12,13]. For a more general form of the process, including for instance, the classical Leray-Schauder approach for quasi linear problems, we refer to [14].

Let X, Y be Banach spaces, let $E : \mathcal{D}(E) \to Y, \mathcal{D}(E) \subset X$, be a linear operator with domain $\mathcal{D}(E) \subset X$, and let $N : X \to Y$ be a nonnecessarily linear operator. Let $X = X_0 + X_1$, $Y = Y_0 + Y_1$ be decompositions of X and Y, and $P : X \to X$, $Q : Y \to Y$ projection operators (i.e., linear, bounded and idempotent), such that $X_0 = PX$, $X_1 = (I - P)X$, $Y_0 = QY$, $Y_1 = (I - Q)Y$, and we assume that the null space of E is contained in X_0 and the image of $X_1 \cap \mathcal{D}(E)$ is Y_1, or $ker\ E \subset X_0$ and $Y_1 = E[X_1 \cap \mathcal{D}(E)]$. Then, the map $E : X_1 \cap \mathcal{D}(E) \to Y$ is one-one and onto, and the partial inverse $H : Y_1 \to X_1 \cap \mathcal{D}(E)$ exists as a linear map. In [13] we have discussed these assumptions together with the natural relations: (k_1) $H(I - Q)E = I - P$, (k_2) $QE = EP$, (k_3) $EH(I - Q) = I - Q$. As proved in [13], the operator H is bounded if both the graph and the range of E are closed. Also, equation

$$Ex = Nx \tag{1}$$

is equivalent to the system of auxiliary and bifurcation equations

$$x = Px + H(I - Q)Nx, \tag{2}$$

$$Q(Ex - Nx) = 0. \tag{3}$$

If $x^* = Px$ and T denotes the map defined by $Tx = x^* + H(I - Q)Nx$, then the auxiliary equation takes the form of a fixed point problem $x = Tx$. For $y = x - Px$, the auxiliary equation takes the form of a Hammerstein equation $(I - K\overline{N})y = 0$, where $K = H(I - Q)$, and $\overline{N}y = N(x^* + y)$.

For every $x^* \in X_0$, we have $PTx = x^* = Px$, that is, T maps the fiber $P^{-1}x^*$ of P through x^* into itself. Thus, if for every x^* the transformation $T : P^{-1}x^* \to P^{-1}x^*$ is a contraction, then it has a unique fixed point $x = Tx = \tau x^*$ which depends on x^*. The bifurcation equation then becomes $Q(E - N)\tau x^* = 0$, an equation in x^*. (See [13] for applications).

For a treatment of auxiliary and bifurcation equations in the case in which $-E$ and N are monotone maps, we refer to [18] and [13].

In general it may be convenient to discuss equations (2) and (3) as a system. For instance, for $X_0 = \ker E$, the bifurcation equation reduces to $QNx = 0$, and the solutions of system (2),(3) can be thought of as the fixed points of the transformation $(x,x^*) \to (\bar{x},\bar{x}^*)$ defined by

$$\bar{x} = x^* + H(I - Q)Nx, \qquad \bar{x}^* = x^* + QNx, \tag{4}$$

or analogously, for $y = x - Px$, as the fixed points of the transformation $(y,x^*) \to (\bar{y},\bar{x})$ defined by

$$\bar{y} = H(I - Q)N(x^* + y), \qquad \bar{x}^* = x^* + QNx,$$

or of analogous transformations.

3. *Forces Oscillations of Liénard Systems.*

The results of this Section have just been obtained by the use of transformations similar to (4), by establishing a priori bounds, and by the use of the Ulam-Borsuk topological theorem.

Let us consider a Lienard system with forcing terms of the type

$$x''(t) + (d/dt)F(x(t) + (d/dt)V(x(t),t) + Ax(t) + g(x(t)) = e(t), \tag{5}$$

where $x(t) = (x_1,\dots,x_n)$, $-\infty < t < +\infty$, $A = [a_{ij}]$, $F(x) = (F_1,\dots,F_n)$, $V(x,t) = (V_1,\dots,V_n)$, $g(x) = (g_1,\dots,g_n)$, $e(t) = (e_1,\dots,e_n)$, where we assume that $e(t)$ is a 2π-periodic continuous function of mean value zero, that A is a constant $n \times n$ matrix, that $V(x,t)$ is 2π-periodic in t for every x and is such that all V_i, $\partial V_i/\partial x_j$ are of class C^1 in R^{n+1}, and that $g(x)$ is continuous in R^n with $g(x)/|x| \to 0$ as $|x| \to 0$.

(3.i) THEOREM. Under the hypotheses above, system (5) has at least one 2π-periodic solution $x(t)$, $-\infty < t < +\infty$, provided that one of the following assumptions hold. Either (α) there exist nonnegative constants $c, d, C, D, C', D', c > 0$, and some integer $p \geq 2$ such that $xF(x) = \Sigma_1^n x_i F_i(x) \geq c|x|^{2p} + d$, $V(x,t) = V_1(x,t) + V_2(x,t)$, $V_1(x,t) = S(t)\ grad\ W(x)$ for some C^2 scalar function $W(x)$ and some 2π-periodic C^1 function $S(t) = (S_1,\ldots,S_n)$, $|V_1(x,t)| \leq C'|x|^{2p-2} + D'$, $|V_2(x,t)| \leq C|x|^p + D$, and moreover $det\ A \neq 0$; or (α') $g \equiv 0$, F and V as in (α) above, A an arbitrary constant matrix; or (β) the same as in (α) with $p = 1$ and $c > C + ||A - A_{-1}||/2$; or (β') $g \equiv 0$, F and V as in (β) above, A an arbitrary constant matrix; or (γ) the same as in (α) with $|V_1(x,t)| \leq C'||x||^{2p-1} + D'$, $c > C'$ for $p > 1$, and $c > C + C' + ||A - A_{-1}||/2$ for $p = 1$; or (γ') $g \equiv 0$, F and V as in (γ), A an arbitrary constant matrix.

This theorem was proved by Cesari and Kannan [19,20] under slightly more restrictive conditions ($V_1 \equiv 0$, $F = grad\ Z(x)$, Z homogeneous of degree $2p$ and constant sign), by alternative method and the considerations mentioned above. De Vries [21] by the same argument extended the theorem to the form stated.

For further results concerning systems analogous to (5) of the Liénard or Rayleigh types under conditions of symmetry, we refer to Bowman [9] and De Vries [21], where use is made of Hale's general concept of symmetry.

4. *Nonlinear Eigenvalues*

Let X, Y be Banach spaces over the reals, and let $C : \mathcal{D}(C) \to Y$, $\mathcal{D}(C) \subset X$, be a linear operator. Let $B : X \to Y$ be an operator, not necessarily linear, with $B(0) = 0$. Thus, the equation $Cx = \mu\ Bx$ has always the solution $x = 0$.

A number μ is said to be an eigenvalue if $Cx = \mu\ Bx$ for some $x \neq 0$, $x \in X$. A number μ_0 is said to be a bifurcation point, or branch point, if for all $\varepsilon > 0$, $\rho > 0$ there are pairs (x,μ), $x \in X$, $x \neq 0$, $|x| < \varepsilon$, $|\mu - \mu_0| < \rho$ with $Cx = \mu\, Bx$.

We assume here that B is of the form $B = D + \Omega$, D linear and bounded, $\Omega(0) = 0$, and there is a monotone continuous function $\beta(\rho) \geq 0$, $0 \leq \rho < +\infty$, with $\rho(0) = 0$ and $||\Omega x - \Omega y|| \leq \beta(\rho)||x - y||$ for all x, $y \in X$, $||x||$, $||y|| \leq \rho$.

(4.i) A necessary condition for μ_0 to be a bifurcation point is that $C - \mu_0 D$ has not a bounded inverse.

Indeed, the equation $Cx = \mu\, Bx$ can now be written in the form $(C - \mu_0 D)x = (\mu - \mu_0)Dx + \mu\Omega x$, and if $(C - \mu_0 D)^{-1}$ exists as a bounded linear operator, then

$$x = (C - \mu_0 D)^{-1}[(\mu - \mu_0)Dx + \mu\Omega x]. \tag{6}$$

For ρ, $\delta > 0$ sufficiently small and $||x|| \leq \rho$, $|\mu - \mu_0| \leq \delta$, then the second member of (6) is a contraction operator, and thus $Cx = \mu\, Bx$ has one and only one solution. Since $x = 0$ is a solution, this is the only one, and μ_0 is not an eigenvalue.

If we take $E = C - \mu_0 D$, $N = (\mu - \mu_0)D + \mu\Omega$, then equation $Cx = \mu\, Bx$ takes the form $Ex = Nx$ of Section 2. Let $X_0 = \ker E$ and let us assume that $X = X_0 + X_1$ is a decomposition of X with $X_0 = PX$, $X_1 = (I - P)X$, where $P : X \to X$ is a projection operator. Analogously, let us assume that $\mathcal{R}E = Y_1$, that $Y = Y_0 + Y_1$ is a decomposition of Y with $Y_0 = QY$, $Y_1 = (I - Q)Y$, where $Q : Y \to Y$ is a projection operator. Then $E : X_1 \cap \mathcal{D}(E) \to Y_1$ is one-one and onto, and the partial inverse $H : Y_1 \to X_1 \cap \mathcal{D}(E)$ exists as a linear operator. We assume that H is bounded. As mentioned in Section 3, the equation $Cx = \mu\, Bx$, or $Ex = Nx$, admits the decomposition into auxiliary and bifurcation equations

$$x = Px + H(I - Q)Nx, \quad QNx = 0,$$

and if we take $x = x^* + z$, $x^* = Px \in X_0$, $z \in X_1$, the same equations become

$$z = H(I - Q)N(x^* + z), \quad QN(x^* + z) = 0.$$

As in [13] let c, d, $0 < c < d$, denote fixed numbers and let $W = [x^* \varepsilon X_0 | \, ||x^*|| \leq c]$, and for every $x^* \varepsilon W$, let $V^* = V(x^*) = [z \varepsilon X_1 | \, |z| \leq d]$.

(4.ii) There exists $\rho > 0$, $\delta > 0$ such that, for $0 < c < d \leq \rho$, $|\mu - \mu_0| \leq \delta$ and $x^* \varepsilon W$, the transformation $z \to Tz = H(I - Q)N(x^* + z)$ is a contraction on $V(x^*)$, and thus the auxiliary equation, or $z = Tz$, has one and only one solution in $V(x^*)$, or $x = x^* + F(x^*,\mu)$ with $F(0,\mu) = 0$, and

(7) $$F(x^*,\mu) = H(I - Q)N[x^* + F(x^*,\mu)].$$

We refer to Cesari [13] or Hale [25] for the simple proof.

Let us assume now that the spaces $X_0 = ker\ E = ker\ (C - \mu_0 D)$ and $Z_0 = QDX_0$ are one dimensional, and therefore, $X_0 = \{a\psi\}$, $Z_0 = \{b\psi\}$, a,b reals, ϕ,ψ not zero, $QD\phi = p\psi$, $p \neq 0$.

(4.iii) For X_0 and Z_0 one dimensional, there always is a bifurcation at $\mu = \mu_0$.

We refer to Hale [25] for a simple proof of this statement. For the equations $x = \mu Ax$, where A is a completely continuous operator with Fréchet derivative at the origin $L = A'(0)$, Krasnoselskii has proved, by topological considerations, that any nonzero eigenvalue of L of odd multiplicity is a bifurcation point.

As an application of the considerations above, Hale [25] has considered the branching of solutions of the second order differential equation

$$(I(t)u_t)_t + f(u,t,\lambda) = 0, \quad t_1 \leq t \leq t_2,$$

with linear homogeneous boundary conditions $B[u(t),\lambda] = 0$, expressible in terms of the values of $u(t)$ and $u'(t)$ at t_1 and t_2 (Cf Keller [28] for a previous treatment). The branching of solutions of differential systems of order n, $u' - A(t)u = f(t,u,\lambda)$ with linear homogeneous boundary conditions $B_1u(t_1) + B_2u(t_2) = 0$, ($A$ an $n \times n$ matrix, B_1, B_2 two $m \times n$ matrices) has been studied by Nagle [30] and Bowman [10]. The same problem for periodic solutions of systems of first and second order ordinary differential equations had been studied, by essentially the same method by Hale and Gambill [26]. The buckling problem of a rod has been studied by Bowman [10].

5. *A Further Decomposition of the Spaces* X *and* Y

As in no. 2 let X, Y be Banach spaces, let $E : \mathcal{D}(E) \to Y$, $\mathcal{D}(E) \subset X$, be a linear operator, and $N : X \to Y$ a nonnecessarily linear operator. Let $X_{00} = \ker E$, $Y_{01} = R(E)$, with the assumption that there are decompositions $X = X_{00} + X_{01}$, $Y = Y_{00} + Y_{01}$ and projection operators $P_0 : X \to X$, $Q_0 : Y \to Y$ such that $X_{00} = P_0 X$, $X_{01} = (I - P_0)X$, $Y_{00} = Q_0 Y$, $Y_{01} = (I - Q\ )Y$. Then, the map $E : X_{01} \cap \mathcal{D}(E) \to Y_{01}$ is one-one and onto, and we denote by H_0 the partial inverse $H_0 : Y_{01} \to X_{01} \cap \mathcal{D}(E)$ as we did in Section 2. Now the relations hold: $Q_0 E = 0$, $EP_0 = 0$, $E(I - P_0) = E$, $H_0(I - Q_0)E = H_0 E = H_0 E(I - P_0) = I - P_0$, in particular relations (k_{123}) of no. 2, and equation $Ex = Nx$ in X is equivalent to the system of auxiliary and bifurcation equations

$$(8) \qquad x = P_0 x + H_0(I - Q_0)Nx, \quad Q_0 Nx = 0.$$

For the next further decomposition of the spaces X, Y, the following lemma will be helpful.

(5.i) Let $S : X \to X$ be any projection operator with $SX \subset X_{01} \cap \mathcal{D}(E)$, $SP_0 = 0$, $P_{0S} = 0$. Then, $P = P_0 + S$ is also a projection operator, and system (8) is equivalent to

$$x = Px + (I - P)H_0(I - Q_0)Nx, \quad Sx = SH_0(I - Q_0)Nx, \quad Q_0 Nx = 0.$$

A proof of this lemma was given in [13, p.25].

Now let Σ' denote a closed subspace of Y_{01} and let Σ be the corresponding subspace $\Sigma = H_0\Sigma' \subset X_{01}$. Let $R' : Y_{01} \to Y_{01}$ be a projection operator with range Σ', or $R'(Y_{01}) = \Sigma'$, and let us define the projection operator $S' : X_{01} \to X_{01}$, by taking $S' = H_0 R' E$. Then, the range of S' is Σ, and for any $\bar{y} \in Y_{01}$ we also have $R'\bar{y} = ES'H_0\bar{y}$. Now we take $S = S'(I - P_0)$,

$R = R'(I - Q_0)$, so that $S : X \to X$, $R : Y \to Y$ are projection operators in X and Y respectively. Moreover, $P_0 S = 0$, $SP_0 = 0$, $Q_0 R = 0$, $RQ_0 = 0$.

(5.ii) Under the assumptions above, then $P = P_0 + S$ is a projection operator in X, $Q = Q_0 + R$ is a projection operator in Y, and, for $X_0 = PX$, $X_1 = (I - P)X$, $Y_0 = QY$, $Y_1 = (I - Q)Y$, we have $X_0 \supset ker\ E = X_{00}$, $Y_1 \subset R(E) = Y_{01}$. Moreover, if H denote the restriction of H_0 to Y_1, or $H = H_0|_{Y1}$, then the relations hold (k_1) $H(I - Q) = I - P$, (k_2) $QE = EP$, (k_3) $EH(I - Q) = I - Q$. Finally, $I - Q = I - Q_0 - R'(I - Q_0) = (I - R')(I - Q_0)$.

A proof of this statement is essentially given in [13, pp.27-28]. Thus, we have the final decompositions

$$X = X_0 + X_1 = X_{00} + \Sigma + X_1, \quad Y = Y_0 + Y_1 = Y_{00} + \Sigma' + Y_1,$$

and with the operators P, Q, H, the equation $Ex = Nx$ is now equivalent to the system of auxiliary and bifurcation equations

$$(9) \qquad x = Px + H(I - Q)Nx, \quad Q(Ex - Nx) = 0,$$

though now we may write the auxiliary equation also in the form

$$(10) \qquad x = Px + H(I - R')(I - Q_0)Nx.$$

6. *The Use of Finite Elements*

First a few general considerations on relations (9). We shall think here that the image $R(N)$ of N is contained in a Banach space $Z \subset Y$, or $N : X \to Z$, $R(N) = NX \subset Z \subset Y$. We denote by $\| \ \|_X$, $\| \ \|_Y$, $\| \ \|_Z$ the norms in X, Y, Z, though we may omit the subscripts when evident from the context.

We shall think of the operators appearing in the auxiliary equation (10) as follows: $N : X \to Z$, $I - Q_0 : Z \to Z$, $I - R' : Z \to Y$, $H_0 : Y_{01} \to X$, $P : X \to X$. Accordingly, we shall introduce corresponding constants as follows:

$$||Nx_1 - Nx_2||_Z \leq L||x_1 - x_2||_X \quad \text{for all} \quad x_1, x_2 \in X;$$

$$||(I - Q_0)z||_Z \leq M||z||_Z \quad \text{for all} \quad z \in Z;$$

$$||(I - R')z||_Y \leq \gamma||z||_Z \quad \text{for all} \quad z \in Z;$$

$$||Hy||_X \leq \Lambda||y||_Y \quad \text{for all} \quad y \in Y_1 \subset Y.$$

Then, for any fixed $x^* \in X$, or $x = x^* + w$, $x^* \in X_0$, $w \in \Sigma$, then

(11) $\quad Tx = x^* + H(I - R')(I - Q_0)Nx \in P^{-1}x^* \subset X$ for all $x \in N$.

Moreover, for any two elements x_1, $x_2 \in P^{-1}x^*$, we also have

$$Tx_1 - Tx_2 = H(I - R')(I - Q_0)(Nx_1 - Nx_2),$$

$$||Tx_1 - Tx_2||_X \leq \Lambda\gamma ML||x_1 - x_2||_X .$$

Thus, T maps $P^{-1}x^*$ into itself, and, if $k = \Lambda\gamma ML < 1$, then T is a contraction map in the norm of X.

The case of $X = H^{m+\sigma}(G)$, $Y = H^{\sigma}(G)$, $Z = H^{m+\sigma-s}(G)$ for some $\sigma \geq 0$, $0 \leq s \leq m - 1$, and a fixed bounded domain G of E^{ν}, $\nu \geq 1$, is of interest.

Now let $\psi_1, \ldots, \psi_N$ be "finite elements" in G of class $H^{\sigma}(G)$, and Σ' be the N-dimensional space spanned by $\psi_1, \ldots, \psi_N$. Let $\phi_j = H_0\psi_j$, $j = 1, \ldots, N$, be the corresponding elements in X_{01}, and Σ the space spanned by $\phi_1, \ldots, \phi_N$; thus, $\Sigma' \subset Y_{01}$, $\Sigma \subset X_{01}$. We conceive $R'z = \Sigma c_j\psi_j$ as an "approximation" of z the fineness of h of the finite system $\psi_1, \ldots, \psi_N$. Indeed, for instance, for $X = H^{m+\sigma}$, $Y = H^{\sigma}$, and $z \in Z = H^{m+\sigma-s}$, $\sigma \geq 0$, $0 \leq s \leq m - 1$, we have ([22],[32])

(12) $\quad ||(I - R')z||_Y \leq Ch^{m-s}||z||_Z ,$

where C is a constant. In this situation then, $k = \Lambda CMLh^{m-s}$, and we can make $k < 1$ by taking h sufficiently small.

(5.i) For $X = H^{m+\sigma}$, $Y = H^{\sigma}$, $Z = H^{m+\sigma-s}$, $\sigma \geq 0$, $0 \leq s \leq m - 1$, the operator $T = Px + H(I - R')(I - Q)Nx$, or $T : P^{-1}x^* \to P^{-1}x^*$ for any $x^* \in X_0$, can be made to be a contraction on the fibers $P^{-1}x^*$ of the projection operator P by taking $h > 0$ sufficiently small, and the auxiliary equation $x = Tx$ has a unique fixed point $x = Tx = \tau x^* \in P^{-1}x^*$ for any $x^* \in X_0$.

This applies to equations $Ex = Nx$ whose underlying linear problem $Ex = 0$ may be nonselfadjoint with possible nontrivial null space $X_{00} = ker\ E$. Even for selfadjoint problems there is no need to compute eigenelements. As usual with the alternative method the process indicated above with the use of finite elements can be applied to the proof of the existence of an exact solution x_0 to a given problem, as well as to the computation of error bounds $||x - x_0||$ for given approximations x of x_0. (Cf. [17]).

The same process above with the use of finite elements also leads to a quasi Newton method of successive approximations (cf. Ku[2]) as an extension of the processes proposed by Banfi [1,2], Casadei [3], Sanchez [31] for the use of proper eigenelements of selfadjoint problems. These are a few of the problems treated by Ku [29] by the quasi Newton method of successive approximations based on finite elements mentioned above:

(a) $x'' - 15x' = \arctan x + \sin \pi t, \quad 0 \le t \le 1,$
$x(0) = x(1), \quad x'(0) = x'(1);$

(b) $x'' - 15x' = x^3 + \sin \pi t, \quad 0 \le t \le 1,$
$x(0) = x(1), \quad x'(0) = x'(1);$

(c) $x_{\xi\xi} + x_{\eta\eta} = \sin x + f(\xi,\eta), \quad (\xi,\eta) \in Q = [0 \le \xi, \eta \le 1],$
$\partial x/\partial n = 0, \quad (\xi,\eta) \in \partial Q.$

7. *Boundary Value Problems for Schauder's Hyperbolic Systems*

Let us consider the quasi linear Schauder's system of equations in a domain $D_a = I_a \times E^r$, $I_a = [0 \le x \le a]$, of the xy

space E^{r+1},

$$(13)\qquad \sum_{j=1}^{m} A_{ij}(x,y,z)[\partial z_j/\partial x + \sum_{k=1}^{r} \rho_{ik}(x,y,z)\partial z_j/\partial y_k]$$

$$= f_i(x,y,z), \quad i = 1,\ldots,m, \quad z = z(x,y) = (z_1,\ldots,z_m)$$

$$(x,y) \in D_a = I_a \times E^r, \quad y = (y_1,\ldots,y_r),$$

where A_{ij}, ρ_{ik}, f_i are functions of (x,y,z) in a domain $D_a \times \Omega$, $\Omega = [z \mid |z_i| \leq \Omega_0, \ i = 1,\ldots,m]$, $\Omega_0 > 0$, $a > 0$, and $det[A_{ij}] \geq \mu > 0$ for $(x,y,z) \in \Omega$. For system (13) we consider the boundary value problem

$$(14)\qquad \sum_{j=1}^{m} b_{ij}(y)z_j(a_i,y) = \psi_i(y), \quad y \in E^r, \quad i = 1,\ldots,m,$$

where $0 \leq a_1,\ldots,a_m \leq a$ are m given numbers between 0 and a, b_{ij}, ψ_i are given functions in E^r, and $det[b_{ij}] \geq \mu_0 \geq 0$ for $y \in E^r$.

Thus, if $b_{ij} = \delta_{ij}$, problem (14) reduces to the one of assigning the values of each function z_i on the hyperplane $x = a_i$; if $b_{ij} = \delta_{ij}$ and $a_i = 0$, then problem (14) reduces to the usual Cauchy problem with all data on the hyperplane $x = 0$.

The following assumptions have been made on the functions A_{ij}, ρ_{ik}, f_i:

$$|A_{ij}(x,y,z)| \leq H,$$

$$|A_{ij}(x,y,z) - A_{ij}(x,\bar{y},\bar{z})| \leq C[|y - \bar{y}| + |z - \bar{z}|],$$

$$|A_{ij}(x,y,z) - A_{ij}(\bar{x},y,z)| \leq \left| \int_{x}^{x} m\,(\alpha)d\alpha \right|,$$

$$|\rho_{ik}(x,y,z)| \leq m(x), \qquad |f_i(x,y,z)| \leq n(x),$$

$$|\rho_{ik}(x,y,z) - \rho_{ik}(x,\bar{y},\bar{z})| \leq \ell(x)[|y - \bar{y}| + |z - \bar{z}|],$$

$$|f_i(x,y,z) - f_i(x,\bar{y},\bar{z})| \leq \ell_1(x)[|y - \bar{y}| + |z - \bar{z}|],$$

for all $0 \le x$, $\bar{x} \le a_0$, y, $\bar{y} \in E^r$, z, $\bar{z} \in \Omega$, for given constants H, C, C' and functions $m(x), m(x), n(x), \ell(x)$, $\ell_1(x) \ge 0$ of class $L_1[0,a_0]$. If $[\alpha_{ij}]$ denote the transpose of the inverse of the $m \times m$ matrix $[A_{ij}]$, then the elements α_{ij} satisfy relations analogous to the ones for A_{ij} above with constants H', C', and a function $\overset{0}{m}'(x) \ge 0$ also L-integrable in $[0,a_0]$. On the functions b_{ij} and ψ_i we assume that

$$|\psi_i(y)| \le \omega_0 < \Omega_0, \qquad |\psi_i(y) - \psi_i(\bar{y})| \le \Lambda_0 |y - \bar{y}|,$$

$$\sum_{j=1}^m |b_{ij}(y) - b_{ij}(\bar{y})| \le \tau_0 |y - \bar{y}|,$$

for all y, $\bar{y} \in E^r$ and constants ω_0, Λ_0, τ_0. Moreover, we need here that both matrices $[A_{ij}]$, $[b_{ij}]$ have "dominant main diagonal". Since equations (13) and (14) can be multiplied by arbitrary nonzero constants, and so the unknowns z_i, we simply assume that

$$A_{ij} = \delta_{ij} + \tilde{A}_{ij}, \quad \alpha_{ij} = \delta_{ij} + \tilde{\alpha}_{ij}, \quad b_{ij} = \delta_{ij} + \tilde{b}_{ij},$$

with $\tilde{A}_{ij}$, $\tilde{\alpha}_{ij}$, $\tilde{b}_{ij}$ "small" in the following sense. First we take

$$\sigma_0 = Max_i \; Sup_y \; \sum_{h=1}^m |\tilde{b}_{ih}(y)|,$$

and as usual we should expect the assumption $0 \le \sigma_0 < 1$. Concerning the matrices $[A_{ij}]$, $[\alpha_{ij}]$ we need slightly more stringent requirements. Let

$$\sigma_1 = Max_i \; Sup \sum_{h=1}^m |\tilde{A}_{ih}(x,y,z)|,$$

$$\sigma_2 = Max_i \; Sup \sum_{h=1}^m |\tilde{\alpha}_{hi}(x,y,z)|,$$

$$\sigma_3 = Max_i \; Sup \sum_{s=1}^m \sum_{h=1}^m |\tilde{\alpha}_{si}(x,y,z)| \, |\tilde{A}_{sh}(x,y,z)|,$$

$$\sigma = \sigma_1 + \sigma_2 + \sigma_3.$$

Our only requirement is now that

$$\sigma + \sigma_0 + \sigma\sigma_0 < 1.$$

Thus, for $[A_{ij}]$ diagonal we have $\sigma = 0$ and we require $\sigma_0 < 1$; for $[b_{ij}]$ diagonal we have $\sigma_0 = 0$ and we require $\sigma < 1$.

Concerning the solution $z = z(x,y) = (z_1,\ldots,z_m)$ we shall demand that it satisfies relations of the form

$$\text{(15)} \qquad \begin{aligned} &|z_i(x,y)| \leq \Omega_0, \qquad |z_i(x,y) - z_i(x,\bar{y})| \leq Q|y - \bar{y}|, \\ &|z_i(x,y) - z_i(\bar{x},y)| \leq \left|\int_{\bar{x}}^{x} \chi(\alpha)d\alpha\right|, \quad i = 1,\ldots,m, \end{aligned}$$

for all $0 \leq x$, $\bar{x} \leq a$, y, $\bar{y} \in E^r$, for a suitable constant Q, and a nonnegative L-integrable function $\chi(x)$, $0 \leq x \leq a$.

(6.i) Theorem. There are positive constants a, ω_0, τ_0, C, C', Q, $0 \leq a \leq a_0$, $0 < \omega_0 < \Omega_0$, and a nonnegative L-integrable function $\chi(x)$, $0 \leq x \leq a$, such that for any system of functions A_{ij}, ρ_{ik}, f_i, ψ_i, b_{ij}, and for any system of numbers $0 \leq a_i \leq a$, $i = 1,\ldots,m$, there is a unique function $z(x,y) = (z_1,\ldots,z_m)$, $(x,y) \in D_a = I_a \times E^r$, satisfying (13) a.e. in D_a, satisfying (14) everywhere in E^r, and satisfying relations (15).

This theorem has been proved by Cesari in [15] by the use of Banach fixed point theorem for contracting maps, and the usual Leray-Schauder device for the study of quasi linear problems. The corresponding method of successive approximations has been streamlined by Bassanini in [4,5] to make it more suitable to numerical computations.

8. A Problem in Nonlinear Optics

If a monocromatic strong laser beam is focused on a thin crystal, say of thickness a, then after emerging the light shows a measurable component of the second harmonic radiation. For instance, a ruby laser, 6940 Å wave length, gives rise to the 3470 Å wave length component. Franken and Ward [23] discuss at great length the experimental and theoretical basis of this phenomenon. (See also Bloembergen [8]). In [24] Graffi considered the main nonlinear problems of electromagnetism, and particularly those of nonlinear optics. From Graffi exposition, as well as from those of other authors, it appears that nonlinear plane waves as monitored by Maxwell equations

$$(16) \qquad -H_x = k^2E_t + \varepsilon EE_t, \quad -E_x = H_t, \quad 0 \le x \le a, \quad t \in E^1,$$

$k > 1$, $\varepsilon > 0$, constants depending on the nature of the crystal, may represent the simplest model for a preliminary study of phenomena of nonlinear optics, otherwise an extremely complex field. The boundary conditions to associate to (16) result from the need to match the values of E and H to those before and after the crystal, that is, to those of the usual linear waves in the empty space $(k = 1,\ \varepsilon = 0)$ for $x < 0$ and $x > a$. The following boundary conditions have been proposed

$$(17) \qquad E(0,t) + H(0,t) = 2w(t), \quad E(a,t) - H(a,t) = 0, \quad t \in E^1,$$

where $w(t)$ represents the incident radiation (cf [16]). In [16] equations (16) and boundary conditions (17) have been transformed into an analogous problem for Schauder's systems with matrices $[A_{ij}]$, $[b_{ij}]$ with "dominant main diagonal" in the sense of previous section, namely,

$$(U_x + \rho_1 U_t) + \theta(V_x + \rho_1 V_t) = 0,$$

$$\theta(U_x + \rho_2 U_t) + (V_x + \rho_2 V_t) = 0,$$

$$2^{-1}(k^{-1} + 1)U(0,t) + 2^{-1}(k^{-1} - 1)V(0,t) = \overline{w}(t),$$

$$2^{-1}(k^{-1} - 1)U(a,t) + 2^{-1}(k^{-1} + 1)V(a,t) = 0,$$

where $\rho_1 = (1 + \varepsilon' u)^{1/2}$, $\rho_2 = -(1 + \varepsilon' u)^{1/2}$, $u = U + V$, $\varepsilon' = \varepsilon k^{-2}$, and $\theta = [(1 + \varepsilon' u)^{1/2} - 1][(1 + \varepsilon' u)^{1/2} + 1]^{-1}$, $|\theta| < 1$, $A_{11} = A_{22} = 1$, $A_{12} = A_{21} = \theta$, $b_{11} = b_{22} = 2^{-1}(k^{-1} + 1)$, $b_{12} = b_{21} = 2^{-1}(k^{-1} - 1)$.

In [6,7] the problem and corresponding transformation are studied in their technical details. In particular, the displacement in function space is made in so that the known solution E_0, H_0 of the corresponding linear problem ((16), (17) with $k = 1$, $\varepsilon = 0$) is taken as a new origin. By this and other devices, and the use of computer, it was shown in [7] for the quartz, and other data as actually encountered in some of the experiments, that a constant $a = cm.0.2$ already guarantees the existence of the solution, and the convergence of the various processes of successive approximations. The actual computation of the solution is at present under way.

REFERENCES

[1] C. Banfi, Sulla determinazione delle soluzioni periodiche di equazioni differentiali periodiche, *Boll. Unione Mat. Ital.* (4)1,1968,608-619.

[2] C. Banfi, Su un metodo di successive approssimazioni per lo studio delle soluzioni periodiche di sistemi debolmente nonlineari, *Atti Accad. Sci. Torino 100*,]965-66, 471-479.

[3] C. Bonfi and G. Casadei, Calcolo di soluzioni periodiche di equazioni differenziali nonlineari, *Calcolo*, 5, Suppl.1, 1968, 1-10.

[4] P. Bassanini, Su una recente dimostrazione circa il problema di Cauchy per sistemi quasi lineari iperbolici, *Boll. Unione Mat. Ital.* (5)13 B, 1976.

[5] P. Bassanini, On a recent proof concerning a boundary value problem for quasi linear hyperbolic systems in the Schauder canonic form, *Boll. Unione Mat. Ital.* to appear.

[6] P. Bassanini, A nonlinear hyperbolic problem arising from a question of nonlinear optics, *I. Journal Applied Math. & Physics* (ZAMP), 27(1976).

[7] P. Bassanini and G. Polidori, A nonlinear hyperbolic problem arising from a question of nonlinear optics, *II, Journal Applied Math. & Physics*, to appear.

[8] N. Bloembergen, *Nonlinear Optics*, Benjamin 1965.

[9] T. T. Bowman, Periodic solutions of Liénard systems with symmetries, To appear.

[10] T. T. Bowman, Perturbation problems for ordinary differential equations, to appear.

[11] L. Cesari, Functional analysis and periodic solutions of nonlinear differential equations, *Contributions to Differential Equations 1*, Wiley 1963, 149-187.

[12] L. Cesari, Alternative methods in nonlinear analysis, *International Conference on Differential Equations* (Antosiewicz, editor), Academic Press, 1975, 95-148.

[13] L. Cesari, Functional analysis, nonlinear differential equations and the alternative method (A Summer Institute at Michigan State University, 1975), *Nonlinear Functional Analysis and Differential Equations*, (Cesari, Kannan, Schuur, editors), Marcel Dekker, New York, 1976, 1-197.

[14] L. Cesari, Nonlinear differential equations, ordinary and partial, in the frame of functional analysis and bifurcation theory, *Upsala 1977 International Conference on Differential Equations*.

[15] L. Cesari, A boundary value problem for quasi linear hyperbolic systems in the Schauder canonic form, *Annali Scuola Normale Sup. Pisa* (4)1, 1974, 311-358.

[16] L. Cesari, Nonlinear oscillations under hyperbolic systems (An international Conference, Providence, R.I. 1974), *Dynamical Systems* (Cesari, Hale, LaSalle, editors), Academic Press, vol. 1, 1976, 251-261.

[17] L. Cesari and T. T. Bowman, Some error estimates by the alternative method, *Quarterly Appl. Mathematics*, to appear.

[18] L. Cesari and R. Kannan, Functional Analysis and Nonlinear Differential Equations, *Boll. Amer. Math. Soc.* 79(1973), 1216-1219.

[19] L. Cesari and R. Kannan, Periodic solutions in the large of nonlinear differential equations, *Rend. Matematica Univ. Roma* (2)8, 1975, 633-654.

[20] L. Cesari and R. Kannan, Solutions in the large of Liénard systems with forcing terms, *Annali Matematica pura appl.*, to appear.

[21] R. De Vries, Periodic solutions of differential systems of Liénard and Rayleigh type, (An international Conference at the University of Florida, Gainesville, 1976), to appear.

[22] G. J. Fix and G. Strand, *An Analysis of the Finite Element Method*, Prentice Hall, 1973.

[23] P. A. Franken and J. F. Ward, Optical harmonics and nonlinear phenomena, *Rev. Mod. Physics* 35(1963), 23-39.

[24] D. Graffi, Problemi nonlineari nella teoria del campo elettromagnetico, *Mem. Acad. Sci. Modena* 11(1967), 172-196.

[25] J. K. Hale, *Applications of alternative problems*, Brown University Lecture Notes, 1971, 1-69.

[26] J. K. Hale and R. A. Gambill, Subharmonic and ultraharmonic solutions for weakly nonlinear systems, *J. Rat. Mech. Anal.*

4(1956),353-398.

[27] M. A. Krasnoselskii, *Topological Methods in the Theory of Nonlinear Integral Equations*, MacMillan 1964.

[28] J. B. Keller, Bifurcation theory for ordinary differential equations, *Bifurcation Theory and Nonlinear Eigenvalue Problems* (Keller and Antman, editors), Benjamin 1969,17-48.

[29] D. Ku, *Boundary value problems and numerical estimates*, Univ. of Michigan thesis, 1976.

[30] K. Nagle, *Boundary value problems for nonlinear ordinary differential equations*, University of Michigan thesis, 1975.

[31] D. A. Sanchez, An iteration scheme for Hilbert space boundary value problems, *Boll. Unione Mat. Ital.* (4)11, 1975, 1-9.

[32] R. S. Varga, *Functional Analysis and Approximation Theory in Numerical Analysis*, Applied Mathematics Series, SIAM, 1971.

A COMPARISON PRINCIPLE FOR STEADY-STATE DIFFUSION OPERATORS

Jagdish Chandra, U. S. Army Research Office
and
Paul Wm. Davis*, Worcester Polytechnic Institute

Many models of steady-state Fickian (or Newtonian) diffusion processes contain the operator

(1) $$div(D(\underline{x},u,\nabla u)\nabla u),$$

in which D is a positive diffusion coefficient, possibly depending on the concentration u and its gradient ∇u. While assuming D to be constant is often an appealing simplification, there are many problems in which the nonlinear, and often complicated, behavior of the diffusion coefficient is a significant feature.

Physical intuition suggests that an analog of the familiar linear comparison principle (inequality on the boundary and satisfaction of a differential inequality implies inequality in the interior) should remain valid even if the diffusion coefficient varies fairly arbitrarily with concentration and its gradient, so long as that coefficient remains positive and thereby preserves the basic character of Fickian diffusion. The comparison principle announced in this note confirms this suggestion.

The need to consider diffusion coefficients with arbitrary nonlinear behavior was suggested by a number of examples, including the nonmonotone thermal diffusion coefficients exhibited in [5] and the gradient-dependent enzyme diffusion coefficients considered in [6]. Operators of similar mathematical structure also occur in the Reynold's equation of gas lubrication theory [3] and in the minimal surface problem. (We make no contribution to the latter in this note, but an important idea in our proof was suggested by a comparison theorem for capillary surfaces in [1].)

* Research supported by the U. S. Army Research Office under grant number DAAG29-76-G-0237.

Since we place only slight restriction on the coefficient $D(\underline{x},u,\nabla u)$, the comparison theorem presented here easily encompasses such a variety of diffusion coefficients.

In view of the lack of restrictions on D, the comparison theorem is most easily stated in a weak setting; cf. [8]. Let $C^{0,1}(\Omega)$ denote the space of functions which are uniformly Lipschitz continuous on compact subsets of the domain $\Omega \subset R^n$. Let $C_0^{0,1}$ denote the subset of $C^{0,1}$ of functions having compact support in Ω. Define $Tu \equiv D(\underline{x},u,\nabla u)\nabla u$ for $u \in C^{0,1}$ and $\underline{x} \in \Omega$. Then the weak inequality

$$div(Tu) - div(Tv) \geq_w 0$$

holds for $u, v \in C_0^{0,1}$ if and only if

$$\int_\Omega [Tu - Tv] \cdot \nabla\phi \, d\underline{x} \leq 0 \tag{2}$$

for all nonnegative functions $\phi \in C_0^{0,1}$.

The derivatives of functions in $C^{0,1}$ are well-defined in the weak sense. Note that the weak inequality (2) is equivalent to the classical inequality if $u, v \in C^2(\Omega)$, $D \in C^1(\Omega \times R \times R^n)$, and Ω is sufficiently well-behaved to permit application of the divergence theorem.

Comparison Theorem: Let Ω be a bounded domain in R^n. Let $u, v \in C^{0,1}(\Omega)$. Let $D(\underline{x},w,\Delta w) > 0$ for $w = u$ or $w = v$ and for all $\underline{x} \in \Omega$. Suppose that D, u, and Ω are such that the linear inequality

$$div(D(\underline{x},u,\nabla u)\nabla\psi) >_w 0, \ \underline{x} \in \Omega, \ \psi(\underline{x}) \leq 0, \ \underline{x} \in \partial\Omega, \tag{3}$$

has a solution $\psi \in C^{0,1}(\Omega)$.

Then the inequalities

$$div(D(\underline{x},u,\nabla u)\nabla u) - div(D(\underline{x},v,\nabla v)\nabla v) \geq_w 0, \ \underline{x} \in \Omega, \tag{4}$$

and

$$u(\underline{x}) - v(\underline{x}) \leq 0, \quad \underline{x} \in \partial\Omega,$$

imply $u(\underline{x}) - v(\underline{x}) \leq 0$ throughout Ω.

Proof. Let $\psi(\underline{x})$ be as in (3). Then (2), (3), and (4) combine to yield

$$\int_{\Omega} [D(u)\nabla(u + \varepsilon\psi) - D(v)\nabla v] \cdot \nabla\phi \, d\underline{x} < 0 \tag{5}$$

for any nonnegative $\phi \in C_0^{0,1}(\Omega)$ and any constant $\varepsilon > 0$. (For convenience, we have written $D(u)$ for $D(\underline{x},u,\nabla u)$ and so forth).

In pursuit of a contradiction, suppose that $u - v > 0$ for some $\underline{x}^* \in \Omega$. If $\varepsilon > 0$ is sufficiently small, then $u + \varepsilon\psi - v > 0$ at $\underline{x}^*$ as well.

For such an ε, define $\phi \equiv max(0,u + \varepsilon\psi - v)$. Then $\phi > 0$ and $\phi \in C_0^{0,1}(\Omega)$. With this choice of ϕ, (5) may be written

$$\int_{\Omega^*} \sum_{i=1}^{n} \begin{pmatrix} u_{x_i} + \varepsilon\psi_{x_i} \\ v_{x_i} \end{pmatrix}^T \begin{bmatrix} D(u) & -D(u) \\ -D(v) & D(v) \end{bmatrix} \begin{pmatrix} u_{x_i} + \varepsilon\psi_{x_i} \\ v_{x_i} \end{pmatrix} d\underline{x} < 0 \tag{6}$$

where Ω^* is the support of ϕ. Since the matrix in the integrant in (6) has the characteristic values zero and $D(u) + D(v) > 0$, the quadratic form appearing in (6) is nonnegative definite.

But inequality (6) cannot hold if the integrand is never negative. Consequently, $u - v \leq 0$ throughout Ω.

While the comparison principle we have just established does not treat a differential operator whose form is as general as the divergence operators considered in [2,7,8] and elsewhere, the result given here complements these in that the expression $D(\underline{x},u,\nabla u)\nabla u$ need not necessarily satisfy the Lipschitz-like conditions or the ellipticity criterion typically imposed in these studies.

To illustrate the utility of this comparison theorem, we consider a model for steady-state diffusion with a nonlinear source

$$div(D(\underline{x},u,\nabla u)\nabla u) + f(\underline{x},u) = 0, \quad \underline{x} \in \Omega, \tag{7}$$

and derive bounds on the concentration u. These bounds which have a simple source-sink physical interpretation, are familiar from the case of the linear diffusion operator; cf. [4]. For simplicity, we suppose that D and Ω are smooth enough to permit (7) to have a classical solution u.

Bounding Theorem: Let u satisfy (7). Let m, M be constants. If $f(\underline{x},u) \geq 0$ for $u < m$ and $\underline{x} \in \Omega$ and if $u \geq m$ on $\partial\Omega$, then $u \geq m$ throughout Ω. Likewise, if $f(\underline{x},u) \leq 0$ for $u > M$ and if $u \leq M$ on $\partial\Omega$, then $u \leq M$ throughout Ω.

Proof. Consider the first case and suppose that $u < m$ at some point in Ω. Let $\Omega^* = \{\underline{x} \in \Omega : u(\underline{x}) < m\}$. Then

$$div(D(\underline{x},u,\nabla u)\nabla u) = -f(\underline{x},u) \leq 0, \quad \underline{x} \in \Omega^*,$$

while $u = m$ on $\partial\Omega^*$. Hence, the comparison principle forces $u \geq m$ in Ω^*, a contradiction. The second case can be treated in the same way.

Sharper comparison results, the time-dependent analog of (1), and other extensions and applications will be considered in more detail elsewhere.

REFERENCES

[1] P. Concus, and R. Finn, *On capillary free surfaces in the absence of gravity*, Acta Math. 132(1974), 177-198.

[2] J. Douglas, T. DuPont, and J. Serrin, *Uniqueness and comparison theorems for nonlinear elliptic equations in divergence form*, Arch. Rational Mech. Anal. 42(1971), 158-168.

[3] W. A. Gross, *Gas Film Lubrication*, John Wiley and Sons, Inc., New York, 1962.

[4] H. B. Keller, *Elliptic boundary value problems suggested by nonlinear diffusion processes*, Arch. Rational Mech. Anal. 35(1969), 363-381.

[5] Y. H. Ma and P. R. Peltre, *Freeze dehydration by microwave energy: Part I. Theoretical Investigation*, AICHE J. (21) (1975), 335-344.

[6] J. D. Murray, *A simple method for obtaining approximate solutions for a class of diffusion-kinetics enzyme problems: I. General class and illustrative examples*, Math. Biosci. 2 (1968), 379-411.

[7] J. Serrin, *On the strong maximum principle for quasilinear second order differential inequalities*, J. Functional Anal. 5(1970), 184-193.

[8] N. S. Trudinger, *On the comparison principle for quasilinear divergence structure equations*, Arch. Rational Mech. Anal. 57(1974), 128-133.

STABILITY OR CHAOS IN DISCRETE EPIDEMIC MODELS

Kenneth L. Cooke, Daniel F. Calef, and Eric V. Level
Pomona College, Claremont, California

Dedicated to Professor Chester G. Jaeger on the occasion of his eightieth birthday.

1. *Introduction*

In a number of recent investigations, it has been observed that very simple recursion relations can give rise to dynamic behavior of startling diversity. For example, the equation

$$x_{n+1} = Ax_n(1 - x_n) \tag{1}$$

has been studied in many papers, including [3], [6], [11]; a good survey has recently been given by May in [9]. The equation has the form

$$x_{n+1} = f(x_n) \tag{2}$$

where $f(x)$ is a quadratic function, and so generates a sequence x_0, $f(x_0), \ldots, f^{(n)}(x_0), \ldots,$ where $f^{(n)}$ denotes the nth iterated function. If $0 \leq A \leq 4$, f maps [0,1] into itself, and all iterates are defined on [0,1]. The behavior of the sequence $f^{(n)}(x_0)$ may be briefly summarized as follows. If $0 < A < 1$, $f^{(n)}(x_0)$ converges to 0 as $n \to \infty$, for any x_0 in [0,1]. If $1 < A < 3$, the sequence converges to $x^* = (A - 1)/A$, a nontrivial fixed point of the mapping f, for all x_0 in (0,1). As A is increased beyond 3, the point x^* becomes unstable and a stable 2-cycle or periodic motion of period 2 appears. Further increase of A results in cycles of period 4, 8, 16, and so on. Above a certain critical value of A, there are infinitely many cycles with all possible integer periods, as well as an uncountable number of asymptotically aperiodic trajectories. Such a state of affairs has been termed "chaos" by Li and Yorke, [6].

For more details, refer to the references below and to the paper by Kaplan and Marotto in these Proceedings.

Various other recursion relations have been studied and found to exhibit similar behavior. For example, Eq. (2) with $f(x) = x\ exp(a - bx)$ has been studied in [8]. See May and Oster, [10], for a catalogue of equations. In most cases, these have arisen as simple models of population growth processes. In addition, a few papers have appeared which deal with Eq. (2) when x is a k-dimensional vector, $k > 1$, and f maps $\mathbb{R}^k$ into $\mathbb{R}^k$. The paper by Beddington, Free, and Lawton, [1], contains interesting numerical studies for a system with $k = 2$ which models a host-parasite interaction. The paper by Marotto, [7], gives a sufficient condition for the appearance of chaos in k-dimensional systems, in terms of the concept of a "snap-back repeller." For a description of this work, see the paper by Kaplan in these Proceedings.

It is the purpose of this paper to describe our recent work on certain systems of difference equations which we have formulated as deterministic epidemic models. The simplest of these is the equation

$$x_{n+1} = (1 - x_n)(1 - e^{-Ax_n}) \tag{3}$$

The model underlying Eq. (3) is a variation of the classical Reed-Frost model, and A is a parameter which can be interpreted as the infectivity of the disease. As we shall see, the global behavior of (3) is entirely different from that of Eq. (1). In fact, there is a strong stability for all $A > 0$, and "chaos" does not appear. At first glance, this is somewhat surprising, in that the linear approximation $1 - Ax$ for e^{-Ax} reduces (3) to (1).

We are actually interested in a hierarchy of models, of which Eq. (3) is the first. The others are, in turn,

$$x_{n+1} = (1 - x_n - x_{n-1})(1 - e^{-Ax_n}) \tag{4}$$

$$x_{n+1} = (1 - x_n - x_{n-1} - x_{n-2})(1 - e^{-Ax_n}) \quad (5)$$

and, generally,

$$x_{n+1} = (1 - \sum_{j=0}^{k-1} x_{n-j})(1 - e^{-Ax_n}) \qquad (6)$$

The questions which we shall address are the following.

(a) For these equations, which arise from plausible physical assumptions, what range of dynamic behavior can occur? In particular, can "chaos" occur?

(b) In what way does the answer to (a) depend, not only on the parameter A, but also on the dimension k of the system?

In the next section, we shall indicate the physical assumptions and derivation of our equations. In Section 3, the Eq. (3) will be analyzed and complete proofs of stability properties will be given. In Section 4, we state the theoretical results which we have obtained so far for the higher dimensional equations (4), (5), and (6). Details of the proofs will be given elsewhere. Also, in Section 4 we report on several numerical experiments which are intended to help answer questions (a) and (b). These simulations indicate that for $k \geq 3$, bifurcation into periodic cycles occurs. Whether these is "chaos" is not yet known.

2. *The Model and Derivation of Equations*

Equations (3) - (6) are special cases of the model equations derived in [1]. However, we shall repeat the formulation of the model here, since this can be done quickly and with greater clarity for the special cases considered here.

In the classical epidemic model of the Reed-Frost type, it is assumed that the infection is spread by effective contact (contact that is sufficiently close to cause transfer of the infecting organism) between an individual who is infective and an individual who is susceptible. Contacts are assumed to be independent random events (the hypothesis of homogeneous mixing). In the deterministic version of the theory, this means that we assume that each individual will have (on the average) the same number of effective contacts with each other individual in a given period of time. Also, it is assumed that there is a constant total population. The infectious period for an individual is assumed to be fixed and is taken as the unit of time. At the end of his infectious period, an individual in the classical theory becomes totally and permanently immune. In our model, however, we shall assume on the contrary that he returns to the susceptible group with no immunity. Thus, our model is of the S-I-S type (susceptible to infectious to susceptible again). There are a number of diseases -- malaria, gonorrhea, the common cold -- for which for various reasons this is at least a reasonable first approximation.

The above assumptions imply that the population can be divided into two disjoint classes, the class of susceptible individuals and the class of infectives. We let

$I(t)$ = number of infectives at time t

$S(t)$ = number of susceptibles at time t

N = total population size

p = probability of effective contact between two specified individuals in one unit of time

$q = 1 - p = e^{-\alpha}$

The probability that a given susceptible does not have effective contact with any of the infectives is q^I and therefore the expected number of new infections (in a unit of time) is

$$S(t)[1 - q^{I(t)}] = S(t)[1 - e^{-\alpha I(t)}]$$

Consequently, the equations for the population dynamics are

$$\begin{aligned} I(t+1) &= S(t)[1 - e^{-\alpha I(t)}] \\ S(t+1) &= I(t) + S(t)e^{-\alpha I(t)} \qquad (1) \\ S(t) + I(t) &= N \end{aligned}$$

Here, $t = 0,1,2,\ldots.$ Introducing normalized variables by

$$I(t) = Nx_1(t), \quad S(t) = Nx_2(t), \quad \alpha N = A \tag{2}$$

we obtain

$$\begin{aligned} x_1(t+1) &= x_2(t)[1 - e^{-Ax_1(t)}] \\ x_2(t+1) &= x_1(t) + x_2(t)e^{-Ax_1(t)} \\ x_1(t) + x_2(t) &= 1 \quad \text{for all } t. \end{aligned} \tag{3}$$

It is perhaps more convenient to think of Eq. (3) as defining a transformation T and to let x be the vector with components x_1 and x_2. Then (3) may be written

$$x(t+1) = Tx(t) \tag{4}$$

Also, since $x_2 = 1 - x_1$, we can eliminate x_2 and replace the system (3) by the scalar equation

$$x_1(t+1) = [1 - x_1(t)][1 - e^{-Ax_1(t)}] \tag{5}$$

Finally, if we write x instead of x_1, and then use the subscript n to denote time, we can replace (5) by

$$x_{n+1} = (1 - x_n)(1 - e^{-Ax_n})$$

This is Eq. (3) of Section 1.

Next, let us alter the model by supposing that after one time unit spent in the infectious state, an individual spends one time unit in an isolated state, and then returns to the susceptible state. Such isolation might occur because the individual withdraws from normal activities after the appearance of symptoms and until recovery. Such a model may be said to be of S-I-R-S (susceptible - infectious - removed - susceptible) type. Define $S(t)$, $I(t)$ as before and let

$R(t)$ = number in the removed state at time t

Then we obtain the equations

$$\begin{aligned} I(t+1) &= S(t)[1 - e^{-\alpha I(t)}] \\ R(t+1) &= I(t) \\ S(t+1) &= R(t) + S(t)e^{-\alpha I(t)} \\ S(t) + R(t) + I(t) &= N \end{aligned} \tag{6}$$

After normalization, these become

$$\begin{aligned} x_1(t+1) &= x_3(t)[1 - e^{-Ax_1(t)}] \\ x_2(t+1) &= x_1(t) \\ x_3(t+1) &= x_2(t) + x_3(t)e^{-Ax_1(t)} \\ x_1(t) + x_2(t) + x_3(t) &= 1 \end{aligned} \tag{7}$$

Elimination of x_3 yields the system

$$\begin{aligned} x_1(t+1) &= [1 - x_1(t) - x_2(t)][1 - e^{-Ax_1(t)}] \\ x_2(t+1) &= x_1(t) . \end{aligned} \tag{8}$$

Equivalent to (8) is the scalar second order difference equation (with $x_1(t)$ replaced by x_n)

$$x_{n+1} = (1 - x_n - x_{n-1})(1 - e^{-Ax_n}) \tag{9}$$

This is Eq. (4) in Section 1.

If we assume that an individual spends one time unit in the infectious state, then two time units in a removed or isolated state, and then returns to the susceptible state, we obtain the system

$$\begin{aligned} x_1(t+1) &= [1 - x_1(t) - x_2(t) - x_3(t)][1 - e^{-Ax_1(t)}] \\ x_2(t+1) &= x_1(t) \\ x_3(t+1) &= x_2(t) \end{aligned} \tag{10}$$

Equivalent to this is the scalar third order difference equation (with $x_1(t)$ replaced by x_n)

$$x_{n+1} = (1 - x_n - x_{n-1} - x_{n-2})(1 - e^{-Ax_n}) \tag{11}$$

which is Eq. (5) in Section 1. Generally, if $(k - 1)$ time units are spent in isolation, Eq. (6) of Section 1 results.

3. *Analysis for One-dimensional Model*

We first consider the model with no removed state, that is, $k = 1$. The equation to be considered is

$$x_{n+1} = (1 - x_n)(1 - e^{-Ax_n}) \equiv f(x_n) \tag{1}$$

$(0 \leq x_n \leq 1)$. The behavior of solutions of this equation is described by the following theorem.

<u>THEOREM 1</u>: For all $A \geq 0$, the function $f(x) = (1 - x)(1-e^{-Ax})$ maps [0,1] into [0,1]. If $0 < A \leq 1$, then for any x_0 in [0,1], the sequence x_n generated by Eq. (1) converges to 0 as $n \to \infty$. If $A > 1$, the map f has a unique nonzero fixed point x^* in (0,0.5) satisfying

$$x^* = (1 - x^*)(1 - e^{-Ax^*}) \tag{2}$$

The point x^* is globally stable in the sense that for any x_0, $0 < x_0 < 1$, the sequence x_n converges to x^*.

<u>Proof</u>: We give an elementary proof. It is obvious that f maps [0,1] into itself. Moreover, since

$$f'(x) = -1 + e^{-Ax}[1 + A(1 - x)],$$

$$f''(x) = -e^{-Ax}[2A + A^2(1 - x)],$$

we see that if $0 < A \leq 1$, the graph of $y = f(x)$ lies under the graph of $y = x$ on $0 < x \leq 1$, is concave down, and has a unique maximum point. Standard arguments then show that if $0 \leq x_0 \leq 1$, the sequence x_n is monotone decreasing to zero.

Now suppose that $A > 1$. Since $f(0) = f(1) = 0$, $f'(0) = A$, $f''(x) < 0$, the graph of $f(x)$ has a unique maximum point and there is a unique intersection of the graph with the line $y = x$. See Figure 1 which shows the graph for various values of A. Let x^* be the intersection. Clearly x^* satisfies (2). Since

$$e^{-Ax^*} = \frac{1 - 2x^*}{1 - x^*}$$

is positive, we have $0 < x^* < 0.5$. Now let $(x_M, f(x_M))$ denote the maximum point. If $x_M \geq x^*$, it is easy to see geometrically that for any x_0, $0 < x_0 < 1$, $x_0 \neq x^*$, we have $x_1 = f(x_0) < x_M$ and thereafter the sequence tends monotonically to x^*.

Now consider the case $x_M < x^*$. We shall first prove that the sequence initiated with $x_0 = x_M$ converges to x^*. In Figure 2, let M denote the maximum point and draw MN horizontally to the point N on the line $y = x$. The abcissa of N is $x_1 = f(x_M)$. Draw a perpendicular from M onto the x-axis, meeting the line $y = x$ at P. Draw PQ parallel to MN and of equal length. Since PN lies on the line of slope 1, MQ has slope -1 and $QPMN$ is a square. Since

$$f'(x) = -1 + e^{-Ax}[1 + A(1 - x)] \geq -1 + e^{-Ax} > -1,$$

the curve $y = f(x)$ must lie above the diagonal MQ for $x_M \leq x \leq 1$, and NQ intersects the curve at a point R above

Q. Therefore, a horizontal line through R intersects $y = x$ at a point S above and to the right of P. Thus, if $x_2 = f(x_1) = f(f(x_M))$, we have $x_M < x_2 < x_1$. It readily follows that the successive iterates satisfy

$$x_M < x_2 < x_4 < \ldots < x_5 < x_3 < x_1$$

Therefore, the even- and odd-numbered iterates must converge to limits, x' and x'' respectively. Suppose that $x' < x''$. Then x' and x'' must be two-periodic points of the iteration, that is, fixed points of $f(f(x))$. However, consider the sequence starting from the initial point x'. By an argument exactly like the above, drawing a square with one corner at $(x', f(x'))$, we see that $f(f(x')) > x'$, so that x' cannot be a two-periodic point. See Figure 3. This proves that $x' = x''$ and thus that the sequence starting with $x_0 = x_M$ converges (necessarily to x^*).

If a sequence is started with $x^* < x_0 \leq 1$, then $x_1 = f(x_0) < x^*$, so without loss of generality we assume that $x_0 < x^*$. If $x_M < x_0 < x^*$, an argument like the above shows that the sequence converges to x^*. If $0 < x_0 < x_M$, then after a few iterations we reach a point x_n such that $x_M < x_n < f(x_M)$. Then by the previous arguments all subsequent iterates lie in $[x_M, f(x_M)]$ and the sequence converges to x^*. This completes the proof of Theorem 1.

Some additional explanation as to why there is global stability rather than chaos for Eq. (1) may be helpful. It has been pointed out by May and Oster, [10], that for one-dimensional systems $x_{n+1} = f(x_n)$, multiple bifurcations and chaos may occur if the graph of $f(x)$ contains a "hump" and if the slope $f'(x)$ can be "tuned" by varying a parameter. For the equation (1) studied here, however, we have seen that $f'(x) > -1$ for x in $[0,1]$ and all $A > 1$. Consequently $|f'(x^*)| < 1$ for all A, and x^* remains stable. This can also be explained in relation to the graph of the second iterate, $f^{(2)}(x) \equiv f(f(x))$. For Eq.

(1) of Section 1, May and Oster show that the graph of $f^{(2)}(x)$ has two humps and that as A increases through the value where the fixed point x^* becomes unstable, two new intersections of $f^{(2)}(x)$ and the 45° line appear. Thus, stability of x^* is exchanged for stability of a new 2-cycle. At a larger value of A, stability of this 2-cycle is exchanged for stability of a new 4-cycle, and so on. In our case, however, a 2-cycle never appears, as suggested by the graphs of $f^{(2)}(x)$ in Figure 4 for various values of A. These graphs intersect the 45° line at only one point.

It is also possible to prove Theorem 1 by Liapunov function methods, as we shall show in a subsequent paper.

4. *Higher Dimensional Systems*

We shall now describe some results which have been obtained for the higher dimensional models derived in Section 2. We let x denote a vector in $\mathbb{R}^k$ with components $x_1, x_2, \ldots, x_k$. The systems under consideration have the form

$$x(t+1) = f(x(t)), \quad t = 0,1,2,\ldots \tag{1}$$

or, in abbreviated notation,

$$x' = f(x) \tag{2}$$

Thus, x' denotes the vector which is the image of x under f, or it is the state vector one time step after x. In our models, the components of f have the form

$$\begin{aligned} f_1(x) &= (1 - x_1 - x_2 - \ldots - x_k)(1 - e^{-Ax_1}) \\ f_j(x) &= x_{j-1} \qquad (j = 2,\ldots,k) \end{aligned} \tag{3}$$

Let

$$S = \{x \text{ in } \mathbb{R}^k: \quad 0 \leq x_j \leq 1 \quad \text{for all } j, \quad \sum_{j=1}^{k} x_j \leq 1\} \quad (4)$$

It is easily seen that $f(S) \subset S$. For, clearly all $x_j' \geq 0$, and moreover

$$\sum_{j=1}^{k} x_j' = (1 - \sum_{j=1}^{k} x_j)(1 - e^{-Ax_1}) + x_1 + \ldots + x_{k-1}$$
$$\leq 1 - \sum_{j=1}^{k} x_j + x_1 + \ldots + x_{k-1} = 1 - x_k \leq 1.$$

Some results on stability can be obtained by using arguments of Liapunov function type. We employ the following theorem of La Salle, [5]. First, a set M is called <u>invariant</u> if $f(M) = M$. Also, we write $f^1(x) = f(x)$ and $f^{(n)}(x) = f(f^{(n-1)}(x))$ for $n = 2,3,\ldots$.

<u>Invariance Principle</u> (La Salle). Consider a system of the form $x' = f(x)$, where $f : \mathbb{R}^k \to \mathbb{R}^k$. Let G be any set in $\mathbb{R}^k$ and let $V : \mathbb{R}^k \to \mathbb{R}$ be a continuous function. Define

$$\dot{V}(x) = V(f(x)) - V(x)$$

and assume that

$$\dot{V}(x) \leq 0, \quad \text{for } x \text{ in } G.$$

Let

$$E = \{x \text{ in } \overline{G}: \quad \dot{V}(x) = 0\},$$

and let M be the largest invariant set in E. If x, $f(x)$, $f^{(2)}(x), \ldots$ is a "solution" of the system which is bounded and in G, then there exists a number c in $\mathbb{R}$ such that

$$f^{(n)}(x) \text{ tends to } M \cap V^{-1}(c) \quad \text{as } n \to \infty.$$

THEOREM 2: Consider the system $x' = f(x)$ where f is given in Eq. (3). If $0 < A \leq 1$, then for any initial vector x^0 in S, the sequence $x^0, f(x^0), f^{(2)}(x^0), \ldots,$ remains in S and converges to the zero vector. In other words, 0 is globally stable if $0 < A \leq 1$.

The proof of Theorem 2 utilizes the Invariance Principle with a function V of the form $V(x) = c^T x$ where c is a suitably chosen constant vector and c^T its transpose. Details will be published elsewhere.

The range of possible behavior is much more complex for $A > 1$. Consider the two dimensional system

$$\begin{aligned} x_1' &= (1 - x_1 - x_2)(1 - e^{-Ax_1}) \\ x_2' &= x_1 \end{aligned} \tag{5}$$

THEOREM 3: If $A > 1$, the system (5) has a unique nonzero equilibrium point $x_1 = x_2 = x^*$, given by

$$x^* = (1 - 2x^*)(1 - e^{-Ax^*}) \tag{6}$$

The number x^* satisfies $0 < x^* < \frac{1}{3}$. This equilibrium is locally stable for all $A > 1$.

It is easy to see that if there is an equilibrium then $x_1 = x_2 = x^*$ where x^* satisfies (6). By rewriting (6) as

$$e^{-Ax^*} = \frac{1 - 3x^*}{1 - 2x^*}$$

and looking at graphs, we see that there is a unique intersection point x^* and it lies in $(0, \frac{1}{3})$. To prove that this equilibrium is locally stable, we linearize (5) around (x^*, x^*). The resulting system has characteristic equation

$$\lambda^2 + (a - b)\lambda + a = 0$$

$$a = 1 - e^{-Ax^*}, \quad b = Ae^{-Ax^*}(1 - 2x^*)$$

Application of the Hermite-Fujiwara stability criterion (see [4]) leads to the symmetric matrix

$$\begin{pmatrix} 1 - a^2 & a - b - a^2 + ab \\ a - b - a^2 + ab & 1 - a^2 \end{pmatrix}$$

which can be shown to be positive definite for all $A > 1$. Details of this proof will be given elsewhere.

Numerical values of x^* for several values of A, as calculated by Newton's method, are given in Table 1.

Table 1

A	1.0	1.5	2.0	5.0	10.0	20.0
x^*	0	.13246	.19741	.30506	.32909	.33319

Numerical simulations strongly indicate that the equilibrium point (x^*,x^*) is globally stable for all $A > 1$. Despite attempts to construct a suitable Liapunov function, we have so far been unable to give a rigorous proof for this. If true, then our models of dimensions one and two exhibit no bifurcation into periodic orbits or "chaos".

The three dimensional system is

$$\begin{aligned} x_1' &= (1 - x_1 - x_2 - x_3)(1 - e^{-Ax_1}) \\ x_2' &= x_1 \\ x_3' &= x_2 \end{aligned} \tag{7}$$

and here the story is very different. For $A > 1$, there is again a unique nonzero equilibrium $x_1 = x_2 = x_3 = x^*$,

$$x^* = (1 - 3x^*)(1 - e^{-Ax^*})$$

and $0 < x^* < \frac{1}{4}$. Calculated values of x^* are given in Table 2.

Table 2

A	1.0	2.0	5.0	10.0	20.0	30.0
x^*	0	.1419	.22278	.24418	.24957	.24997

For system (7), the characteristic equation can be shown to be

$$\lambda^3 + (a - b)\lambda^2 + a\lambda + a = 0$$
$$a = 1 - e^{-Ax^*}, \quad b = Ae^{-Ax^*}(1 - 3x^*)$$

Application of the Hermite-Fujiwara criterion leads to a three-by-three symmetric matrix. Although we have not been able completely to settle the question of positive definiteness for this matrix, we have been able to prove that it is not positive definite for $A > A_1$ where

$$A_1 = \frac{4e - 3}{e - 1} = 4.58 \ldots$$

Therefore, we conjecture that local stability of the equilibrium is lost at $A = A_1$.

This conjecture is supported by numerical studies. Figure 5 shows the graph of $x_1(t)$ versus t for $t = 0,1,\ldots,1000$, starting from the initial values $x_1(0) = .3$, $x_2(0) = .2$, $x_3(0) = .3$, with $A = 4.55$. Each individual point $(t,x_1(t))$ is plotted. The three separate "curves" which appear in the plot are the loci of points $x_1(3n)$, $x_1(3n + 1)$, and $x_1(3n + 2)$, $n = 1,2,3,\ldots,$ respectively. It seems clear that there is slow convergence of $x_1(t)$ to the equilibrium value x^*.

Figure 6 shows the graph of $x_1(t + 1)$ against $x_1(t)$ for $A = 6$. After an initial settling down, the points appear to fall on a continuous curve. For $A = 10$, the solution after a fairly short transient phase seems to be asymptotic to a stable 5-cycle. For $A = 20$, the behavior is similar except that the transient phase is longer. Figure 7 shows $x_1(t)$ plotted against t for $A = 20$ and 1000 time steps. For $A = 30$, there appears to be convergence to a 10-cycle after 7000 or 8000 time steps, and for $A = 40$ there seems to be a stable 20-cycle reached after some 30000 steps. Limitations on numerical accuracy have made it difficult to extend these studies to still larger values of A.

Similar considerations of the four-dimensional system show that there is a unique nonzero equilibrium x^*, $0 < x^* < \frac{1}{5}$, which is locally unstable for $A > A_2 = 1 + A_1$.

In conclusion, it appears that for the one- and two-dimensional models, the solutions always approach stable equilibria. For higher dimensional models, at least for Eq. (7), stability of the equilibrium is lost for sufficiently large A, and periodic cycles appear. Further increase of A tends to produce bifurcation into cycles of larger period. Whether there is a regime of chaos has not yet been determined. We hope to present further explanation for these phenomena at a later time.

REFERENCES

[1] J. R. Beddington, C. A. Free, and J. H. Lawton, *Dynamic complexity in predator prey models framed in difference equations*, Nature 255(1975), 58-60.

[2] K. L. Cooke, *A discrete-time epidemic model with classes of infectives and susceptibles*, Theor. Population Biology 7(1975), 175-196.

[3] F. C. Hoppensteadt and J. M. Hyman, *Periodic solutions of a logistic difference equation*. A Preprint.

[4] R. E. Kalman, *On the Hermite-Fujiwara theorem in stability theory*, Quart. Appl. Math. 23(1965), 279-282.

[5] J. P. LaSalle, *Stability theory for difference equations*, in *A Study of Ordinary Differential Equations*, J. K. Hale (editor), Math. Association of America, forthcoming.

[6] T. Y. Li, and J. A. Yorke, *Period three implies chaos*, Amer. Math. Monthly 82(1975), 985-992.

[7] F. R. Marotto, *Snap-back repellers imply chaos in* R^n, J. Math.

Anal. Appl., to appear.

[8] R. M. May, *Biological populations with nonoverlapping generations; stable points, stable cycles, and chaos*, Science 186(1974), 645-647.

[9] R. M. May, *Simple mathematical models with very complicated dynamics*, Nature 261(1976), 459-467.

[10] R. M. May, and G. F. Oster, *Bifurcations and dynamic complexity in simple ecological models*, Amer. Nat. 110, in press.

[11] S. Smale, and R. F. Williams, *The qualitative analysis of a difference equation of population growth*, J. Math. Biology 3(1976), 1-4.

ACKNOWLEDGMENT

This work was partially supported by N.S.F. Grant No. MCS72-04965 A04.

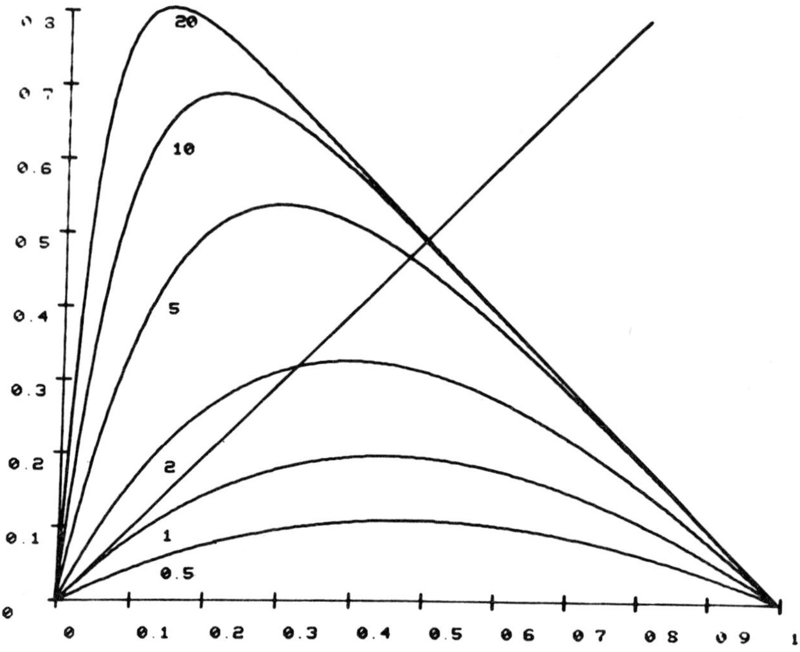

Figure 1. Graph of $f(x) = (1 - x)(1 - e^{-Ax})$ for $A = 0.5$, 1, 2, 5, 10, 20.

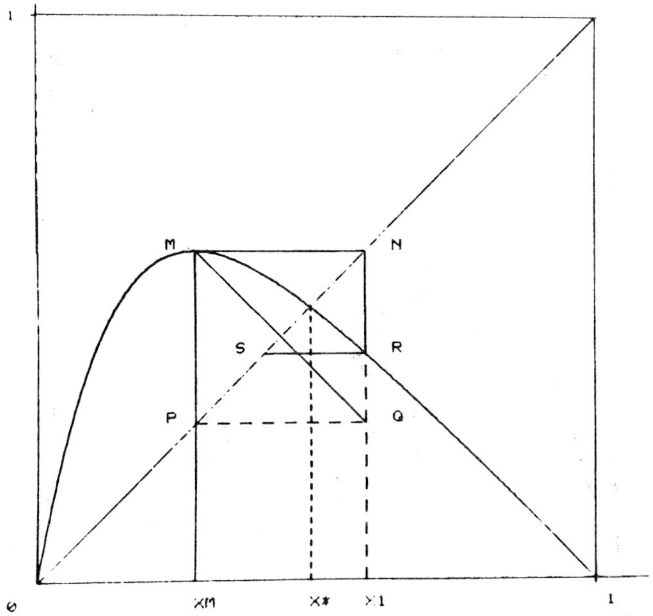

Figure 2. Construction of Iterates Starting from Maximum Point.

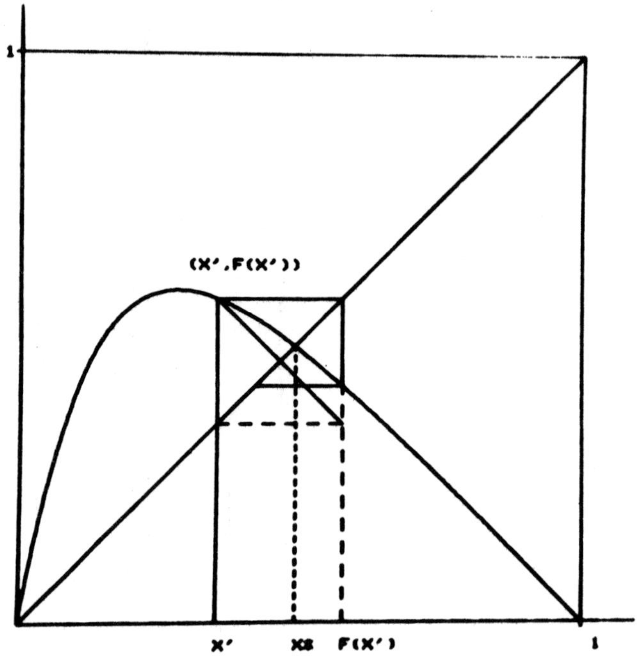

Figure 3. Demonstration that there is no Two-periodic Point.

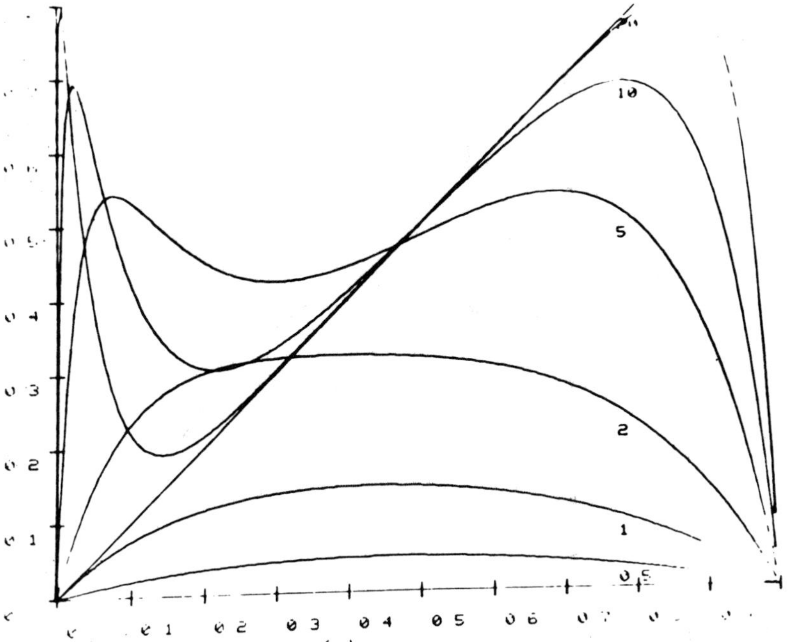

Figure 4. Graph of $f^{(2)}(x)$ for $A = 0,5,1,2,5,10,20$, showing only one intersection with 45° line.

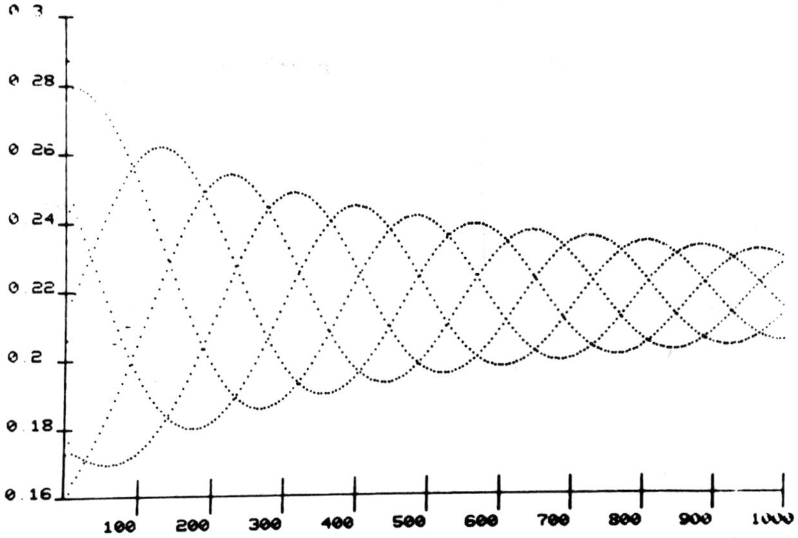

Figure 5. Graph of $x_1(t)$ versus t for model of dimension 3 with $A = 4.55$.

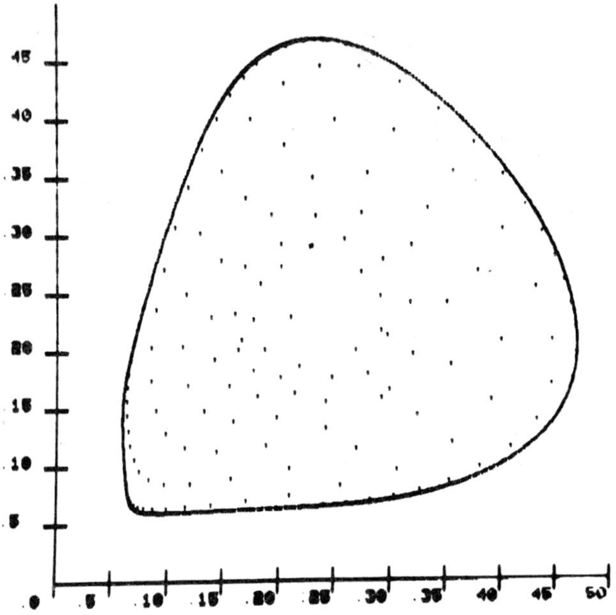

Figure 6. Graph of $x_1(t + 1)$ versus $x_1(t)$ for model of dimension 3 with $A = 6$.

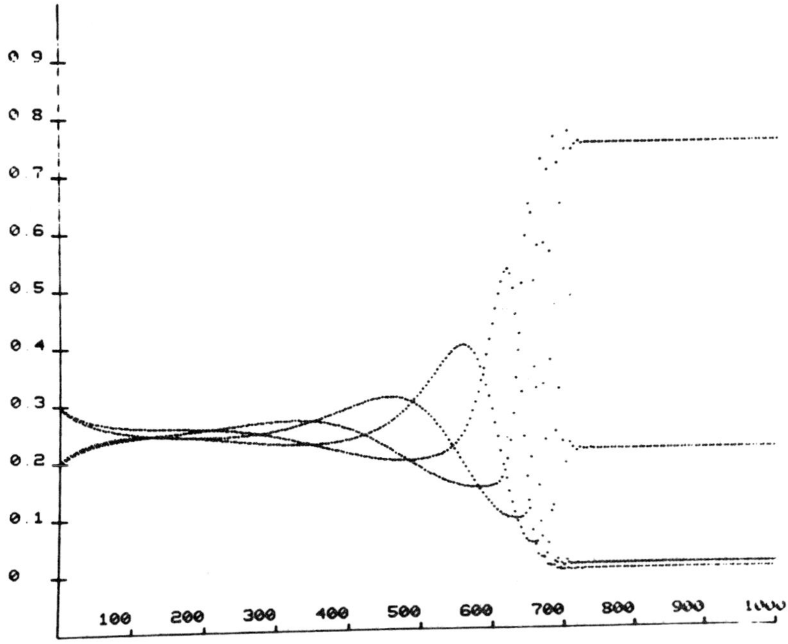

Figure 7. Graph of $x_1(t)$ versus t for model of dimension 3 with $A = 20$.

ANALYTIC METHODS FOR APPROXIMATE SOLUTION OF ELLIPTIC FREE BOUNDARY PROBLEMS

by

B. A. Fleishman*, Rensselaer Polytechnic Institute

and

T. J. Mahar, Courant Institute of Mathematical Sciences

We illustrate here the application of perturbation and iteration methods to certain BVP's for elliptic PDE's of the form

$$\Delta u + f(x,u) = 0 \tag{1}$$

where $x = (x_1,\ldots x_n)$ is a point in R^n, Δ denotes the Laplacian operator, u is a real scalar variable, and f is a piecewise-continuous function of $x_1,\ldots x_n$ and u. When f has jump discontinuities with respect to u, among the interfaces across which f changes abruptly there may be "free boundaries" which are not known a priori but must be found along with the solution $u = u(x)$.

Suppose f is a step-function in u and depends also on m of the independent variables, say, $x_1,\ldots x_m$, where $0 \leq m < n$. Let D be a fixed region in R^n whose bounding surfaces are independent of $x_{m+1},\ldots,x_n$.

Now consider a BVP for (1) on D, denoted by $P(\varepsilon)$, in which a small parameter ε occurs in the boundary conditions in such a way that the "reduced" problem $P(0)$ does not involve $x_{m+1},\ldots,x_n$.

Suppose a non-trivial solution $u_0 = u_0(x_1,\ldots,x_m)$ of $P(0)$ is obtainable. In both the perturbation and iteration treatments presented here we seek a solution of $P(\varepsilon)$ in a neighborhood of the known solution u_0 of the nonlinear "reduced" problem $P(0)$, with the free boundary close to that of u_0.

The perturbation technique is applied in a formal manner to

* Research supported by U.S. Army Research Office.

a BVP for the PDE $\Delta u + f(u) = 0$ (where f is a step-function); the analysis is developed in detail. Then a different BVP (for the same PDE) is attacked by a constructive iterative method. An existence proof for a classical solution of the free boundary problem is outlined; details will be given elsewhere.

Free boundary problems for equations similar to (1) occur in plasma physics; namely, equations of the form $Lu + f(x,u) = 0$, where L is an elliptic operator and f is, however, piecewise-linear in u, not discontinuous. Free boundary problems for equations of the form div $(K \; grad \; u) = 0$, where $K = K(x,u)$ is a piecewise-continuous function, arise in the equilibrium Stefan problem and govern certain diffusion and metallurgical processes. Such problems are also being investigated by the methods illustrated here.

Besides occurring naturally, problems with discontinuous nonlinearities are sometimes introduced as approximations to problems with smooth nonlinearities (which, in general, can not be solved explicitly).

NONLINEAR BOUNDARY VALUE PROBLEMS FOR ELLIPTIC SYSTEMS IN THE PLANE*

R. P. Gilbert†
University of Delaware

Introduction

In this paper we present some new results on nonlinear elliptic systems in the plane. We are interested, in particular, with semilinear equations where the principal part consists of an elliptic system of $2r$ equations. These may be put into a normal form using the hypercomplex algebra of Douglis [5] or the matrix notation of Bojarski [4]. When $r = 1$ the normal form for the principal part reduces to the Cauchy-Riemann operator.

In Section I the recent work of Begehr-Gilbert [2] concerning the nonlinear Cauchy-Riemann equations is reviewed. Sections II-III presents the results of Begehr-Gilbert [3] concerning generalized hyperanalytic function theory, that is the function theoretic study of the linear elliptic systems of $2r$ equations in the plane. Since this theory is at present reasonably well developed, an extension of the use of function theoretic methods to nonlinear problems is possible. This is the content of Section IV, which presents two approaches to the problem of existence and uniqueness of these nonlinear, boundary-value problems, namely a Schauder-fixed-point method, and also an imbedding method approach based upon a priori estimates.

I. *Nonlinear Cauchy-Riemann Problems*

In this section we present the investigations of Begehr and Gilbert [2] on a boundary value problem associated with a nonlinear Cauchy-Riemann equation,

$$w_{\bar{z}} = f(z,w). \tag{1.1}$$

* An invited address at the International Conference on Nonlinear Systems and Applications.

† This research was supported in part by the Air Force Office of Scientific Research through Grant AF-AFOSR 76-2879.

Here the $\bar{z}$ derivative is meant to be understood in a Sobolev sense. The interest in such problems stems from the earlier investigations of Bers [1], Haack [13], and Vekua [15], concerning the linear equations

$$(1.2) \qquad w_{\bar{z}} = Aw + B\bar{w} + C,$$

which are the complex form of the Hilbert normal form for a linear elliptic system of first order,

$$(1.3) \qquad \begin{aligned} u_x - v_y &= au + bv + c, \\ u_y + v_x &= \alpha u + \beta v + \gamma. \end{aligned}$$

Let G be a simply-connected region in $\mathbb{C}$ that has at least two boundary points. Then there exists a conformal mapping ϕ of this region on the unit disk, and the Green's functions of the 1st and 2nd kind are given by

$$(1.4) \qquad G^I(\zeta,z) := \frac{1}{2\pi} \log \left| \frac{\phi(\zeta) - \phi(z)}{1 - \overline{\phi(\zeta)}\phi(z)} \right|, \quad (\zeta,z \in D),$$

and

$$(1.5) \qquad G^{II}(\zeta,z) := \frac{1}{2\pi} \log |(\phi(\zeta) - \phi(z))(1 - \overline{\phi(\zeta)}\phi(z))|, \quad (\zeta,z \in D).$$

If $\dot{D}$ is smooth then there exists a constant c which is determined by

$$(1.6) \qquad c := \sup_{\zeta,z \in G} \left| \frac{\phi'(z)(\zeta - z)}{\phi(\zeta) - \phi(z)} \right| \geq 1;$$

consequently,

$$(1.7) \qquad |G^K_\zeta(\zeta,z)| \leq \frac{1}{2\pi} \left| \frac{\phi'(\zeta)}{\phi(\zeta)-\phi(z)} \right| \leq \frac{c}{|\zeta-z|}, \quad (\zeta,z \in D), \quad (K=I,II).$$

By means of the above Green's functions we may represent all functions $w(z) \in \mathcal{C}^1(G) \cap \mathcal{C}(\bar{G})$ as [13]

$$(1.8) \qquad w(z) = -\theta(z) + (\underset{\sim}{P}w)(z), \quad \text{where}$$

$$\underset{\sim}{P}(z) := i\int_G \{w_{\bar\zeta}(\zeta)[G^I_\zeta(\zeta,z) + G^{II}_\zeta(\zeta,z)] \tag{1.9}$$
$$+ \overline{w_{\bar\zeta}(\zeta)}[G^I_{\bar\zeta}(\zeta,z) - G^{II}_{\bar\zeta}(\zeta,z)]\}d\zeta d\bar\zeta, \quad (z \in D).$$

Here $\theta(z)$ is given by

$$\theta(z) := \int_{\dot G} \{Re\, w(\zeta)[d_n G^I(\zeta,z) - idG^{II}(\zeta,z)] \tag{1.10}$$
$$+ i\, Im\, w(\zeta) d_n G^{II}(\zeta,z).$$

Recalling the properties of the Green's functions

$$G^I(\zeta,z) = 0, \quad d_n G^{II}(\zeta,z) = -\frac{1}{2\pi}|d\phi(\zeta)|, \qquad \zeta \in G,\ z \in G,$$

we recognize that $\theta = \tilde\phi + iC$, where $\tilde\phi$ is a holomorphic function satisfying the boundary conditions

$$Re\,\tilde\phi|_{\dot G} = Re\, w(\zeta), \quad \int_{\dot G} Im\,\tilde\phi(\zeta)|d\phi(\zeta)| = 0. \tag{1.11}$$

The constant C is determined by the "Boundary normalization" of $Im\, w$ namely

$$C := \frac{1}{2\pi}\int_{\dot G} Im\, w(\zeta)|d\phi(\zeta)|.$$

We investigate the solvability of the nonlinear boundary value problem

$$w_{\bar z} = f(z,w), \quad Re\, w|_{\dot G} = \phi, \quad \frac{1}{2\pi}\int_{\dot D} Im\, w(\zeta)|d\phi(\zeta)| = C, \tag{1.12}$$

under the hypotheses

H-1 $f(z,w(z))$ <u>is defined in</u> $\bar G \times \mathbb{C}$, <u>and for all</u> $w \in \mathcal{C}^0(\bar G)$ $f(z,w(z)) \in L_p(\bar G)$, $p > 2$.

H-2 $|f(z,w) - f(z,\omega)| \le \frac{1}{8c} g(z,|w-\omega|)$ $(z \in \bar G;\ w,\omega \in \mathbb{C})$ <u>where</u> $g(z,x)$ <u>is defined on</u> $\bar G \times [0,\infty]$ <u>and has the properties</u>

(i) $g(z,0) = 0, \quad g(z,x) \leq g(z,y), \quad (z \in G, \quad 0 \leq x \leq y)$

(ii) $g(z,x(z)) \in L_p(\overline{G}), \quad (p > 2)$ for each nonnegative, continuous function $x(z)$.

(iii) There exists a positive K such that

$$\int_G g(\zeta,K) \frac{d\xi d\eta}{|\zeta - z|} \leq K, \tag{1.13}$$

(iv) On the set of continuous, non-negative functions, bounded above by K, the integral operator $\underset{\sim}{I}$,

$$(\underset{\sim}{I}x)(z) := \int_G g(\zeta,x(\zeta)) \frac{d\xi d\eta}{|\zeta - z|}, \quad (z \in G, \ x \in \mathcal{C}(\overline{G}), \ \xi = \xi + i\eta) \tag{1.14}$$

does not have 1 as an eigenvalue.

Lemma 1 [2]: Under conditions (i)-(iv) above on $g(z,x(z))$ the integral inequality

$$x(z) \leq (\underset{\sim}{I}x)(z), \tag{1.15}$$

$$x \in \mathcal{B}_K := \{y(z) : y(z) \in \mathcal{C}^0(\overline{G}), \quad 0 \leq y(z) \leq K\},$$

has only the zero-function for a solution.

Proof: Let $\mathcal{L}_K$ be the class of all solutions of (1.15). Then because of the monotonicity of

$$\underline{\overline{X}}_0(z) := \sup\{x(z) : x(z) \in \mathcal{L}_K, \quad z \in \overline{G}\}$$

is also a solution of (1.15). The sequence of functions defined by

$$\underline{\overline{X}}_n(z) := (\underset{\sim}{I}\,\underline{\overline{X}}_{n-1})(z), \quad (n = 1,2,\ldots)$$

are non-negative, continuous, monotone increasing and bounded above by K. Consequently

$$\underline{\overline{X}}(z) := \lim_{n\to\infty} \underline{\overline{X}}_n(z)$$

exists and belongs to $\mathcal{L}_K$. Because of (iv) $\underline{\overline{X}}(z) \equiv 0$ from which our conclusion follows.

We are now able to prove an existence and uniqueness

THEOREM 1: If the constant K of (1.13) fulfills the condition

$$\|\theta\| + M[L_p(f_0,\overline{G}) + L_p(g_{2K},\overline{G})] \leq K \tag{1.16}$$

where $f_0(z) := 8c\, f(z,0)$, $g_{2K}(z) := g(z,2K)$,

$$M = M(p,G) := \left(\frac{2}{\alpha q}\right)^{1/q} d^{\alpha}(G), \quad \left(\alpha = \frac{p-2}{p}, \quad \frac{1}{p} + \frac{1}{q} = 1\right),$$

$$d(G) := diam\ (G),$$

then there exists one and only one solution of the nonlinear boundary value problem (1.12) in the set

$$\mathcal{A} := \{w : w \in (\overline{G}), \quad \|w\| \leq K\}$$

Proof: $\mathcal{A}$ is a subset of the Banach space $\mathcal{B}$ of continuous functions with the maximum norm, $\|w\| := \max_{z\in\overline{G}} |w(z)|$. The operator $\underset{\sim}{F} := -\theta + \underset{\sim}{P}$ takes the compact, convex subset $\mathcal{A}$ into itself; hence, from the Schauder fixed point theorem there exists a solution in $\mathcal{A}$. This solution is Hölder continuous when $\tilde{\phi}$ is Hölder continuous. That it satisfies the boundary conditions may be shown as in [13].

We show uniqueness by noting that if w_1, w_2 are two solutions, then from the Hölder inequality

$$|w_1(z)-w_2(z)| = |\underset{\sim}{P}w_1 - \underset{\sim}{P}w_2| \leq \int_G g(\zeta, |w_1(\zeta)-w_2(\zeta)|)\cdot\frac{d\xi d\eta}{|\zeta - z|} \tag{1.17}$$

$$\leq ML_p(g_{2K},\overline{G}) \leq K, \quad z \in D.$$

Since $x(z) := |w_1(z) - w_2(z)|$ is a solution of (1.15) our Lemma implies $w_1(z) \equiv w_2(z)$.

Remark: If $f(z,w(z))$ is Hölder continuous for Hölder continuous functions $w(z)$, then $\frac{\partial w}{\partial \bar{z}}$ exists in the classical sense [2] and we have a classical solution.

Remark: It is clear that instead of condition (1.16) it is sufficient to demand that

$$(1.18) \qquad |\theta(z)| + \int_G (|f_0(\zeta)| + g(\zeta,K)) \frac{d\xi d\eta}{|\zeta - z|} \le K \quad z \in \overline{G}.$$

and

$$(1.19) \qquad \int_G g(\zeta, 2K) \frac{d\xi d\eta}{|\zeta - z|} \le K, \quad z \in \overline{G},$$

hold. One must only choose K such that

$$(1.20) \qquad 2||\theta|| \le K, \quad 2ML_p(f\ ,\overline{G}) \le K(1 - 4ML_p(g,\overline{G})),$$

under the assumption that

$$4ML_p(g,\overline{G}) < 1,$$

which is surely satisfied for a sufficiently small region. We list a result which follows quickly from the above discussion.

THEOREM 2: The boundary value problem (1.12) has in $\mathcal{B}(\overline{G})$ a unique generalized solution if

(i) $f(z,w(z)) \in L_p(\overline{G})$, $(p > 2)$, $w \in \mathfrak{C}(\overline{G})$,

(ii) $|f(z,w_1) - f(z,w_2)| \le g(z)|w_1 - w_2|$, where $g \in L_p(\overline{G})$,

$$4\int_G g(\zeta) \frac{d\xi d\eta}{|\zeta - z|} \le 1.$$

Remark: The conditions (1.16) and (1.18), (1.19) are clearly satisfied by (i), (ii) for any K (providing $||\Theta|| < K$) if $d(\overline{G})$ is sufficiently small.

II. *Elements of Hypercomplex Function Theory*

A. Douglis [5] showed that elliptic systems of the first order in two independent variables, with sufficient smoothness requirements on the coefficients of the first order terms, can be decomposed into the cannonical subsystems

$$
\begin{aligned}
&u_{0,x} - v_{0,y} + \dots &&= g_0,\\
&u_{0,y} + v_{0,x} + \dots &&= h_0,\\
&u_{k,x} - v_{k,y} + au_{k-1,x} + bu_{k-1,y} + \dots &&= g_k,\\
&u_{k,y} + v_{k,x} + av_{k-1,x} + bv_{k-1,y} + \dots &&= h_k,\\
&(k = 1,2,\dots,r-1),
\end{aligned}
$$

where the dots represent terms without derivatives. In order to treat these subsystems Douglis introduces a hypercomplex function algebra

$$w := \sum_{k=0}^{r-1} e^k w_k(x,y), \quad e^r = 0, \tag{2.2}$$

with which the principal part of this subsystem could be written in a particularly simple form†

$$Dw = 0, \quad D := \frac{\partial}{\partial \bar{z}} + q(x,y)\frac{\partial}{\partial z}, \tag{2.3}$$

where $q(z) := q(x,y)$ is the nilpotent††

$$q(z) := \sum_{k=1}^{r-1} e^k q_k(z). \tag{2.4}$$

The power series development of analytic functions may be used to obtain hyperanalytic functions, i.e. $\mathcal{C}^1$ functions $\phi(z)$ satisfying $D\phi = 0$. Examples of these are

$$exp\ z := exp\ z_0 \sum_{k=0}^{r-1} \frac{1}{k!}(z - z_0)^k$$

and

† This normalization is a variation on Douglis' form. See Bojarski [4], Kühn [14], and Gilbert-Hile [11].

†† The notation $q(z)$ means here that q is a function of the point z, not the variable z.

$$log\ z := log\ z_0 + \sum_{k=1}^{r-1} \frac{(-1)^{k-1}}{k} \left(\frac{z - z_0}{z_0}\right)^k, \quad (z_0 \neq 0),$$

where $z_0 :=$ complex part of z (c.p. [z]), and $(z - z_0) :=$ nilpotent part of z (n.p. [z]). We remark that the addition theorem for hypercomplex quantities

$$exp\ a\ exp\ b = exp\ (a + b),$$

and the identities

$$log\ exp\ z = z,\ exp\ log\ z = z, \quad (z_0 \neq 0),$$

easily follow. In order to discuss convergence of sequences of hyperanalytic functions, the "algebraic norm", $|w| := \sum_{\ell=0}^{r-1} |w_\ell|$ is frequently employed.

A hypercomplex quantity with nonvanishing complex part permits a hypercomplex polar representation. Let

$$z := \sum_{k=0}^{r-1} e^k z_k, \quad \bar{z} := \sum_{k=0}^{r-1} e^k \bar{z}_k, \quad \rho := \sum_{k=0}^{r-1} e^k \rho_k.$$

Then under the assumption that $z_0 \neq 0$, the equation

$$z\bar{z} = \rho^2$$

may be solved for real-hypercomplex ρ, i.e. the real components ρ_k may be obtained from

$$\sum_{\ell=0}^{k} z_\ell \bar{z}_{k-\ell} = \sum_{\ell=0}^{k} \rho_\ell \rho_{\kappa-\ell} \quad (0 \leq k \leq r - 1).$$

Indeed there are exactly two solutions which are uniquely determined by the sign of ρ_0.

Let us write $z = \rho\ exp(i\theta)$, with hypercomplex θ; then it follows that

$$\theta = arg\ z_0 + 2n\pi + i \sum_{k=1}^{r-1} \frac{(-1)^k}{k}\ exp(ik\ arg\ z_0) N^k,$$

$(n = 0, \pm 1, \pm 2, \ldots)$, and $N := \text{n.p.}[\rho^{-1}z]$. The complex part is only determined modulo 2π, i.e. $\theta_0 = \arg z_0 + 2n\pi$, whereas the nilpotent part is uniquely determined. For nilpotent hypercomplex numbers θ does not exist.

From the equation

$$\rho^2 = z\bar{z} = \rho^2 \exp i(\theta - \bar{\theta}),$$

it follows that θ is real hypercomplex. We note that in general $|\rho| \neq |z|$.

In [8], [9], [10] the differential equations

$$(2.5) \qquad Dw = aw + b\bar{w} \qquad (a,b \text{ hypercomplex functions}),$$

and more generally,

$$(2.6) \qquad Dw = \sum_{k,\ell=0}^{r-1} e^k (a_{k\ell} w_\ell + b_{k\ell} \bar{w}_\ell)$$

are investigated and an extension of the Bers-Haack-Vekua theory [1], [13], [15] is developed. We present below some of the results of [3] associated with the linear, boundary-value problem

$$(2.7) \qquad Dw = aw + b\bar{w}, \quad Re(\bar{\gamma}w)|_{\dot{G}} = \phi.$$

In what follows we assume that the coefficients a, b are Hölder continuous in a simply connected, bounded region G, with Hölder-smooth boundary $\dot{G}$. On $\bar{G}$ a, b are to be continuous with the exception of jump discontinuities. q is to be once Hölder-continuously differentiable on $\bar{G}$. γ and ϕ are to be respectively once Hölder-continuously differentiable, and Hölder continuous on $\dot{G}$. Finally $\gamma_0 := \text{c.p.}[\gamma]$ is to have no zeros on $\dot{G}$, and the Index of $\bar{\gamma}$, designated by

$$(2.8) \qquad Ind[\bar{\gamma}] := Ind[\bar{\gamma}_0] := \frac{1}{2\pi} \int_{\dot{G}} d \arg \bar{\gamma}_0,$$

is to be an integer.

Remark: By using generalized derivatives we may clearly weaken our regularity conditions on the coefficients.

As in Section I we will attempt to represent solutions of boundary value problems by means of the operator $\underset{\sim}{P}$ (1.9). We recall [3], that if f is Hölder continuous in $\overline{G}$, then $\underset{\sim}{P}f \in \mathcal{C}^1$, and

$$\frac{\partial}{\partial \bar{z}} (\underset{\sim}{P}f) = f, \quad \underset{\sim}{\pi}f := \frac{\partial}{\partial z} (\underset{\sim}{P}f),$$

the singular integral operator $\underset{\sim}{\pi}$ may be expressed through

(2.9) $$(\underset{\sim}{\pi}f)(z) :=$$

$$\frac{\phi'(z)}{2\pi i} \int_G \left[f(\zeta) \frac{\phi'(\zeta)}{[\phi(\zeta) - \phi(z)]^2} + \overline{f(\zeta)} \frac{\overline{\phi'(\zeta)}}{[1 - \overline{\phi(\zeta)}\phi(z)]^2} \right] d\zeta d\bar{\zeta}$$

The first integral is interpreted as a Cauchy principal value; whereas, the second integral represents a function holomorphic in G and continuous in $\overline{G}$. The operator $\underset{\sim}{P}$ may be represented by means of the Pompeiu operator [3]

(2.10) $$(\underset{\sim}{T}_G f)(z) := \frac{1}{2\pi i} \int_G f(\zeta) \frac{d\zeta d\bar{\zeta}}{\zeta - z}, \quad (z \in \mathbb{C}),$$

as

(2.11) $$(\underset{\sim}{P}f)(z) := (\underset{\sim}{T}_E f(\psi)\overline{\psi'})(\phi(z)) - \phi(z)(\underset{\sim}{T}_E f_1(\psi_1)\psi_1')(\phi(z)),$$

where $\psi := \phi^{-1}$ is the inverse mapping of ϕ and takes the unit circle E onto G. ψ_1 and $f_1(\psi_1)$ are defined by

(2.12)
$$\psi_1(z) := \psi\left(\frac{1}{\bar{z}}\right), \quad z \in E' := \mathbb{C} - E,$$

$$f_1(\psi_1(z)) := \frac{1}{z} \overline{f(\psi_1(z))}, \quad z \in E'.$$

From Vekua [15] we recall that $\underset{\sim}{T}_G$ is a compact operator in $\mathcal{C}_\alpha(\overline{G})$ and imbeds $\mathcal{C}_\alpha(\overline{G})$ in $\mathcal{C}_\alpha(\overline{G})$. Hence $\underset{\sim}{P}f \in \mathcal{C}^1_\alpha(\overline{D})$ for each subdomain D, such that $\overline{D} \subset G$. In particular, one has for the

Hölder norm $\mathcal{C}_\alpha(f,D) := \sup_D |f| + \sup_{D\times D} \frac{|f(z) - f(z')|}{|z - z'|^\alpha}$, the inequalities [3]

$$\begin{aligned} &\mathcal{C}_\alpha(\underset{\sim}{P}f,G) \leq M_1(\alpha,G)\mathcal{C}_\alpha(f,G), \\ (2.13) \quad & \\ &\mathcal{C}_\alpha(\underset{\sim}{\pi}f,D) \leq M_2(\alpha,D)\mathcal{C}_\alpha(f,D). \end{aligned}$$

The operators $\underset{\sim}{P}$ and $\underset{\sim}{\pi}$ may be applied to hypercomplex functions providing the componentwise application is defined. For a hypercomplex, Hölder-continuous function in $\overline{G}$, we define the norm

$$(2.14) \qquad \mathcal{C}_\alpha(f,\overline{G}) := \sum_{k=0}^{r-1} \mathcal{C}_\alpha(f_k,\overline{G})$$

III. *Boundary Value Problems for Generalized Hyperanalytic Functions (Linear Systems).*†

It is sufficient for the case of index zero to consider the problem

$$(3.1) \qquad Dw = \tilde{a}w + \tilde{b}\overline{w}, \quad Re\, w|_{\dot{G}} = \phi.$$

We show this by noting that when $Ind\,\overline{\gamma} = Ind\,\overline{\gamma}_0 = 0$ it is possible to harmonically continue the hypercomplex R,θ defined by

$$(3.2) \qquad \gamma = R\,exp(i\theta), \quad R_0 \neq 0$$

into G such that $R_0 = |\gamma_0|$ remains positive. In this way γ is continued into G as a continuously differentiable function such that its complex part does not vanish; consequently, γ^{-1} exists in G. If w is a solution of the boundary value problem

$$(3.3) \qquad Dw = \tilde{a}w + \tilde{b}\overline{w}, \quad Re(\overline{\gamma}w)|_{\dot{G}} = \phi,$$

† The material presented in this section was obtained by the author and Heinrich Begehr [3].

then

(3.4) $$\omega := \bar{\gamma} w$$

satisfies

(3.5) $$D\omega = a\omega + b\bar{\omega}, \quad Re\,\omega|_{\dot{G}} = \phi$$

where

$$a := \bar{\gamma}^{-1} D\bar{\gamma} + \tilde{a}, \quad b := \bar{\gamma}\tilde{b}\gamma^{-1}.$$

Conversely, if ω is a solution of (3.3) then $w = \bar{\gamma}^{-1}\omega$ is a solution of the original boundary value problem. The problem (3.4) may be expressed in terms of its components as

(3.6) $$\frac{\partial \omega_k}{\partial \bar{z}} = a_0\omega_k + b_0\bar{\omega}_k + f_k, \quad Re\,\omega_k|_{\dot{G}} = \phi_k \quad (0 \leq k \leq r-1)$$

where

(3.7) $$f_0 := 0, \quad f_k := \sum_{\ell=0}^{k-1} (a_{k-\ell}\omega_\ell + b_{k-\ell}\bar{\omega}_\ell - q_{k-\ell}\frac{\partial}{\partial z}\omega_\ell),$$

$$(0 \leq k \leq r-1).$$

From Section I, it is clear that the boundary-value problem (3.6)-(3.7) is equivalent to the system of integral equations

(3.8) $$\omega_k = \Omega_k + \underset{\sim}{P}(a_0\omega_k + b_0\bar{\omega}_k + f_k),$$

where Ω_k is the analytic function

(3.9) $$\Omega_k(z) := -\int_{\dot{G}} \phi_k(\zeta)\left[d_n G^I(\zeta,z) - i dG^{II}(\zeta,z)\right] + C_k.$$

C_k is an arbitrary constant, which we may make definite by setting

(3.10) $$C_k = -i\int_{\dot{G}} I_m\omega_k(\zeta) d_n G^{II}(\zeta,z).$$

The integral equations are uniquely solvable in the inhomogeneous case $(f_k \neq 0)$, modulo the "boundary normalization" (3.10). The homogeneous problem $f_k = 0$, $\phi_k = 0$ has a nontrivial solution, more specifically it has as a solution a function which vanishes nowhere and such that $C_k \neq 0$. Furthermore we may obtain all solutions of the problem by multiplying the nontrivial solution by real constants. These solutions are with the exception of the discontinuities of the coefficients differentiable, satisfy the differential equation, and the boundary data.

Starting with

$$\omega_0 = \Omega_0 + \underset{\sim}{P}(a_0\omega_0 + b_0\overline{\omega}_0),$$
$$\frac{\partial \omega_0}{\partial z} = \Omega_0' + \underset{\sim}{\pi}(a_0\omega_0 + b_0\overline{\omega}_0),$$

and successively eliminating the derivatives $\frac{\partial \omega_\ell}{\partial z}$ in the expression for ω_k leads to the integral equation†

$$\omega_k = \Omega_k + \underset{\sim}{P}\left[\sum_{\ell=0}^{k} (-q\underset{\sim}{\pi})^\ell (a\omega + b\overline{\omega}) + \sum_{\ell=0}^{k-1} (-q\underset{\sim}{\pi})^\ell (-q\Omega')\right]_k \tag{3.11}$$

Since q is a nilpotent we may write the system (3.11) as

$$\omega_k = \Omega_k + \underset{\sim}{P}\left[\sum_{\ell=0}^{r-1} (-q\underset{\sim}{\pi})^\ell (a\omega + b\overline{\omega}) + \sum_{\ell=0}^{r-1} (-q\underset{\sim}{\pi})(-q\Omega')\right]_k \tag{3.12}$$

so that for $\omega = \sum_{k=0}^{r-1} e^k \omega_k$ we have

$$\omega = \Omega + \underset{\sim}{P}\left[\sum_{\ell=0}^{r-1} (-q\underset{\sim}{\pi})^\ell (a\omega + b\overline{\omega} - q\Omega')\right], \tag{3.13}$$

or

$$\omega = \Omega - \underset{\sim}{R}(q\Omega') + \underset{\sim}{R}(a\omega + b\overline{\omega}), \tag{3.14}$$

where

$$\underset{\sim}{R} := \underset{\sim}{P}(1 + q\underset{\sim}{\pi})^{-1} = \underset{\sim}{P}\left(\sum_{k=0}^{r-1} (-q\underset{\sim}{\pi})^k\right). \tag{3.15}$$

† Here the index k means the k-th component of the corresponding hypercomples quantity.)

Recalling (2.13), (2.14) we have the following imbedding inequality

$$(3.16)\qquad \mathcal{C}_\alpha(\underset{\sim}{R}f,\overline{D}) \leq M_1(\alpha,G)\,\frac{M_3^{\,r}(\alpha,D)-1}{M_3(\alpha,D)-1}\,\mathcal{C}_\alpha(f,\overline{D}),$$

where

$$M_3(\alpha,D) := (r-1)M_2(\alpha,D)\,\mathcal{C}_\alpha(q,\overline{G}),$$

for each $\overline{D} \subset G$ and each hypercomplex $f \in \mathcal{C}_\alpha(\overline{D})$.

We next turn to the case of negative index, namely the problem,

$$(3.17)\qquad Dw = \tilde{a}w + \tilde{b}\overline{w},\quad Re(\overline{\gamma}w)\,|_{\dot{G}} = \phi,$$

where

$$(3.18)\qquad Ind\,\overline{\gamma} = \frac{1}{2\pi}\int_{\dot{G}} d\,arg\,\overline{\gamma}_0 = -n (n \in N).$$

By means of a transformation of the form

$$v := \psi^{-1}w,\quad \psi(z) := \prod_{\nu=1}^{n}\,(t(z)-t(z_\nu))$$

where $z_\nu (1 \leq \nu \leq n)$ are n distinct points, and $t(z)$ is the generating solution [5], [8] for the operator D, the problem is reduced to one of index zero, namely

$$(3.19)\qquad Dv = av + b\overline{v},\quad Re(\overline{p}v)\,|_{\dot{G}} = \phi,$$

with

$$a := \tilde{a},\quad b := \tilde{b}\,\overline{\psi}/\psi,\quad p = \gamma\overline{\psi},\quad Ind\,\overline{p} = 0.$$

Let v_h be a nontrivial solution of the homogeneous problem $(\phi = 0)$, with nonvanishing complex part v_{h_0}. Then the constant

$$C_h = -i\int_{\dot{G}} I_m v_h(\zeta)\,d_n G^{II}(\zeta,z)$$

The $c.p.[v_h] = v_{h_0}$ is a solution of a type problem which was investigated by Haack-Wendland [13] and Vekua [15], namely

$$\text{(3.20)}\qquad \frac{\partial}{\partial \bar{z}} v_{h0} = a_0 v_{h0} + b_0 \prod_{\nu=1}^{n} \overline{\frac{z - z_\nu}{z - z_\nu}}\, \overline{v_{h0}},$$
$$Re(\bar{p}_0 v_{h0})|_{\dot{G}} = 0, \quad Ind\, \bar{p}_0 = 0.$$

For the system (3.20) it is known [13] (11.1) that $|v_{h0}| > 0$, and hence $|v_h| := \sum_{k=0}^{r-1} |v_{hk}| > 0$.

The function ψv_h is a solution of the corresponding homogeneous system

$$\text{(3.21)}\qquad Dw = aw + b\bar{w}, \quad Re(\bar{\gamma} w)|_{\dot{G}} = 0,$$

and each solution of this homogeneous boundary value problem may be written in the form

$$w_h = \lambda \psi v_h, \; \lambda = \sum_{k=0}^{r-1} e^k \lambda_k (\lambda_k \in I\!R, \; 0 \le k \le v - 1, \; \lambda_0 \ne 0).$$

Furthermore, for every real hypercomplex λ w_h is a solution. Since $v_{h0} \ne 0$ in $\bar{G}$ we may represent

$$v_h = exp(\omega_h), \quad \text{where} \quad \omega_h := log\, v_h.$$

If w_1 is a solution of (3.21) whose complex part does not vanish identically and w_2 another solution with the same zeros $z_\nu (1 \le \nu \le n)$ as w_1 in G, then $w_2 = \lambda w_1$, where λ is real hypercomplex. Each continuous solution with $n + 1$ different zeros in $\bar{G}$, of which n lie in G, is identically zero.

If v_0 is a particular solution of (3.19) with $\phi \ne 0$, then $w_0 = \psi v_0$ is a particular solution of (3.17) with $\phi \ne 0$, and consequently $w = \psi(v_0 + \lambda w_h)$ is the general solution of the boundary value problem (3.17) with negative characteristic. In this representation of the solution we have $2n + r$ free, real-parameters which arise from the appearance of the λ and z_ν $(1 \le \nu \le n)$ terms.

In order to investigate the soltuions w_h of (3.21) which

have nonvanishing complex part, we consider the algebra $\mathcal{A}_r$ of real hypercomplex quantities. As usual we call m hypercomplex functions $w_\mu (1 \le \mu \le m)$ over $\mathcal{A}_r$, having nonidentically vanishing complex parts $w_{\mu 0}$ $(1 \le \mu \le m)$ linear independent, when

$$\sum_{\mu=1}^{m} \lambda_\mu w_\mu = 0 \quad (\lambda_\mu \in \mathcal{A}_r),$$

implies that $\lambda_\mu = 0$ $(1 \le \mu \le m)$. This concept is identical with the linear independence over R of the complex parts $w_{\mu 0}$ $(1 \le \mu \le m)$.

THEOREM 4: The homogeneous boundary value problem (3.21) with Index $-n$ $(n \in N)$ has exactly $2n + 1$ linearly independent solutions over $\mathcal{A}_r$ with non-identically vansihing complex part, and exactly $(2n + 1)r$ linearly independent solutions over R.

Proof: We show first that at most $2n + 1$ linearly independent solutions over $\mathcal{A}_r$ exist. To this end let there be given n separate points z_ν $(1 \le \nu \le n)$ in G, and $2n + 1$ linearly independent solutions w_μ $(0 \le \mu \le 2n)$ of (3.21) over $\mathcal{A}_r$. Furthermore, these solutions are to have nonvanishing complex part. Using a nontrivial solution.

$$\lambda_\mu^{(0)} = \sum_{k=0}^{r-1} e^k \lambda_{\mu k}^{(0)} \qquad (0 \le \mu \le 2n),$$

of the linear system,

$$\sum_{\mu=0}^{2n} \lambda_\mu^{(0)} \tilde{w}_{\mu 0}(z_\nu) = 0 \qquad (1 \le \nu \le n),$$

that is the real system of $2nr$ equations

$$\sum_{\mu=0}^{2n} \lambda_\mu^{(0)} \tilde{w}_{\mu 0}(z_\nu) = 0 \qquad (1 \le \nu \le n),$$

$$\sum_{\mu=0}^{2n} \sum_{\ell=0}^{k} \lambda_{\mu 1}^{(0)} \tilde{w}_{\mu, k-\ell}(z_\nu) = 0 \qquad (1 \le k \le r - 1, \ 1 \le \nu \le n),$$

we define the hypercomplex function

$$w_0 := \sum_{\mu=0}^{2n} \lambda_\mu^{(0)} \tilde{w}_\mu$$

Without loss of generality, we may assume $\lambda_{00}^{(0)} \neq 0$. The system $\{w_0, \tilde{w}_\mu (1 \leq \mu \leq 2n)\}$ is clearly linear independent over $\mathcal{A}_r$. For each $\mu (1 \leq \mu \leq 2n)$ we set

$$w_\mu := \sum_{k=1}^{2n} \lambda_k^{(\mu)} \tilde{w}_k$$

Where we determine the real hypercomplex quantities $\lambda_k^{(\mu)}$ $(1 \leq k, \ \mu \leq 2n)$ from the systems of equations

$$w_{2\mu}(z_\nu) = \delta_{\mu\nu}, \quad w_{2\mu-1}(z_\nu) = i\delta_{\mu\nu} \quad (1 \leq \mu, \ \nu \leq n),$$

that is

$$\sum_{k=1}^{2n} \sum_{\ell=0}^{m} \lambda_{k\ell}^{(2\mu)} w_{k,m-\ell}(z_\nu) = \delta_{0m}\delta_{\mu\nu}$$

$$(3.22) \qquad (0 \leq m \leq r-1, \quad 1 \leq \mu, \quad \nu \leq n)$$

$$\sum_{k=1}^{2n} \sum_{\ell=0}^{m} \lambda_{k\ell}^{(2\mu-1)} w_{k,m-\ell}(z_\nu) = i\delta_{0m}\delta_{\mu\nu} .$$

The unique solvability of the two inhomogeneous systems (3.22) follows from the observation that the corresponding homogeneous system

$$(3.23) \qquad \sum_{k=1}^{2n} \kappa_k \tilde{w}_k(z_k) = 0 \quad (1 \leq \nu \leq n; \ \kappa_k \in \mathcal{A}_r, \quad 1 \leq k \leq n),$$

has only the trivial solution since otherwise the left hand side of (3.23) would be of the form $\kappa w_0(z_k)$, which negates the linear independency of $\{w_0, \tilde{w}_\mu (1 \leq \mu \leq 2n)\}$. See Haak-Wendland [13].

If w is an arbitrary solution of (3.21) with not identically vanishing complex part, then w may be represented as a linear combination over $\mathcal{A}_r$ of the $w_\mu (0 \leq \mu \leq 2n)$.

$$\lambda_{2\nu} := Re\, w(z_\nu), \quad \lambda_{2\nu-1} := Im\, w(z_\nu) \quad (1 \leq \nu \leq n)$$

we notice that the solution $w - \sum_{\mu=1}^{2n} \lambda_\mu w_\mu$ must be a multiple of w_0 since they both have the zeros $z_\nu (1 \leq \nu \leq n)$.

We may show analogously that there are at most $(2n + 1)r$ linear independent solutions over R; we refer the reader to Begehr-Gilbert [3] for these details.

What remains is to show that there are exactly $2n + 1$ over $\mathcal{Q}_r$ [and $(2n + 1)r$ over R] linearly independent solutions of (3.21). We consider here only the case over $\mathcal{Q}_r$ [3].

Let $z_k (1 \leq k \leq n)$ be n distinct points in G and w_0 a nontrivial continuous solution of (3.21) which vanishes at these points, but whose complex part does not vanish identically. Suppose we have $2n$ further solutions $w_\mu (1 \leq \mu \leq 2n)$ with the conditions

$$w_{2k}(z_\ell) = \delta_{k\ell}, \quad w_{2k-1}(z_\ell) = i\delta_{k\ell} \quad (1 \leq k, \quad \ell \leq n)$$

Since these solutions do not have identically vanishing complex part we can form with w_0 a linear independent set.

As in [3] (see also [13] pg. 283.) we may show, for each fixed k, the existence of two solutions of (3.21) $\{\omega_1(z), \omega_2(z)\}$ such that

$$Im\{\omega_{10}(z_k)\overline{\omega_{20}(z_k)}\} \neq 0.$$

This means we may uniquely determine real hypercomplex numbers $\{\lambda_{2k}^{(1)}, \lambda_{2k}^{(2)}\}$ respectively $\{\lambda_{2k-1}^{(1)}, \lambda_{2k-1}^{(2)}\}$, that are solutions of the linear systems

$$\lambda_{2k}^{(1)} \omega_1(z_k) + \lambda_{2k}^{(2)} \omega_2(z_k) = 1,$$

$$\lambda_{2k-1}^{(1)} \omega_1(z_k) + \lambda_{2k-1}^{(2)} \omega_2(z_k) = i.$$

With these hypercomplex quantities we define the solutions of (3.21).

$$w_\mu = \lambda_\mu^{(1)} w_1 + \lambda_\mu^{(2)} w_2 \quad (\mu = 2k - 1,\ 2k),$$

with the properties

$$w_{2k}(z_\ell) = \delta_{k\ell},\quad w_{2k-1}(z_\ell) = i\delta_{k\ell} \quad (1 \le \ell \le n)$$

Repeating this procedure for each $i(1 \le k \le n)$ provides $2n + 1$ linear independent solutions of (3.21) over $_r$.

We remark that for the case of positive index, namely

$$(3.23) \qquad Dw = \tilde{a}w + \tilde{b}\overline{w},\quad Re(\overline{\gamma}w)|_{\dot{G}} = \phi,\quad Ind\ \overline{\gamma} = n \in N,$$

a similar treatment is possible. We introduce in this instance a function $\psi(z)$ defined by

$$\frac{1}{\psi(z)} := \sum_{\nu=1}^{n} (t(z) - t(z_\nu))^{-1}.$$ Setting $\omega = \psi w$

takes (3.23) into the same type of problem, however again of index zero,

$$(3.24) \qquad \begin{aligned} &D\omega = a\omega + b\overline{\omega},\quad a := \tilde{a},\quad b := \tilde{b}\psi/\overline{\psi}, \\ &Re(\overline{\gamma}\omega/\psi)|_{\dot{G}} = \phi. \end{aligned}$$

This case may now be treated as the negative index problem was in the preceeding material [3].

IV. *Boundary Value Problems for Nonlinear Systems in the Plane.*

We consider below the nonlinear elliptic system

$$(4.1) \qquad Dw = f(z,w) := \sum_{k=0}^{r-1} e^k f_k(z,w),$$

for the case of homogeneous boundary data with index zero, namely $Re\ w|_{\dot{G}} = 0$. Equation (4.1) may be written in component form as

$$(4.2) \qquad \frac{\partial w_k}{\partial \overline{z}} = f_k(z,w) - \sum_{\ell=0}^{k-1} q_{k-\ell} \frac{\partial w_\ell}{\partial z} \quad (0 \le k \le r - 1),$$

$$f_k(z,w) := f_k(z,w_0,\ldots,w_{r-1}). \tag{4.3}$$

Following the procedure of Section III we may obtain an integral equation formulation for the solution of (4.1). Namely, setting

$$w_0 = \Omega_0 + \underset{\sim}{P}(f_0(z,w)), \tag{4.4}$$

$$\frac{\partial}{\partial z} w_0 = \Omega_0' + \underset{\sim}{\pi}(f_0)(z),$$

etc., we are finally led to

$$\begin{aligned} w(z) &= \Omega(z) + \underset{\sim}{P} \sum_{\ell=0}^{r-1} (-q\underset{\sim}{\pi})^{\ell} f(z,w) - \underset{\sim}{P} \sum_{\ell=0}^{r-1} (-q\underset{\sim}{\pi})^{\ell} q\Omega' \\ &= \Omega(z) - (\underset{\sim}{R}\, q\Omega')(z) + (\underset{\sim}{R}\, f)(z). \end{aligned} \tag{4.5}$$

If $Re\, w|_{\dot{G}} = 0$, then from Section I, we have instead of (4.5)

$$w(z) = (\underset{\sim}{R}\, f)(z) := (\underset{\sim}{Q}\, w)(z), \tag{4.6}$$

with the boundary normalization

$$c := \frac{1}{2\pi} \int_{\dot{G}} Im\, w(\zeta)\, |d\phi(\zeta)|.$$

The operator $\underset{\sim}{Q}$ has the form

$$\begin{aligned} (\underset{\sim}{Q}\, w)(z) := i \int_G \Big\{ & f(\zeta,w(\zeta)) \left[G_\zeta^I(\zeta,z) + G_\zeta^{II}(\zeta,z) \right] \\ & + \overline{f(\zeta,w(\zeta))} \left[G_{\bar{\zeta}}^I(\zeta,z) - G_{\bar{\zeta}}^{II}(\zeta,z) \right] \Big\}\, d\zeta d\bar{\zeta}, \end{aligned} \tag{4.7}$$

and $(\underset{\sim}{\chi}\, w)(z) := (\underset{\sim}{\pi}\, f)(z)$ may be written as

(4.8) $(\underset{\sim}{\chi}\, w)(z) := $

$$= \frac{\phi'(z)}{2\pi i} \int_G \left\{ \frac{f(\zeta,w(\zeta))\phi'(\zeta)}{(\phi(\zeta) - \phi(z))^2} + \frac{\overline{f(\zeta,w(\zeta))}\,\overline{\phi'(\zeta)}}{(1 - \overline{\phi(\zeta)}\phi(z))^2} \right\} d\zeta d\overline{\zeta}$$

We make the following assumptions concerning $f(z,w(z))$ in order to apply the methods of Section I to the present case.

H-1 $f(z,w(z))$ is defined in $\overline{G} \times \not{C}^r$, and for all $w \in C^0(\overline{G})$ $f(z,w(z)) \in L_p(\overline{G})$, $p > 2$

H-2 $|f(z,w) - f(z,\omega)| := \sum_{k=0}^{r-1} |f_k(z,w) - f_k(z,\omega)|$

(4.9) $\leq k\, g(z, |w - \omega|), \quad (z \in \overline{G}\,;\, w, \omega \in \not{C}^r),$

where $g(z,x)$ is defined on $\overline{G} \times [0,\infty]$ and has the properties (i)-(iv) of Section I. The constant k, moreover, is defined as

$$k := \frac{1}{8c} \left(\frac{M_3(\alpha,G) - 1}{M_3^r(\alpha,G) - 1} \right),$$

where c is given by (1.6), and M_3 is the imbedding constant of (3.16).

Let B be the Banach space of continuous hypercompelx functions defined on $\overline{G} \subset \not{C}$, and equipped with the norm

$$||w|| := \max_{z\in\overline{G}} |w|, \quad |w| := \sum_{k=0}^{r-1} |w_k| .$$

Using the results of Section I the following theorem may be easily established.

THEOREM 5: If the constant K of (1.13) fulfills

(4.10) $M[L_p(f_0,\overline{G}) + L_p(g_{2k},\overline{G})] \leq K,$

where

$$f_0(z) := 8c \left(\frac{M_3^r(\alpha,G) - 1}{M_3(\alpha,G) - 1} \right) f(z,0) ,$$

$$g_{2k}(z) := g(z,2k),$$

$$M = \left(\frac{2\pi}{\alpha q}\right)^{1/q} d^{\alpha}(G) \qquad \left(\alpha = \frac{p-2}{p},\ \frac{1}{p}+\frac{1}{q}=1\right)$$

then there exists one and only one solution of the nonlinear boundary value problem,

$$Dw = f(z,w),\quad Re\, w|_{\dot{G}} = 0,\ \frac{1}{2\pi}\int_{\dot{G}} Im\, w\ |d\phi(\zeta)| = c,$$

(where c is a real, hypercomplex constant,

in the ball of radius K in B.

Remark: Here $L_p(f_0) := \left(\int_G |f_0|^p dxdy\right)^{1/p}.$

We next turn to the nonhomogeneous boundary value problem of index zero. We treat this case using a variant by Wendland [17] of an imbedding method due to Wacker [16]. To illustrate how this method may be adapted to the hypercomplex case we must obtain apriori estimates for the linear problem.

(4.11) $$Dw = aw + b\overline{w} + c$$

(4.12) $$Re\, w|_{\dot{G}} = 0,\quad \int_{\dot{G}} (Im\, w)\ \sigma\ ds = 0,$$

which we also may rewrite componentwise as

$$\frac{\partial w_k}{\partial \overline{z}} = a_0 w_k + b_0\overline{w}_k + f_k$$

$$f_k(z) := c_k + \sum_{\ell=0}^{k-1}\left(a_{k-\ell}w_\ell + b_{k-\ell}\overline{w}_\ell - q_{k-\ell}\frac{\partial w_\ell}{\partial z}\right)$$

$$Re\, w_k|_{\dot{G}} = \phi_k,\quad \int_{\dot{G}} (Im\, w_k)\ \sigma\ ds = 0.$$

The apriori estimate we need is a generalization appropriate for the hyperanalytic operator of Wendland's estimate. [17] (pg. 442, Lemma 5) for the Cauchy-Riemann operator.

Remark: If the coefficients a, b are in $C^\alpha(G)$ then Wendland's lemma [17] may be written with Hölder-norms instead of sup-norms. The proof is analogous to the original one. We use this fact below.

Lemma 2: Let w be a solution of

$$Dw = aw + b\overline{w} + c$$

$$Re\, w|_{\dot{G}} = 0, \quad \int_{\dot{G}} (Im\, w)\sigma ds = 0,$$

where $a, b, c \in C^\alpha(G)$ and $C_\alpha(a,G)$, $C_\alpha(b,G) \leq K$. Then there exists a constant λ depending only on G, K, σ such that

$$C_\alpha(w,G) \leq \lambda C_\alpha(c,G), \tag{4.13}$$

where $$C_\alpha(w,G) := \sum_{k=0}^{r-1} C_\alpha(w_k, G).$$

Remark: For the Cauchy-Riemann operator (complex case) Wendland has an estimate of the type (4.13) with $\alpha = 0$. It is not difficult to show that such an estimate also holds for the C_α-norm by making a slight variation in his argument.

Proof of Theorem: From (4.11), (4.12) we compute †

$$C_\alpha(w_1,G) \leq \lambda\, C_\alpha(c_1 + a_1 w_0 + b_1\overline{w}_0 - q_1 \frac{\partial w_0}{\partial z}, G)$$

$$\leq \lambda \left[C_\alpha(c_1,G) + k\, C_\alpha(w_0,G) + C_\alpha(q_1,G) C_\alpha\left(\frac{\partial w_0}{\partial z}, G\right)\right].$$

Since $\frac{\partial w}{\partial z} = \underset{\sim}{\pi}(a_0 w_0 + b_0 w_0 + c_0)$, we have, using the estimate (2.13) (where we may replace D by G), that

† In what follows λ will always be taken to mean a generic constant.

$$(4.14)\qquad C_\alpha\left(\frac{\partial w_0}{\partial z},\ G\right) \le M_\alpha C_\alpha(a_0 w_0 + b_0\overline{w}_0 + c_0, G)$$

$$\le M_\alpha[2k\lambda C_\alpha(w_0,G) + C_\alpha(c_0,G)].$$

Combining (4.14) and (4.15) yields, taking into account [15] (pg. 442, Lemma 5)

$$C_\alpha(w_1,G) \le \lambda[C_\alpha(c_1,G) + (2k\lambda + M_\alpha(2k\lambda + 1)C_\alpha(q_1,G))\cdot C_\alpha(c_0,G)]$$

$$\le \lambda[C_\alpha(c_0,G) + C_\alpha(c_1,G)].$$

Proceeding in this manner we obtain a bound of the form

$$C_\alpha(w_k,G) \le \lambda[C_\alpha(c_0,G) + C_\alpha(c_1,G) + \ldots + C_\alpha(c_k,G)]$$

$$\le \lambda\ C_\alpha(c,G),$$

from which our lemma follows.

We now turn to the semilinear problem

$$(4.15)\qquad Dw = f(z,w)\ :=\ H(z,\overline{z},w_0,\ldots,w_{r-1},\overline{w}_0,\ldots,\overline{w}_{r-1}).$$

Let $\delta_z^h f = \delta f(z;h)\ :=\ \lim\limits_{\zeta\to}\frac{1}{\zeta}[f(z+\zeta h) - f(z)]$ designate the Gateaux differential of a function f with respect to z. We ask that the following Gateaux differentials exist,

$$\delta_w^h f,\quad \delta_{\overline{w}}^{\overline{h}} f,\quad \left(\delta_w^h\right)^2 f,\quad \delta_w^h\,\delta_{\overline{w}}^{\overline{h}} f,\quad \left(\delta_{\overline{w}}^{\overline{h}}\right)^2 f.$$

These functions along with $f(z,w)$ are to be continuous (and hence bounded in norm by k) in $\overline{G}\times \mathbb{C}^r$; furthermore for each fixed value of $w \in \mathbb{C}^r$, $f(z,w) \in C^\alpha(\overline{G})$.

We consider the following imbedding:

$$(4.16) \qquad (Dw)(z,t) = tf(z,w) \qquad z \in G,\ w \in \not{C}^r),\quad t \in [0,1],$$

$$Re\ w|_{\dot{G}} = \chi, \qquad \int_{\dot{G}} (Im\ w)\sigma\ ds = C.$$

We note that for $t = 0$ (4.6) reduces to the differential equation for hyperanalytic functions, namely $Dw = 0$. Once the Green's function is known, solutions for this case may be represented using (3.14) as

$$w(z) = \Omega(z) - \underset{\sim}{R}(q\Omega')(z).$$

Otherwise, a solution is available using a layer method, such as in Fichera [6] [7], Gilbert-Hsiao [12], or Wendland [17]. Having obtained a solution for $t = 0$, we use it as the initial approximation in a Newton's method for $t = t_1 > 0$. The solution obtained by the Newton iteration for $t = t_1$ then becomes the initial approximation for $t = t_2 > t_1$, and so on. Eventually we reach $t = t_j > t_{j-1}$. The Newton iteration is then given by

$$(Dw_{n+1})(z) = t_j(\delta_w^h f)(z,w_n) + t_j(\delta_{\overline{w}}^{\overline{h}} f)(z,w_n)$$

$$(4.17) \qquad + t_j f(z,w_n),$$

$$h := w_{n+1} - w_n, \quad \overline{h} := \overline{w}_{n+1} - \overline{w}_n,$$

and

$$Re\ w_{n+1}|_{\dot{G}} = \chi, \quad \int_{\dot{G}} Im\ w_{n+1}\ \sigma\ ds = C, \cdot \ (n = 0,1,\ldots).$$

We remark that the iteration (4.17) is a linear equation for w_{n+1}. To obtain an estimate of the rate of convergence we investigate the difference $(w_{n+1} - w_n)$ which up to second order accuracy in the nonhomogeneous term satisfies

$$(w_{n+1} - w_n)_{\bar{z}} = t_j(\delta_w^h f)(z,w_n) + t_j(\delta_{\bar{w}}^{\bar{h}} f)(z,w_n)$$

$$+ t_j(\delta_{(w,w)}^{(h,h)} f)(z) + 2t_j(\delta_{(w,\bar{w})}^{(h,\bar{h})} f)(z) \tag{4.18}$$

$$+ t_j(\delta_{(\bar{w},\bar{w})}^{(\bar{h},\bar{h})} f)(z),$$

where

$$(\delta_{(w,w)}^{(h,h)} f)(z) := \int_0^1 (\delta_w^h)^2 f(z, tw_n + (1-t)w_{n-1})dt,$$

and other terms are analogously defined. Using the apriori estimates (4.13) and our regularity assumptions about the Gateaux differentials of f we have that

$$C_\alpha(w_{n+1} - w_n, G) \leq 4Kt_j\lambda\, C_\alpha(w_n - w_{n-1}, G)^2, \tag{4.19}$$

since $|(w_n - w_{n-1})\overline{(w_n - w_{n-1})}| \leq |w_n - w_{n-1}|^2$

It is clear that <u>if</u>
$4Kt_j\lambda\, C_\alpha(w_1 - w_0, G) < 1$ then the iteration method converges. We estimate this term by noticing that the difference $(w_1 - w_0)$ must satisfy an equation of the form

$$(w_1 - w_0)_{\bar{z}} = t_j(\delta_w^h f)(z,w_0) + t_j(\delta_{\bar{w}}^{\bar{h}} f)(z,w_0) + (t_j - t_{j-1})f(z,w_0)$$

it follows from the apriori estimates that

$$C_\alpha(w_1 - w_0, G) \leq (t_j - t_{j-1})\lambda\, C_\alpha(f(z,w_0), G) \leq (t_j - t_{j-1})\lambda K.$$

If t_j is chosen such that

$$t_j(t_j - t_{j-1}) < \frac{1}{4K^2\lambda^2} ,$$

the Newton iteration method will converge for $t = t_j$. Since the right hand side provides a uniform bound, the imbedding method may be extended to $t = 1$.

Uniqueness and regularity of the solution w follow by the usual arguments [17].

REFERENCES

[1] Bers, L., *Theory of Pseudo-Analytic Functions*, Lecture Notes, Courant Institute, New York, 1953.

[2] Begehr, H., and Gilbert, R. P., *Über das Randwert-Normproblem für ein nichtlineares elliptisches System*, (to appear).

[3] Begehr, H., and Gilbert, R. P., *Randwertaufgaben ganzzahliger Charakteristik für verallgemeinerte hyperanalytische Funktionen*, (to appear), Applicable Analysis.

[4] Bojarski, B. B., *The theory of generalized analytic vectors*, Annales Polonici Mathematici, 17(1966), 281-320.

[5] Douglis, A., *A function-theoretic approach to elliptic systems of equations in two variables*, Comm. Pure Applied Math., VI, (1953), 259-289.

[6] Fichera, G., *Linear elliptic equations of higher order in two independent variables and singular integral equations*, Proc. Conf. on Partial Differential Equations and Continum Mechanics, Univ. Wisconsin Press, Madison, Wisc., 1961.

[7] _______, *The single layer potential approach in the theory of boundary value problems for elliptic equations*, Conference on Function Theoretic Methods for Partial Differential Equations.

[8] Gilbert, R. P., *Constructive Methods for Elliptic Equations*, Lecture Notes in Math. No. 365, Springer-Verlag, Berlin, 1974.

[9] Gilbert, R. P., and Hile, G. N., *Generalized hypercomplex function theory*, Trans. Am. Math. Soc. 195(1974), 1-29.

[10] _______, *A function theory in the sense of Bers for hypercomplex functions*, Mathematische Nachrichten (to appear).

[11] _______, *Hilbert function modules with reproducing kernels*, to appear, Journal Nonlinear Analysis.

[12] Gilbert, R. P., and Hsiao, G. C., *On Dirichlet's problem for*

quasi-linear elliptic equations, Lecture Notes in Math. No. 430, Springer-Verlag, Gerlin, 1975.

[13] Haack, W., and Wendland, W., *Lectures on Partial and Pfaffian Differential Equations*, Pergamon Press, Oxford, 1972.

[14] Kühn, E., *Über die Funktionentheorie und das Ähnlichkeitsprinzip einer Klasse elliptischer Differentialgleichungssysteme in der Ebene*, Dissertation, Dortmund, 1974.

[15] Vekua, I. N., *Generalized Analytic Functions*, Pergamon Press, London, 1962.

[16] Wacker, H. J., *Eine Lösungsmethode zur Behandlung nichtlinearer Randwertprobleme*, Iterationsverfahren, Numerische Mathematik, Approximationstheorie ISNM, Vol. 15, Birkhauser, 1970.

[17] Wendland, W., *An integral equation method for generalized analytic functions*, Lecture Notes in Math. No. 430, Springer-Verlag, serlin, 1975.

ERROR PROPAGATION AND CATASTROPHES IN PROTEIN SYNTHESIZING MACHINERY

by

Narendra S. Goel*
School of Advanced Technology
State University of New York
Binghamton, New York]3901

Introduction

It is commonly known that proteins are very essential for life; they supply not only carbon, hydrogen, and oxygen (also done by carbohydrates and fats) but also nitrogen, sulfur and often phosphorus. Some of the proteins also act as super-catalysts for many chemical reactions necessary for maintaining life in all its aspects. These proteins are known as enzymes. Almost all of the proteins are made up of 20 types of small molecules, the amino acids. The number of amino acids of various types and their arrangement (known as primary structure) differ from proteins to proteins. For some proteins, minor change in the arrangements causes drastic change in their function while for others it is not so. To assure ample supply of proteins with right functions, cells synthesize proteins by attaching amino acids to each other in the right order. This protein synthesizing machinary (PSM) gets the basic information about the order of arrangement of amino acids by DNA molecule. This machinery also has a self-replicating aspect to it; some of the proteins formed by the machinery are necessary for its proper functioning in the sense of rate of formation of proteins and proper arrangement of the amino acids. If these proteins have errors in their primary sequences, they may not function properly and make PSM unreliable, possibly leading to the formation of the same next generation proteins with more errors. This process of increase of errors may continue and finally lead to an "error catastrophe" in protein synthesis and thus death

* Also at Center for Theoretical Biology, Vestal, New York.

of the cell. (This has been suggested to be a mechanism of aging by Orgel (1963, 1973)). On the other hand, some of the errors may not be catastrophic-evolution is a well known example of it. In this paper we discuss various models for propagation of errors in the protein synthesizing machinery and attempt to delineate the conditions under which errors may increase, leading to an error catastrophe, and may decrease.

For the benefit of those readers who are not familiar with the protein synthesizing machinery, we give a brief description of PSM and refer them to two books for details (Yčas, 1969; Watson, 1975).

The DNA molecule, with its famous double helical structure, is a polymer molecule with each strand composed of four types of small molecules known as nucleotides. They are denoted by A, T, C and G. There is an important complimentarity between the sequences of the nucleotides in two strands. In the double helical form A and T, and G and C are always opposite each other. First a copy of one of the two strands is transcribed. This copy is known as messenger RNA (m-RNA) which is a single-stranded poly-nucleotide, also consisting of four types of nucleotides similar to DNA with the exception that T is replaced by another nucleotide U. This transcription process is aided by an enzyme known as transcriptase. Parts of m-RNA carry the information for the sequence in which twenty types of amino acids will be attached in the form of sequence of non-overlapping triplets (known as codons). Since there are $4^3 = 64$ triplets which code for 20 types of amino acids, for some amino acids there are many codons (up to six). The specific relationship is now established and is known as genetic code. The translation of information from m-RNA into sequence of amino acids i.e. proteins, is carried out in a somewhat complex function. First each of the amino acids gets attached to a small RNA molecule, known as transfer RNA (t-RNA) with the help of an enzyme known as amino acyl synthetase (t-RNA) molecules are also transcribed from DNA very much like m-RNA molecules). Each transfer RNA molecule acts like an adaptor molecule in the sense that each t-RNA molecule has a triplet of nucleotide on it (known as anticodon) that recognizes the relevant codon on the m-RNA by pairing of nucleotides. The amino acid-tRNA complexes diffuse to certain spherical particles (known as ribosomes) attached to m-RNA. Ribosomes contain both RNA (known as r-RNA) and proteins molecules and have specific surfaces that bind m-RNA, aa-tRNA and the growing protein chain in a suitable sterochemical position. Once a ribosome is attached to m-RNA and protein synthesis has commences, the ribosome moves along the m-RNA and the amino acid corresponding to the next triplet on the m-RNA gets attached to the already formed protein chain. Each protein is coded by a part of the mRNA molecule known as cistron. Each cistron has codons at the two ends for starting and terminating the synthesis of the proteins.

The overall mechanism of protein synthesis is schematically shown in Figure 1.

Let us now briefly describe where and how errors could be introduced in the protein synthesizing machinery.

(a) Errors Due to Synthetase Molecules

As noted above, an amino acyl synthetase (synthetase for short) catalyzes the attachment of an amino acid with t-RNA. Let us denote by a_i, $i = 1,\ldots,20$, the twenty amino acids and by t-RNA$_j$ those t-RNA molecules which recognizes codons for jth ($j = 1,\ldots,20$) amino acids. In an error free system, these exist synthetase molecules which attach amino acid a_i to t-RNA$_i$. However, if the primary sequences of one or more of the synthetases are not correct, they may not attach the correct amino acid to a given t-RNA molecule. Since t-RNA molecule recognizes the codon, in the growing protein chain a wrong amino acid will be inserted in the protein being synthesized, including that (transcriptase) which will be used for transcription from DNA to m-RNA.

(b) Errors Due to Transcriptase Molecules

If transcriptase molecules have wrong sequences, the mRNA and t-RNA molecules, transcribed from DNA molecule, may have wrong sequences. This means codons and anticodons may be wrong. Hence even if there are no errors in the synthetase molecule i.e. even if amino-acid-tRNA complexes are correct, the protein produced (including synthetase and transcriptase) may have wrong sequences.

(c) Errors Due to Replicase Molecules

DNA molecules are continually replicated in almost all the cells. This process, necessary for production of new cells, is assisted by an enzyme known as replicase. Once again if its sequence of amino acids is wrong, the new DNA molecule may have a wrong sequence, which even in the presence of error free protein synthesizing machinery, will lead to proteins with altered primary sequences.

It should be emphasized that in replication and transcription, it is generally believed that the major determinants of sequences of DNA and RNA produced are not the enzymes, replicase and transcriptase, but complimentarity of nucleotide bases $(A \to T, A \to U, C \to G)$. If this is indeed so, erroneous enzymes should only be able to introduce a certain and perhaps quite limited amount of noise into replication and transcription.

In the next section we review a simple model proposed earlier by the author (Goel and Ycas, 1975, 1976 Goel and Islam, 1977) for propagation of errors in synthetase molecules initially introduced (by genetic mutations or by artificial means). The model leads to linear difference or equivalent linear differential equations which can easily be analyzed. In the final Section 3, we discuss a more general model, which leads to nonlinear difference equations. We also make some general remarks, especially about the other models under investigations.

2. *Simple Basic Model*

This simple model was introduced and analyzed in Goel and Ycas (1975). Further analysis has been carried out by Goel and Islam (1977). In the following we outline the model and the analysis and give the key results.

A given synthetase molecule will have a certain number of amino acid sites which must not be changed if it is to retain normal specificity (i.e. ability to attach right amino acid to right t-RNA molecules) and activity (i.e. rate with which it does so). These sites may be different for different synthetases. Let $x_{ij} (i,j = 1,2,\ldots,N)$ be the number of such sites of an amino acid a_j in the ith synthetase. Of these sites let $y_{ij} (0 \le y_{ij} \le x_{ij})$ be ones that must be occupied by incorrect amino acids for the synthetase to become erroneous (to attach wrong amino acid to t-RNA). The remaining $x_{ij} - y_{ij}$ sites must not change, otherwise the synthetase will suffer still another change in specificity or becomes inactive (does not attach any amino acid).

In principle, several different sequences, coded by the same cistron, may have normal (or the same erroneous) activity with respect to a given tRNA molecule, and the number of critical sites x_{ij} and y_{ij} may differ in these sequences. In such cases x_{ij} and y_{ij} are the average numbers, so that in general they may be non-negative fractions.

Let q_i, q_i', and q_i'' denote the fractions of normal, erroneous and inactive ith synthetase molecules. Bearing in mind that the effects of an erroneous synthetase molecule are not site specific, the normal fraction $q_1(t + 1)$ of normal synthetase for amino acid a_1, S_1, in generation $(t + 1)$ is given by

$$\left(\frac{q_1}{q_1+q_1'+q_1''}\right)_{t+1} = \left(\frac{q_1}{q_1+q_1'}\right)_t^{x_{11}} \left(\frac{q_2}{q_2+q_2'}\right)_t^{x_{12}} \cdots \left(\frac{q_N}{q_N+q_N'}\right)_t^{x_{1N}} \quad (2.1a)$$

This follows because errors will be binomially distributed and therefore the fraction of molecules with amino acid a_1 at the specified x_{11} locations is given by the first term in the right hand side, the fraction of molecules with amino acid a_2 at specified x_{12} locations is given by the second term, etc.

The corresponding equation for the fraction of erroneous synthetase depends upon the assumption made about its specificity and the nature of errors in the primary sequence of normal synthetase which produce erroneous synthetase. In a simple case, if the occupation of y_{ij} sites by <u>any</u> of the incorrect amino acid produces erroneous synthetase which can insert any of the incorrect amino acids, then the fraction $q_1'(t+1)$ of erroneous synthetase S_1 in generation $(t+1)$ is given by

$$\left(\frac{q_1'}{q_1+q_1'+q_1''}\right)_{t+1} = \left(\frac{q_1}{q_1+q_1'}\right)_t^{x_{11}-y_{11}} \left(\frac{q_1'}{q_1+q_1'}\right)_t^{y_{11}} \cdots \left(\frac{q_N}{q_N+q_N'}\right)_t^{x_{1N}-y_{1N}} \left(\frac{q_N}{q_N+q_N'}\right)_t^{y_{1N}} \tag{2.1b}$$

Equations similar to the above two equations hold for the synthetases for other amino acids.

The above equation represents a system in which the erroneous synthetases are non-specific (except that they do not insert correct amino acids) and their number is very large because y_{ij} can be occupied by any of incorrect amino acids in a large number of ways.

Dividing equation (2.1a) by (2.1b) and defining

$$Q_i = ln\{q_i(t)/q_1'(t)\}, \quad i = 1,2,\ldots,N \tag{2.2}$$

we get

$$Q_i(t+1) = \sum_{j=1}^{N} y_{ij}\, Q_j(t), \quad i = 1,2,\ldots,N \tag{2.3}$$

In the continuous limit of the discrete process discussed above, Eq. (2.3) becomes

$$\frac{dQ}{dt} = BQ \tag{2.4}$$

where

$$\frac{dQ_i}{dt} = Q_i(t+1) - Q_i(t) \tag{2.5a}$$

B is a matrix and Q is a vector defined by

$$B = \begin{pmatrix} y_{11}-1 & y_{12} & \cdots & y_{1N} \\ \vdots & & & \\ y_{N1} & y_{N2} & \cdots & y_{NN}-1 \end{pmatrix}, \quad Q = \begin{pmatrix} Q_1 \\ Q_2 \\ \vdots \\ Q_N \end{pmatrix} \tag{2.5b}$$

Equation (2.4) describes the evolution of errors in PSM. Its formal solution is

$$Q(t) = e^{Bt}Q(0) \tag{2.6}$$

where $Q(0)$ is the initial value of the vector Q. Since all the non-diagonal elements of the matrix B are greater than or equal to zero, the matrix B is an "essentially non-negative" matrix. Therefore, the theory of such matrices (Bellman, 1970; Seneta, 1973) provides the necessary tool for analyzing the consequences of equation (2.4).

For the purpose of evolution of errors, it is convenient to divide the discussion into four cases according to the <u>initial conditions</u>.

(A) THE QUANTITIES OF ALL THE NORMAL SYNTHETASE MOLECULES ARE GREATER THAN THOSE OF ERRONEOUS ONES

For this case, the initial vector $Q(0)$ is positive. Therefore, from equation (2.6), it will either remain positive forever $(Q(t) > 0)$ or will eventually become equal to zero (this will occur when the largest characteristic value of B is <0. Zero value corresponds to equal concentrations of normal and erroneous synthetase molecules). Thus at all times, quantities of all normal synthetase molecules will equal or exceed those of the erroneous ones.

(B) THE QUANTITIES OF ALL THE NORMAL SYNTHETASE MOLECULES ARE SMALLER THAN THOSE OF ERRONEOUS ONES.

Similar to the case (A) above, at all times, quantities of all erroneous synthetase molecules will equal or exceed those of the normal ones.

(C) NORMAL AND ERRONEOUS SYNTHETASE MOLECULES ARE PRESENT IN EQUAL QUANTITIES.

For this case $Q(0) = 0$ and from equation (2.6), $Q(t) = 0$. Thus at all times the initial condition of equal normal and erroneous synthetase molecules will exist.

(D) SOME SYNTHETASES HAVE NORMAL MOLECULES MORE THAN, SOME LESS THAN, AND SOME EQUAL TO THE ERRONEOUS ONES.

For this mixed initial condition, the evolution of system can lead to many asymptotic outcomes depending upon the characteristics of the system as embedded in the matrix B. It is convenient to divide further analysis according to reducibility and decomposibility of the matrix B.

(i) B is an irreducible (and essentially non-negative) matrix B.

For such matrices, as shown in Goel and Islam (1976), asymptotically either the system will completely recover itself from errors or will totally suffer an error catastrophe or will have equal quantities of normal and erroneous synthetase molecules. What happens is determined by the following theorem: There exists a time $\tau \geq 0$ such that for any time $t > \tau$, $Q(t) \geq 0$ if and only if $(Q(0),P) \geq 0$ where $Q(0)$ is the initial condition, P is the Perron vector (characteristic vector corresponding to the largest characteristic value of an essentially non-negative matrix) of the matrix B^T, the transpose of B, and $(\ ,\ )$ denotes the inner product of the two vectors i.e.

$$(Q(0),P) = Q_1(0)P_1 + \ldots + Q_N(0)P_N \tag{2.7}$$

Thus if:

$$(Q(0),P) > 0 \tag{2.8a}$$

the system will completely recover itself from errors,

$$(Q(0),P) < 0 \tag{2.8b}$$

the system will eventually suffer from an error catastrophe and

$$(Q(0),P) = 0 \tag{2.8c}$$

the system will have equal concentrations of normal and erroneous synthetase molecules.

As an example consider the matrix

$$Y = \begin{pmatrix} 1 & 1 & 1 \\ 1 & 1 & 1 \\ 1 & 1 & 1 \end{pmatrix}, \qquad B = \begin{pmatrix} 0 & 1 & 1 \\ 1 & 0 & 1 \\ 1 & 1 & 0 \end{pmatrix}$$

The matrix Y describes three amino acids -- three synthetases system for which each of the synthetases one additional error in each of the three amino acid locations beyond those tolerable for normal synthetases will create erroneous synthetase. The Perron vector, P, of B^T for this system is (1,1,1). Thus from Eqs. (2.7) and (2.2) the quantity determining the asymptotic behavior is

$$\begin{aligned} I_Q &\equiv Q_1(0) + Q_2(0) + Q_3(0) \\ &= \left[\ln \left(\frac{q_1}{q_1'}\right) \left(\frac{q_2}{q_2'}\right) \left(\frac{q_3}{q_3'}\right) \right]_{t=0} . \end{aligned} \tag{2.10}$$

$I_Q \gtreqless 0$ correspond to eventual recovery from errors, error catastrophe, and equal amounts of normal and erroneous synthetase molecules respectively.

In Figure 2.1 is plotted $Q(t)$ as a function of t for various initial vectors $Q(0)$. For cases A, B, and C, the initial vectors are

$$Q_A(0) = \begin{pmatrix} 3 \\ 1 \\ -4.01 \end{pmatrix}, \quad Q_B(0) = \begin{pmatrix} 3 \\ 1 \\ -3.99 \end{pmatrix}, \quad Q_C(0) = \begin{pmatrix} 3 \\ 1 \\ -4 \end{pmatrix} \tag{2.11}$$

$$(Q_A(0),P) = -.01, \quad (Q_B(0),P) = +.01, \quad (Q_C(0),P) = 0 \qquad (2.12)$$

Thus for these cases, the quantities of erroneous synthetase 3 molecules are $e^{4.01} \simeq 55.15$, $e^{3.99} \simeq 54.06$, and $e^4 \simeq 54.598$ times that of normal molecules, respectively. In spite of only the small differences in the initial concentrations, in case A, the system will suffer an eventual error catastrophe, in case B, it will recover itself from all the erroneous synthetase molecules and in case C, eventually $Q(t) \to 0$ i.e. for all synthetases, eventually the concentrations of normal and erroneous molecules will become the same.

Thus there is a threshold phenomena for recovery. If the relative concentrations of normal and erroneous molecules are below some threshold values (defined by $(Q(0),P) = 0$), the system suffers an error catastrophe. If they are more, it recovers from the errors. If they are at threshold, the system continues to operate in the presence of equal concentrations of normal and erroneous synthetase molecules.

(ii) B is a decomposible (and essentially non-negative) matrix. Such a matrix, by definition, can be reduced to the form $\begin{pmatrix} A & 0 \\ 0 & D \end{pmatrix}$, when A and D are irreducible square matrices, by simultaneous permutation of its rows and columns. For each of the matrices A and D, the preceding discussion for irreducible matrix is separately applicable. The system is essentially decomposed into two clusters of synthetases (and corresponding amino acids) and the initial relative concentrations of normal and erroneous synthetases belonging to one cluster do not affect the asymptotic behavior of synthetases in the other cluster.

(iii) B is a reducible but not decomposible (and essentially non-negative) matrix. In this case if the

Perron characteristic value is simple (multiplicity = 1), then the results for irreducible matrix B are still true. For greater than one multiplicity, we have not been able to prove any general theorem and the asymptotic outcome could only be determined by solving basic equation (2.4).

2. *General Model*

In this section we will present a general model in which each synthetase can have more than one type of erroneous synthetase molecules. Each type is produced by inserting a specific set of erroneous amino acids in the primary sequence of synthetase. Each type of erroneous synthetase also has a specified specificity for attaching various amino acids to t-RNA molecules i.e. for inserting various amino acids in the growing protein chain synthesized by the machinery. For simplicity, we will only consider the corresponding system of three amino acids and three synthetases; the equations and the results, however, are general.

Let q_i, q_i', and p_i' denote the fractions of three active th synthetase molecules and q_i'' that of inactive synthetase molecules, all produced by cistron for ith synthetase molecules $(q_i + q_i' + p_i' + q_i'' = 1)$. Let α_{ij}, β_{ij}, and γ_{ij} be the relative specificities of three types of active ith synthetase molecules for attaching amino acid, a_j, to $(\text{t-RNA})_i$ for ith amino acid. Thus if $\alpha_{ii} = 1$ and zero otherwise, q_i will denote the fraction of normal ith synthetase molecule. Let us choose the following requirements for the production of active ith synthetase molecules.

(i) Type 1: x_{i1}, x_{i2} and x_{i3} sites of amino acids 1, 2, and 3, respectively must be same as for the naturally occuring synthetase 1 molecule.

(ii) Type 2 (or 3): Of the x_{i1} sites normally occupied by amino acid a_1, y_{i1q} (or y_{i1p}) sites must be replaced by a_2, and z_{i1q} (or z_{i1p}) sites must be replaced by a_3. Of the x_{i2} sites normally occupied by a_2, y_{i2q} (or y_{i2p}) sites must be replaced by a_1 and z_{i3q} (or z_{i3p}) sites must be replaced by a_2. Of the x_{i3} sites normally occupied by a_3, y_{i3q} (or y_{i3p}) sites must be replaced by a_1 and z_{i3p} sites must be replaced by a_2.

Any other replacement causes synthetase to become inactive.

Then, as in the preceding section, at time $t + 1$, the fractions of Type 1 active synthetase 1 molecules is given by

$$\left(\frac{q_1}{q_1+q_1'+p_1'+q_1''}\right)_{t+1} = \left(\frac{\alpha_{11}q_1+\beta_{11}q_1'+\gamma_{11}p_1'}{q_1+q_1'+p_1'}\right)_t^{x_{11}} \left(\frac{\alpha_{22}q_2+\beta_{22}q_2'+\gamma_{22}p_2'}{q_2+q_2'+p_2'}\right)_t^{x_{12}} \left(\frac{\alpha_{33}q_3+\beta_{33}q_3'+\gamma_{33}p_3'}{q_3+q_3'+p_3'}\right)_t^{x_{13}} \tag{3.1}$$

For brevity, let us introduce the following notation.

$$U_{iq} \equiv q_i'/q_i, \quad U_{ip} = p_i'/p_i, \quad i = 1,2,3. \tag{3.2}$$

$$S_i \equiv 1 + U_{iq} + U_{ip}, \quad i = 1,2,3. \tag{3.3}$$

$$A_{ij} \equiv \alpha_{ij} + \beta_{ij}U_{iq} + \gamma_{ij}U_{ip}, \quad i,j = 1,2,3. \tag{3.4}$$

Eq. (3.1) then becomes

$$\left(\frac{q_1}{q_1+q_1'+p_1'+q_1''}\right)_{t+1} = \left(\frac{A_{11}}{S_1}\right)_t^{x_{11}} \left(\frac{A_{22}}{S_2}\right)_t^{x_{12}} \left(\frac{A_{33}}{S_3}\right)_t^{x_{13}} \tag{3.5a}$$

Keeping in mind the above mentioned requirements for producing Type 2 active synthetase 1 molecules, its fraction is given by

$$\left(\frac{q_1'}{q_1+q_1'+p_1'+q_1''}\right)_{t+1} = \left(\frac{A_{11}}{S_1}\right)_t^{x_{11}-y_{11q}-z_{11q}} \left(\frac{A_{12}}{S_1}\right)_t^{y_{11q}} \left(\frac{A_{13}}{S_1}\right)_t^{z_{11q}}$$

$$\left(\frac{A_{33}}{S_3}\right)_t^{x_{13}-y_{13q}-z_{13q}} \left(\frac{A_{31}}{S_3}\right)_t^{y_{13q}} \left(\frac{A_{32}}{S_3}\right)_t^{z_{13q}} \qquad (3.5b)$$

The fraction of Type 3 active synthetase 1 molecules is given by a similar equation except that the subscript q is replaced by p in y's and z's. Dividing Eq. (3.5b) by Eq. (3.5a), we get

$$(U_{1q})_{t+1} = \left(\frac{A_{12}}{A_{11}}\right)_t^{y_{11q}} \left(\frac{A_{13}}{A_{11}}\right)_t^{z_{11q}} \left(\frac{A_{23}}{A_{22}}\right)_t^{y_{12q}} \left(\frac{A_{21}}{A_{22}}\right)_t^{z_{12q}}$$

$$\left(\frac{A_{31}}{A_{33}}\right)_t^{y_{13q}} \left(\frac{A_{32}}{A_{33}}\right)_t^{z_{13q}} \qquad (3.6)$$

The other U's are given by equations similar to (3.6), except that subscripts for y and z are appropriately replaced. The six equations for six U's can be compactly written as

$$(U_{iq})_{t+1} = (\tilde{A}_{12})_t^{y_{i1q}} (\tilde{A}_{13})_t^{z_{i1q}} (\tilde{A}_{23})_t^{y_{i2q}} (\tilde{A}_{21})_t^{z_{i2q}} (\tilde{A}_{31})_t^{y_{i3q}} (\tilde{A}_{32})_t^{z_{i3q}}$$

$$i = 1,2,3 \qquad (3.7a)$$

$$(U_{ip})_{t+1} = (\tilde{A}_{12})_t^{y_{i1p}} (\tilde{A}_{13})_t^{z_{i1p}} (\tilde{A}_{23})_t^{y_{i2p}} (\tilde{A}_{21})_t^{z_{i2p}} (\tilde{A}_{31})_t^{y_{i3p}} (\tilde{A}_{32})_t^{z_{i3p}}$$

$$i = 1,2,3 \qquad (3.7b)$$

where

$$\tilde{A}_{ij} = A_{ij}/A_{ii}; \quad i,j = 1,2,3. \qquad (3.7c)$$

For given set of values of y's and z's i.e. the matrix

$$\tilde{y} = \begin{pmatrix} y_{11q} & z_{11q} & y_{12q} & z_{12q} & y_{13q} & z_{13q} \\ y_{11p} & z_{11p} & y_{12p} & z_{12p} & y_{13p} & z_{13p} \\ y_{21q} & z_{21q} & y_{22q} & z_{22q} & y_{23q} & z_{23q} \\ y_{21p} & z_{21p} & y_{22p} & z_{22p} & y_{23p} & z_{23p} \\ y_{31q} & z_{31q} & y_{32q} & z_{32q} & y_{33q} & z_{33q} \\ y_{31p} & z_{31p} & y_{32p} & z_{32p} & y_{33p} & z_{33p} \end{pmatrix} \tag{3.8}$$

set of α_{ij}, β_{ij}, γ_{ij} i.e. the matrices

$$\alpha = \begin{pmatrix} \alpha_{11} & \alpha_{12} & \alpha_{13} \\ \alpha_{21} & \alpha_{22} & \alpha_{23} \\ \alpha_{31} & \alpha_{32} & \alpha_{33} \end{pmatrix}, \quad \beta = \begin{pmatrix} \beta_{11} & \beta_{12} & \beta_{13} \\ \beta_{21} & \beta_{22} & \beta_{23} \\ \beta_{31} & \beta_{32} & \beta_{33} \end{pmatrix},$$

$$\gamma = \begin{pmatrix} \gamma_{11} & \gamma_{12} & \gamma_{13} \\ \gamma_{21} & \gamma_{22} & \gamma_{23} \\ \gamma_{31} & \gamma_{32} & \gamma_{33} \end{pmatrix} \tag{3.9}$$

and the initial relative concentrations of q_i, q_i' and p_i' i.e. the initial value of the vector

$$U = \begin{pmatrix} U_{1q} \\ U_{1p} \\ U_{2q} \\ U_{2p} \\ U_{3q} \\ U_{3p} \end{pmatrix} \equiv \begin{pmatrix} q_1'/q_1 \\ p_1'/q_1 \\ q_2'/q_2 \\ p_2'/q_2 \\ q_3'/q_3 \\ p_3'/q_3 \end{pmatrix} \tag{3.10}$$

nonlinear difference equations can be solved to determine the evolution of q_i, q_i' and p_i' i.e. the evolution of errors.

As for the case discussed in the preceding section, one finds threshold phenomenon for recovery of errors and error catastrophe. We will now discuss two specific cases and quantify threshold phenomenon. These cases are as follows.

CASE 1

$$\alpha = \begin{pmatrix} 1 & 0 & 0 \\ 0 & 1 & 0 \\ 0 & 0 & 1 \end{pmatrix}, \quad \beta = \begin{pmatrix} 0 & 1 & 0 \\ 0 & 0 & 1 \\ 1 & 0 & 0 \end{pmatrix}, \quad \gamma = \begin{pmatrix} 0 & 0 & 1 \\ 1 & 0 & 0 \\ 0 & 1 & 0 \end{pmatrix} \tag{3.11}$$

Here the synthetases denoted by fractions q_1, q_1', p_1', all coded by cistron 1, have specificity for amino acids 1 (normal), 2 and 3, respectively. Similarly, those denoted by fractions q_2, q_2', p_2' all coded by cistron 2, have specificity for amino acids 2 (normal), 3 and 1, respectively, and those denoted by fractions q_3, q_3', p_3', all coded by cistron 3, have specificity for amino acids 3 (normal), 1 and 2, respectively.

CASE 2

$$\alpha = \begin{pmatrix} 1 & 0 & 0 \\ 0 & 1 & 0 \\ 0 & 0 & 1 \end{pmatrix}, \quad \beta = \begin{pmatrix} 0 & 0.5 & 0.5 \\ 0.3 & 0 & 0.7 \\ 0.5 & 0.5 & 0 \end{pmatrix},$$

$$\gamma = \begin{pmatrix} 0 & 0.6 & 0.4 \\ 0.8 & 0 & 0.2 \\ 0.6 & 0.4 & 0 \end{pmatrix} \tag{3.12}$$

Here the synthestases denoted by q_1, q_2 and q_3 are specific for amino acids 1,2, and 3, respectively i.e. they are the normal synthetases. The synthetases denoted by q_1' and p_1' do not insert amino acid 1 but insert amino acids 2 and 3 in the ratio 0.5: 0.5 and 0.6:0.4, respectively. The synthetases denoted by q_2' and p_2' do not insert amino acid 2 but insert amino acids 1 and 3 in the ratio 0.3:0.7 and 0.8:0.2, respectively. The synthetases denoted by q_3' and p_3' do not insert amino acid 3 but most amino acids 1 and 2 in the ratio 0.5:0.5 and 0.6:0.4 respectively.

For both cases we choose the $\tilde{y}$ matrix as

$$\tilde{y} = \begin{pmatrix} 1 & 2 & 1 & 2 & 1 & 2 \\ 2 & 1 & 2 & 1 & 2 & 1 \\ 1 & 2 & 1 & 2 & 1 & 2 \\ 2 & 1 & 2 & 1 & 2 & 1 \\ 1 & 2 & 1 & 2 & 1 & 2 \\ 2 & 1 & 2 & 1 & 2 & 1 \end{pmatrix} \tag{3.13}$$

Here for each of the synthetases three errors on each of the three amino acid sites must be introduced (beyond those for normal synthetases) for them to become erroneous. The exact nature of error determine the type (see definition of y's and z's in the beginning of this section).

For simplicity, let us consider two different types of initial conditions

INITIAL CONDITION (a)

Here initially

$$q_1'/q_1 = p_1'/q_1 = r, \quad q_2'/q_2 = p_3'/q_3 = r$$

and

$$p_2'/q_2 = q_3'/q_3 = s \tag{3.14}$$

In other words, for the two types of erroneous synthetase 1, the relative concentration is equal (to r). The relative concentration of first type of erroneous synthetase 2 and second type of erroneous synthetase 3 is also equal to r. However, the relative concentration of second type of erroneous synthetase 2 and first type of erroneous synthetase 3 is both equal to s.

INITIAL CONDITION (b)

Here initially

$$q_1'/q_1 = p_1'/q_1 = q_2'/q_2 = p_2'/q_2 = p_3'/q_3 = r$$

and

$$q_3'/q_3 = s \tag{3.15}$$

In Table 3.1, we have given the threshold values of s for recovery for various values of r. For values of s exceeding the threshold value, the system will suffer an error catastrophe. Also for case 2, there is no threshold for recovery i.e. error catastrophe is ensured for $r \geq 1.27$ for initial condition (b). Note that case 1, for both initial conditions, is less prone to error catastrophe than case 2.

To compare the results of Table 3.1 with the corresponding

results for model of the preceding section (for simplicity, let us call it case 3), we should choose the Y and B matrices of equations (2.3) and (2.5b) as

$$Y = \begin{pmatrix} 3 & 3 & 3 \\ 3 & 3 & 3 \\ 3 & 3 & 3 \end{pmatrix}, \quad B = \begin{pmatrix} 2 & 3 & 3 \\ 3 & 2 & 3 \\ 3 & 3 & 2 \end{pmatrix} \tag{3.16}$$

The corresponding transpose of Perron vector, P, of transpose of B is (1,1,1). Since in the model of preceding section, we only had one type of erroneous molecules for each of three synthetases, for the purpose of comparison we don't differentiate between q_i' and p_i' and set

$$(q_i'/q_i) + (p_i'/q_i) = U_i, \quad i = 1,2,3 \tag{3.17}$$

for the ratio of erroneous to normal molecules for ith synthetase. From Eq. (2.10), the threshold for error catastrophe is given by

$$U_1(0)\ U_2(0)\ U_3(0) = 1$$

In Table 3.2 we have given the threshold for recovery from errors for cases 1 and 2 for both initial conditions and for case 3, corresponding to model of the preceding section. For initial condition (*a*) $U_3(0) = U_2(0)$ while for initial condition (b)$U_1(0)=U_2(0)$.

From this table one can see that in general, when a threshold exists, case (3) gives lower threshold than the other cases. Also the threshold is lower for the initial condition (a) than for (b). Thus initial condition (b) is less prone to error catastrophe than initial condition (a).

The biological significance of the above remarks is as follows. As the mechanisms of production of erroneous synthetase molecules become specific (case 3 least specific while cases 1 and 2 most specific), the chances of error catastrophe decreases. This is so because only certain sets of errors will qualify for production of erroneous synthetase molecules and thus the production of such molecules will be reduced. Also as the specificity of erroneous synthetase molecules increases (from case 3 to case 2 to case 1), the chances of error catastrophe also decrease.

In the above models we limited ourselves to errors at translation level due to synthetase molecules. As a first approximation this is a good assumption. The models in which one incorporates the errors at replication and transcription level are more complicated and nonlinear. They are currently being investigated in detail in collaboration with S. Islam and M. Yčas. We hope to report the results in the near future.

ACKNOWLEDGMENTS

The author gratefully acknowledges the discussions with Sirajul Islam. Some of the numerical computations reported here were carried out by him using the facilities of SUNY-Binghamton Computer Center. This research was supported under NIH Grant No. 5R01 HD05 13605 and an award from the Biomedical Research Support Grant awarded by NIH to SUNY-Binghamton.

REFERENCES

[1] Bellman, R., *Introduction to Matrix Analysis*, New York, McGraw Hill, 1970.

[2] Goel, N. S. and Islam, S., *J. Theor. Biol.*, (in press), 1977.

[3] Goel, N. S. and Yčas, M., *J. Theor. Biol.* 55(1975), 245.

[4] Goel, N. S. and Yčas, M., *J. Math. Biol.* 3(1976), 121.

[5] Orgel, L. E., *Proc. Natn. Acad. Sci.*, 49(1963), 517, U.S.A.

[6] Orgel, L.E., *Nature, Lond.* 243(1973), 441.

[7] Seneta, E., *Non-negative Matrices*, New York, John Wiley, 1973.

[8] Watson, J. D., *Molecular Biology of the Gene*, Menlo Park, California, W. A. Benjamin, 1975.

[9] Yčas, M., *The Biological Code*, New York, John Wiley, 1969.

TABLE 3.1

THRESHOLDS FOR RECOVERY FROM ERRORS

Threshold s

	Initial Condition (a)		Initial Condition (b)	
r	Case 1	Case 2	Case 1	Case 2
0.1	100	6.76	100,000	202.28
0.2	25	4.62	3125	50.40
0.3	11.1	3.6	411.5	22.10
0.4	6.25	2.94	97.66	12.25
0.5	4.0	2.46	32.0	7.6
0.6	2.78	2.08	12.86	5.02
0.7	2.04	1.76	5.95	3.43
0.8	1.56	1.48	3.05	2.36
0.9	1.23	1.24	1.69	1.60
1.0	1.0	1.02	1.0	1.03
1.1	.83	.82	.62	.58
1.2	.69	.63	.40	.22
1.3	.59	.46	.27	--
1.4	.51	.30	.19	--
1.5	.44	.15	.13	--
1.6	.39	.015	.095	--

Initial Conditions

a) $q_1'/q_1 = p_1'/q_1 = q_2\ q_2 = p_3'/q_3 = r,\ q_3'/q_3 = p_2'/q_2 = s.$

b) $q_1'/q_1 = p_1'/q_1 = q_2'/q_2 = p_2'/p_2 = p_3'/q_3 = r,\ q_3'/q_3 = s.$

Case 1(a), s is given by $s = r^{-2}$.
1(b), s is given by $s = r^{-5}$.

TABLE 3.2

THRESHOLDS FOR RECOVERY FROM ERRORS

	Initial Condition (a) $\frac{U_2(0) = U_3(0)}{U_3(0)}$			Initial Condition (b) $\frac{U_1(0) = U_2(0)}{U_3(0)}$		
$U_1(0)$	Case 1	Case 2	Case 3	Case 1	Case 2	Case 3
0.2	100.1	6.86	2.24	100,000.1	202.38	25
0.4	25.2	4.82	1.58	3,125.2	50.60	6.25
0.6	11.4	3.9	1.29	411.8	22.49	2.78
0.8	6.65	3.34	1.12	98.06	12.65	1.56
1.0	4.5	2.96	1.0	32.50	8.1	1.0
1.2	3.38	2.68	.91	13.46	5.62	0.69
1.4	2.74	2.46	.85	6.65	4.13	0.51
1.6	2.36	2.28	.79	3.85	3.16	0.39
1.8	2.13	2.14	.75	2.59	2.50	0.31
2.0	2.0	2.02	.71	2.0	2.03	0.25
2.2	1.93	1.92	.67	1.72	1.68	0.21
2.4	1.89	1.83	.65	1.60	1.42	0.17
2.6	1.89	1.76	.62	1.57	--	0.15
2.8	1.91	1.70	.60	1.59	--	0.13
3.0	1.94	1.65	.58	1.63	--	0.11
3.2	1.99	1.62	.56	1.7	--	0.10

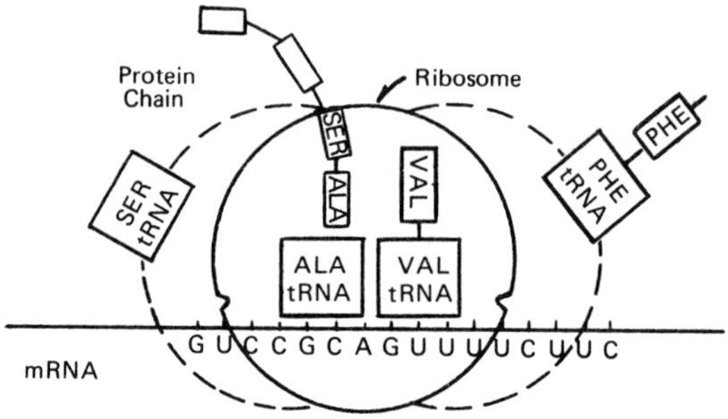

FIGURE 1.1

1.1 Schematics of mechanism of protein synthesizing machinery. UCC, GCA, GUU and UUC, respectively code for amino acids SER, ALA, VAL, PHE. Bigger rectangles represent t-RNA molecules while smaller rectangles represent amino acids.

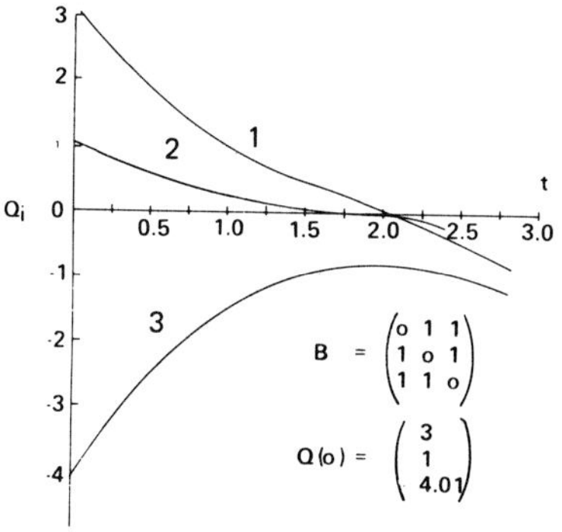

Fig. 2.1A

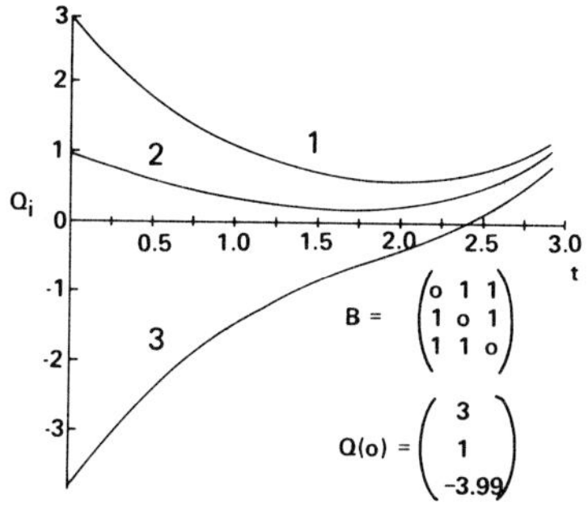

Fig. 2.1B

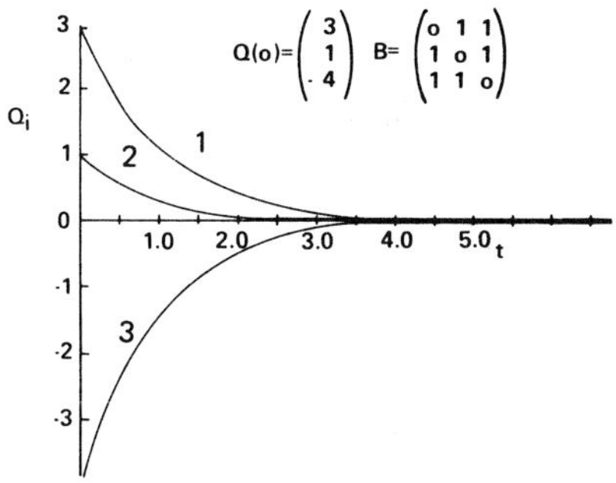

Fig. 2.1C

2.1 Behavior of the protein synthesizing machinery defined by the B-matrix $\begin{pmatrix} 0 & 1 & 1 \\ 1 & 0 & 1 \\ 1 & 1 & 0 \end{pmatrix}$ with Perron vector $\begin{pmatrix} 1 \\ 1 \\ 1 \end{pmatrix}$ for different initial conditions given by $Q(0)$. The first element of $Q(0)$ is the natural logarithm of the ratio of normal and erroneous molecules for synthetase 1, the second and third elements are the corresponding numbers for synthetase 2 and 3. A, B, and C correspond to the initial vectors with transposes (3,1,-4.01), (3,1,-3.99) and (3,1,-4.0), respectively. 1, 2, and 3 denote the three components of $Q(t)$. The time is in terms of generations. From Goel and Islam (1976).

SOME NEW APPLICATIONS OF POPOV'S FREQUENCY-DOMAIN METHOD

A. Halanay
Bucharest University

This report may be considered as a continuation of a similar one presented at the Equadiff Conference in Brussels in 1973 [1]. Three years ago I mentioned the important progress that has been made in Bucharest concerning the applications of frequency-domain methods to differential-delay equations, mainly in the contributions of Dr. Vladimir Rasvan. We want now to describe some new results we obtained in joint work with Dr. Rasvan.

1. *Frequency-Domain Criteria for Nuclear Reactor Stability*

The first results on frequency domain stability criteria for nonlinear reactor models have been obtained by Popov [2]. More recently Kappel and Di Pasquantonio [3] considered a general model for the uncontrolled nuclear reactor and by using a Liapunov functional obtained a condition of stability and estimates of the domain of attraction. The results of Kappel and Di Pasquantonio have been extended by considering more general situations and the case of controlled reactors in [4] and [5]. We shall sketch here the general result for controlled reactors.

We consider the following model

$$\begin{cases} \dot{x}(t) = Ax(t) + \sum_1^r B_k x(t - \theta_k) + b_0\rho(t) + \sum_1^r b_k\rho(t - \theta_k) \\ \dot{\rho}(t) = -\frac{1}{\Lambda}\sum_1^M \beta_k[\rho(t) - \eta_k(t)] - \frac{P}{\Lambda}[1 + \rho(t)]\nu(t) \\ \dot{\eta}(t) = \lambda_k[\rho(t) - \eta_k(t)], \quad k = 1,\ldots,M \\ \nu(t) = c'x(t) + \sum_1^r c_k' x(t - \theta_k) + \sum_0^\infty \alpha_k\rho(t - \tau_k) \\ \qquad + \int_{-\infty}^t K(t - s)\rho(s)ds, \end{cases} \tag{1}$$

with the basic assumptions $\beta_k > 0$, $\Lambda > 0$, $P > 0$, $\sum_{k=0}^{\infty} |\alpha_k| < \infty$

$K \in L_1 \ (0,\infty)$.

Here x is the state vector of the control system, ρ is the reactor power, η_k the flow of the delayed neutrons. All these variables are normalized around some nominal values which represent the steady state of the reactor and control. The linear functional ν represents the reactivity feedback (both internal and external). The input in the control system is the power and the delays θ_k are due to the fact that the measurement devices of the regulating system are situated at a certain distance from the active zone of the reactor. The regulating system may influence only the reactivity ν and this explains the structure of the model considered; the infinite delays in the functional ν correspond to the description of various effects in the reactor. With corresponding notations system (1) may be written in the compact form

$$\dot{z}(t) = Az(t) + \sum_{1}^{\infty} B_k z(t - t_k) + \int_0^t k(t - s)z(s)ds + f(t,z)$$

$$z = col(x,\rho,\eta_1 \dots \eta_M), \quad t_k = \begin{cases} \theta_k & k \leq r \\ \tau_{k-r} & k > r \end{cases} \tag{2}$$

$$f(t,z) = \begin{pmatrix} 0 \\ -\frac{P}{\Lambda}\left[\int_{-\infty}^{0} K(t - s)\rho_0(s)ds + \rho(t)\nu(t)\right] \\ 0 \end{pmatrix}$$

where $\rho_0(t)$, $t \leq 0$ is the initial condition for ρ. We associate to the problem the transfer functions γ_R (corresponding to the effect of the delayed neutrons), γ_ρ corresponding to the internal reactivity feedback, γ_r corresponding to the external reactivity feedback

$$\gamma_R(\sigma) = \frac{1}{\sigma\left(1 + \frac{1}{\Lambda}\sum_1^M \frac{\beta_k}{\sigma + \lambda_k}\right)} \quad \gamma_\rho(\sigma) = \sum^\infty \alpha_k e^{-\sigma\tau_k} + \tilde{K}(\sigma)$$

$$\gamma_r(\sigma) = c'(\sigma)R(\sigma)b(\sigma), \quad \tilde{K}(\sigma) = \int_0^\infty e^{-\sigma t}K(t)dt$$

$$c(\sigma) = c_0 + \sum_1^r c_k e^{-\sigma\theta_k}$$

$$R(\sigma) = (\sigma I - A - \sum B_k e^{-\sigma\theta_k})^{-1}$$

$$b(\sigma) = b_0 + \sum b_k e^{-\sigma\theta_k}$$

$$\gamma_f(\sigma) = \gamma_r(\sigma) + \gamma_\rho(\sigma), \quad \gamma_1(\sigma) = \frac{\gamma_R(\sigma)}{1 + \frac{P}{\Lambda} h_1 \gamma_R(\sigma)\gamma_f(\sigma)},$$

$$\gamma_2(\sigma) = \frac{P}{\Lambda} h_1 \gamma_f(\sigma)\gamma_1(\sigma), \quad h_1 > 0.$$

THEOREM. Assume

1°. $det(\sigma I - A - \sum_1^r B_k e^{-\sigma\theta_k}) \neq 0$ for $Re\ \sigma \geq 0$

2°. There exists h_1, h_2, $0 < h_1 < 1 < h_2$ and $\delta_0 \geq 0$, $\delta_2 \geq \delta_1 \geq 0$ with $\delta_1 + \delta_2 \neq 0$ such that:

(i) The linear system

$$\dot{y}(t) = Ay(t) + \sum_1^r B_k y(t - \theta_k) + b_0\xi(t) + \sum b_k\xi(t - \theta_k)$$

$$\dot{\xi}(t) = -\frac{1}{\Lambda}\sum_1^M \beta_k[\xi(t) - \zeta_k(t)] - \frac{P}{\Lambda} h[c_0'y(t) + \sum_1^r C_k' y(t - \theta_k)$$

$$+ \sum_0^\infty \alpha_k\xi(t - \tau_k) + \int_{-\infty}^t K(t - s)\xi(s)ds]$$

$$\dot{\zeta}_k(t) = \lambda_k[\xi(t) - \zeta_k(t)]$$

is exponentially stable for $h = h_1$ and $h = h_2$

(ii) $h_1 \leq 1 - \gamma_0$ where $\gamma_0 \in (0,1)$, $\hat{\gamma}_0 \in (0, \sqrt{h_2} - 1)$ are such that $\phi(\xi) \leq \phi(\sqrt{h_2} - 1)\ \min\{\ , \frac{\beta_1}{\lambda_1 \Lambda}, \dots, \frac{\beta_M}{\lambda_M \Lambda}\} \stackrel{\Delta}{=} l$

for $-\gamma_0 \leq \xi \leq \hat{\gamma}_0$,

$$\phi(\xi) \stackrel{\Delta}{=} \xi - \ln(1 + \xi) - \frac{1}{2h_2}\xi^2$$

(iii) $Re\ H(i\omega) > 0$ for all $\omega \in R$ where

$$H(i\omega) = \frac{\delta_0}{h_1}\left[\frac{h_1}{h_2 - h_1} + \gamma_2(i\omega)\right] + \delta_1 \gamma_1(i\omega) + $$
$$+ \delta_2 \frac{P}{\Lambda}(h_2 - h_1)|\gamma_1(i\omega)|^2 \gamma_f(i\omega);$$

If $\delta_0 = 0$ we have to assume moreover

$$\delta_1 \frac{1}{\Lambda}\sum_1^M \beta_k + [\delta_1 h_1 + \delta_2(h_2 - h_1)](\alpha_0 - \sum_1^\infty |\alpha_k|) > 0.$$

Then the zero solution of the system is asymptotically stable. Before describing the method of proof let us discuss some special cases.

a) For $\delta_0 > 0$, $\delta_1 = 0$, $\delta_2 = \frac{\Lambda}{P}\frac{1}{h_2 - h_1}$ one gets an extension of Popov's criterion [2] for the case of functional differential equations.

b) for $\delta_0 = 0$, $\delta_1 = \delta_2 = 1$ one gets an extension to the controlled reactor of Kappel - Di Pasquantonio result in [3].

Consider the linear block

$$\dot{y}(t) = Ay(t) + \sum_1^r B_k y(t - \theta_k) + b_0 \xi(t) + \sum_1^r b_k \xi(t - \theta_k)$$

$$\dot{\xi}(t) = -\frac{1}{\Lambda}\sum_{1}^{M} \beta_k[\xi(t) - \zeta_k(t)] - \frac{P}{\Lambda} h[c_0'y(t) + \sum_{1}^{r} c_k'y(t - \theta_k) +$$

$$+ \sum_{0}^{\infty} \alpha_k \xi(t - \tau_k) + \int_{-\infty}^{t} K(t - s)\xi(s)ds] + \mu(t)$$

$$\dot{\zeta}_k(t) = \lambda_k[\xi(t) - \zeta_k(t)] \qquad k = 1,\dots,M$$

$$\lambda(t) = c_0'y(t) + \sum_{1}^{r} c_k'y(t - \theta_k) + \sum_{0}^{\infty} \alpha_k \xi(t - \tau_k) + \int_{-\infty}^{t} K(t - s)\xi(s)ds$$

with input μ and output λ.

Take $h = h_1$, define the control μ_T by

$$\mu_T(t) = \begin{cases} -\frac{P}{\Lambda}(1 + \rho(t) - h_1 \nu(t) & 0 \leq t \leq T \\ -\frac{P}{\Lambda}(h_2 - h_1)\hat{\nu}(t) & t > T \end{cases}$$

where $\nu(t)$ is the one in (1) and

$$\hat{\nu}(t) = c_0'\hat{y}(t) + \sum_{1}^{k} c_k'\hat{y}(t - \theta_k) + \sum_{0}^{\infty} \alpha_k \hat{\xi}(t - \tau_k) + \int_{-\infty}^{t} K(t - s)\hat{\xi}(s)ds$$

with $\hat{y}$, $\hat{\xi}$, $\hat{\zeta}_k$ the solution of (3) for $h = h_1$ defined for $t > T$ by the initial conditions

$$\hat{y}_0(t) = \begin{cases} x_0(t) & -\theta \leq t \leq 0 \\ x(t) & 0 \leq t \leq T \end{cases} \qquad \hat{\xi}_0(t) = \begin{cases} \rho_0(t) & t \leq 0 \\ \rho(t) & 0 \leq t \leq T \end{cases}$$

$$\hat{\zeta}_k(T) = \eta_k(T),$$

x, ρ, η_k being the solution of (1) defined by the initial data x_0, ρ_0, $\eta_k(0)$.

Associate the Popov index

$$X(T) = \delta_0 \int_0^T \mu_T(t) \left[\frac{1}{h_1 - h_2} \mu_T(t) + \frac{P}{\Lambda} (h_2 - h_1)\lambda(t) \right] dt +$$

$$+ \int_0^T \xi(t)[\delta_1 \mu_T(t) + \delta_2 \frac{P}{\Lambda} (h_2 - h_1)\lambda(t)]dt$$

This index can be rewritten in the form

$$X(T) = \delta_0 \int_0^\infty \mu_T(t) \left[\frac{1}{h_2 - h_1} \mu_T(t) + \frac{P}{\Lambda} (\lambda_T(t) + \check{\lambda}_0(t)) \right] dt +$$

$$+ \int_0^\infty \xi_T(t) \left[\delta_1 \mu_T(t) + \delta_2 (h_2 - h_1) \frac{P}{\Lambda} (\lambda_T(t) + \check{\lambda}_0(t)) \, dt - \right.$$

$$- (\delta_2 - \delta_1)(h_2 - h_1) \frac{P}{\Lambda} \int_T^\infty \xi_T(t)(\lambda_T(t) + \check{\lambda}_0(t))dt$$

where

$$\lambda_T(t) = c_0' y_T(t) + \sum_1^r c_k' y_T(t - \theta_k) + \sum_0^\infty \alpha_k \xi_T(t - \tau_k) + \int_0^t K(t - s)\xi_T(s)ds$$

$$\check{\lambda}_0(t) = \begin{cases} \sum_1^r c_k' \check{y}_0(t - \theta_k) + \sum_1^\infty \alpha_k \check{\xi}_0(t - \tau_k) + \int_{-\infty}^0 K(t - s)\rho_0(s) & t > 0 \\ 0 & t < 0 \end{cases}$$

$$\check{y}_0(t) = \begin{cases} x_0(t) & -\theta \le t \le 0 \\ 0 & t < -\theta, \quad t > 0 \end{cases}, \quad \check{\xi}_0(t) = \begin{cases} \rho_0(t) & t \le 0 \\ 0 & t > 0 \end{cases}$$

y_T, ξ_T, ξ_{kT} being defined by

$$\dot{y}_T(t) = Ay_T(t) + \sum_1^r B_k y_T(t - \theta_k) + b_0 \xi_T(t) + \sum_1^r b_k \xi_T(t - \theta_k)$$

$$+ \sum_1^r B_k \check{y}_0(t - \theta_k) + \sum_1^r b_k \check{\xi}_0(t - \theta_k)$$

$$\dot{\xi}_T(t) = -\frac{1}{\Lambda}\sum \beta_k[\xi_T(t) - \xi_{kT}(t)] - \frac{P}{\Lambda} h_1[c_0' y_T(t) + \sum_1^r c_k' y_T(t-\theta_k) +$$

$$+ \sum_1^\infty \alpha_k \xi_T(t-\tau_k) + \int_0^t K(t-s)\xi_T(s)ds] - \frac{P}{\Lambda} h_1[\sum_1^r c_k' \check{y}_0(t-\theta_k) +$$

$$+ \sum_1^\infty \alpha_k \check{\xi}_0(t-\tau_k) + \int_{-\infty}^0 K(t-s)\rho_0(s)ds] + \mu_T(t)$$

$$\dot{\xi}_{kT}(t) = \lambda_k[\xi_T(t) - \xi_{kT}(t)], \quad k = 1,\ldots,M.$$

Using the assumption that (3) is stable for $h = h_2$ one gets that ξ_T, λ_T, $\check{\lambda}_0$, μ_T belong to $L_1(0,\infty) \cap L_2(0,\infty)$, hence the integrals in $X(T)$ are convergent and for the first ones we may use Parseval's Theorem.

After several manipulations we get for $X(T)$, if $\delta_0 \neq 0$

$$X(T) = \delta_0\left(\frac{P}{\Lambda}\right)^2 \frac{1}{h_2 - h_1}\int_0^T (h_1 - 1 - \rho(t))(h_2 - 1 - \rho(t))\nu^2(t)dt -$$

$$- \delta_1 h_1[\Omega(T)-\Omega(0)] - \delta_1 \frac{1}{\Lambda}\sum_1^M \beta_k \int_0^T (\rho-\eta_k)^2\left[\frac{h_2}{(1+\rho)(1+\eta_k)} - 1\right] dt -$$

$$- (\delta_2 - \delta_1)(h_2 - h_1)\{\Omega_1(T) - \Omega_1(0) - \frac{1}{\Lambda}\sum_0^M \beta_k \int_0^T \frac{(\rho-\eta_k)^2}{(1+\rho)(1+\eta_k)} dt\} =$$

$$= \frac{1}{2\Pi}\int_{-\infty}^\infty \left|\sqrt{Re\ H(i\omega)}\ \tilde{\mu}_T(i\omega) + \frac{1}{2}\frac{\tilde{q}(i\omega)}{\sqrt{Re\ H(i\omega)}}\right|^2 d\omega -$$

$$- \frac{1}{8\Pi}\int_{-\infty}^{+\infty} \frac{|\tilde{q}(i\omega)|}{Re\ H(i\omega)} d\omega + \frac{1}{4\Pi}\int_{-\infty}^\infty [\tilde{q}_1(-i\omega)\tilde{q}_2(i\omega) + \tilde{q}_1(i\omega)\tilde{q}_2(-i\omega)]d\omega -$$

$$- (\delta_2 - \delta_1)(h_2 - h_1)\frac{1}{2h_2}\left[\rho^2(T) + \frac{1}{\Lambda}\sum_1^M \frac{\beta_k}{\lambda_k}\eta_k^2(T)\right] +$$

$$+ \frac{1}{\Lambda h_2} \sum_1^M \beta_k \int_T^\infty [\hat{\xi}(t) - \hat{\zeta}_k(t)]^2 dt,$$

$$\Omega(t) = \phi(\rho(t)) + \frac{1}{\Lambda} \sum_1^M \frac{\beta_k}{\lambda_k} \phi(\eta_k(t))$$

$$\Omega_1(t) = \psi(\rho(t)) + \frac{1}{\Lambda} \sum_1^M \frac{\beta_k}{\lambda_k} \psi(\eta_k(t))$$

$$\psi(\xi) = \xi - ln\ (1 + \xi).$$

From this equality we obtain the fundamental inequality

$$\Omega(T) + \frac{1}{[\delta_2(h_2-h_1)+\delta_1 h_1]\Lambda} \sum_1^M \beta_k \int_0^T (\rho-\eta_k)^2 \left[\frac{h_2}{(1+\rho)(1+\eta_k)} - 1 \right] dt +$$

$$+ \frac{1}{h_2-h_1} \frac{\delta_0}{\delta_2(h_2-h_1)+\delta_1 h_1} \left(\frac{P}{\Lambda}\right)^2 \int_0^T (1 + \rho(t) - h_1)(h_2 - 1 - \rho(t))\nu^2(t) dt \leq$$

$$\leq \Omega(0) + \frac{(\delta_2 - \delta_1)(h_2 - h_1)}{\delta_2(h_2 - h_1) + \delta_1 h_1} \cdot \frac{1}{2h_2} \left[\rho^2(0) + \frac{1}{\Lambda} \sum \frac{\beta_k}{\lambda_k} \eta_k^2(0) \right] +$$

$$+ \frac{1}{\delta_2(h_2 - h_1) + \delta_1 h_1} \frac{1}{8\Pi} \int_{-\infty}^{+\infty} \frac{|\tilde{q}(i\omega)|^2}{Re\ H(i\omega)} d\omega +$$

$$+ \frac{1}{\delta_2(h_2-h_1)+\delta_1 h_1} \frac{1}{4\Pi} \int_{-\infty}^{+\infty} (\tilde{q}_1(-i\omega)\tilde{q}_2(i\omega) + \tilde{q}_1(i\omega)\tilde{q}_2(-i\omega)) d\omega \overset{\Delta}{=} E(0);$$

(the last two intervals depend only on the initial conditions and on the parameters of the problem).

From this inequality it is seen that if

$$-\gamma_0 \leq \rho(0) \leq \hat{\gamma}_0, \quad -\gamma_k < \eta_k(0) < \hat{\gamma}_k, \quad E(0) < l$$

then $\rho(t)$, $\eta_k(t)$ satisfy the same inequalities for all $t > 0$, hence we have global existence and Liapunov stability for all solutions in the described domain.

From the same fundamental inequalities one gets

$$\int_0^T (1 + \rho(t) - h_1)(h_2 - 1 - \rho(t))\nu^2(t)dt \leq constant$$

where the constant does not depend on T, hence $\nu \in L_2(0,\infty)$, $(1 + \rho)\ \nu \in L_2(0,\infty)$; by using a lemma of Popov we deduce that $\lim_{t\to\infty} \rho(t) = \lim_{t\to\infty} \eta_k(t) = 0$. After that the equations for x give $\lim_{t\to\infty} x(t) = 0$. The case $\delta_0 = 0$ is considered in the same way. To show that all manipulations involved in the proof are allowed we had to use results of Corduneanu [6] concerning properties of systems that have the form (2). (See also [1], pages 397-405).

2. *Absolute Stability and Forced Oscillations for Some Distributed Parameter Systems*

Let us start with the problem (motivated by examples in several domains including electrotechniques).

$$\frac{\partial u_1}{\partial t} + \tau(\lambda)\frac{\partial u_1}{\partial \lambda} = 0 \qquad \frac{\partial u_2}{\partial t} - \tau(\lambda)\frac{\partial u_2}{\partial \lambda} = 0$$

$$u_1(0,t) + \alpha_1 u_2(0,t) = \nu_1(t) - \beta_1\psi(\nu(t)) + f_1(t)$$

$$\alpha_2 u_1(l,t) + u_2(l,t) = \nu_2(t) - \beta_2\psi(\nu(t)) + f_2(t)$$

$$\dot{x}(t) = Ax(t) + b_{11}u_1(0,t) + b_{12}u_2(0,t) + b_{21}u_1(l,t) + {}$$

$$+ b_{22}u_2(l,t) - b_0\psi(\nu(t)) + f_3(t)$$

$$\nu_1(t) = c_1'x(t), \quad \nu_2(t) = c_2'x(t), \quad \nu(t) = c_0'x(t)$$

$$u_i(\lambda,0) = \gamma_i(\lambda), \quad 0 \leq \lambda \leq l, \quad i = 1,2, \quad x(0) = x_0 .$$

By using the characteristic method as in [7] and [8] the problem is reduced to

$$\dot{x}(t) = (A + b_{11}c_1' + b_{22}c_2')x(t) + (b_{21} - \alpha_2 b_{22})\eta_1(t - h) +$$

$$+ (b_{12} - \alpha_1 b_{11})\eta_2(t - h) - (b_0 + \beta_1 b_{11} + \beta_2 b_{22})\psi(\nu(t)) + g_1(t)$$

$$\eta_1(t) = c_1'x(t) - \alpha_1\eta_2(t - h) - \beta_1\ \psi(\nu(t)) + f_1(t)$$

$$\eta_2(t) = c_2'x(t) - \alpha_2\eta_1(t - h) - \beta_2\ \psi(\nu(t)) + f_2(t)$$

$$\nu(t) = c_0'x(t).$$

The connection between the two problems is obtained from

$$\eta_1(t) = u_1(0,t), \quad \eta_2(t) = u_2(l,t)$$

and

$$u_1(\lambda,t) = \eta_1(t_1(0;\lambda,t)), \quad u_2(\lambda,t) = \eta_2(t_2(l;\lambda,t))$$

t_1, t_2 being solutions of the equations of the characteristics $\frac{dt}{d\lambda} = \frac{1}{\tau(\lambda)}$, $\frac{dt}{d\lambda} = -\frac{1}{\tau(\lambda)}$. These considerations motivate the study of systems of the form

$$\dot{x}(t) = A_0x(t) + A_1y(t - h) - b_0\psi(\nu(t)) + f(t)$$

$$y(t) = A_2x(t) + A_3y(t - h) - b_1\psi(\eta(t)) + g(t)$$

$$\nu(t) = c_0'x(t).$$

Absolute stability conditions in frequency-domain form for such systems have been obtained by Rasvan ([1] pages 381-396). New derivations of the results of Rasvan have been given by Hale and Martinez.

Let us state without proof the results of Rasvan:

1. Assume $0 \leq \frac{\psi(\nu)}{\nu} \leq k \leq \infty, \quad \psi(0) = 0$
2. Let the eigenvalues of A_3 lie inside the unit circle and the roots of $\det H(\sigma) = 0$ be in the half-plane $Re\ \sigma \leq -\alpha < 0$, where

$$H(\sigma) = \begin{pmatrix} \sigma I - A_0 & -A_1 e^{-\sigma h} \\ -A_2 & I - A_3 e^{-\sigma h} \end{pmatrix}$$

3. Let $\gamma(\sigma) = (c_0' \ 0)\, H^{-1}(\sigma) \begin{pmatrix} b_1 \\ b_2 \end{pmatrix}$; assume $q \geq 0$ exists such that $\frac{1}{k} + Re(1 + i\omega q)\nu(i\omega) > 0$ for all $-\infty \leq \omega \leq \infty$. Then the system is absolutely stable.

The critical case

$$\dot{x}(t) = A_0 x(t) + A_1 y(t - h) + d_0 \xi(t) - b_0 \psi(\nu(t))$$

$$y(t) = A_2 x(t) + A_3 y(t - h) + d_1 \xi(t) - b_1 \psi(\nu(t))$$

$$\dot{\xi}(t) = -\psi(\nu(t)), \quad \nu(t) = c_0' \xi(t) + \gamma \xi(t)$$

is also considered by Rasvan in [8]. He proved absolute stability under the following assumptions:

1. $0 < \frac{\psi(\nu)}{\nu} < k < \infty, \quad \psi(0) = 0.$
2. The eigenvalues of A_3 lie inside the unit circle and the roots of $\det H(\sigma) = 0$ lie in $Re\ \sigma \leq -\alpha < 0$.
3. $\gamma + \gamma_2(0) > 0$
4. There exists $q \geq 0$ such that

$$\frac{1}{k} + \gamma q + Re(1 + i\omega q)\gamma_1(i\omega) + q\, Re\gamma_2(i\omega) + \frac{1}{\omega} Im\gamma_2(i\omega) \geq 0$$

$$\gamma_2(\sigma) = (c_0' \ 0)H^{-1}(\sigma)\begin{pmatrix} d_0 \\ d_1 \end{pmatrix},$$ and γ_1 is the same as above.

The most recent results obtained in a joint work with Rasvan concerns frequency domain conditions for the existence of periodic solutions. Let us state the result; consider the system

$$\dot{x}(t) = A_0x(t) + A_1y(t-h) - B_0\psi(\nu(t)) + f(t)$$

$$y(t) = A_2x(t) + A_3y(t-h) - B_1\psi(\nu(t)) + g(t)$$

$$\nu(t) = Cx(t).$$

Assume $\psi(\nu) = col(\psi_i(\nu_i))$. Let $H(\sigma)$ be defined as above and

$$Y(\sigma) = (C \ 0)H^{-1}(\sigma)\begin{pmatrix} B_0 \\ B_1 \end{pmatrix}.$$

<u>THEOREM</u>: Let $det\ H(\sigma) \neq 0$ for $Re\ \sigma > -\alpha$, $\alpha > 0$ $det\ (\sigma I - A_3) \neq 0$ for $|\sigma| \geq 1$; assume there exists $\tau_j \geq 0$, $\delta > 0$ such that

$$\tau_d\mu_d^{-1} + Re\ \tau_d\ Y(iw) \geq \delta I \quad \text{for all real } w$$

($\tau_d = diag(\tau_1..\tau_m)$, $\mu_d = diag(\mu_1,\ldots,\mu_m)$). Then for all ψ such that

$$0 \leq \frac{\psi_j(\alpha_1) - \psi_j(\alpha_2)}{\alpha_1 - \alpha_2} \leq \mu_j, \quad \psi_j(0) = 0$$

and for all f, g with $|f(t)| \leq M$, $|g(t)| \leq M$ for all real t, the system has a bounded on the real axis solution which is periodic, almost periodic respectively if f, g are periodic, almost periodic respectively; moreover this solution is exponentially stable. The proof of this theorem is similar to the one we had in a paper published several years ago.

The system considered will define a flow in the Hilbert space $R^n \times L^2(-h,0;R^p)$; we shall write $c(t;\tilde{t},\tilde{c})$ for $(x(t;\tilde{t},\tilde{x}(\tilde{t}),y_{\tilde{t}}(\cdot)), \ y_t(\tilde{t},x(\tilde{t}),y_{\tilde{t}}(\cdot))(\cdot))$ where $y_{\tilde{t}} \in L^2(-h,0;R^p)$. It is easy to check that $c(\tilde{t};\tilde{t},\tilde{c}) = \tilde{c}$, $c(t_2;t_1,c(t_1;\tilde{t},\tilde{c})) = c(t_2 + t_1;\tilde{t},\tilde{c})$, $t_2 \geq t_1 \geq \tilde{t}$. We prove then usual continuity properties; it is important to mention the uniform L^2-continuity of y_t with respect to t.

$$||y_{t_2} - y_{t_1}||^2 = \int_{-h}^{0} |y_{t_2}(s) - y_{t_1}(s)|^2 ds \leq X(t_2 - t_1)$$

$$\lim_{\rho \to 0} X(\rho) = 0.$$

This inequality is obtained by careful computation using the representation formula.

$$y(t) = A_3^k y_{\tilde{t}}(t - \tilde{t} - kh) + \sum_{0}^{k-1} A_3^j F(t - jh), \quad \tilde{t}+(k-1)h \leq t < \tilde{t}+kh$$

where $F(t) = A_2 \, x(t) - B_1 \psi(\nu(t)) + g(t)$.

The next task is to obtain estimates of the form

$$||c(t;\tilde{t},\tilde{c})|| \leq a_1 + a_2 e^{-\mu(t-\tilde{t})} ||\tilde{c}||$$

$$||c(t;\tilde{t},\tilde{c}_1) - c(t;\tilde{t},\tilde{c}_2)|| \leq a_3 e^{-\mu(t-\tilde{t})} ||\tilde{c}_1 - \tilde{c}_2||$$

Indeed, such estimates allow, by using an idea essentially due to Kurzweil, to prove the existence of a motion which is bounded on the whole axis and exponentially stable; if the flow has some special properties corresponding to periodicity of f, g, almost periodicity respectively, this motion is periodic, almost periodic respectively.

To get the basic estimates a variation of constants formula is first used to get

$$\gamma(t) = \zeta(t) - \int_0^t K(t - s)\psi(\gamma(s))ds$$

for $\gamma(t) = \nu(t + \tilde{t})$; $\zeta(t)$ is written with the initial data and the forcing functions f, g. A direct computation shows that the Laplace transform of the kernel K is just $Y(\sigma)$. The frequency domain condition in the statement gives the inequality $\tau_d\mu_d^{-1} + Re\tau_d Y(i\omega - \mu) \geq \frac{\delta}{2} I$ for all ω and $\mu > 0$ sufficiently small. Then, in the same way as in [9] we obtain

$$\sum_1^m \int_0^T [e^{\mu t}\psi_j(\gamma_j(t))]^2 dt \leq \tilde{k} \left[\sum_1^m \tau_j \left(\int_0^T e^{2\mu t}\zeta_j^2(t)dt \right)^{1/2} \right]^2$$

and since $|\zeta(t)| \leq k_1 e^{-\lambda t}(|x(\tilde{t})| + ||y_{\tilde{t}}||) + k_2$ the estimate

$$|x(t)| \leq k_3 + k_4{}^{-\mu(t-\tilde{t})} (|x(\tilde{t})| + ||y_{\tilde{t}}||)$$

is obtained.

In the same way we obtain the corresponding estimate for $|x_1(t) - x_2(t)|$. We use then the representation formula for y to write

$$y(t + s) = A_3^{k-1} y_{\tilde{t}}(t + s - \tilde{t} - (k - 1)h) + \sum_0^{k-2} A_3^j F(t + s - jh)$$

for $-h \leq s < \tilde{t} + (k - 1) h - t$

$$y(t + s) = A_3^k y_{\tilde{t}}(t + s - \tilde{t} - kh) + \sum_0^{k-1} A_3^j F(t + s - jh)$$

$\tilde{t} + (k - 1) h - t \leq s < 0$.

We get the estimate

$$||y_t|| \le k_5\rho^{k-1}||y_{\tilde{t}}|| + k_6 \sum_0^{k-1} \rho^j ||F_{t-jh}||$$

where $\rho < 1$ is larger than the spectral radius of A_3.

We use then the estimate

$$|F(t)| \le M + M_1|x(t)| \le k_7 + k_8 e^{-\mu(t-\tilde{t})}(|x(\tilde{t})| + ||y_{\tilde{t}}||)$$

to obtain for y_t the estimate of the form

$$||y_t|| \le k_9 + k_{10} e^{-\mu(t-\tilde{t})}(|x(\tilde{t})| + ||y_{\tilde{t}}||).$$

We may thus see the specific features of the frequency-domain method; the condition in frequency-domain form is used just to derive some basic estimates which are then used. It is an art to associate the quadratic functional that gives the good estimates. Let us finally remark that in the last 15-16 years, it has been a continuous competition between frequency-domain and Liapunov function approach. For the models described above the typical frequency-domain approach proved itself up to now more convenient, but it is possible that recent results of Yakubovic [10] concerning extension of the Yakubovic-Kalman-Popov main lemma to the Hilbert space case will lead to a simple way to get results of the above type by Liapunov function methods.

REFERENCES

[1] P. Janssens, J. Mawhin, N. Rouche, *editors Equations differentielles et fonctionnelles nonlineares Hermann*, Paris, 1973, 357-380, 381-396, 397-405.

[2] V. M. Popov, A *new criterion for the stability of systems containing nuclear reactors*, Rev. T'Electrotechnique d'Energetique Serie A(1963), Vol. 8, No. 1, 113-130.

[3] F. Di Pasquantonio, F. Kappel, A *stability criterion for the kinetic reactor equations with linear feedback*, Energia Nucleare (1972), Vol. 19, No. 3, 162-175.

[4] A. Halanay, Vl. Rasvan, *Frequency domain criteria for nuclear reactor stability I. The case of the uncontrolled reactor*, Rev. Roum. Sci. Techn. Electrotechn. et Energ., Vol. 19, No. 2, 367-378 (1974).

[5] A. Halanay, Vl. Rasvan, *Frequency-domain criteria for nuclear reactor stability II. The controlled reactor*, ibidem, Vol. 20, No. 2, 233-250 (1975).

[6] C. Corduneanu, *Some differential equations with delay*, Proc. Equadiff, Brno, 1972.

[7] K. L. Cooke, R. W. Krumme, *Differential-difference equations and nonlinear initial-boundary value problems for linear partial differential equations*, J. Mat. Anal. Appl. 24, No. 2, 372-387 (1968).

[8] Vl. Rasvan, *Stabilitatea absoluta a sistemelor automate cu intirziere*, Editura Academiei, Bucharest, 1975.

[9] A. Halanay, *Almost periodic solutions for a class of nonlinear systems with time lag*, Rev. Roum. Math. Pures Appl., Vol. 14 (1969), 1269-76.

[10] V. A. Yakubovic, *The frequency domain theorem for the case the spaces of states and controls are* Hilbert and *applications to some problems of optimal control systhesis II*, Sibirskii Mat. J. T. XVI, No. 5, 1975 (Russian).

BOUNDS FOR SOLUTIONS OF REACTION-DIFFUSION EQUATIONS

Thomas G. Hallam and Eutiquio C. Young
Florida State University

1. *Introduction*

Reaction-diffusion equations have been employed as models for problems in chemical morphogenesis: Turing [15]; in chemical reactors: Cohen and Laetch [4]; in genetics: Aronson and Weinberger [1], Fisher [6], Fleming [7], Hoppensteadt [9]; in molecular recombination theory: Gray and Kerr [5], Rosen [12]; in ecological population theory: Rosen and Fizell [13], Segel and Levin [14]; as well as in classical heat transfer processes: Cohen [3], Friedman [8], Keller and Olmstead [10].

Reaction processes are usually nonlinear in character; hence, approximate solutions to the dynamical equations representing these phenomena are important for understanding behavior and can often be profitably exploited in the quantitative and qualitative study of such processes. The main objective of this paper is to indicate a general method for constructing upper and lower bounds for the solutions of initial-boundary value problems associated with nonlinear reaction-diffusion equations. The germ of our procedure is contained in a paper by Rosen [11].

Our nonlinear analysis approximation technique employs Lagrange's theory of first order characteristics, a transformation which is required to be monotone in the solution argument, and differential inequalities. The procedure is described in detail in Section 2. An example illustrating the construction and relationship to a previous result is presented in Section 3.

2. *The Nonlinear Analysis Approximation Technique*

Let Γ denote an open connected subset of Euclidean n-space E^n, $R_+ \equiv [0,\infty)$, and $u : \Omega \equiv \Gamma \times R_+ \to E^1$. The symbol $\overline{A}$ will denote the closure of the set A and ∂A will denote the boundary of A.

Consider the equation

$$Lu = -g(u,t) \tag{1}$$

where the operator L is defined by

$$Lu \equiv \left(\frac{\partial}{\partial t} - D\nabla^2 \right) u;$$

the diffusivity D is a positive constant; ∇^2 denotes the Laplacian operator; and, g is continuous from $E^1 \times R_+$ into E^1. Equation (1) coupled with the nonnegative initial distribution

$$u(x,0) = \alpha(x), \quad x \in \Gamma \tag{2}$$

and the impermeability condition

$$\frac{\partial u}{\partial n} = 0 \quad \text{on} \quad \partial\Gamma \times R_+ \tag{3}$$

where n denotes the unit vector normal to the surface $\partial\Gamma$, constitutes the mathematical model of interest. For convenience we shall refer to this model consisting of equations (1), (2), and (3) as M. It is tacitly assumed that M has a solution; see Chan [2] for existence and uniqueness results.

The Green's function, $G(\underset{\sim}{x},\underset{\sim}{y},t)$, which satisfies the linear equation

$$L\, G(\underset{\sim}{x},\underset{\sim}{y},t) = \delta(\underset{\sim}{x} - \underset{\sim}{y})\, \delta(t)$$

and the Neumann boundary condition

$$\frac{\partial G}{\partial n}(\underset{\sim}{x},\underset{\sim}{y},t) = 0$$

can be used to obtain an integral equation representation of the solutions of M.

Specifically, we employ the notation

$$\alpha_1(\underset{\sim}{x},t) \equiv \int_\Gamma G(\underset{\sim}{x},\underset{\sim}{y},t)\, \alpha(\underset{\sim}{y})\, dV_y,$$

$$(G * g)(x,t) \equiv \int_0^t \int_\Gamma G(\underset{\sim}{x},\underset{\sim}{y};t-s)\, g(u(\underset{\sim}{y},s),s)\, dV_y\, ds,$$

and recall that finding a solution of M is equivalent to solving the integral equation

$$u(\underset{\sim}{x},t) = \alpha_1(\underset{\sim}{x},t) - (G * g)(\underset{\sim}{x},t), \quad (\underset{\sim}{x},t) \in \Omega. \tag{4}$$

We now describe the procedure which generates bounds for the problem M. To capitalize on the ordinary differential equation structure present when diffusion is absent, we find an integral, c, of a solution of the ordinary differential equation

$$dr/dt = -g(r,t).$$

Utilizing Lagrange's method for finding characteristics, we note that $c = c(u,t)$ is a solution of the first order partial differential equation

$$\frac{\partial c}{\partial t} - g(u,t)\,\frac{\partial c}{\partial u} = 0 \tag{5}$$

It is our objective to relate the integrals associated with (5) to solutions of equation (1). To this end, we define function w by $w \equiv \phi(c)$, where ϕ is an arbitrary C^2-function and observe that w is a solution of the equation

$$Lw = -D|\nabla w|^2\,\psi \tag{6}$$

where
$$\psi \equiv \frac{\partial^2 c}{\partial u^2}\,\phi'(c) + \left(\frac{\partial c}{\partial u}\right)^2 \phi''(c).$$

The functions $\partial c/\partial u$ and $\partial^2 c/\partial u^2$ are prescribed through (5). The flexibility of our approximation technique resides in the arbitrary choice of the function ϕ. It is convenient to consider situations where ϕ is selected so that ψ has a constant sign. When ψ is of constant sign, the right hand side of (6) is also of constant sign. This leads to two possible initial-boundary value, differential inequality problems

$$P_-: \qquad Lw_- \leq 0, \; w_-(\underset{\sim}{x},0) = \phi(c(\alpha(\underset{\sim}{x}),0)), \quad \underset{\sim}{x} \in \Gamma, \quad t \in R_+;$$

$$\frac{\partial w_-}{\partial n}(\underset{\sim}{x},t) = 0, \quad \underset{\sim}{x} \in \partial\Gamma, \quad t \in R_+;$$

$$P_+: \qquad Lw_+ \geq 0, \quad w_+(\underset{\sim}{x},0) = \phi(c(\alpha(\underset{\sim}{x}),0)), \quad \underset{\sim}{x} \in \Gamma, \quad t \in R_+;$$

$$\frac{\partial w_+}{\partial n}(\underset{\sim}{x},t) = 0, \quad \underset{\sim}{x} \in \partial\Gamma, \quad t \in R_+;$$

associated with M. It is also reasonable to compare solutions of P_- and P_+ with solutions of the problem composed of the homogeneous equation together with similar initial-boundary data

$$P_0: \qquad Lv = 0, \; v(\underset{\sim}{x},0) = \phi(c(\alpha(\underset{\sim}{x}),0)), \quad \underset{\sim}{x} \in \Gamma, \quad t \in R_+;$$

$$\frac{\partial v}{\partial n}(\underset{\sim}{x},t) = 0, \quad \underset{\sim}{x} \in \partial\Gamma, \quad t \in R_+.$$

It follows from (4) and the nonnegativity of G that if $v = v(x,t)$ is a solution of P_0 and w_- is a solution of P_- then

$$w_-(\underset{\sim}{x},t) \leq v(\underset{\sim}{x},t), \quad \underset{\sim}{x} \in \overline{\Gamma}, \quad t \in R_+.$$

Similarly, the relationship between solutions of problems P_+ and P_0 yields

$$v(\underset{\sim}{x},t) \leq w_+(\underset{\sim}{x},t), \quad \underset{\sim}{x} \in \overline{\Gamma}, \quad t \in R_+.$$

The transformation $w = \phi(c) = \phi(c(u,t))$ contains the solution u to problem M implicitly. A basic requirement we shall impose is that this implicit relationship can be used to define a function $\mu = \mu(\underset{\sim}{x},t)$ by the equation

$$v = \phi(c(\mu,t)), \quad \underset{\sim}{x} \in \Gamma, \quad t \in R_+,$$

where v is the solution of problem P_0. Comparison of w and v shows that properties of c and ϕ share in the determination of μ as an upper or lower bound for u. The subsequent situations indicate outcomes generated by different choices of the function ϕ. In the following, R_0 represents a subset of the (u,t) plane which will be determined by the choice of ϕ and by c (see the example in Section 3).

I. *Let the following conditions hold:* $\psi \geq 0$; $w = w(u,t) \equiv \phi(c(u,t))$ *is nondecreasing as a function of* u *for each* $t \in R_+$, $(u,t) \in R_0$. *Then,* μ *is a pointwise upper bound for* u *as long as* (u,t) *is in* R_0.

The nonnegativity of ψ implies that $w = \phi(c)$ is a solution of problem P_- and, as such, satisfies

$$w = \phi(c(u,t)) \leq \phi(c(\mu,t)) = v, \ \underset{\sim}{x} \in \Gamma, \ t \in R_+, \ (u,t) \in R_0.$$

Since w is nondecreasing as a function of u, it follows that

$$u(\underset{\sim}{x},t) \leq \mu(\underset{\sim}{x},t), \ \underset{\sim}{x} \in \Gamma, \ t \in R_+$$

as long as (u,t) is in R_0; that is, when these hypotheses are satisfied, μ is an upper bound for u.

II. *Let* $\psi \leq 0$ *and* w *be nonincreasing as a function of* u *for each* $t \in R_+$, $(u,t) \in R_0$; *then,* μ *is an upper bound for* u *as long as* (u,t) *is in* R_0.

In this situation, we conclude that w is a solution of problem P_+; this leads to the conclusion

$$w = \phi(c(u,t)) \geq \phi(c(\mu,t)) = v, \ t \in R_+, \ \underset{\sim}{x} \in \Gamma, \ (u,t) \in R_0.$$

The nonincreasing property of w implies that

$$u(\underset{\sim}{x},t) \leq \mu(\underset{\sim}{x},t), \ \underset{\sim}{x} \in \Gamma, \ t \in R_+;$$

hence, again μ is an upper bound for u. In an analogous manner the following situations can occur.

III. *Let* $\psi \geq 0$ *and* w *be a nondecreasing function of* u *for each* $t \in R_+, (u,t) \in R_0$; *then* μ *is a lower bound for* u *as long as* (u,t) *remains in* R_0.

IV. *Let* $\psi \leq 0$ *and* w *be a nondecreasing function of* u *for each* $t \in R_+$, $(u,t) \in R_0$; *then* μ *is a lower bound for* u *as long as* (u,t) *remains in* R_0.

3. *Illustration of the Method*

To demonstrate the technique, we consider a second order process wherein g is given by

$$g(u) = ku^2 \quad (k \text{ a positive constant})$$

and the spatial domain is the entire space E^3. Rosen [11] has considered this specific problem; we shall compare the two lower bounds. The Green's function for this problem is given by

$$G(\underset{\sim}{x} - \underset{\sim}{y}), t) = (4\pi Dt)^{-3/2} exp(-|\underset{\sim}{x} - \underset{\sim}{y}|/4Dt).$$

Equation (5) becomes

$$c_t - ku^2 c_u = 0,$$

so that the characteristics are integrals of the equation

$$dt = \frac{du}{-ku^2};$$

that is, $$c(u,t) = 1/ku - t.$$

We shall first determine a lower bound for the solution u. To this end, we choose $\phi = \phi(\xi) = -\sqrt{\xi}$, $\xi \geq 0$. As long as $(u,t) \in R_0 = \{(u,t): 1/u \geq kt\}$, the characteristic c is nonnegative. The function $w = -\sqrt{\frac{1}{ku} - t}$, satisfies P_+ with the initial condition

$$w(\underset{\sim}{x},0) = -\frac{1}{\sqrt{k\alpha(\underset{\sim}{x})}} \equiv \beta(\underset{\sim}{x})$$

since ψ is nonpositive for $1/u \geq 4/3kt$. The solution v of the associated homogeneous problem P_0: $Lv = 0$, $v(\underset{\sim}{x},0) = \beta(\underset{\sim}{x})$; is given by

$$v(\underset{\sim}{x},t) = -\int_{E^3} G(\underset{\sim}{x} - \underset{\sim}{y},t)\ \beta(\underset{\sim}{y})\ dV_y =$$

$$- \frac{1}{(4\pi Dt)^{3/2} k^{1/2}} \int_{E^3} \frac{1}{[\alpha(\underset{\sim}{y})]^{1/2}} e^{-|\underset{\sim}{x}-\underset{\sim}{y}|/Dt} dV_y$$

Defining $\mu_- = \mu_-(\underset{\sim}{x},t)$ implicitly by

$$v = -\sqrt{\frac{1}{k\mu_-} - t}$$

we obtain a lower bound for the solution u given by

$$\mu_- = \left\{ kt + \frac{1}{(4\pi Dt)^3} \left[\int_{E^3} \frac{1}{[\alpha(y)]^{1/2}} e^{-|\underset{\sim}{x}-\underset{\sim}{y}|/Dt} dV_y \right]^2 \right\}^{-1} .$$

If the initial function α is specialized to a Gaussian distribution

$$\alpha(\underset{\sim}{x}) = ae^{-b|\underset{\sim}{x}|^2} \quad (a,b \text{ positive constants})$$

then a lower bound for u is

$$\mu_-(\underset{\sim}{x},t) = \begin{cases} \{kt + (1 - 2bDt)^{-3}a^{-1}\exp[b|\underset{\sim}{x}|^2/(1 - 2bDt)]\}^{-1}, & 0 \le t \le (2Db)^{-1}; \\ 0 & t \ge (2Db)^{-1}. \end{cases}$$

Rosen's approximation to this problem essentially uses $\phi(\xi) = -\xi$ and results in a nonzero μ_- defined only on the interval $0 \le t \le (4Db)^{-1}$.

An upper bound for this solution may be generated by choosing the function ϕ as

$$\phi(\xi) = \xi^{-1}, \quad \xi > 0$$

A direct computation (for convenience, we took $k = 1$) shows that ψ is nonnegative for $c = 1/u - t > 0$ and that $w(u,t) \equiv [1/u - t]^{-1}$ is nondecreasing in u. This places the example in category I; hence, an upper bound for u is found as

$$\mu_+ = [t + v^{-1}]^{-1}$$

where v is the solution of the equation $Lv = 0$ which satisfies

$v(\underset{\sim}{x},0) = \alpha(\underset{\sim}{x})$ and the boundary condition $\partial v/\partial n = 0$.

From the representation formula (4), the positivity of the Green's function, and the existence of a lower (or upper) bound for a solution, it is often possible to generate an upper bound directly. For example, taking advantage of the monotone property of $g(u) = ku^2$, we find an upper bound, μ_+, of the solution is given by

$$\mu_+ \equiv \int_{E^3} G(\underset{\sim}{x} - \underset{\sim}{y}, t)\alpha(\underset{\sim}{y})dVy - \int_0^t \int_{E^3} G(\underset{\sim}{x} - \underset{\sim}{y}, t - s)k[\mu_-(y,s)]^2 dV_y ds$$

Preliminary investigations of the case where $g = g(u,x,t)$ are underway. We plan to discuss this more complicated situation in a future publication.

Acknowledgment. The research of T. G. Hallam was supported in part by the Office of Naval Research under contract N00014-76-C-0006. The authors would like to acknowledge helpful discussions with Professor C. Y. Chan.

REFERENCES

[1] D. G. Aronson, and H. F. Weinberger, *Multidimensional nonlinear diffusion arising in population genetics*, Advances in Mathematics, to appear (1977).

[2] C. Y. Chan, *Positive solutions for nonlinear parabolic second order initial boundary value problems*, Quart. Appl. Math., 32(1974), 443-454.

[3] D. S. Cohen, *Generalized radiation cooling of a convex solid*, J. Math. Anal. Appl., 35(1971), 503-511.

[4] D. S. Cohen, and T. W. Laetch, *Nonlinear boundary value problems suggested by chemical reactor theory*, J. Differential Eqs., 7(1970), 217-226.

[5] E. P. Gray, and Donald E. Kerr, *The diffusion equation with a quadratic loss term applied to electron-ion volume recombination in a plasma*, Ann. Physics, 17(1962), 276-300.

[6] R. A. Fisher, *Gene frequencies in a cline determined by selection and diffusion*, Biometrics 6(1950), 353-361.

[7] W. H. Fleming, *A selection migration model in population genetics*, J. Math. Biol., 2(1975), 219-233.

[8] A. Friedman, *Generalized heat transfer between solids and gases under nonlinear boundary conditions*, J. Math. Mech., 8(1959), 161-183.

[9] F. C. Hoppensteadt, *Analysis of a stable polymorphism arising in a selection-migration model in population genetics*, J. Math. Biol., 2(1975), 235-240.

[10] J. B. Keller, and W. E. Olmstead, *Temperature of a nonlinear radiating semi-infinite solid*, Quart. Appl. Math., 9(1951), 163-184.

[11] Gerald Rosen, *Approximate general solutions to nonlinear reaction-diffusion equations*, Math. Biosciences, 17(1973), 367-370.

[12] Gerald Rosen, *Solutions to the nonlinear recombination equation for the infinite spatial domain*, SIAM J. Appl. Math., 29(1975), 146-151.

[13] Gerald Rosen, and Richard G. Fizell, *Bounds for the total population for species governed by reaction-diffusion equations in arbitrary two-dimensional regions*, Bull. Math. Biol., 37(1975), 71-78.

[14] L. Segel, and S. Levin, *Application of nonlinear stability theory to the study of the effects of diffusion on predator prey interactions*, to appear.

[15] A. M. Turing, *The chemical basis of morphogenesis*, Phil. Trans. Roy. Soc. B, 237(1952), 37-72.

IMBEDDING METHODS FOR BIFURCATION PROBLEMS AND POST-BUCKLING BEHAVIOR OF NONLINEAR COLUMNS

R. Kalaba[1], K. Spingarn[2], and E. Zagustin[3]

1. *Introduction*

In this paper we consider anew the classical problem of a uniform column built in vertically at the bottom and subject to a load at the top, which is free. When the column is sufficiently short, the column will stand upright in the equilibrium position. As the length is increased, eventually a length will be reached for which the column sags over into an equilibrium position which is different from the upright one. One task is to determine this critical length, and a second is to examine post-buckling equilibrium configurations [1,2]. Normally these problems are treated as bifurcation problems for nonlinear two-point boundary value problems or integral equations [3].

Our approach is different. We shall study the potential energy of the column as a function of the angle of inclination at the base and the length. When that angle is zero and the length is sufficiently small, the column stands upright and its potential energy is simply the product of the length and the weight. When, however, the length exceeds the critical length the upright position no longer provides a minimum of potential energy. For some sagged configuration the potential energy is less than the product of length by weight. Thus by determining the potential energy function we can determine the critical length at which the non-straight configuration bifurcates from upright equilibrium configurations.

We shall also study the moment at the base and other features of the column as functions of the length and angle at the base.

[1] Department of Biomedical Engineering, University of Southern California, Los Angeles, California 90007

[2] Hughes Aircraft Company, Los Angeles, California.

[3] Department of Civil Engineering, California State University, Long Beach, California 90840.

All of these functions satisfy certain initial value problems which, as numerical experiments have shown, are feasible and stable.

2. The Potential Energy Function (2)

Consider a column of length L bearing a weight b. The end carrying the weight is free, and the other end is built into a horizontal support at an angle ψ with the vertical. We suppose that the equilibrium configuration is the one in which the potential energy of the rod-weight system is a minimum. Let us denote arc length along the rod, measured from the free end, by s and by $\theta = \theta(s)$, $0 \le s \le L$, the angle of inclination with respect to the vertical of the column at s. We take the energy of bending to be

$$E_b = a \int_0^L \left(\frac{d\theta}{ds}\right)^2 ds \,, \tag{1}$$

where a is an elastic constant depending on the cross section and the material, $\left(\frac{d\theta}{ds}\right)^2$ is, of course, the square of the curvature. In conventional terms we have

$$a = \frac{EI}{2} \tag{2}$$

where E is the Young's modulus and I is the moment of inertia of the column. The potential energy of the weight is simply the product of the weight b and its altitude above the horizontal support,

$$E_w = b \int_0^L (\cos\,\theta)\; ds \,. \tag{3}$$

Thus the total potential energy of the system is, for any suggested configuration $\theta(s)$,

$$E_t = \int_0^L \left[a\left(\frac{d\theta}{ds}\right)^2 + b\,\cos\,\theta\right] ds \,. \tag{4}$$

The actual configuration is the one for which E_t is minimized, the condition at $s = L$ being $\theta(L) = \psi$, the other end being free.

Let us note that the minimum of potential energy is a function of ψ and L, $-\infty < \psi < \infty$, $0 \leq L$,

$$f(\psi,L) = \min_{\theta} \int_0^L \left[a\left(\frac{d\theta}{ds}\right) + b \cos \theta \right] ds . \tag{5}$$

We shall now obtain a partial differential equation for this function, using Bellman's principle of optimality (4). Divide the rod into two parts: one from 0 to $L - h$ and the other from $L - h$ to L, h being small. At $s = L$ the value of θ is given by $\theta(L) = \psi$. We denote the curvature there by κ

$$\kappa = - \left.\left(\frac{d\theta}{ds}\right)\right|_{s = L} \geq 0 . \tag{6}$$

We then write the basic functional equation

$$f(\psi,L) = \min_{\kappa} \left[(a\kappa^2 + b \cos \psi)h + f(\psi + \kappa h, L - h) + 0(h) \right]. \tag{7}$$

The first term in the brackets is the contribution to the total potential energy of the section from $L - h$ to L, and the second is the minimal contribution of the remaining $L - h$ units of the rod, where we have observed that its initial angle is $\psi + \kappa h$ and its length is $L - h$. The function $0(h)$ has the property that

$$\lim_{h \to 0} 0(h)/h = 0 . \tag{8}$$

Then κ must be chosen so that the term in brackets assumes its minimum value. Using Taylor's theorem we find

$$f(\psi,L) = \min_{\kappa}\left[(a\kappa^2 + b\cos\psi)h + f(\psi,L) + \kappa h f_{\psi} - h f_L + 0(L)\right], \tag{9}$$

where the subscripts denote partial derivatives. Passing to the limit we see that the function f satisfies the partial differential equation

$$f_L(\psi,L) = \min_{\kappa}\left[a\kappa^2 + b\cos\psi + \kappa f_{\psi}(\psi,L)\right]. \tag{10}$$

The initial condition for $L = 0$ is

$$f(\psi,0) = 0, \quad -\infty < \psi < +\infty, \tag{11}$$

as is seen from Eq. (5).

Since the term in brackets is simply a quadratic function of the decision variable κ, we can perform the minimization by setting its derivative with respect to κ equal to zero, which yields

$$2\,a\,\kappa + f_{\psi}(\psi,L) = 0, \tag{12}$$

or

$$\kappa = -\left(\frac{1}{2a}\right) f_{\psi}. \tag{13}$$

Substituting this minimizing value of κ (which is a function of ψ and L) into Eq. (10) shows that the potential energy function f satisfies the partial differential equations

$$f_L = a\left(\frac{1}{4a^2}\right) f_{\psi}^2 + b\cos\psi - \left(\frac{1}{2a}\right) f_{\psi}^2, \tag{14}$$

$$f_L = b\cos\psi - \left(\frac{1}{4a}\right) f_{\psi}^2, \quad -\infty < \psi < +\infty, \quad 0 \le L. \tag{15}$$

Later we shall see that this partial differential equation, subject to the initial condition in Eq. (11) is readily integrated numerically using a simple finite difference scheme.

Observe from Eq. (15) that the function f has the properties

$$f(-\psi,L) = f(\psi,L), \tag{16}$$

and

$$f(\pi,L) = -bL \ . \tag{17}$$

Also f is clearly periodic in ψ with period 2π and

$$lim\ f(\psi,L) = \begin{cases} bL & 0 \le L \le L_{crit.}, \\ <bL & L > L_{crit.}, \end{cases} \tag{18}$$

where L_{crit} is the critical length of the column, the least upper bound of the lengths for which the column remains upright. The numerical determination of f will show that this critical length really exists and allow us to estimate its value.

3. *The Euler Equation*

The usual treatment of the variational problem in Eq. (5) consists in writing the Euler equation for the optimizing function $\theta = \theta(s)$, $0 \le s \le L$,

$$\ddot{\theta} = -\ \omega\ sin\ \theta, \quad 0 \le s \le L, \tag{19}$$

where

$$\ddot{\theta} = \frac{d^2\theta}{ds^2}\ , \tag{20}$$

and

$$\omega = \frac{b}{2a}\ . \tag{21}$$

The boundary conditions are

$$\theta(L) = \psi, \tag{22}$$

and

$$\dot{\theta}(0) = 0 \ , \tag{23}$$

a free boundary condition.

The buckling problem consists in determining the least upper bound of the values of L for which the nonlinear boundary value problem in Eqs. (21), (22), and (23), with $\psi = 0$, has only the solution

$$\theta(s) \equiv 0, \quad 0 \leq s \leq L.$$

For this value of L the solution $\theta \equiv 0$ bifurcates: for larger values of L the solution $\theta \equiv 0$ represents an unstable vertical equilibrium configuration and the solution $\theta \not\equiv 0$ represents the sagged minimum potential energy configuration.

The linearized version of Eq. (19) is

$$\theta = -\omega\theta \ , \tag{24}$$

with

$$\theta(L) = 0, \quad \dot{\theta}(0) = 0. \tag{25}$$

From this linear eigenvalue problem we see that L_{crit} satisfies the condition

$$L_{crit.} = \frac{\pi}{2}\left(\frac{2a}{b}\right)^{1/2} . \tag{26}$$

That this is the critical length can also be seen by using the closed form solution of Eqs. (19), (22) and (23), with $\psi = 0$ which involves elliptic functions.

4. *The Moment Function*

Now let us consider the moment at the base of the column,

$$M = M(\psi, L), \quad -\infty < \psi < +\infty, \quad 0 \leq L. \tag{27}$$

Allow this moment to work to increase the base angle from ψ to $\psi + \Delta\psi$. The work done is $M \, \Delta \, \psi$. On the other hand, this is the negative of the change in potential energy of the rod in going from the one equilibrium configuration to the other,

$$M \, \Delta \, \psi = -\Delta f. \tag{28}$$

Thus we take as our analytical definition of the moment at the base

$$M(\psi, L) = -\left(\frac{\partial f}{\partial \psi}\right). \tag{29}$$

Later we shall adopt a more conventional viewpoint. Referring back to Eq. (13), we see that the curvature at the base and the moment there are related by the formula

$$\kappa = \frac{1}{2a} M, \tag{30}$$

which is a form of a well-known formula of Euler

$$E \, I \, \kappa = M. \tag{31}$$

We now obtain the partial differential equation and the initial condition for the function M. Differentiating both sides of Eq. (15) with respect to ψ we find that

$$(f_\psi)_L = -(b \sin \psi) - \left(\frac{1}{2a}\right) f_\psi (f_\psi)_\psi . \tag{32}$$

Then, using the definition in Eq. (29) we obtain the partial differential equation for M.

$$M_L = b \sin \psi + \frac{1}{2a} M \, M_\psi . \tag{33}$$

For the rod of length zero we see that the initial condition is

$$M(\psi,0) = 0, \quad -\infty < \psi < +\infty, \tag{34}$$

a relation which follows from Eq. (11) through differentiation. Also the function M has the properties

$$M(\pi,L) = 0, \quad 0 \leq L, \tag{35}$$

which follows from Eq. (17), and

$$M(\psi + 2\pi,L) = M(\psi,L), \tag{36}$$

$$M(-\psi,L) = -M(+\psi,L). \tag{37}$$

It is also clear that

$$\lim_{\psi\to 0} M(\psi,L) = \begin{cases} 0, & 0 \leq L \leq L_{crit}. \\ > 0, & L_{crit} < L, \end{cases} \tag{38}$$

since with base angle $\psi = 0$ the column will be in a vertical position for $L < L_{crit}$, with no moment at the base, and the moment there will be greater than zero when the rod has sagged.

By numerically integrating the system in Eqs. (33) and (34), we can examine the results, and, in light of the condition in Eq. (38), determine the critical length, which gives the bifurcation value of L.

Lastly, let us introduce the function $\delta = \delta(\psi,L)$, $-\infty < \psi < \infty$, $0 \leq L$, to be the horizontal displacement of the free end of the column of length L built in at the angle ψ at the base. Analytically we have

$$\delta(\psi,L) = \int_0^L \sin\,\theta(s)ds. \tag{39}$$

But from the Euler differential equation in Eq. (19) we have

$$\begin{aligned} \delta(\psi,L) &= -(1/\omega)\int_0^L \ddot{\theta}(s)ds \\ &= -(1/\omega)[\dot{\theta}(L) - \dot{\theta}(0)] \\ &= -(1/\omega)\dot{\theta}(L) \\ &= +(1/\omega)\kappa(\psi,L). \end{aligned} \tag{40}$$

As we have seen, though, in Eq. (13)

$$\kappa = -\left(\frac{1}{2a}\right) f\psi, \tag{41}$$

so that

$$\delta(\psi,L) = -\left(\frac{1}{2a\omega}\right) f\psi, \tag{42}$$

or

$$\delta(\psi,L) = -\left(\frac{1}{b}\right) f\psi. \tag{43}$$

Then, in view of Eq. (29),

$$M(\psi,L) = b\delta(\psi,L), \tag{44}$$

remember also

$$M(\psi,L) = -f\psi(\psi,L), \tag{45}$$

so that M is clearly the moment transmitted through the rod to the base by the weight b at the free end which has been displaced by a horizontal distance δ. Thus M is characterized in three equivalent ways:

$$M(\psi,L) = b\delta(\psi,L), \tag{46}$$

$$M(\psi,L) = -f\psi(\psi,L), \tag{47}$$

$$M(\psi,L) = 2a\kappa(\psi,L). \tag{48}$$

3. Numerical Results

In our first experiment we solve the partial differential equation for the potential energy function f in Eq. (15), with $a = 0.5$ and $b = 4.0$. The critical length is $L_{crit} = \pi/4 \cong 0.785$. We limit ourselves to the interval $0 < \psi \leq \pi$ and as initial conditions we have

$$f(0,L) = 0, \tag{49}$$

also when $\psi = \pi$ we have

$$f(\pi,L) = -bL, \quad 0 \leq L \leq L_{max}. \tag{50}$$

We divide the interval $[0,\pi]$ into R parts and put

$$\Delta\psi = \frac{\pi}{R}; \tag{51}$$

also we divide the interval $[0,L_{max}]$ into N parts and put

$$\Delta L = \frac{L_{max}}{N}. \tag{52}$$

We wish to determine the values of the function f at the grid points

$$\psi = i\Delta\psi, \quad i = 0,1,2,\ldots.,R; \tag{53}$$

$$L = j\Delta L, \quad j = 0,1,2,\ldots.,N. \tag{54}$$

The values of $f(0,0)$, $f(\Delta\psi,0)\ldots f(R\Delta\psi,0)$ are known. Assume now that we know the values of $f(0,j\Delta L)$, $f(\Delta\psi,j\Delta L)$, $f(2\Delta\psi,j\Delta L)\ldots f(R\Delta\psi,j\Delta L)$. To obtain the values of f when j is replaced by $j+1$, we employ simple finite difference formulas. For $i = 1, 2,\ldots,R-1$ we write

$$\frac{\partial f(i\Delta\psi,j\Delta L)}{\partial\psi} \cong \frac{f((i+1)\Delta\psi,j\Delta L) - f(i\Delta\psi,j\Delta L)}{\Delta\psi}. \tag{55}$$

Then we obtain an approximation to f_L at the point $\psi = i\Delta\psi$, $L = j\Delta L$ through use of the partial differential equation (15),

$$\frac{\partial f(i\Delta\psi,j\Delta L)}{\partial L} = b\cos(i\Delta\psi) - \frac{1}{4a}\left[\frac{\partial f(i\Delta\psi,j\Delta L)}{\partial\psi}\right]^2. \tag{56}$$

And for $f(i\Delta\psi,(j+1)\Delta L)$ we write

$$f(i\Delta\psi,(j+1)\Delta L) = f(i\Delta\psi,j\Delta L) + \left[\frac{\partial f(i\Delta\psi,j\Delta L)}{\partial L}\right] \Delta L. \quad (57)$$

For $i = R$, we have the known value

$$f(R\Delta\psi,(j+1)\Delta L) = -b(j+1)\Delta L. \quad (58)$$

We determine $f(0,(j+1)\Delta L)$ through use of a quadratic extrapolation formula,

$$f(0,(j+1)\Delta L) = f(3\Delta\psi,(j+1)\Delta L)$$

$$+ 3[f(2\Delta\psi,(j+1)\Delta L) - f(\Delta\psi,(j+1)\Delta L)].$$

We choose to do this since the function f is singular at $\psi = 0$ and we wish to be able to integrate beyond this length to obtain the postbuckling behavior. Calculations are done on an IBM System 360/44 in single precision with $R = 90$, $L_{max} = 0.80$, $\Delta L = 0.01$. The entire run takes 14 sec., including compile time.

Some results are presented in Table 1.

TABLE 1. SOME VALUES OF $f(0,L)$ and bL

L	$f(0,L)$	bL
0.10	.4000	.4000
0.50	1.9998	2.0000
0.75	2.9940	3.0000
0.76	3.0322	3.0400
0.77	3.0699	3.0800
0.78	3.1067	3.1200
$\pi/4$	--	--
0.79	3.1423	3.1600
0.80	3.1767	3.2000

Though our primary interest in this calculation is feasibility and stability, we see that for values of L near the critical length, which is $\pi/4$, the computed values of $f(0,L)$ begin to become less than the corresponding values of bL, the potential energy that the column has in the upright position, indicating that buckling occurs near that value. More accurate results could be obtained using the finer grid, centered differences, and so on. But we shall pass on to a second experiment.

We now wish to determine the moment at the base M by solving the partial differential equation (33) together with the initial condition in Eq. (34). We also use the property in Eq. (35). As before we consider the case for which $a = 0.5$ and $b = 4.0$, so that $L_{crit} \cong 0.785$. We employ the same type of simple finite difference scheme as before, and, indeed, we find that values of $M\ (0,L)$ begin rising above zero at about $L = \pi/4$. We continue the integration until $L = 3.2$, well beyond the critical length, using $\Delta\psi = \pi/360$ and $L = 0.0025$. We put

$$x = \frac{b}{b_{crit}} = 4bL^2/(2\pi^2 a), \tag{60}$$

and

$$y = \frac{\delta(0,L)}{L} = M(0,L)/\ L\ . \tag{61}$$

The value of b_{crit} is obtained from Eq. (26), so that x is a non-dimensional load applied, and y is a non-dimensional horizontal displacement of the free end. Both x and y are viewed as functions of L. Vol'mir [1] shows that for the universal curve representing y as a function of x, y is identically zero for $0 \leq x \leq 1$. Then as x increases further, y increases rapidly to a maximum and then decreases. Vol'mir, using tables of elliptic integrals, gives the coordinates of the maximum as (1.75, .806). The nearest point in our calculation is (1.7618, .80418), which shows that good accuracy is obtained from a value of the load which is seventy-five percent beyond the critical load. The last point we obtained is $x = 2.3343$ and $y = .76867$, which to graphical accuracy, falls on Vol'mir's curve. This indicates that treatment of the post-buckling behavior presents no special difficulty for the imbedding method. Total elapsed time for the second experiment is 1 min. 41 sec.

6. *The Curvature and Free-End Angle Functions*

Lastly, let us consider κ, the curvature at the bottom and Ψ, the angle with the vertical at the free-end, as functions of ψ and L. Write Eq. (19) in the form of a first order system [5]

$$\dot{\kappa} = \omega \sin \theta, \quad \kappa(0) = 0, \tag{62}$$

$$-\dot{\theta} = \kappa, \quad \theta(L) = \psi . \tag{63}$$

Actually, of course, we have

$$\kappa = \kappa(s,\psi,L), \tag{64}$$

and

$$\theta = \theta(s,\psi,L), \quad -\infty < \psi < +\infty, \quad 0 \leq s \leq L. \tag{65}$$

Differentiate Eqs. (62) and (63) with respect to ψ and then with respect to L to obtain

$$\dot{\kappa}_\psi = (\omega \cos \theta)\theta_\psi, \quad \kappa_\psi(0,\psi,L) = 0, \tag{66}$$

$$-\dot{\theta}_\psi = \kappa_\psi, \quad \theta_\psi(L,\psi,L) = 1, \tag{67}$$

and

$$\dot{\kappa}_L = (\omega \cos \theta)\psi_L, \quad \kappa_L(0,\psi,L) = 0, \tag{68}$$

$$-\dot{\theta}_L = \kappa_L, \quad \dot{\theta}_L(L,\psi,L) + \theta_L(L,\psi,L) = 0. \tag{69}$$

In Eq. (69) we have used the chain rule; the dot represents differentiation with respect to the first argument, the subscript differentiation with respect to the third argument. Then, assuming that there is a unique solution for Eqs. (66) and (67), we see that the solutions are related by the equations

$$\kappa_L(s,\psi,L) = -\dot{\theta}(L,\psi,L)\kappa_\psi(s,\psi,L), \tag{70}$$

$$\theta_L(s,\psi,L) = -\dot{\theta}(L,\psi,L)\theta_\psi(s,\psi,L), \quad L \geq s. \tag{71}$$

But

$$-\dot{\theta}(L,\psi,L) = \kappa(\psi,L), \tag{72}$$

the curvature at the bottom, so Eqs. (70) and (71) become the two partial differential equations, valid for $L \geq s$, where s is viewed as fixed,

$$\kappa_L(s,\psi,L) = \kappa(\psi,L)\kappa_\psi(s,\psi,L), \tag{73}$$

and

$$\theta_L(s,\psi,L) = \kappa(\psi,L)\theta_\psi(s,\psi,L), \quad s \geq L. \tag{74}$$

To round out this system we obtain the partial differential equation for the function $\kappa(\psi,L) = \kappa(L,\psi,L)$.

By definition

$$\kappa(\psi,L) = \kappa(L,\psi,L), \tag{75}$$

so that, using Eqs. (62) and (73),

$$\kappa_L(\psi,L) = \dot{\kappa}(L,\psi,L) + \kappa_L(L,\psi,L) \tag{76}$$

$$\kappa_L(\psi,L) = \omega \sin \psi + \kappa(\psi,L)\kappa_\psi(\psi,L), \tag{77}$$

which is the desired partial differential equation for the function $\kappa(\psi,L)$. Putting

$$\theta(0,\psi,L) = \Psi(\psi,L), \tag{78}$$

we see from Eq. (74) that Ψ satisfies the partial differential equation

$$\Psi_L(\psi,L) = \kappa(\psi,L)\Psi_\psi(\psi,L). \tag{79}$$

The equations (77) and (79) together with their initial conditions at $L = 0$, serve to determine the functions $\kappa(\psi,L)$ and $\Psi(\psi,L)$. Furthermore, in conjunction with Eqs. (73) and (74), for a fixed s, the functions $\kappa(s,\psi,L)$ and $\theta(s,\psi,L)$ may be determined for $L \geq s$. Their initial conditions at $L = s$ are

$$\kappa(s,\psi,s) = \kappa(\psi,s), \tag{80}$$

and

$$\theta(s,\psi,s) = \psi. \tag{81}$$

We have now shown how to determine the functions $\kappa(\psi,L)$, $\kappa(s,\psi,L)$ and $\Psi(\psi,L)$ as solutions of initial value problems.

7. *Discussion*

We have presented initial value problems for determining critical lengths and the post-buckling behavior of columns with large deflections. Some numerical results have demonstrated the feasibility, stability and simplicity of this approach. Easy generalizations are to columns of non-uniform cross sections. Beyond this, we wish to apply such methods to the analysis of arches in the nonlinear region.

The problem of determining higher bifurcation points by methods such as these remains open.

REFERENCES

[1] A. S. Vol'mir, *Stability of Elastic Systems*, translation by Foreign Technology Division, Wright-Patterson Air Force Base, Ohio, 1965.

[2] R. Kalaba, *Dynamic Programming and the Variational Principles of Classical and Statistical Mechanics*, pp. 1-9 of Developments in Mechanics, Vol. 1, Plenum Press, New York, 1961.

[3] J. Keller and S. Antman (eds.), *Bifurcation Theory and Nonlinear Eigenvalue Problems*, W. A. Benjamin, Inc., New York, 1969.

[4] R. Bellman and R. Kalaba, *Dynamic Programming and Modern Control Theory*, Academic Press, New York, 1965.

[5] H. Kagiwada and R. Kalaba, *Derivation and Validation of an Initial Value Method for Certain Nonlinear Two Point Boundary Value Problems*, Journal of Optimization Theory and Applications, Vol. 2, No. 6, pp. 378-385, 1968.

MONOTONICITY AND MEASURABILITY

R. Kannan
University of Missouri at St. Louis

1. We consider here the question of existence of random solutions of random nonlinear operator equations. In general the method for proving the randomness of the solution of the operator equation $T(x,\omega) = 0$ is to solve the associated deterministic problem $T(x) = 0$ obtained by fixing the ω and then establishing the randomness of the map which associates with each ω the set of solutions of $T(x) = 0$. The difficulties in the latter step are as follows: for the deterministic problem the solution need not be unique and iterative techniques need not always be available. This necessitates the study of measurability of multivalued maps. In this paper we will establish the existence of a random solution for a particular nonlinear boundary value problem where the solution need not be unique to each ω and further we do not make use of any iterative techniques. As will be obvious from the proof, the importance of the selected nonlinear boundary value problem is that it gives rise to some special properties for the set of solutions which enables us to prove the randomness of the solution map. Hence our method of proof is not special to the random nonlinear boundary value problem chosen here but might be applicable to more general classes of random nonlinear operator equations.

2. Throughout this paper let $(\Omega,\mathcal{B},\mu)$ be a complete probability space and let D be a bounded smooth domain in R^n with boundary ∂D. Let $\beta : R \times \Omega \to R$ be a function that is monotone increasing and continuous for each ω. In other words let β have a maximal monotone graph. We consider the boundary value problem

$$(1)\qquad \begin{aligned} -\Delta u + \beta(u,\omega) &= f(x,\omega) \quad \text{in } D \\ \frac{\partial u}{\partial n} &= 0 \qquad\quad \text{on } \partial D \end{aligned}$$

for the existence of random solutions.

For each fixed $\omega \in \Omega$, the associated deterministic problem has been recently studied by several authors ([5],[8],[12],[18]) in various situations. An interesting feature is the existence of a non-trivial null-space for the associated linear boundary value problem. In this paper we will assume hypotheses on f and β such that the associated deterministic problem is solvable. The question of existence of random solutions for random nonlinear boundary value problems involving monotonic nonlinearities has been studied in [13] where we resort to the iterative techniques that are available. In this paper we will follow in the lines of [8] and by using a result [15] on randomness of generalized inverses we will reduce the problem to one of proving the randomness of a multivalued correspondence. As remarked above, we hope that the ideas in the proof of the randomness of the multivalued map may be used to establish the existence of more general random nonlinear operator equations. Before we consider the problem (1) as a general abstract operator equation, we will assimilate some concepts from the theory of random operator equations. However we will assume familiarity with the concepts from the theory of monotone operators. A good reference to this theory is [4].

3. Let Ω be a measure space and $\mathcal{B}$ be the σ-algebra of subsets of Ω. For any metric space Y, B_Y denotes the σ-algebra of subsets of Y generated by closed subsets of Y. Then a function g from Ω into Y is called a Y-valued random variable if the inverse image, under g, of each Borel set $B \in B_Y$ belongs to $\mathcal{B}$. For any arbitrary set, the mapping T from $\Omega \times \Gamma$ into Y is called a random operator if for each fixed $\gamma \in \Gamma$, the function $T(\cdot,\gamma)$ is a Y-valued random variable. An equation of the type $T[\cdot,x(\cdot)] = y(\cdot)$ where T is a random operator from $\Omega \times X$ into Y and y is a given random variable with values in Y is called a random operator equation. Any mapping $x(\omega) : \Omega \to X$ which satisfies $T(\cdot,x(\cdot)) = y(\cdot)$ a.e. is said to be a wide-sense

solution of the above equation. However any X-valued random variable $x(\omega)$ which satisfies $\mu\{\omega : T(\cdot,x(\cdot)) = y(\cdot)\} = 1$ is said to be a random solution of the above equation. For a detailed survey of the theory of random operator equations and some of the recent developments in the theory of random fixed-point theorems the reader is referred to Bharucha-Reid [2], [3]. A random operator T from $\Omega \times X$ into Y is said to be separable if there exists a countable set $S \subset X$ and a negligible set $N \in \mathcal{B}$, $\mu(N) = 0$, such that

$$\{\omega : T(\omega,x) \in K,\ x \in F \cap S\} \,\Delta\, \{\omega : T(\omega,x) \in K,\ x \in F\} \subset N$$

(where X is a separable metric space, Y a metric space) for every closed set K in B_Y and every F in B_X. Simply stated the above definition of separability implies that there exists a negligible set $N \in \mathcal{B}$ and a countable set $S \subset X$ such that for each $\omega \notin N$ and each $x \in X$ there exists a sequence $\{x_i\}$ such that $x_i \to x$ and $T(\omega,x_i) \to T(\omega,x)$. This concept of separability may be seen in the monographs of Doob [10], Gikhman and Skorohod [11], Neveu [16], and Bharucha-Reid [3]. An important theorem on the randomness of the inverse (single-valued) of random operators is the one due to Nashed and Salehi [15] which states that: let $(\Omega,\mathcal{B},\mu)$ be a complete probability space, X a separable metric space and Y a metric space. Let T be a separable random operator from $\Omega \times X$ into Y such that a.s. $T(\omega,\cdot)$ is invertible and its inverse $T^{-1}(\omega,\cdot)$ is continuous. Then T^{-1} is also a random operator from $\Omega \times Y$ into X.

A multivalued map $T : \Omega \to X$ where X is another set, is a subset of $\Omega \times X$. The set of all $x \in X$ such that $(\omega,x) \in T$ for a given ω is denoted by $T(\omega)$. In this paper we will be concerned only with multivalued maps $T : \Omega \to X$ that are closed valued in the sense that $T(\omega)$ is a closed subset of X for every $\omega \in \Omega$, where X is a separable metrizable space. A multivalued map $T : \Omega \to X$ is said to be a random map if and only if

$T^{-1}(B)$ is measurable for each closed subset B of X. Multivalued measurable maps have been studied extensively in recent years by numerous authors ([1],[6],[9],[17]). Much of these works assumes some extra properties on the measurable space Ω; and in many of these results X is assumed to be Euclidean. We will now state three fundamental results from the theory of multivalued measurable maps.

a) For a closed-valued multifunction $T : S \to R^n$, the following conditions are equivalent:

1) T is measurable;
2) $domT$ is measurable, and there is a countable (or finite) family $(x_i | i \in I)$ of measurable functions $x_i : domT \to R^n$, such that

$$T(s) = cl\{x_i(s) | i \in I\} \quad \text{for all} \quad s \in domT.$$

b) If $T : S \to R^n$ is a measurable closed-valued multifunction, there exists at least one measurable selection, i.e., a measurable function $x : domT \to R^n$ such that $x(s) \in T(s)$ for all $s \in domT$ [cf. 14].

Both of these results are true when the space R^n is replaced by a separable complete metrizable space [6].

c) Let $T : S \to R^n$ be closed-valued. Then the following are equivalent when $(\Omega, \mathcal{B}, \mu)$ is a complete measure space:

1) T is a measurable multifunction;
2) graph of T is a $\mathcal{B} \otimes \mathcal{B}$-measurable subset of $\Omega \times R^n$ (where B is the algebra of Borel sets);
3) $T^{-1}(C)$ is measurable for all Borel sets $C \subset R^n$.

4. We will now outline the method of proving the existence of a solution to the associated deterministic problem. For the details of the arguments one can look at [7]. We consider here the general abstract version of problem (1). Let E be a random positive self-adjoint differential operator mapping a separable reflexive

Banach space X^* into a real Hilbert space S where (X,S,X^*) is in normal position. Also let N be a random nonlinear operator from X^* to S. We assume that the associated eigenvalue problem of E i.e., $Ex + \lambda x = 0$ together with the boundary conditions has a countable system of eigenvalues $[\lambda_i(\omega)]$: $\lambda_i(\omega) \geq 0$, $\lambda_i(\omega) \to \infty$ and a corresponding complete orthonormal system of eigenfunctions $\{\emptyset_i(\omega)\}$ which form a basis for S. Let $S_0(\omega) = \mathcal{N}(E)$ which is a finite dimensional $= \{\emptyset_i;\ldots,\emptyset_m\}$ and let $P : S \to S_0(\omega)$ be the idempotent projection operator. Also let $H : S \to X^*$ be a linear operator such that

$$H(I - P)Ex = (I - P)x \quad \text{for all} \quad x \in X^*.$$

Nor the operator $E : X^* \subseteq S \to S$ is continuous and random. Hence it is separable with respect to X^*. Also H is bounded. We can then conclude by the theorem of Nashed and Salehi that H is a random operator. Further, this implies that $\mathcal{R}(H)$ is measurable, but $\mathcal{R}_H = \eta(E)^1$. Hence $S_0 = \mathcal{N}(E)$ is measurable. Thus $H(I - P)$ is measurable. (The measurability of P and H follows from the fact that under our hypothesis λ_i's are measurable and the ϕ_i's can be measurably instead.) Then H satisfies $\langle -H\{I - P\}x,x\rangle \geq \lambda_{m+1}(\omega)||-H(I - p)x||^*$ for all $x \in X^*$. Clearly H is the generalized inverse of E and by virtue of the theorem of Nashed and Salehi, $-H(I - P) : S \to X^*$ is a random operator.

By applying the operator $H(I - P)$ to the random differential equation $Ex = Nx$ we get

$$(I - P)x = H(I - P)Nx$$

(2) or

$$x - H(I - P)Nx = Px.$$

Thus every solution of the random differential equation is a solution of (2). If $x \in X^*$ is a solution of (2) then by applying E to both sides of (2),

$$Ex - (I - P)Nx = EPx = PEx.$$

Thus

$$P(Ex - Nx) = Ex - Nx.$$

Hence any solution of (2) is a solution of (1) if and only if

(3) $$P(Ex - Nx) = 0.$$

But $P : S \to \mathcal{N}(E)$ and thus (3) reduces to

(4) $$PNx = 0 .$$

If, however, the random Hammerstein equation

(5) $$x - H(I - P)Nx = x^*(\omega),$$

for any aribtrary $x^*(\omega) \in S_0$, has a unique solution $x(\omega)$ then equation (4) can be rewritten as

(6) $$PN[I - H(I - P)N]^{-1} x^*(\omega) = 0.$$

We have thus, under the above circumstances, reduced the random nonlinear differential equation (1) to the equivalent system of operator equations (5) and (6). This form of splitting (1) is due to Cesari and Kannan [8]. In order to establish the existence of a solution of the equation $Ex = Nx$, we proceed as follows. As may be seen in [7], equation (6) is of the type

$$T_2 x^* = 0$$

where T_2 is a nonlinear random operator from the finite dimensional space S_0 into itself and T_2 is maximal monotone. Hence if T_2 is coercive, (6) can be solved for x^*, for a fixed ω. For each solution x^* of (6) (keeping ω fixed) we can solve (5) from the theory of maximal monotone operators and Hammerstein integral equations. Each such solution x of (5) will then be a solution of the associated deterministic problem.

5. In this section we will discuss the question of existence of random solutions of the operator equation $Ex = Nx$. We will first establish the existence of a multivalued random solution of (6), then we extract a single-valued random selection and then from (5) we obtain a random solution of the abstract problem. For brevity's sake we omit the details of the proofs here. A nonlinear operator T from a Banach space X to its dual X^* is said to be <u>demi-continuous</u> if for every sequence $\{x_n\}$ which converges to x_0 in X and for every choice $y_n \in T(x_n)$ there exists a subsequence $\{y_{n_i}\}$ which is weak* convergent to some $v_0 \in T(x_0)$. Equation (6) is of the type $T_2 x^* = 0$ where T_2 is a maximal monotone operator from S_0 into itself. We will now show that T_2 is demicontinuous. It must be noted here that we do not use the finite dimensionality of S_0, but rather the separability only and this remark is of importance in considering questions of existence of random solutions for more general random nonlinear operator equations. Let $\{x_n\}$ be convergent to $x_0 \in S_0$ and let $y_n = T_2 x_n$. It can be proved, by the maximality of T_2, that $\{y_n\}$ is bounded. Hence it suffices to show that if $\{y_n\}$ is weak* convergent to y_0 then $y_0 = T_2 x_0$. But this will follow from the maximality of T_2. Since T_2 is maximal monotone, so is T_2^{-1}. Thus in order to prove the randomness of T_2^{-1} it suffices to prove that T_2^{-1} is demicontinuous (in other words T_2^{-1} has a closed graph property). Now the nonlinear operator $T_2 : S_0 \to S_0$ being single-valued and defined over all of S_0 is demicontinuous and since S_0 is finite dimensional, it follows that T_2 is continuous. Hence the randomness of T_2 and the separability of S_0 imply that T_2 is a separable random operator. Thus $T_2 : \Omega \times S_0 \to S_0$ is a separable random operator such that T_2^{-1} exists a.s. and is demicontinuous. Proceeding analogously to the proof of Nashed and Salehi [15] we can now show that the map $T_3 : \omega \to T_2^{-1}(0)$ is also a random operator. But by the theorem of Kuratowsi and Ryll-Nardzewski [14], T_3 would have a single-valued random selection if $T_3 : \Omega \to S_0$

is closed for each ω. Once again this is true because of the maximal monotonicity of T_2^{-1} and the fact that for a maximal monotone operator $T_2^{-1}x$ is closed and convex for each x in the domain of T_2^{-1}. Thus we have proved that there exists a single-valued random solution $x^*(\omega)$ of (6). In order to prove the existence of a random solution to the nonlinear abstract equation $Ex = Nx$ we will now apply the theorem of Nashed and Salehi [15]. The operators $-H(I - P)$ and N being random, so is $I - H(I - P)N$. Also for each fixed ω, $[I - H(I - P)N]^{-1}$ is defined over all of X because of the maximal monotonicity of $-H(I - P)$ and N over X. It remains to show $[I - H(I - P)N]^{-1}$ is continuous over X. This has been established in [7], [13] under the additional hypothesis that N is bounded. Thus it follows from the theorem of Nashed and Salehi [15] that $[I - H(I - P)N]^{-1}$ is random. Hence we have proved the existence of a random solution of the random abstract operator equation $Ex = Nx$ when $N : X \to X^*$ is a maximal monotone. Since the nonlinear random boundary value problem (1) is of the type $Ex = Nx$ as considered in Sections 4 and 5, we have thus established the existence of a random solution of (1).

6. We conclude this paper with two remarks:

a) The crucial step in the proof was the demicontinuity of T_2^{-1}. Since such a property is true for other classes of nonlinear operators, it would be interesting to extend the ideas of this paper to wider classes of random nonlinear problems. In fact multi-valued nonexpansive maps are an example and this would answer a question raised by Bharucha-Reid [2].

b) It has been proved in Brézis [5] that the solution of the associated deterministic problem of (1) can be treated as a limit of solutions of a sequence of nonlinear problems. It would be of interest to study a stochastic approximation scheme for a solution of the random problem (1).

[

REFERENCES

[1] R. J. Aumann, *Integrals of set-valued functions*, J. Math. Anal. Appl. 12(1965), 1-12.

[2] A. T. Bharucha-Reid, *Fixed point theorems in probabilistic analysis*, Bull. Amer. Math. Soc. 82(1976), 641-657.

[3] A. T. Bharucha-Reid, *Random Integral Equations*, Academic Press, 1972.

[4] H. Brézis, *Opérateurs Maximaux Monotones*, North Holland, 1973.

[5] H. Brézis, *Quelques propriétés des opérateurs monotones et des semi-groups non linéaires, Nonlinear operators and the calculus of variations*, Bruxelles 1975, Springer Verlag, Lecture Notes in Mathematics, 543.

[6] C. Castaing, *Sur les multi-applications mesurables*, Rev. Francaise d'Informatique et de Recharche Opérationelle, 1(1967), 91-126.

[7] L. Cesari, *Nonlinear oscillations in the frame of alternative methods*, Int. Conf. Dynamical Systems, Providence, 1974, Academic Press 1976, vol. 1, 29-50.

[8] L. Cesari and R. Kannan, *Functional analysis and nonlinear differential equations*, Bull. Amer. Math. Soc. 79(1973), 1216-1219.

[9] G. Debrey, *Integration of correspondences*, Proc. Fifth, Berkeley Symp. on Math. Stat. and Prob., Vol. II, Part I, 351-372.

[10] J. L. Doob, *Stochastic Processes*, John Wiley, New York, 1953.

[11] I. I. Gikman and A. V. Skorohod, *Introduction to the Theory of Random Processes*, W. B. Saunders, Philadelphia, 1969.

[12] R. Kannan and John Locker, *Nonlinear boundary value problems and operators* TT^*, Jour. Diff. Eqns. (to appear).

[13] R. Kannan and H. Salehi, *Random nonlinear equations and monotonic nonlinearities*, Jour. Math. Anal. Appl. (to appear).

[14] K. Kuratowski and C. Ryll-Nardzewski, *A general theorem on selectors*, Bull. Acad. Polon. Sci. Sér. Math. Sci. Astronom. Phys. 13(1965), 397-403.

[15] M. Z. Nashed and H. Salehi, *Measurability of generalized inverses of random linear operators*, SIAM J. Appl. Math. 25(1973), 681-692.

[16] J. Neveu, *Mathematical Foundations of the Calculus of Probability*, Holden-Day, San Francisco, 1965.

[17] R. T. Rockafellar, *Measurable dependence of convex sets and functions on parameters*, Jour. Math. Anal. Appl. 28(1969), 4-25.

[18] M. Schatzman, *Problémes aux limites, nonlinéaires, non coercifs*, Ann. Scuola Norm. Sup. Pisa, 27(1973), 641-686.

CHAOTIC BEHAVIOR IN DYNAMICAL SYSTEMS

James L. Kaplan and Frederick R. Marotto
Boston University

In recent years numerous researchers have investigated dynamical systems which, for certain parameter values, exhibit "chaotic" behavior; that is, for such parameter values these systems possess an infinite number of different periodic orbits of arbitrarily long period, as well as an uncountable number of initial points for which the trajectories, although bounded, do not settle into any cycle. These studies were originally motivated by the brilliant paper by E. Lorenz [1], who considered the forced dissipative system:

$$\begin{aligned} x' &= -\sigma x + \sigma y \\ y' &= -xz + rx - y \qquad (1) \\ z' &= xy - bz \; . \end{aligned}$$

These equations are an approximation to a system of partial differential equations describing finite amplitude convection in a fluid layer heated from below. Ruelle and Takens [2] have proposed that the chaotic behavior exhibited by the Lorenz system (1) is a mathematical analogue of turbulence in the flow of fluids.

In studying the system (1), Lorenz attempted to reduce the dimensionality of the system by determining some single scalar feature of a trajectory which would characterize its behavior. Such a feature appeared to be the maximum value of z on any circuit of the trajectory about one of the non-zero equilibrium solutions. (See [1] for details.) For a given trajectory, let M_n denote the maximum value of z on the nth circuit around either of these fixed points. Lorenz plotted the collection of all ordered pairs (M_n, M_{n+1}) and obtained a relation which implicitly determined a mapping $f : \mathbb{R} \to \mathbb{R}$. Thus, Lorenz was led to consider first order, scalar difference equations of the type:

$$M_{n+1} = f(M_n).$$

With Lorenz's work in mind, Li and Yorke [3] established conditions under which an arbitrary first order, scalar difference scheme of the form:

$$x_{k+1} = f(x_k) \qquad f : \mathbb{R} \to \mathbb{R} \tag{2}$$

will behave chaotically. They proved the following remarkable result.

<u>THEOREM 1</u>. (Li and Yorke). Let J be an interval and let $f : J \to \mathbb{R}$ be continuous. Suppose there exists a point of period 3 for (2). Then:

(i) for each $k = 1,2,\ldots$ there is a point of (2) having period k.

Furthermore,

(ii) there is an uncountable invariant set $S \subset J$, called a <u>scrambled set</u>, containing no periodic points, which satisfies

(a) for every $p, q \in S$, $p \neq q$,

$$\limsup_{n\to\infty} |f^n(p) - f^n(q)| > 0,$$

(b) for every $p \in S$ and periodic point $q \in J$,

$$\limsup_{n\to\infty} |f^n(p) - f^n(q)| > 0.$$

Additionally,

(c) there is an uncountable subset $S_0 \subset S$ such that,

$$\liminf_{n\to\infty} |f^n(p) - f^n(q)| = 0 \quad \text{for all } p, q \in S_0.$$

(Note: A point p is said to have period n for (1) if $f^n(p) = p$ but $f^j(p) \neq p$ if $j < n$, where f^j represents the composition of the function f with itself j times.)

This result has been applied to numerous difference equations modelling population growth in seasonally breeding populations with nonoverlapping generations. (See May [4].) Additional applications may be found in such diverse fields as epidemiology [5], economics [6], [7], [8], and social science [9].

Difference schemes describing population growth are by no means restricted to single-species models. There are also prey-predator systems and competitive systems where generations do not overlap. (Anthropod systems provide many such examples.) In a very interesting paper Beddington, Free and Lawton [10] investigate a particular host-parasite system. Numerical studies show regions (in the parameter space) of stable points, stable limit cycles, and chaos. The paper is illustrated by fascinating photographs of oscilloscope trajectories.

In another paper Guckenheimer, Oster and Ipaktchi [11] discuss another discrete model (the so-called Leslie model). Here the population has been divided into two age classes. Again, numerical investigation of the model shows that there are stable points, stable cycles, stable invariant curves, and chaotic orbits, depending upon the values of the parameters. The authors give a heuristic explanation of the reasons for this behavior.

It is significant to note that the behavior of these multi-dimensional systems cannot be described or explained in terms of Theorem 1. First of all, the proof of this theorem relies heavily upon the fact that f is a real valued function of a single real variable. Its proof uses the following simple fixed point theorem.

LEMMA 1. Let $G : \mathbb{R} \to \mathbb{R}$ be continuous. Let $I \subset \mathbb{R}$ be a compact interval, and assume $I \subset G(I)$. Then there exists a fixed point of G in I.

Of course Lemma 1 is no longer true if $G : \mathbb{R}^n \to \mathbb{R}^n$ for $n > 1$. (See [3] for a proof of Lemma 1 as well as for a complete proof of Theorem 1.)

Secondly, simple examples show that the existence of a periodic point of period 3 does not suffice to produce chaos in higher dimensions. For instance, the difference scheme:

$$\begin{aligned} x_{k+1} &= (ax_k + by_k)(1 - ax_k - by_k) \\ y_{k+1} &= x_k \end{aligned} \tag{3}$$

has a stable cycle of period 3 when $a = 1.9$ and $b = 2.1$, although chaotic behavior does not occur for those parameter values. (It does occur for different values, however.)

A theorem which provides a basis for a rigorous analysis of the system (3), as well as a large class of other higher dimensional difference schemes, has been given by Marotto [12].

Consider the general first order difference equation:

(4) $\quad x_{k+1} = f(x_k)$ where $f : \mathbb{R}^n \to \mathbb{R}^n$, f continuous.

DEFINITION 1. Let $f : \mathbb{R}^n \to \mathbb{R}^n$ be continuously differentiable in some ball $B_r(z)$ about some fixed point z. We shall say that z is a repelling fixed point of f if all eigenvalues of $Df(x)$ exceed 1 in norm for all $x \in B_r(z)$.

The repelling fixed point z will be called a snap-back repeller if there exists $x_0 \in B_r(z)$, with $x_0 \neq z$, and an integer M such that:

$$f^M(x_0) = z$$

and $|Df^M(x)| \neq 0$.

THEOREM 2. (Marotto). Snap-back repellers imply chaos in $\mathbb{R}^n$. More precisely, suppose z is a snap-back repeller for f. Then,

(i) there is an integer N such that for every $k > N$ there is a periodic point of (4) having period k. Furthermore, there exists an uncountable set $S \subset B_r(z)$ satisfying (ii) of Theorem 1 (with the absolute value replaced by the Euclidean norm in $\mathbb{R}^n$).

Sketch of proof of Theorem 2(i): Without loss of generality we can assume that $x_0 \in B_r(z)$ but $f(x_0) \not\in B_r(z)$. Let $x_k = f^k(x_0)$. Now $f^M(x_0) = z$ and $|Df^M(x_0)| \neq 0$. Thus, by the inverse function theorem, there exists a ball $B_\varepsilon(z)$ and a function $G : B_\varepsilon(z) \to \mathbb{R}^n$ such that $G(z) = x_0$ and $G|_{B_\varepsilon(z)} = f^{-M}$.

Now since G is a repelling fixed point, it follows that there must exist some integer $\tilde{J}$ such that for $J > \tilde{J}$:

$$f^{-J}(Q) \subset B_\varepsilon(z)$$

where $Q = G(B_\varepsilon(z))$. But:

$$\begin{aligned} F^{-J}(Q) &= f^{-J}(f^{-M}(B_\varepsilon(z))) \\ &= f^{-(J+M)}(B_\varepsilon(z)) \subset B_\varepsilon(z). \end{aligned}$$

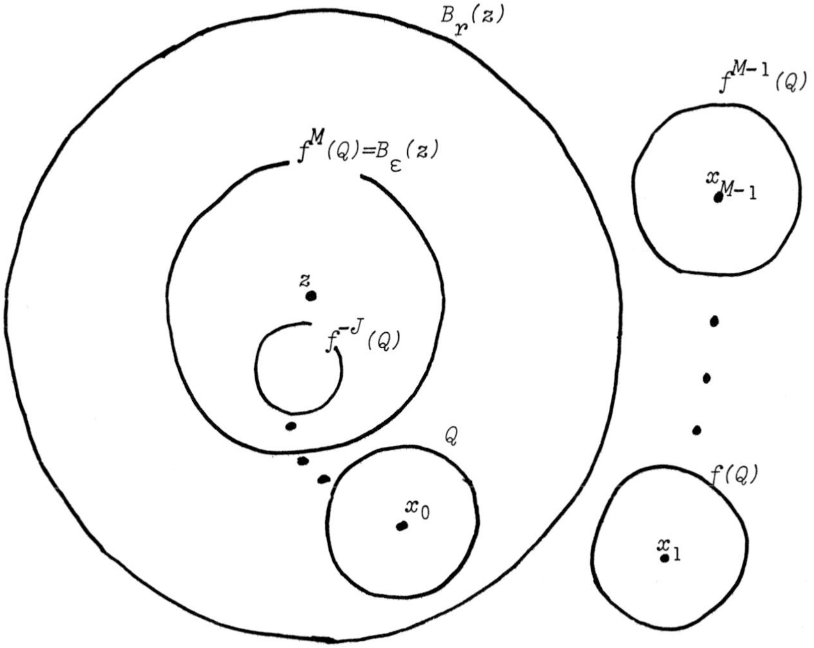

It follows from the Brouwer fixed point theorem that there exists a fixed point of $f^{-(J+M)}$ in $B_\varepsilon(z)$ for all $J > \tilde{J}$. Such a fixed point is automatically a fixed point of f^{J+M}. Thus (i) is established with $N = M + \tilde{J}$.

It is interesting to note that for the special case of (4) with $n = 1$, the existence of a snap-back repeller of f, and of a point of period 3 of some iterate of f, are equivalent notions. (See Marotto [13].) Thus Theorem 2 is a true generalization of Theorem 1.

Heuristically, the idea behind the preceeding proof is as follows. If I is a compact, convex subset of $\mathbb{R}^n$ for $n > 1$, and if $I \subset f(I)$, then we would like to be able to infer the existence of a fixed point of f in I (as was done in Lemma 1 for $n = 1$). If f is invertible then $f^{-1}(I) \subset f^{-1}(f(I)) = I$, and so the existence of a fixed point of f^{-1} (and hence of f) is assured by the Brouwer fixed point theorem. The condition $|Df^M(x_0)| \neq 0$ insures that for some sufficiently large iterate of f we can define an appropriate inverse, even though f itself may be a many-to-one mapping.

It is possible to weaken the hypothesis that z be a repelling fixed point to require only that z be conditionally stable. In this case our proof of part (i) proceeds as above to construct $Q = G(B_\varepsilon(z))$. Now, however, we can no longer resort to the Brouwer fixed point theorem. Instead, we must use degree theory. It is easily seen that in the figure shown $deg\ f^{-(J+M)}(B_\varepsilon(z)) \neq 0$, so that $f^{-(J+M)}$ must have a fixed point in $B_\varepsilon(z)$ for each sufficiently large J.

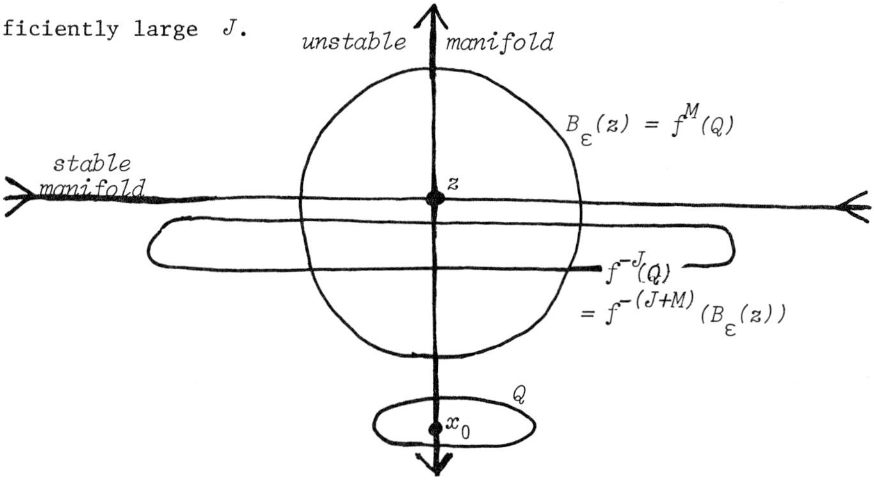

The previous arguments are reminiscent of Smale's famous "horseshoe" example [14] for diffeomorphisms $f : \mathbb{R}^n \to \mathbb{R}^n$. Smale has shown that the existence of a homoclinic orbit for which the intersection of the stable and unstable manifolds is transverse is sufficient to imply the existence of an infinite number of periodic orbits, as well as an uncountable number of aperiodic ones. In our case the snap-back repeller is an analogue of a homoclinic orbit, while the hypothesis $|Df^M(x_0)| \neq 0$ is analogous to Smale's transversality condition.

Moreover, there is a natural extension of Smale's result to flows, rather than discrete dynamical systems, by considering the suspension of a diffeomorphism. Significantly, in studies of the Lorenz system (1), Kaplan and Yorke [15] have shown that turbulence first occurs subsequent to the appearance of a homoclinic orbit, although in this system a "horseshoe" is not present. Instead, the analysis relies upon a configuration which has been called a "broken horseshoe". (We remark, however, that chaos can occur for differential equations only for systems of three or more equations, while it can occur in even one-dimensional difference schemes.)

Consideration of these findings suggests that the phenomenon of chaos is associated with a single feature, namely, the homoclinic orbit. It is not unreasonable to suspect, therefore, that similar processes may be occuring within more general types of dynamical systems which have not as yet been investigated with respect to chaotic behavior. In particular, we would like to suggest that just such a phenomenon is occuring for the differential-delay equation:

$$x(t) = \int_{t-L_1-L_2}^{t-L_1} bx(s)(1 - x(s))\, ds \tag{5}$$

which has been used to model the spread of infectious diseases. (For a discussion see Cooke and Yorke [16].) If we let $a = b/L_2$ then (5) becomes:

$$x(t) = \frac{1}{L_2} \int_{t-L_1-L_2}^{t-L_1} bx(s)(1 - x(s))\, ds \,. \tag{6}$$

Taking the limit of this expression as $L_2 \to 0$, (6) reduces to the widely studied finite difference equation:

$$x(t) = bx(t - L_1)(1 - x(t - L_1)).$$

(See [3], [4], [11], [12], [17], for example.) In this equation chaos occurs for $b > 3.57$. This suggests that for L_2 sufficiently small (or, equivalently, for $L_1 >> L_2$) we might encounter chaos in the trajectories of (5) for appropriate choices of the parameter a. In fact, preliminary numerical investigations seem to corroborate this conjecture.

We would like to describe a possible mechanism for interpreting this phenomenon.

First, note that the system (6) has the non-trivial equilibrium solution:

$$x^*(t) = \frac{b - 1}{b}$$

If we linearize (6) about x^* and substitute $e^{\lambda t}$ to determine the eigenvalues λ of the linearized equation, we find that λ must satisfy:

$$\frac{2 - b}{\lambda} [exp(-\lambda L_1) - exp(-\lambda(L_1 + L_2))] = 1.$$

Since there are only a finite number of conjugate pairs of roots with positive real part, the dimension of the unstable manifold at x^* is finite. (In fact, the unstable manifold is always finite dimensional for differential-delay equations. See Hale [18].) More importantly, in the complex plane these eigenvalues $\lambda = \alpha + \beta i$ appear to lie along a "parabola-like" curve as depicted below:

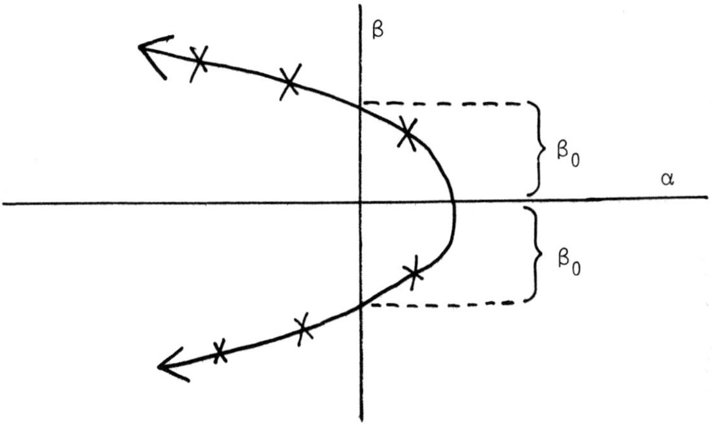

Thus, for any $\lambda = \alpha + \beta i$:

$$\text{if} \quad \alpha > 0 \quad \text{then} \quad |\beta| < \beta_0$$

and

$$\text{if} \quad \alpha < 0 \quad \text{then} \quad |\beta| > \beta_0$$

for some $\beta_0 > 0$. Intuitively then, the stable manifold at x^* contains trajectories of (6) which oscillate quickly around x^*, and the unstable manifold, those which oscillate slowly.

Another way to see this is as follows. Suppose we have a trajectory $x(t)$ of (6) which is quickly oscillating around x^* with small amplitude. If L_2 is large then $x(t)$ is integrated over an interval in which it is sometimes above and sometimes below x^*.

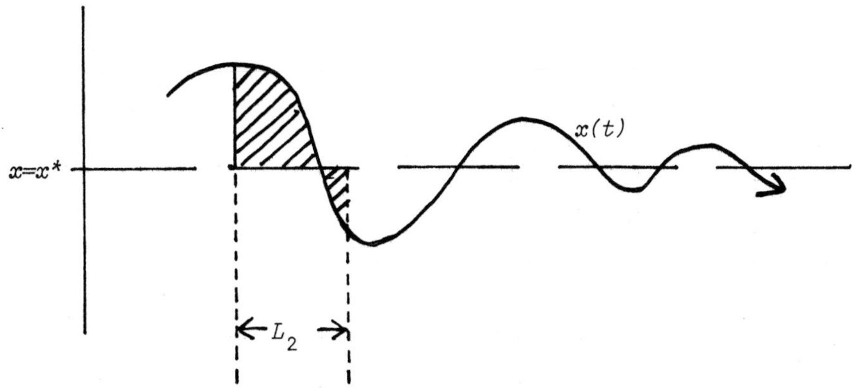

This will produce damped oscillations of $x(t)$ because of cancellation. On the other hand, if $x(t)$ is slowly oscillating and L_2 is relatively small, then very little cancellation will occur and thus (for b large enough) $x(t)$ will oscillate with increasing amplitude.

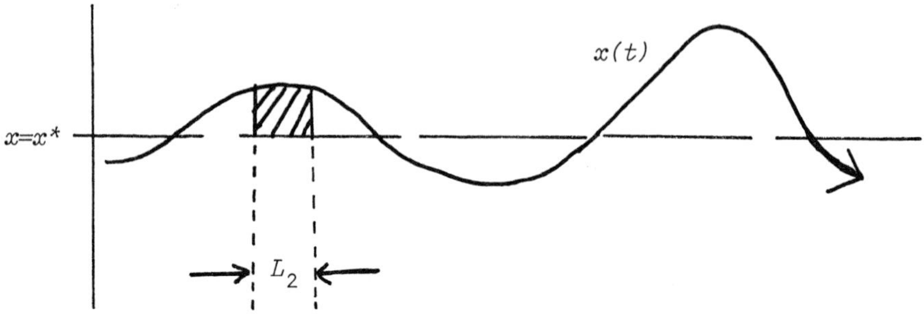

Note, therefore, that a homoclinic orbit of (6) must be a trajectory which begins as a slowly oscillating function arbitrarily close to x^*, and ultimately becomes quickly oscillating. One way in which such an orbit may arise is demonstrated by the following heuristic argument.

Suppose $x(t)$ is a slowly oscillating trajectory about x^*. As $x(t)$ increases in amplitude, it will eventually cross the line $x = 1/2$ at which the function $bx(1 - x)$ assumes its maximum. Thus, $L_1 + L_2$ units of time later $x(t)$ will behave as though it were "folded over". That is, if $0 < x(t) < 1/2$ for $t_1 < t < t_2$, then for $t_1 + L_1 + L_2 < t < t_2 + L_1 + L_2$ there will be an extra "hump" in $x(t)$ (as though $x(t)$ were folded horizontally from the dotted to the solid curve in the figure below). If this folding is sufficiently great, this will result in an increase in the frequency of oscillation around x^*. This process may be repeated, each time increasing the frequency of oscillation. If the increased frequency thus produced is sufficiently large, damping may occur and $x(t)$ will then decay to x^*.

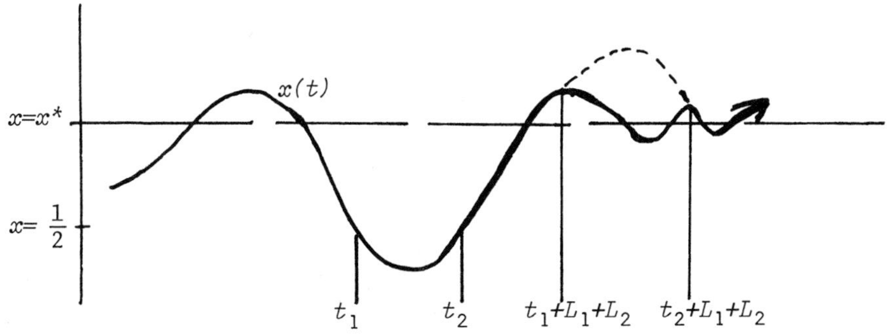

Hence, a function $x_t(\cdot)$ initially lying in the unstable manifold of x^*_t may eventually be an element of the stable manifold. (Note: $x_t \in C\,[-L_1-L_2,-L_1],\mathbb{R})$ is defined by $x_t(s) = x(t+s)$ for $-L_1-L_2 \le s \le -L_1$. See Hale [18] for a further discussion of functional differential equation notation.) In this manner it seems that a homoclinic orbit in the function space can arise.

Numerical evidence confirms that this type of "folding over" process does occur for this equation. In addition, the resulting increase in the frequency of oscillation becomes more pronounced as L_1/L_2 is increased. However, rigorously demonstrating the existence of a homoclinic orbit remains to be investigated.

A careful proof that such an orbit actually implies chaos for a differential-delay equation also merits further investigation. Hopefully, some analogue of Smale's "horseshoe" can be developed which will include infinite dimensional problems of this type.

REFERENCES

[1] E. N. Lorenz, *Deterministic Nonperiodic Flow*, J. Atmos. Sci., 20(1963), p. 130-141.

[2] D. Ruelle, and F. Takens, *On the Nature of Turbulence*, Commun. Math. Phys. 20(1971), p. 167-192.

[3] T. Y. Li, and J. A. Yorke, *Period Three Implies Chaos*, Amer. Math. Monthly 82(1975), p. 985-992.

[4] R. M. May, *Simple Mathematical Models with Very Complicated Dynamics*, (preprint).

[5] F. C. Hoppensteadt, *Mathematical Theories of Populations:*

Demographis, Genetics, and Epidemics, SIAM Pubs., Philadelphia (1975).

[6] P. A. Samuelson, *Foundations of Economic Analysis*, Harvard University Press, Cambridge, Massachusetts (1947).

[7] R. E. Goodwin, *The nonlinear accelerator and the persistence of business cycles*, Econometrica 19(1951), p. 1-17.

[8] W. J. Baumol, *Economic Dynamics, 3rd edition*, Macmillan, New York (1970).

[9] J. Kemeny, and J. L. Snell, *Mathematical Models in the Social Sciences*, MIT Press, Cambridge, Massachusetts (1972).

[10] J. R. Beddington, C. A. Free, and J. H. Lawton, *Dynamic Complexity in Predator Prey Models Framed in Difference Equations*, Nature 255(1975), p. 58-60.

[11] J. Guckenheimer, G. Oster, and A. Ipaktchi, *The Dynamics of Density Dependent Population Models*, (Preprint).

[12] F. R. Marotto, *Snap-Back Repellers Imply Chaos in* R^n, J. Math. Anal. Appl. (To appear).

[13] ______, Doctoral Thesis, Boston University, Boston, Massachusetts (In preparation).

[14] S. Smale, *Differentiable Dynamical Systems*, Bull. Amer. Math. Soc. 73(1967), p. 747-817.

[15] J. L. Kaplan, and J. A. Yorke, *On the transition to turbulence in the Lorenz system*, (In preparation).

[16] K. L. Cooke, and J. A. Yorke, *Some equations modelling growth processes and gonorrhea epidemics*, Math. Biosci. 16(1973), p. 75-101.

[17] R. M. May, *Biological populations with nonoverlapping generations: stable points, stable cycles, and chaos*, Science 186(1974), p. 645-647.

[18] J. Hale, *Functional Differential Equations*, Springer-Verlag, New York (1971).

STABILITY TECHNIQUE AND THOUGHT PROVOCATIVE DYNAMICAL SYSTEMS*

G. S. Ladde
The State University of New York at Potsdam

1. Since 1940, there is a continuous growth in stability conditions of linear differential systems. As far as known, no attempt has been made to unification of these conditions. However, a systematic attempt in the direction of comparing the existing stability conditions for linear systems has been made by Newman [15].

Very recently, based on Rosen's striking work and some of his very interesting remarks [16], the stability analysis of chemical systems [5], compartmental systems [6], hereditary competitive systems [9,11], random competitive systems [8], and diffusion systems [10] has been made in a systematic and unified way to cover a wide range of dynamic processes in biological, physical and social sciences.

In the present work, we present a very simple stability conditions that includes the earlier stability conditions for linear systems. In addition, it is also applicable for special type of nonlinear systems that are mathematical models of thought provocative phenomena, such as chemical systems, compartmental systems, eco-systems, economic systems, social systems, and etc. Furthermore, the presented stability conditions illustrate several intrinsic problems that are of practical interest, for example, complexity vs. stability, measurability of complexity, sensitivity of stability, etc. In short, it bridges the gap between the two mathematical stability conditions and the gap between the mathematical conditions and certain real life problems.

In Section 2, we present certain basic notations and definitions. In Section 3, we give a basic stability result that unifies the earlier results in a systematic and unified way. In Sec-

*The research reported herein was supported by SUNY Research Foundation Faculty Fellowship No. 0629-01-030-75-0.

tion 4, we state certain observations that illustrate the practical significance of the stability conditions.

2. Let us consider a system of differential equations

$$\dot{x} = A(t,x)f(x) \tag{2.1}$$

where $x \in R^n$; $A \in C[R^+ \times D(0,\rho), R^{n^2}]$; $f \in C[R^n, R^n]$ and $f(x) = (f_1(x_1), f_2(x_2), \ldots, f_n(x_n))^T$; $D(0,\rho) = \{x \in R^n: \ 0 \leq |x_i| < \rho_i,\ i \in \{1,2,\ldots,n\} = I,\ \rho_i > 0\}$. Further assume that the function A, f are smooth enough to assure the existence and uniqueness of solutions of (2.1). Furthermore, without loss in generality, we assume that $x \equiv 0$ is unique solution of (2.1).

(H_1) Assume that $\frac{\partial f(x)}{\partial x} = f_x(x)$ exists and it is continuous and $f_x(x) > 0$ on $D(0,\rho)$. Furthermore, $f(0) = 0$.

(H_2) Let us suppose that

$$\mu(f_x(x)A(t,x)) = \lim_{h \to 0^+} \frac{1}{h}[||I + hf_x(x)A(t,x)|| - 1] \tag{2.2}$$

exists and satisfies the relation

$$\mu(f_x(x)A(t,x)) < -\alpha \tag{2.3}$$

for $(t,x) \in R^+ \times D(0,p)$ and some $\alpha > 0$.

3. In this section, we shall present a basic stability result which plays an important role in unifying and systemizing the stability analysis of (2.1).

THEOREM 3.1: Let the hypotheses (H_1) and (H_2) be satisfied. Then, the trivial solution $x \equiv 0$ of (2.1) is exponentially stable.

PROOF: Define $v(x) = ||f(x)||$, where $f(x)$ is as defined in (2.1). From the hypothesis (H_1), it is obvious that $v(x)$ is positive definite and locally Lipschitzian in . From (H_1), for small $h > 0$, we have

$$v(x + hA(t,x)f(x)) = ||f(x + hA(t,x)f(x))||$$
$$= ||f(x) + hf_x(x)A(t,x)f(x) + 0(h)||.$$

This together with the hypothesis (H_2) and the properties of the norm, yields

$$D^+_{(2.1)}v(x) \le \mu(f_x(x)A(t,x))v(x), \tag{3.1}$$

for $(t,x) \in R^+ xD(0,\rho)$. From (3.1) and applying result in [13], we have

$$v(x(t,t_0,x_0)) \le u_0 exp -\alpha(t - t_0) , \quad t \ge t_0, \tag{3.2}$$

whenever $v(x_0) \le u_0$. By applying the standard technique [13], the proof of the theorem follows immediately. We omit the details.

<u>*REMARK 3.1*</u>: The following special type of stability conditions will justify our remark regarding the unification and systemization of the above presented stability conditions.

I. If we take $v(x) = (f^T(x)\cdot f(x))^{1/2}$ and $||A(t,x)||$ = square root of the largest eigenvalue of $A^T(t,x)A(t,x)$, then the $\mu(f_x(x)A(t,x))$ in (H_2) is equal to the largest eigenvalue of $\frac{1}{2}[(f_x(x)A(t,x))^T + f_x(x)A(t,x)]$, where T stands for transpose of a matrix or a vector. Thus the stability condition includes the stability condition due to Wintner [19] for linear time-varying system as a very special case.

II. If we take $v(x) = (f^T(x)Qf(x))^{1/2}$ and $||A(t,x)||$ = square root of the largest eigenvalue of $A^{-1}A^TQA$, then the $\mu(f_x(x)A(t,x))$ in (H_2) is equal to the largest eigenvalue of $\frac{1}{2}Q^{-1}[f_x(x)A(t,x))^TQ + Qf_x(x)A(t,x)]$.

III. If we take $v(x) = \sum_{i=1}^{n} |f_i(x_i)|$ and $||A(t,x)|| =$

$\sup_j [\sum_{i=1}^{n} |a_{ij}(t,x)|]$, then

$$\mu(f_x(x)A(t,x)) = \sup_j [b_{jj} + \sum_{\substack{i=1 \\ i\neq j}}^{n} |b_{ij}(t,x)|] \tag{3.3}$$

where $(b_{i_j}(t,x)) = (f_x(x)A(t,x))$. From (3.3) and (2.2), it is obvious that the matrix $f_x(x)A(t,x)$ is row dominant diagonal matrix [15] with diagonal elements negative. Such a stability condition has been utilized in studying stability properties of economic systems [1,15], compartmental systems [3,4,6], chemical systems [3,5], ecosystems [6,12,17] and recently in stability analysis of hereditary competitive processes [8] and competitive processes of diffusion type [10]. In addition, if $f_x(x) = D$ constant matrix, then the stability condition (2.2) in view of (3.3) implies that the matrix $A(t,x)$ is quasi-dominant diagonal matrix [1,15] with diagonal elements negative. Again such a stability conditions have been used in investigating the stability properties of chemical systems [5], compartmental systems [6], economic systems [18], ecosystems [12,17], competitive processes [8,9,10, 11].

IV. If we take $v(x) = \sup_k\{|f_k(x_k)|\}$ and $||A(t,x)|| = \sup_j [\sum_{i=1}^{n} |a_{i_\gamma}(t,x)]$, then the $\mu(f_x(x)A(t,x) =$

$$\sup_i [b_{ii} + \sum_{\substack{j=1 \\ k\neq i}}^{n} |b_{ij}|] \tag{3.4}$$

where b_{ij} is as defined in III. From (3.4) and (2.2), it is obvious that the matrix $f_x(x)A(t,x)$ is column dominant diagonal matrix [1,15] with diagonal elements negative. Such a stability conditions have been used economic systems [1,15].

REMARK 3.2: For comparative study of the stability conditions of type I, III and IV, see [15].

4. In this section, we briefly interpret the special form of the

systems of differential equations (2.1). Furthermore, we will point out that how our stability conditions provides the structural insight of the biological, physical and social systems in a coherent way.

The system of differential equations (2.1) represents the dynamical mathematical model of any thought provocative system in biological, physical and social sciences of n interacting species.

x = state vector of species.

$A(t,x)$ = n x n process rate matrix that depends on the time, and state of the species.

$f_j(x_j)$ = the law by means of which the state of the i-th species is affected by the state of the j-th species.

For example in chemical system, x = concentration vector of the chemical reactants, $A(t,x)$ = reaction rate matrix, $f_j(x_j)$ = the law of mass action.

As we observed in the preceeding section, Theorem 3.1 gives very general sufficient condition for stability analysis in a systematic and unified way. In addition, it provides the following significant observations of the stability conditions in terms of the parameters or rate coefficients of the system.

(O_1) It has been noted [15] that the stability conditions (I) does not relate to its coefficients in a simplified way. In addition, as the size n of the system increases the computation of eigenvalues of corresponding matrices is very difficult.

(O_2) An observation similar to (O_1) can be made relative to the II in the Section 3.

Even though, the special type of stability conditions associated with I and II are interesting in analytical point of view, but practical point of view are very difficult to test. However, the special type of stability conditions associated with III and

IV are very interesting.

(O_3) The stability conditions associated with III and IV are algebraically simple and easy to compute. Furthermore, they are expressed in terms of parameters or rate coefficients of the system.

(O_4) The stability conditions associated with III and IV imply that the weighted self-inhibitory effects are larger than the activatory effects.

(O_5) These conditions also partially resolve the complexity vs. stability problem in a natural way. For details see [5,6,7,8,9, 10,11,12,14,17,18].

(O_6) These stability conditions are invariant under structural perturbations, for example, a shut down of a pathway between i-th to j-th species, shut down of path-ways between i-th and j-th species, a complete isolation of a p-th species, a complete isolation and connection of species in the system. For further details, see [5,6,8,9,10,11,12,17,18].

(O_7) We note that the stability conditions are only sufficient. However, this drawback of the stability conditions is balanced by the reliability of the stability that follows from the quasi-dominant diagonal property of rate matrix $A(t,x)$. For more details, see [5,6,8,9,10,11,12,17,18].

(O_8) The stability conditions associated with III and IV are insensitive of the signs of cross-coupling coefficients. In other words, so long as the self-inhibitory effects are strong enough to overcome the sum of the magnitude of interactions, the system (2.1) is stable regardless of the nature of interactions (activational, inhibitory, neutral, etc.) among the species. For more details, see [8,9,10,11].

(O_9) These conditions also determine the tolerance of the complexity by the stable isolated systems. Furthermore, they also determine the measurability of the complexity of the system. For details, see [8,9,10,11].

REFERENCES

[1] Arrow, K. J., and Hahn, F. H., *General Competitive Analysis*, San Francisco, Holden-Day, 1971.

[2] Coppel, W. A., *Stability and Asymptotic Behavior of Differential Equations*, Heath, Boston, 1965.

[3] Hearon, J. Z., *Theorems on linear systems*, Ann. N.Y. Acad. Sci. 108(1963), 36-67.

[4] Jacquez, J. A., *Compartmental Analysis in Biology and Medicine*, American Elseview Pub. Co., New York, 1972.

[5] Ladde, G. S., *Cellular systems I: Stability of chemical systems*, Math. Biosci. 29(1976), 309-330.

[6] Ladde, G. S., *Cellular systems II: Stability of compartmental systems*, Math. Biosci. 30(1976), 1-21.

[7] Ladde, G. S., *Stability of model ecosystems with time-delay*, J. Theor. Biol. 61(1976).

[8] Ladde, G. S., *Competitive processes I: Stability of random systems*, (to appear).

[9] Ladde, G. S., *Competitive processes II: Stability of hereditary systems*, (to appear).

[10] Ladde, G. S., *Competitive processes III: Stability of diffusion systems*, (to appear).

[11] Ladde, G. S., *Competitive processes IV: Stability of stochastic hereditary systems*, (in preparation).

[12] Ladde, G. S., and Siljak, D. D., *Stability of multispecies communities in randomly varying environment*, J. Math. Biol. 2(1975), 165-78.

[13] Lakshmikantham, V., and Leela, S., *Differential and Integral Inequalities, Theory and Applications, Vol. I & II*, Academic Press, New York, 1969.

[14] May, R. M., *Stability and Complexity in Model Ecosystems*, Princeton University Press, Princeton, N.J., 1973.

[15] Newman, P. K., *Some notes on stability conditions*, Rev. Econ. Stu. 72(1959), 1-9.

[16] Rosen, R., *Dynamical System Theory in Biology, Vol. I*, Wiley-Interscience, New York, 1970.

[17] Siljak, D. D., *When is a complex ecosystem stable?*, Math. Biosci. 25(1975), 25-50.

[18] Siljak, D. D., *Connective stability of competitive equili-*

brium, Automatica 11(1975), 389-400.

[19] Wintner, *Asymptotic integration constant*, American J. Math. 68(1946), 553-59.

THE CURRENT STATUS OF ABSTRACT CAUCHY PROBLEM

by

V. Lakshmikantham
The University of Texas at Arlington, Texas

The study of abstract Cauchy problems for differential equations has taken two different directions. One is to find compactness type conditions that assure the existence of solutions and the corresponding results are extensions of the classical Peano's theorem. The other approach is to employ dissipative type conditions that guarantee existence as well as uniqueness of solutions and the corresponding results may be regarded as extensions of the classical Picard's theorem.

Let E be a real Banach space with norm $||\cdot||$. Consider the differential equation

(1) $$x' = f(t,x),\ x(t_0) = x_0,\ t_0 \geq 0,$$

where x, $f \in E$.

Generally speaking, the methods of proving existence of solutions of (1) consist of three steps, namely:

(a) constructing a sequence of approximate solutions of some kind for (1);

(b) showing the convergence of the constructed sequence;

(c) proving that the limit function is a solution of (1). If f is continuous, the steps (a) and (c) are standard and straight forward. It is the step (b) that deserves attention. This, in turn, leads to two possibilities:

(i) to show that the sequence of approximate solutions is a Cauchy sequence;

(ii) to show that the sequence of approximate solutions is relatively compact so that one can appeal to Ascoli's theorem.

These two possibilities respectively lead to dissipative and compactness type conditions. If, on the other hand, f is not continuous, the steps (a) and (c) also require closer attention and the notion of solution has to be appropriately modified. Also, the construction of approximate solutions has to be changed. Let us first deal with the case where f is continuous.

A general result in this direction is the following [15-18, 25,27].

THEOREM 1. Assume that

(i) $f \in C[R_0,E]$ and a, $b > 0$, $M \geq 1$ are chosen such that $||f(t,x)|| \leq M$ on R_0 where $R_0 = [t_0,t_0 + a] \times B[x_0,b]$ and $B[x_0,b] = \{x \in E : ||x - x_0|| \leq b\}$;

(ii) $g \in C[[t_0,t_0 + a] \times R_+,R]$, $g(t,0) \equiv 0$ and $u \equiv 0$ is the only solution of

(2) $$u' = g(t,u),\ u(t_0) = 0;$$

(iii) $V \in C[[t_0,t_0 + a] \times B[x_0,b] \times B[x_0,b],R_+]$, $V(t,x,x) \equiv 0$, $V(t,x,y) > 0$ if $x \neq y$, $|V(t,x_1,y_1) - V(t,x_2,y_2)| \leq L[||x_1 - x_2|| + ||y_1 - y_2||]$, if $\{x_n\}$, $\{y_n\}$ are sequences in $B[x_0,b]$ such that $\lim_{n\to\infty} V(t,x_n,y_n) = 0$ then $\lim_{n\to\infty} ||x_n - y_n|| = 0$ and for $t \in [t_0,t_0 + a]$, x, $y \in B[x_0,b]$,

$$D_-V(t,x,y) = \liminf_{h\to 0^-} \frac{1}{h} [V(t + h,\ x + hf(t,x),\ y + hf(t,y)) - V(t,x,y)]$$

$$\leq g(t,V(t,x,y)).$$

Then, there exists a unique solution for the problem (1) on $[t_0,t_0 + a]$.

Consider the special case $V(t,x,y) = ||x - y||$ [24]. The condition (iii) of Theorem 1 reduces to

(3) $$\liminf_{h\to 0^-} \frac{1}{h}\ ||x - y + h(f(t,x) - f(t,y))|| - ||x - y||$$

$$\leq g(t,||x - y||)$$

which is clearly satisfied when one assumes Perron's type uniqueness condition

$$(4) \qquad ||f(t,x) - f(t,y)|| \le g(t, ||x - y||).$$

In fact, the convexity of $||x||$ implies that the limit in (3) exists as does the case with $V(t,x,y) = ||x - y||^2$ or $||x - y||^p$, $p \ge 1$. The case $V(t,x,y) = ||x - y||^2$ leads to the dissipative conditions

$$(x - y, f(t,x) - f(t,y))_- \le \frac{1}{2} g(t, ||x - y||^2)$$

or

$$(x - y, f(t,x) - f(t,y))_- \le g(t, ||x - y||)||x - y||,$$

where $(x,y)_\pm$ are the generalized inner products defined by

$$(x,y)_\pm = \begin{matrix} sup \\ inf \end{matrix} \{x^*(y) : x^* \in J(x)\},$$

J being the duality map such that

$$J(x) = \{x^* \in E^* : ||x^*|| = ||x|| \text{ and } x^*(x) = ||x||^2\}.$$

See for details [1,9,18,24]. Thus, the generality of Theorem 1 is clear.

Let $F \subset E$ be a locally closed set, that is, for each $x \in F$ there exists a $b > 0$ such that $F_0 = F \cap B[x_0,b]$ is closed in E. Suppose that we consider the problem (1) with $x_0 \in F$. Assume, in place of (i) of Theorem 1, that

(i*) $f \in C[[t_0,t_0 + a] \times F,E]$ and the numbers a, $b > 0$, $M \ge 1$ are chosen such that $||f(t,x)|| \le M$ on $[t_0,t_0 + a] \times F_0$.

Suppose also that

(iv) $$\liminf_{h \to 0^+} \frac{1}{h} [d(x + hf(t,x),F)] = 0,$$

$$(t,x) \in [t_0,t_0 + a] \times F.$$

The condition (iv) is closely related to the notion of flow-invariant sets. See [11,14,28,30,31]. Obviously, (iv) holds at all interior points of F. It is therefore a boundary condition. The difficulty that arises here is due to the fact that the closed F may have empty interior. Nonetheless, the conclusion of Theorem 1 remains true if (i*), (ii) and (iv) hold and the condition (iii) is replaced by the stronger assumption (4). This we shall call Theorem 1*.

If F_0 is assumed convex, Theorem 1* is true under the conditions (i*), (ii), (iii) and (iv). However, in the general case, that is, when F_0 is not assumed to be convex, the conditions of Theorem 1* are not strong enough. But, if the condition (iii) is restricted to

$$(x - y,\ f(t,x) - f(t,y))_+ \leq L||x - y||^2,$$

Theorem 1* is true [23]. The main difficulty is that the proof needs transfinite induction which crucially depends on the fact that $(x,y)_+$ is upper semicontinuous and the linearity of the comparison function $g(t,u) = Lu$. The failure of this upper semicontinuity property by $D_-V(t,x,y)$ is the cause for Theorem 1* not being valid with the assumptions (1*), (ii), (iii) and (iv). However, one could use a Lyapunov-like function V and the general comparison function $g(t,u)$ to prove Theorem 1* with slightly stronger assumptions in place of (iii).

For details, see [19].

For any bounded set $A \subset E$, let $\alpha(A)$ denote the Kuratowski measure of noncompactness. Let $\Omega = \{A : A$ is a subset of $B[x_0,b]\}$. An existence theorem that employs a general compactness type condition is the following [10,15,16,18].

<u>THEOREM 2</u>. Assume the hypotheses (i) and (ii) of Theorem 1 with the additional condition that f is uniformly continuous on R_0.

Suppose further that

(iii*) $V : [t_0, t_0 + a] \times \Omega \to R_+$, $V(t,A)$ is continuous in t and α-continuous in A, $V(t,A) = 0$ *iff* A is relatively compact, $|V(t,A) - V(t,B)| \leq L|\alpha(A) - \alpha(B)|$ for $t \in [t_0, t_0 + a]$, A, $B \in \Omega$ and for $t \in [t_0, t_0 + a]$, $A \in \Omega$,

$$D\_V(t,A) = \lim_{h\to 0^-} \inf \frac{1}{h} [V(t + h, A_h(f)) - V(t,A)]$$

$$\leq g(t, V(t,A)),$$

where $A_h(f) = \{y : y = x + hf(t,x),\ x \in A\}$.
Then, there exists a solution for the problem (1) on $[t_0, t_0 + a]$.

Taking $V(t,A) = \alpha(A)$, $A \in \Omega$, we see that (iii)*) is satisfied with $\alpha(A_h(f)) \geq \alpha(A) + hg(t,\alpha(A))$, for $h < 0$ sufficiently small and the corresponding version of Theorem 2 is a result in [22].
A stronger assumption is to require that

$$\alpha(f\ (t,A)) \leq g(t,\alpha(A)).$$

If, on the other hand, we assume that

$$\lim_{h\to 0^+} \alpha(f(I_h, A)) \leq g(t,\alpha(A)),$$

where $I_h = [t, t + h]$, then the uniform continuity assumption on f can be dispensed with [9,29]. In the general situation it is not known whether we can dispense with the uniform continuity assumption. Theorem 2 does not pose any severe problem if one considers solutions in closed sets. Theorem 2* can be proved with (1*), (ii), (iii*) and the boundary condition (iv). See for details [9,18,24].

Consider the equation (1) in weak topology. If E is a reflexive Banach space, Peano's existence theorem is valid [32]. If E is not assumed to be reflexive and the notion of measure of weak-noncompactness is available, one could prove existence results under weak-compactness and weak-dissipative type of conditions on f. See for details [4].

V. *Maximal Solution*

Let K be a solid cone in E and K^* be the set of all continuous linear functionals c on E such that $c(x) \geq 0$ for all $x \in K$. The cone K induces the order relations on E. A function $f : E \to E$ is said to be quasimonotone non-decreasing if $x \leq y$ and $c(x) = c(y)$ for some $c \in K^*$, then $c(f(x)) \leq c(f(y))$. Existence of the maximal solution for the problem (1) relative to the cone K can be shown under the conditions of Theorem 1, if in addition, f is quasimonotone nondecreasing with respect to K. Having such a result at our disposal, one could easily prove a general comparison theorem in this set up. See [9,18,24] for further details.

VI. *Delay Differential Equations*

In proving the corresponding existence and uniqueness results for differential equations of retarded type in Banach space E, the main difficulty is in imposing assumptions, since, in this case, the domain and the range of the function involved are not in the same Banach space. To overcome this difficulty, one needs to impose conditions over a subset of the domain in a suitable way and employ weaker forms of the theory of differential inequalities. Utilizing this idea, only recently the existence and uniqueness results have been proved under dissipative and compactness type conditions [20,21].

VII. *Set-valued Differential Equations*

Let us consider the set-valued differential equation

(5) $$x' \in Ax, \quad x(0) = x_0,$$

in the Banach space E.

If E is reflexive, A maps E into $K_0(E)$, where $K_0(E)$ consists of all nonvoid compact, convex subsets of E, A is upper semicontinuous and γ-Lipschitzian, γ being the Hausdorff measure of noncompactness, then it is shown in [8] that there exists a solution of the problem (5). Here, proving that the limit function is a solution requires E to be reflexive and the proof is not straight-forward as in the case when f is continuous.

Let us identify the operator A with its graph. So that we write $[x,y] \in A$ if $y \in Ax$. The operator A is said to be ω-quasi dissipative if, for any $[x_i,y_i] \in A$, $i = 1,2$,

$$(x_1 - x_2, y_1)_- + (x_2 - x_1, y_2)_- \leq \omega||x_1 - x_2||^2.$$

Obviously, quasi-dissipativity implies dissipativity and the converse is not true in general.

Suppose also that

$$\lim_{h\to 0^+} \inf \frac{1}{h} (d[R(I - hA),x]) = 0, \quad \text{for} \quad x \in \overline{D(A)}, \tag{6}$$

where $D(A)$ denotes the domain of A and $R(A)$ is the range of A. Let $x_0 \in E$. An E-valued continuous function $x(t)$ on $[0,T]$ is said to be an integral solution of type ω of (5) on $[0,T]$ if it satisfies

(i) $x(0) = x_0$;

(ii) for every s, $t \in [0,T]$ such that $s \leq t$ and $[x,y] \in A$,

$$e^{-2\omega t}||x(t) - x||^2 \leq e^{-2\omega s}||x(s) - x||^2 + \int_s^t e^{-2\omega\tau}(x(\tau) - x, y)_+ \, d\tau .$$

The definition of integral solution is introduced in [2]. Clearly, a solution is an integral solution. The notion of integral solution is appropriate only if A is dissipative.

Let $x(t)$ be continuous on $[0,T]$ and let $x_0 \in \overline{D(A)}$. Then,

$x(t)$ is said to be a DS-limit solution of (5) on $[0,T]$ if there exists a backward difference scheme approximate solution (DS-approximate solution) $x_n(t)$ of (5) on $[0,T]$ such that $x_n(t)$ converges uniformly to $x(t)$ on $[0,T]$.

Let $\{x_n(t)\}$ be a sequence of E-valued simple functions on $[0,T]$ defined by

$$x_n(t) = \begin{cases} x_0^n, & t = 0 \\ x_i^n, & t \in (t_{i-1}^n, t_i^n] \cap [0,T], \quad i = 1,2,\ldots,N_n, \end{cases}$$

and $n \geq 1$, where $\{t_i^n\}$ represents a partition $\Delta_n = \{0 = t_0^n < t_1^n < \ldots < t_{N_n-1}^n < T \leq t_{N_n}^n\}$ of $[0,T]$ such that $|\Delta_n| = \max_{1\leq i\leq N_n} (t_i^n - t_{i-1}^n) \to 0$ as $n \to \infty$. The sequence $\{x_n(t)\}$ is said to be a DS-approximate solution of (5) on $[0,T]$ if

$$\frac{x_i^n - x_{i-1}^n}{t_i^n - t_{i-1}^n} - \varepsilon_i^n \in Ax_i^n, \quad n = 1,2,\ldots,N_n, \quad n \geq 1,$$

$$x_0^n \to x_0 \quad \text{as} \quad n \to \infty$$

and $\varepsilon_n = \sum_{i=1}^{N_n} ||\varepsilon_i^n||(t_i^n - t_{i-1}^n) \to 0$ as $n \to \infty$.

The following important theorem is proved in [12]. See also [6] for an entirely different method of proof of the same result.

THEOREM 3. Let A be a ω-quasi-dissipative operator in E satisfying the condition (6). Then,

(a) there exists a DS-approximate solution $x_n(t)$ on $[0,T]$;

(b) $x_n(t)$ converges uniformly to a DS-limit solution $x(t)$ on $[0,T]$ such that if $x_0 \in \overline{D(A)}$, $x(t) \in \overline{D(A)}$ for $t \in [0,T]$ and $x(0) = x_0$;

(c) a DS-limit solution is a unique integral solution of (5);

(d) there exists a unique semigroup $\{T(t) : t \geq 0\}$ of type ω on $\overline{D(A)}$ such that for $x \in \overline{D(A)}$, $x(t) = T(t)x$ is the unique DS-limit solution of (5) on $[0,\infty]$ with $x(0) = x$.

The proof of the result is elementary. It is here that all the three steps (a), (b) and (c) of proof in the existence theory require changes (from the continuous case).

Let A be a ω-quasi-dissipative operator satisfying the condition

$$R(I - \lambda A) \supset \overline{D(A)} \quad \text{for} \quad 0 < \lambda \leq \lambda_0 . \tag{7}$$

Then, the condition (6) is clearly satisfied and hence the well known result in [7] is a special case of Theorem 3. The influence of linear theory led to the emphasis on the exponential formula $T(t) = \lim_{n\to\infty} (I - \frac{t}{n} A)^{-n}$ given in [7] rather than an equivalent and more fundamental idea of backward-difference scheme as given in [12].

If A is ω-quasi-dissipative, $D(A)$ is closed and A is continuous on $D(A)$, then the condition (6) is equivalent to the boundary condition $\liminf_{h\to 0^+} \frac{1}{h} [d(x + hf(t,x), D(A))] = 0$ for $x \in D(A)$. Hence Theorem 1* is also included which is a well known result in [23].

There are now some books and lecture notes available which in-include much information on this subject. See [1,3,9,18,24]. The survey paper [5] gives an introduction to evolution equations governed by dissipative operators and has several references. We have therefore given only those references which are pertinent to our development and are new. For more recent work, see [13,26,33].

REFERENCES

[1] Barbu, V., *Nonlinear semigroups and differential equations in Banach spaces*, Noordhoff Publishing, The Netherlands, 1976.

[2] Benilan, Ph., *Equations d'evolution daus un espace de Banach quelconque et applications*, Thesis, Univ. Paris XI, Orsay, 1972.

[3] Brezis, H., *Operateurs maximaux monotones et semigroupes de contractions daus les espace de Hilbert*, North-Holland Publishing, Amsterdam, 1973.

[4] Bronson, E., Lakshmikantham, V., and Mitchell, A. R., *Existence of weak solutions of differential equations in arbitrary Banach space*, to appear.

[5] Crandall, M.G., *An introduction to evolution governed by accretive operators*, Proc. Intern. Symp. on Dynamical Systems, Brown Univ., Providence, R.I., 1974, 94-119.

[6] Crandall, M. G., and Evans, L. C., *On the relation of the operator* $\frac{\partial}{\partial s} + \frac{\partial}{\partial \tau}$ *to evolution governed by accretive operators*, Israel Jour. Math., Vol. 21 (1975), 261-278.

[7] Crandall, M. G., and Liggett, T. M., *Generation of semigroups of nonlinear transformations in general Banach spaces*, Amer. Jour. Math. 93(1970), 265-298.

[8] DeBlasi, F. S., *Existence and stability of solutions for autonomous multivalued differential equations in Banach space*, to appear in Pacific Jour. Math.

[9] Deimling, K., *Ordinary differential equations in Banach spaces*, to appear in Springer Verlag Lecture Notes Series.

[10] Eisenfeld, J., and Lakshmikantham, V., *On the existence of solutions of differential equations in a Banach space*, to appear in Ann. Polon. Math.

[11] Hartman, P., *On invariant sets and on a theorem of Ważewski*, Proc. Amer. Math. Soc. 32(1972), 511-520.

[12] Kobayashi, Y., *Difference approximation of Cauchy problems for quasi-dissipative operators and generation of nonlinear semigroups*, Jour. Math. Soc. Japan 27(1975), 640-665.

[13] Kobayashi, Y., and Kobayasi, K., *On the perturbation of nonlinear dissipative operators in Banach spaces*, to appear.

[14] Ladde, G. S., and Lakshmikantham, V., *On flow invariant sets*, Pacific Jour. Math. 51(1974), 215-220.

[15] Lakshmikantham, V., *Stability and asymptotic behavior of*

solutions of differential equations in a Banach space, CIME, Edizioni Cremonese, Roma 1974, 39-98.

[16] Lakshmikantham, V., *Existence and comparison results for differential equations*, Proc. Intern. Conf. on Diff. Eqs., Academic Press, 1975, 459-473.

[17] Lakshmikantham, V., and Leela, S., *Differential and integral inequalities, Vol. II*, Academic Press, 1969.

[18] Lakshmikantham, V., and Leela, S., *Theory of nonlinear differential equations in abstract spaces*, Lecture Notes.

[19] Lakshmikantham, V., Mitchell, R. W., and Mitchell, A. R., *Differential equations on closed subsets of a Banach space*, Trans. Amer. Math. Soc. 220(1976), 103-113.

[20] Lakshmikantham, V., and Mitchell, A. R., *On the existence of solutions of differential equations of retarded type in a Banach space*, To appear in Ann. Polon. Math.

[21] Leela, S., and Moauro, V., *Existence of solutions of differential equations with delay in Banach spaces*, to appear.

[22] Li, T. Y., *Existence of solutions for ordinary differential equations in Banach spaces*, Jour. Diff. Eqs.

[23] Martin, R. H., *Differential equations on closed subsets of a Banach space*, Trans Amer. Math. Soc. 179(1973), 399-414.

[24] Martin, R. H., *Nonlinear operators and differential equations in Banach spaces*, John Wiley & Sons, 1976.

[25] Martin, R. H., *Lyapunov functions and autonomous differential equations in a Banach space*, Math. Sys. Theory 7(1973), 66-72.

[26] Miyadera, I., and Kobayashi, Y., *Convergence and approximation of nonlinear semigroups*, To appear.

[27] Murakami, H., *On nonlinear ordinary and evolution equations*, Funk. Ekvac., 9(1966), 151-162.

[28] Nagumo, M., *Uber die laga der integralkurven geuohnlicher differential gleichungen*, Proc. Phy. - Math. Soc. Japan, 24(1942), 551-559.

[29] Pianigiani, G., *Existence of solutions for ordinary differential equations in Banach spaces*, Bull. Acad. Polon. Sci., Vol. 23(1975), 853-857.

[30] Redheffer, R. M., *The theorems of Bony and Brezis on flow-invariant sets*, Amer. Math. Monthly 79(1972), 740-747.

[31] Redheffer, R. M., and Walter, W., *Flow-invariant sets and differential inequalities in normed spaces*, Applicable Analysis 5(1975), 149-161.

[32] Szepy, A., *Existence theorem for weak solutions of ordinary differential equations in reflexive Banach spaces*, Studia. Sci. Math. Hungarica 6(1971), 197-203.

[33] Takahashi, T., *Semigroups of nonlinear operators and invariant sets*, to appear.

ON THE METHOD OF CONE-VALUED LYAPUNOV FUNCTIONS

V. Lakshmikantham, University of Texas at Arlington

and

S. Leela, SUNY at Geneseo

It is well known that the technique of employing vector Lyapunov functions and the theory of vector differential inequalities offers a great deal of flexibility in studying nonlinear systems [6-10]. Particularly, this has been a very effective method to study large scale interconnected dynamical systems which occur in many diverse areas like chemical kinetics, pharmacokinetics, multimarket economics, ecosystems and various other competitive systems which involve interaction between several agents [1,3-5,11, 12]. Also, vector Lyapunov functions arise naturally in the study of such systems by the aggregation-decomposition method.

However, for the success of this method, it is required that the comparison system be quasimonotone nondecerasing. In the case of linear comparison system, this requirement implies that the nondiagonal elements of the comparison matrix be all nonnegative. But quasimonotonicity of the matrix is not a necessary condition for the matrix to be stable. Thus, the limitation of the application potential of this general and effective method is due to the fact that comparison systems may have a desired property like positivity or stability of solutions without being quasimonotone.

It was observed in [7] that this difficulty is due to the choice of the cone relative to the comparison system, namely R_+^n, the cone of nonnegative elements of R^n and that a possible approach to overcome this limitation is to choose an appropriate cone other than R_+^n. In this report, this idea is investigated and by developing the theory of differential inequalities through cones, the method of cone-valued Lyapunov functions is shown to be a useful tool in applications. We only state the main results in this direction. The details will appear elsewhere.

Let R^n denote the n-Euclidean space with scalar product $(\ ,\ )$ and norm $||\cdot||$. A closed subset K of R^n with nonvoid interior is called a positive cone in R^n if (i) $u, v \in K$, α, $\beta \in R_+$ implies $\alpha u + \beta v \in K$ and (ii) $u, -u \in K$ implies $u = 0$. The cone K induces the order relations in R^n given by

$$u \underset{K}{\leq} v \quad \text{iff} \quad v - u \in K,$$

$$u \underset{K^o}{<} v \quad \text{iff} \quad v - u \in K^o,$$

where K^o = interior of K. Let

$$K^* = \{\phi \in R^n : (\phi, u) \geq 0 \ \text{ for all } \ u \in K\}$$

be the adjoint cone. We shall denote by ∂K and K_0 the boundary of K and the set $K - \{0\}$ respectively. Following Elsner [2] we define a function $f : D \to R^n$, $D \subset R^n$, to be quasimonotone relative to the cone K if $u, v \in D$ and $v - u \in \partial K$ imply that there exists $\phi \in K_0^*$ such that $(\phi, v - u) = 0$ and $(\phi, f(v) - f(u)) \geq 0$. In the special case when $f(u) = Au$, A being a $n \times n$ matrix and $K = R_+^n$, this definition implies that the nondiagonal elements of A are nonnegative.

THEOREM 1. Assume that $g : R_+ \times R^n \to R^n$ is continuous and $g(t,u)$ is quasimonotone in u relative to K for each $t \in R_+$. Let $r(t,t_0,u_0)$ be the maximal solution of (1) $u' = g(t,u)$, $u(t_0) = u_0$ relative to K existing on $[t_0,\infty)$. Suppose that $m : R_+ \to R^n$ is continuous and satisfies for $t \geq t_0$,

$$Dm(t) \underset{K}{\leq} g(t,m(t)).$$

Then $m(t_0) \underset{K}{\leq} u_0$ implies $m(t) \underset{K}{\leq} r(t,t_0,u_0)$, $t \geq t_0$.

A few useful remarks are in order.

REMARKS. (i) The flow-invariance of the cone K follows from Theorem 1 with the additional assumptions of $g(t,0) \equiv 0$ and the uniqueness of solutions of (1).

(ii) If $K = R_+^n$, Theorem 1 gives the componentwise estimates

$m(t) \leq r(t,t_0,u_0)$.

(iii) Whenever P, Q, $P \subset Q$ are two cones in R^n, the order relations in P imply the same order relations in Q even though quasimonotonicity of the function g relative to P need not imply quasimonotonicity of g relative to Q.

In view of Remark (iii), we state below a typical result concerning the stability of solutions of the differential system

$$(2) \qquad x' = f(t,x), \; x(t_0) = x_0,$$

using a cone-valued Lyapunov function and two cones P, Q, with respect to which certain relevant conditions are satisfied.

<u>THEOREM 2.</u> Let P, Q be two cones in R^n such that $P \subset Q$. Assume that (i) $f : R_+ \times S(\rho) \to R^n$, $S(\rho) = \{x \in R^n : ||x|| < \rho\}$, is continuous and $f(t,0) \equiv 0$;

(ii) $V : R_+ \times S(\rho) \to Q$, $V(t,x)$ is continuous and satisfies a local Lipschitz condition in x relative to P and for $(t,x) \in R_+ \times S(\rho)$,

$$D^+V(t,x) \equiv \limsup_{h \to 0^+} \frac{1}{h}[V(t + h, x + hf(t,x)) - V(t,x)]$$

$$\underset{P}{\leq} g(t,V(t,x));$$

(iii) $g : R_+ \times Q \to R^n$ is continuous, $g(t,u)$ is quasimonotone in u with respect to P for each $t \in R_+$ and $g(t,0) \equiv 0$;

(iv) for some $\phi_0 \in Q_0^*$ and $(t,x) \in R_+ \times S(\rho)$,

$$b(||x||) \leq (\phi_0, V(t,x)) \leq a(t,||x||),$$

where $b(u)$ and $a(t,u)$ for each $t \in R_+$ are standard K-class functions;

(v) the trivial solution $u \equiv 0$ of (1) is ϕ_0-equistable, i.e., given $\varepsilon > 0$ and $t_0 \in R_+$, there exists a $\delta = \delta(t\ ,\varepsilon) > 0$ such that $(\phi_0,u_0) < \delta$ implies $(\phi\ ,r(t,t_0,u_0)) < \varepsilon, t \geq t_0$. Then the trivial solution $x \equiv 0$ of (2) is equistable.

REFERENCES

[1] Bailey, F. N., *The application of Lyapunov's second method to interconnected systems*, J. SIAM Control 27(1971), 543-48.

[2] Elsner, L., *Quasimonotonie und ungleichungen in halbgeordneten Raiimen*, Linear Algebra and its Applications 8(1974), 249-61.

[3] Grujic, L. T., and Šiljak, D. D., *Asymptotic stability and instability of large scale systems*, IEEE Trans. Ac 18(1973), 636-45.

[4] Ladde, G. S., *Competitive systems and comparison differential systems*, to appear in Trans. Amer. Math. Soc.

[5] ______, *Cellular systems I; stability of chemical systems*, Math. Biosci., 29(1976), 309-30.

[6] Lakshmikantham, V., *Several Lyapunov functions*, Proc. Intern. Conf. on Nonlinear Oscillations, Kiev, U.S.S.R. (1970).

[7] ______, *Vector Lyapunov functions*, Proc. 12th Annual Allerton Conference on circuit and system theory, (1974), 71-77.

[8] ______ and Leela, S., *Differential and integral inequalities, Vol I*, Academic Press, New York, 1969.

[9] Matrosov, V. M., *Vector Lyapunov function in the analysis of nonlinear interconnected systems*, Symp. Math. Roma, 6 Mechanica Nonlineare Stabilita, 1970 (1971), 209-42.

[10] ______, *Comparison method in system's dynamics*, Diff. Uravn. 10(1974), 1547-59; ibid. 11(1975), 403-17, (Russian).

[11] Šiljak, D. D., and Ladde, G. S., *Stability of multispecies communities in randomly varying environment*, Jour. Math. Bio. 2(1975), 165-78.

[12] Šiljak, D. D., *Competitive economic systems; stability, decomposition and aggregation*, IEEE Trans. Ac. 21(1976), 149-60.

PERIODIC SOLUTIONS OF SOME INTEGRAL EQUATIONS FROM THE THEORY OF EPIDEMICS

Roger D. Nussbaum*
Rutgers University

1. *Introduction*

In a recent paper [2] Cooke and Kaplan have studied the integral equation

$$x(t)=\int_{t-\tau}^{t} f(s,x(s))ds \qquad (1)_\tau$$

which they suggested as a simplified model for the spread of diseases with seasonal dependence (the dependence of f on s). Cooke and Kaplan considered τ as a parameter, assumed f is periodic of period ω in the s variable and searched for positive, periodic solutions of $(1)_\tau$ of period ω. They proved the existence of a number $\beta(f)$ (which they called a "periodicity threshold") such that for $\tau > \beta(f)$ the equation (1) has positive periodic solutions. However, numerical studies for the case $f(s,x) = (1+\frac{1}{2}\sin 2\Pi s)x(1-x)$ for $0 \leq x \leq 1$, $f(s,x) = 0$ otherwise, suggested that the Cooke-Kaplan number $\beta(f)$ is not best possible: $\beta(f) = 2$ in this case and the correct periodicity threshold appeared to be 1. Furthermore, this example suggested that, at least for a wide variety of examples, it should be possible to obtain uniqueness results for solutions of $(1)_\tau$.

In [6] and [7] this author has studied the equation $(1)_\tau$ and the more general equations

$$x(t) = \int_{t-\beta(\tau)}^{t-\alpha(\tau)} f(s,x(s))ds \qquad (2)$$

$$x(t) = \int_{t-\beta(\tau)}^{t-\alpha(\tau)} P(t-s,\tau)f(s,x(s))ds \qquad (3)$$

* Partially supported by a National Science Foundation Grant.

as bifurcation problems. In particular we have given a best possible value τ_0 for the periodicity threshold ($\tau_0 = 1$ for the example mentioned). Furthermore, we have studied the question of uniqueness of solutions and some associated questions in linear operator theory which are raised by the problem of existence and uniqueness of solutions.

In sections 2-4 below we shall summarize the above-mentioned results from [6] and [7]. A key role in the abstract theory underlying these results is played by the fixed point index. In the fifth section we prove some new results about the fixed point index of cone maps. For example, we prove a nonlinear version of the Krein-Rutman theorem and show that it implies the linear Krein-Rutman theorem. The Krein-Rutman theorem has been used in the literature [8] to obtain results above the fixed point index of positive linear operators. Our results come full circle: elementary arguments yield information about the fixed point index which in turn yields the Krein-Rutman theorem.

2. *A Global Bifurcation Theorem*

Let X denote a real Banach space. By a "wedge" $K \subset K$ we shall mean a closed, convex subset of X such that $x \in K$ implies $tx \in K$ for all $t \geq 0$. By a "cone" K (with vertex at 0) we shall mean a wedge such that $x \in K - \{0\}$ implies $-x \notin K$. Let $J = (a,\infty)$ denote an interval of real numbers with $-\infty \leq a < \infty$ and $Y = X \times \mathbb{R}$ with $||(x,\lambda)|| = ||x|| + |\lambda|$ for $(x,\lambda) \in Y$.

Suppose now that $K \subset K$ is a cone, and $F : K \times J \to K$ is a continuous map. We wish to investigate the nontrivial solutions of $F(x,\lambda) = x$. We make the following assumptions about F.

<u>H1</u>. $F : K \times J \to K$ is a continuous map such that $F(0,\alpha) = 0$ for all $\alpha \in J$. It is a local strict-set-contraction with respect to the norms on Y and X (see [9], Section A, for definitions). If B is any bounded subset of $K \times J$, then $\{(x,\alpha) \in B : F(x,\alpha) = x\}$ has compact closure in Y.

In all our applications the map will be such that $F(0,\alpha) = 0$ for $\alpha \in J$ and F takes bounded sets to precompact sets, and H1 is trivially valid in this case.

H2. If $a > -\infty$ and $F(x_k,\alpha_k) = x_k$ for a sequence (x_k,α_k) such that $x_k \neq 0$ and $\alpha_k \to a$, then $\lim_{k\to\infty} ||x_k|| = \infty$.

Our next hypothesis is the crucial one. Note that in contrast to the work in [6] it is not necessary to assume that the set Λ below has no accumulation points. This is important for applications in Section 3.

H3. For each $\lambda \in J = (a,\infty)$ there exists a compact linear operator L_λ such that $L_\lambda(K) \subset K$. The map $\lambda \to L_\lambda$ is continuous in the norm topology and if $R(x,\lambda) = F(x,\lambda) - L_\lambda(x)$, then $\lim_{||x||\to 0} (||x||^{-1})||R(x,\lambda)|| = 0$ uniformly on bounded λ intervals. If $\Lambda = \{\lambda \in J : L_\lambda(x) = x$ for some $x \in K - \{0\}\}$, there exist $\alpha > a$ and $\beta < \infty$ such that $\Lambda \subset [\alpha,\beta]$. Furthermore, if $r(L) = \lim_{n\to\infty} ||L^n||^{\frac{1}{n}}$ denotes the spectral radius of a bounded linear operator L, then one has $r(L_\lambda) < 1$ for all $\lambda < \alpha$ and $r(L_\lambda) > 1$ for all $\lambda > \beta$.

Our next theorem is proved in [7]. Recall that a cone in a Banach space X is "total" if $X =$ closure $\{x - y : x, y \in K\}$.

THEOREM 1. Let K be a total cone in a Banach space X, $J = (a,\infty)$ an interval of reals, and $F : K \times J \to K$ a continuous map which satisfies H1, H2 and H3. If Λ is as in H3, write $Q = \{(0,\lambda) : \lambda \in \Lambda\}$ and $S = \{(x,\lambda) \in K \times J : x \neq 0, F(x,\lambda) = x\} \cup Q$. Then S is closed and there exists $(0,\lambda_0) \in Q$ such that if S_0 is the maximal connected component of S which contains $(0,\lambda_0)$, S_0 is unbounded.

3. *Existence and Uniqueness of Solutions to Some Nonlinear Integral Equations*

In this section we wish to describe how the results of the previous section can be applied to the integral equations mentioned in the introduction. Thus we discuss the equation

$$x(t) = \int_{t-\beta(\tau)}^{t-\alpha(\tau)} P(t - s,\tau)f(s,x(s))ds \qquad (4)_\tau$$

We are interested in periodic solutions of period ω so we shall always denote by X the Banach space of real-valued periodic functions of period ω (in the sup norm). About the function f we assume

H4. $f : \mathbb{R} \times [0,\infty) \to [0,\infty)$ is a continuous function such that $f(s + \omega,x) = f(s,x)$ for all (s,x) and $f(s,0) = 0$ for all s. There is a strictly positive, continuous function $a(s)$ such that $\lim_{x\to 0^+} (x^{-1})(f(s,x)) = a(s)$ uniformly in s. We suppose that $\lim_{x\to\infty} (x^{-1})(f(s,x)) = 0$ uniformly in s.

About the functions $\alpha(\tau)$ and $\beta(\tau)$ we suppose

H5. $\alpha, \beta : [0,\infty) \to [0,\infty)$ are continuous, nonnegative functions such that $\beta(\tau) - \alpha(\tau) > 0$ for all τ, $\lim_{\tau\to 0} (\beta(\tau) - \alpha(\tau)) = 0$ and $\lim_{\tau\to\infty} (\beta(\tau) - \alpha(\tau)) = \infty$.

We suppose about P that

H6. $P : \mathbb{R} \times [0,\infty) \to [0,\infty)$ is a nonnegative continuous function such that $P(u,\tau)$ is strictly positive for $u \in (\alpha(\tau),\beta(\tau))$. If $a(s)$ is the function in H4, we also assume

$$\lim_{\tau\to\infty} (\inf_t \int_{t-\beta(\tau)}^{t-\alpha(\tau)} P(t - s,\tau)a(s)ds) > 1 \qquad (5)$$

Let K denote the cone of nonnegative functions. Then for each $\tau > 0$ we have a continuous, compact map $F_\tau : K \to K$ defined by

$$(F_\tau x)(t) = \int_{t-\beta(\tau)}^{t-\alpha(\tau)} P(t - s,\tau)f(s,x(s))ds \qquad (6)$$

We also have a compact, linear map L_τ defined by

$$(L_\tau x)(t) = \int_{t-\beta(\tau)}^{t-\alpha(\tau)} P(t - s,\tau)a(s)x(s)ds \qquad (7)$$

If $\Lambda = \{\tau \in [0,\infty) : L_\tau x = x$ for some $x \in K - \{0\}\}$, one can show (assuming H4–H6) that there exist $c > 0$ and $d < \infty$ such that $\Lambda \subset [c,d]$ and that $\Lambda = \{\tau : r(L_\tau) =$ spectral radius of $L_\tau = 1\}$. By applying Theorem 1 of Section 2 one can prove

THEOREM 2. (See [7]). Assume that K, X and Λ are as defined above, that F_τ is defined by equation (5) and that assumptions H4–H6 hold. Define $S = \{(x,\tau) \in K \times (0,\infty) : F_\tau(x) = x$ and $x \neq 0\} \cup \{(0,\tau) : \tau \in \Lambda\}$. Then there exists $\tau_0 \in \Lambda$ such that if S_0 is the connected component of S containing $(0,\tau_0)$, S_0 is unbounded. In fact, for each $\tau > sup\{\lambda : \lambda \in \Lambda\}$, there exists $x \in K - \{0\}$ such that $(x,\tau) \in S_0$.

Theorem 2 immediately implies that $(4)_\tau$ has a positive periodic solution of period ω for $\tau > sup\{\lambda : \lambda \in \Lambda\}$. In certain cases (for example, if $\alpha(\tau) = 0$, $\beta(\tau) = \tau$ and $P(u,\tau) \equiv 1$), the map $\tau \to r(L_\tau)$ is strictly increasing and Λ consists of a single point. However, even if $P(u,\tau) \equiv 1$ and $\beta(\tau) - \alpha(\tau)$ is increasing, it is by no means clear, in general, that Λ has no accumulation points.

Theorem 2 says nothing about the detailed structure of the set S. In many cases one expects that for each $\tau > 0$ there is at most one nonzero periodic solution, and we shall now describe some uniqueness results. Our model equation is the example

$$x(t) = \int_{t-\tau}^{t} (1 + \frac{1}{2} \sin 2\Pi s) g(x(s)) ds \tag{8}$$

where $g(x) = x(1 - x)$ for $0 \leq x \leq 1$ and 0 otherwise. Motivated by this example we study equations of the form

$$x(t) = \int_{t-\tau}^{t} a(s) h(x(s)) ds \tag{9}$$

Our techniques apply equally to the problem of uniqueness of positive, periodic solutions of

$$x(t) = \int_{t-\beta}^{t-\alpha} a(s)h(x(s))ds \tag{10}$$

where $0 < \alpha < \beta$. For convenience we shall (in this section) write L_τ and F_τ for the maps of X into X defined by

$$(L_\tau x)(t) = \int_{t-\tau}^{t} a(s)x(s)ds$$
$$(F_\tau x)(t) = \int_{t-\tau}^{t} a(s)h(x(s))ds \tag{11}$$

We suppose about $a(s)$ and $h(x)$ that

<u>H7</u>. The function a is strictly positive, continuous and periodic of period ω. The map $h : [0,\infty) \to [0,\infty)$ is continuous and $h(0) = 0$. There exists $M > 0$ such that $h(x) > 0$ for $0 < x < M$, $h(x) = 0$ for $x \geq M$ and h is continuously differentiable on $[0,M]$ with $h'(0) = 1$. Furthermore, there exists M_0 with $0 \leq M_0 \leq M$ such that $h'(x)$ is <u>strictly</u> decreasing for $0 \leq x \leq M_0$, $h'(M_0) = 0$ and $h'(x) \leq 0$ for $M_0 \leq x \leq M$.

As usual, let $r(L_\tau)$ denote the spectral radius of L_τ. Define a function $\phi(u)$ for $u > 0$ by

$$\phi(u) = \inf_{0 \leq y \leq u} (y^{-1})h(y) \tag{12}$$

and define $\phi(0) = h'(0) = 1$. The next lemma provides some crude information about the size of positive, periodic solutions of (9).

<u>LEMMA 1</u>. (See [7]). Assume that a and h satisfy H8 and define τ_0 to be the unique value of τ such that $r(L_\tau) = 1$. Then equation (9) has no solution $x \in K - \{0\}$ for $0 < \tau \leq \tau_0$ and at least one solution $x \in K - \{0\}$ for each $\tau > \tau_0$. If $x \in K - \{0\}$ is a solution of (9) and if ϕ is defined by (12), then $0 < x(t) < M$ for all t and

$$\phi(||x||)r(L_\tau) \leq 1 \tag{13}$$

If $j\omega \leq \tau < (j+1)\omega$ and $x \in K - \{0\}$ is a solution of (9), then

$$x(t) \geq \frac{j}{j+1} ||x|| \tag{14}$$

It is not hard to see that if we define

$$\gamma(\tau) = \inf_t \int_{t-\tau}^{t} a(s)ds$$
$$\delta(\tau) = \sup_t \int_{t-\tau}^{t} a(s)ds \tag{15}$$

then $\gamma(\tau) \leq r(L_\tau) \leq \delta(\tau)$ and consequently $\lim_{\tau\to\infty} r(L_\tau) = \infty$. It follows from this and from the equations (13) and (14) that for τ large enough (how large can be estimated crudely using (13) and (14)) every solution $x \in K - \{0\}$ of equation (9) satisfies $x(t) \geq M_0$ for all t.

THEOREM 3. Assume that the functions a and h satisfy H8 and that τ_0 is the unique value of τ such that $r(L_\tau) = 1$. Define $\tau_1 = \sup \{\sigma \geq \tau_0$: every solution $x \in K$ of equation (9) for $\tau_0 \leq \tau \leq \sigma$ satisfies $||x|| \leq M_0\}$; one can prove that $\tau_1 > \tau_0$. Let τ_2 be a number such that if $x \in K - \{0\}$ is a solution of equation (9) for some $\tau \geq \tau_2$, then $x(t) \geq M_0$ for all t. For every $\tau > 0$ there is at most one solution $x \in K - \{0\}$ of equation (9) such that $||x|| \leq M_0$; in particular equation (9) has a unique solution $x \in K - \{0\}$ for $\tau_0 < \tau < \tau_1$. If $C = \sup\{|g'(x)| : M_0 \leq x \leq M\}$, assume that

$$\sup_t C \int_t^{t+\frac{1}{2}\omega} a(s)ds < 1 \tag{16}$$

Then it follows that for every $\tau \geq \tau_2$, equation (9) has a unique solution $x \in K - \{0\}$.

Uniqueness for the range $\tau_0 < \tau < \tau_1$ has been obtained independently by Hal Smith [14] who used some ideas of Krasnoselskii. However, the Krasnoselskii approach fails in the range $\tau > \tau_1$, so a different method was used in [7]. The key lemma in [7] is the following.

LEMMA 2. [7]. Assume that a and h satisfy H8 and that $x \in K - \{0\}$ is a solution of equation (9). Define a linear operator B (the Frechet derivative of F_τ at x) by

$$(Bu)(t) = \int_{t-\tau}^{t} a(s)h'(x(s))u(s)ds$$

and let I denote the identity operator. Assume one of the following three cases hold:

case i. $||x|| \leq M_0$

case ii. $x(t) \geq M_0$ for all t and if σ denotes the distance of τ to the nearest integral multiple of ω,

$$\sup_t \int_{t-\sigma}^{t} a(s)|h'(x(s))|ds < 1$$

case iii. If σ denotes the distance of τ to the nearest integral multiple of ω,

$$\sup_t \int_{t-\sigma}^{t} a(s)|h'(x(s))|ds < \frac{1}{2}$$

$$\sup_t \int_{t-\tau}^{t} a(s)h'(x(s))ds < 1$$

Then it follows that $I - B$ is one-to-one and onto and if V is an open (in the relative topology on K) neighborhood in K of x which contains no other fixed points of F_τ, $i_K(F_\tau, V) = 1$.

By using Lemma 2, one can give more refined versions of Theorem 3. In particular, the gap in uniqueness results for the range $\tau_1 \leq \tau \leq \tau_2$ can sometimes be filled. For example, the uniqueness question can be completely settled for the Cooke-Kaplan example $a(s) = (1 + \frac{1}{2} \sin 2\Pi s)$, $h(x) = x(1 - x)$ for $0 \leq x \leq 1$, $h(x) = 0$ otherwise: for each $\tau > \tau_0 = 1$ the equation (9) has (in this case) a unique, positive periodic solution.

We believe that further progress in analyzing the structure

of the solution set of equation (9) or (10) involves better understanding of the compact linear operator $L : X \to X$ defined by

$$(Lx)(t) = \int_{t-\beta}^{t-\alpha} b(s)x(s)ds$$

where $0 \leq \alpha < \beta$ and $b \in X$. For example, are all eigenvalues of L simple?

4. *Some Linear Spectral Theory*

In the previous section we encountered the linear map of X to X given by

$$(L_\tau x)(t) = \int_{t-\tau}^{t} a(s)x(s)ds \tag{17}$$

and it was important to determine that value of τ for which $r(L_\tau) = 1$. Since simple formulas for $r(L_\tau)$ are not available, it becomes important to give a method of determining $r(L_\tau)$ to a specified accuracy. We should remark that the technique of proof of the results we shall describe applies equally to some other equations, eg

$$(Lx)(t) = \int_{t-\beta}^{t-\alpha} a(s)x(s)ds \tag{18}$$

for $0 < \alpha < \beta$.

As a first step we define a sequence $\{P_n\}$ of finite dimensional, linear projections on the space X of continuous, ω-periodic functions. For each $n \geq 1$, define $t_j = \frac{j\omega}{n}$ for $0 \leq j \leq n$ and define $P_n x = y$, where $y(t_j) = x(t_j)$ and

$$y(t) = \left(\frac{t_{j+1} - t}{t_{j+1} - t_j}\right) y(t_j) + \left(\frac{t - t_j}{t_{j+1} - t_j}\right) y(t_{j+1})$$

for $t_j \leq t \leq t_{j+1}$. Notice that $P_n(K) \subset K$ and $P_n(X)$ is n-dimentional.

With this notation the following lemma can be proved.

LEMMA 3. Suppose $\tau > 0$ and $a(s)$ is a continuous, strictly positive ω-periodic function. Then L_τ has a unique positive eigenvector $x(t)$ of norm 1 and for each $n \geq 1$, $P_n L_\tau$ has a unique positive eigenvector $x_n(t)$ of norm one (both x and x_n depend on τ).

DEFINITION. If x and x_n are as above, $(k(\tau) = \inf_t x(t))^{-1}$ and $k_n(\tau) = (\inf_t x_n(t))^{-1}$

LEMMA 4. Let $a(s)$ be as in Lemma 3 and suppose that $\gamma(\tau)$ is defined by equation (15) of the previous section. Then one has

$$k(\tau) \leq exp\left(\frac{||a||}{\gamma(\tau)}\,\omega\right) \tag{19}$$

and if $\frac{||a||\omega}{n\gamma(\tau)} < \frac{1}{2}$ one has

$$k_n(\tau) \leq exp\left(\frac{2||a||}{\gamma(\tau)}\,\omega\right) \tag{20}$$

THEOREM 4. Let a be a strictly positive, periodic function of period ω and assume that a is Lipschitzian with Lipschitz constant C. Then if $r(P_n L_\tau)$ denotes the spectral radius of $P_n L_\tau$ and $r(L_\tau)$ the spectral radius of L_τ one has

$$r(P_n L_\tau) \geq r(L_\tau) - \left(C + \frac{||a||^2}{\gamma(\tau)}\right)\left(\frac{\omega}{n}\right)^2 k(\tau)$$

$$r(L_\tau) \geq r(P_n L_\tau) - \left(C + \frac{||a||^2}{\gamma(\tau)}\right)\left(\frac{\omega}{n}\right)^2 k_n(\tau)$$

where $k(\tau)$ and $k_n(\tau)$ can be estimated by (19) and (20).

REMARK 1. Theorem 4 reduces the problem to that of finding the spectral radius of $P_n L_\tau$, which is the same as finding the spectral radius of an $n \times n$ matrix.

REMARK 2. Theorem 4 is a corrected version of Theorem 4 in [6]. The erroneous assertion that $P_n x \leq x$ for $x \in K$ leads to an error in the proof in [6]. A corrected proof will appear in a revised version of [6].

It is natural to ask what differentiability properties the map $\tau \to r(L_\tau)$ possesses if, for example, L_τ is defined by equation (7). Our next theorem answers this question. This theorem is not quite proved in the stated generality in [6], but a similar proof gives the result.

THEOREM 5. (Compare [6]). Assume $a(s)$ is a strictly positive, ω-periodic function which is C^n (n times continuously differentiable) for some $n \geq 0$. Let $P(u,\alpha,\beta)$ be a real-valued map defined on $D = \{(u,\alpha,\beta) : (u,\alpha,\beta) \in \mathbb{R}, \ \alpha < \beta\}$ such that $\left(\frac{\partial^j}{\partial u^j}\right)\left(\frac{\partial^k}{\partial \alpha^k}\right)\left(\frac{\partial^l}{\partial \beta^l}\right)P$ exists and is continuous for $j + k + l \leq n + 1$ and $j \leq n$. Assume also that $P(u,\alpha,\beta) > 0$ for $u \in (\alpha,\beta)$. Define a linear operator $L_{\alpha,\beta} : X \to X$ by the formula

$$(L_{\alpha,\beta}x)(t) = \int_{t-\beta}^{t-\alpha} P(t - s,\alpha,\beta)\, a(s)x(s)ds \tag{22}$$

Then the map $(\alpha,\beta) \to r(L_{\alpha,\beta})$ = the spectral radius of $L_{\alpha,\beta}$ is C^{n+1} on its domain.

REMARK 3. The difficulty in proving results like Theorem 5 stems from the fact that even if L_τ is given by equation (17) and $a(s) \equiv 1$, the map $\tau \to L_\tau$ is not Frechet differentiable.

If $P(u,\alpha,\beta) = P(u)$ is independent of α and β and everywhere positive and $L_{\alpha,\beta}$ is as above, it is not hard to see that for each $\alpha > 0$ there is a unique $\beta = \phi(\alpha)$ such that $r(L_{\alpha,\beta}) = 1$. If P and a are C^n for $n \geq 0$, one can prove by using Theorem 5 and the implicit function theorem that the map $\alpha \to \phi(\alpha)$ is C^{n+1}. In the notation of Section 3, the set $\Lambda = \{\tau : \phi(\alpha(\tau)) = \beta(\tau)\}$. We should remark that the map ϕ also arises naturally in a model considered by Hal Smith [13].

5. *The Fixed Point Index of Some Cone Maps*

In this section we want to indicate some elementary means of calculating the fixed point index and some fixed point theorems and eigenvalue theorems which are consequences. We are motivated by the crucial role played by the fixed point index in much of the previous work.

First we need some definitions and notation. K will always denote a total cone in a real Banach space X. The cone induces a partial ordering by $x \leq y$ iff $y - x \in K$. The cone is called "normal" if there exists an equivalent norm $||\cdot||_1$ on X such that $0 \leq x \leq y$ implies $||x||_1 \leq ||y||_1$; the cone is "repreducing" if $X = \{x - y : x, y \in K\}$.

If $f : D \subset K \to K$ is a map, we shall say that f is "increasing" if $x, y \in D$ and $x \leq y$ imply that $f(x) \leq f(y)$. We shall say that f is "sublinear" if whenever $x \in D$, $tx \in D$ for $0 \leq t \leq 1$ and $f(tx) \geq tf(x)$ for $0 \leq t \leq 1$. If $e \in K - \{0\}$ the function f is called "e-increasing" if whenever $x, y \in D$, $x \leq y$ and $x \neq y$, there exist constants α and β (dependent on x and y) with $\alpha > 0$ and $\beta > 0$ such that

$$\alpha e \leq f(y) - f(x) \leq \beta e$$

The function f is called "e-sublinear" if whenever t is a real number with $0 < t < 1$ and $x \in D - \{0\}$, there exists a number $\delta = \delta(t,x) > 0$ such that

$$f(tx) - tf(x) \geq \delta e$$

Recall that if A is any bounded subset of a Banach space X, then $\gamma(A)$, the measure of noncompactness of A, is given by $\inf\,\{d > 0 : A$ equals a finite union of sets of diameter less than $d\}$. If μ is a function which assigns to each bounded set A a nonnegative real number $\mu(A)$, μ will be called a generalized measure of noncompactness if (1) there exist positive constants m and M such that $m\mu(A) \leq \gamma(A) \leq M\mu(A)$ for every bounded set A, (2) $\mu(\overline{co}(A)) = \mu(A)$ for every bounded set A ($\overline{co}(A)$ = convex closure of A), (3) $\mu(A \cup B) = \max(\mu(A),\mu(B))$ and (4) $\mu(A + B)$

$\leq \mu(A) + \mu(B)$. A map $f : D \quad X \to X$ will be called a "k-set-contraction with respect to μ" if f is continuous and $\mu(f(A)) \leq k\mu(A)$ for every bounded set $A \subset D$. We define $\mu(f) = inf\{k > 0 : f$ is a k-set-contraction w.r.t.$\mu\}$.

If L is a bounded linear map of X into X such that $L(K) \quad K$, it will be convenient to define $||L||_K = sup\{||Lx|| : x \in K, ||x|| \leq 1\}$ and $\mu_K(L) = \mu(L|K)$, where μ denotes a generalized measure of noncompactness. It will be convenient to make the following definitions

$$\begin{aligned} r(L) &= \lim_{n\to\infty} ||L^n||^{\frac{1}{n}}, \quad r_K(L) = \lim_{n\to\infty} ||L^n||_K^{\frac{1}{n}} \\ \rho(L) &= \lim_{n\to\infty} (\gamma(L^n))^{\frac{1}{n}}, \quad \rho_K(L) = \lim_{n\to\infty} (\gamma_K(L^n))^{\frac{1}{n}} \end{aligned} \tag{23}$$

We refer the reader to [10] for more details on the significance of $r(L)$ and $\rho(L)$ in spectral theory. In equation (23), γ denotes the measure of noncompactness; the numbers $\rho(L)$ and $\rho_K(L)$ are the same if a generalized measure of noncompactness is used in the formula. The limits exist by the same argument which proves $\lim_{n\to\infty} ||L^n||^{\frac{1}{n}}$ exists. It is not hard to prove that

$$\begin{aligned} &\rho_K(L) \leq \rho(L) \leq r(L) \\ &r_K(L) \leq r(L) \end{aligned} \tag{24}$$

If the cone K is reproducing, it is known (see, for example, [15]), that $r_K(L) = r(L)$; and the same sort of argument also gives that $\rho_K(L) = \rho(L)$ in this case.

We shall need a uniqueness result essentially due to Krasnoselskii [4] but proved in the stated generality by Amann [1].

PROPOSITION 1. (See [4,1]). Let K be a cone in a Banach space X, D a subset of K such that $x \in D$ implies $tx \in D$ for $0 \leq t \leq 1$ and $f : D \to K$ a continuous map such that $f(0) = 0$. Assume that there exists $e \in K - \{0\}$ such that f is e-sublinear and e-increasing. Then there exists at most one nonzero

$x \in D$ such that $f(x) = x$.

Our next proposition, while quite elementary, will prove to be useful.

PROPOSITION 2. Let K be a cone in a Banach space X, $V = \{x \in K : ||x|| < r\}$ and $f : \overline{V} \to K$ a continuous map which is a k-set-contraction, $k < 1$, with respect to μ. Define $S = \{x \in K : ||x|| = r\}$ and let $C : [0,\infty) \times S \to K$ be a map which takes bounded sets to precompact sets and has the properties that $\lim_{t\to\infty} (\inf\{||C(t,x)|| : x \in S\}) = \infty$ and $C(0,x) = 0$ for all $x \in S$. Assume that $f(x) + C(t,x) \neq x$ for $||x|| = r$ and $t \geq 0$. Then it follows that $i_K(f,V) = 0$. If there exists a number $R > r$ such that $x \neq tf(x)$ for $||x|| = R$ and $0 \leq t \leq 1$, then it follows that there exists $x \in K$ with $r < ||x|| < R$ such that $f(x) = x$.

PROOF: It is not hard to construct a continuous retraction $\phi : \overline{V} \to S$ (see [11], Lemma 2.2). Thus we can define C on $[0,\infty) \times \overline{V}$ by $C(t,x) = C(t,\phi(x))$. If we define $f_t(x) = f(x) + C(t,x)$ for $x \in \overline{V}$, the homotopy property for the fixed point index implies that $i_K(f,V) = i_K(f_t,V)$ for all $t \geq 0$. Since f is bounded on $\overline{V}$ and $\lim_{t\to\infty} (\inf\{||C(t,x)|| : x \in \overline{V}\}) = \infty$, the equation $f_t(x) = x$ has no solutions $x \in \overline{V}$ for t large enough, so $i_K(f_t,V) = i_K(f,V) = 0$.

If $B = \{x \in K : ||x|| < R\}$ and $U = \{x \in K : r < ||x|| < R\}$, then the homotopy $f_t(x) = tf(x)$ for $0 \leq t \leq 1$ and $x \in \overline{B}$ shows that

$$i_K(f_1,B) = i_K(f_0,B) = 1$$

The additivity property of the fixed point index now gives that $i_K(f,U) = 1$, which implies the existence of a fixed point x with $r < ||x|| < R$. ■

As an immediate corollary we obtain the following result, which Grafton proves (but does not explicitly state) by a lengthy argument in [3]. Corollary 1 is the basic abstract tool Grafton uses in studying periodic solutions of functional differential equations; the result that $i_K(f,U) = 1$ seems new.

COROLLARY 1. (See [3]). Let K be a cone in a Banach space X, $U = \{x \in K : r < ||x|| < R\}$ and $f : \overline{U} \to K$ a continuous, compact map. Assume that $tf(x) \neq x$ for $||x|| = r$ and $t \geq 1$ and $sf(x) \neq x$ for $||x|| = R$ and $0 \leq s \leq 1$. Finally, suppose $inf\{||f(x)|| : x \in K, ||x|| = r\} = \delta > 0$. Then it follows that $i_K(f,U) = 1$ and f has a fixed point in U.

PROOF. By using the continuous retraction ϕ in Proposition 2, we can extend f to a compact map on $\{x \in K : ||x|| \leq R\}$. Define $C(s,x) = sf(x)$ for $s \geq 0$ and $||x|| = r$ and apply Proposition 2. ∎

PROPOSITION 3. Let K be a cone in a Banach space X and $D \subset K$ a closed set such that $y \in D$ and $0 \leq x \leq y$ imply $x \in D$. Let $f : D \to K$ be an increasing, sublinear map such that $f(0) = 0$ and suppose there exists $r > 0$ such that $V = \{x \in K : ||x|| < r\} \subset D$ and $f(x) \neq x$ for $0 < ||x|| \leq r$. Assume there exists $x_0 \in D - \{0\}$ and an integer $p \geq 1$ such that $f^p(x_0)$ is defined and $f^p(x_0) \geq x$. Finally, suppose there is a generalized measure of noncompactness μ such that f is a k-set-contraction, $k < 1$, with respect to μ. Then it follows that $i_K(f,V) = 0$.

PROOF: Let $A = \{y \in K : y \geq x_0\}$, so A is a closed set not containing 0. It follows that by decreasing r (which does not change the fixed point index) we can assume $A \cap \overline{V}$ is empty. In the notation of Proposition 2, let $C(t,x) = tx_0$. It suffices to prove that $f(x) + tx_0 \neq x$ for $0 \leq t$ and $||x|| = r$, or equivalently for $0 < t < 1$ and $||x|| = r$. If $f(x) + tx_0 = x$, we have $tx_0 \leq x$. To obtain a contradiction it suffices to show that whenever $\sigma x_0 \leq x$ for $0 < \sigma \leq 1$ and $||x|| = r$, then $(\sigma + t)x_0 \leq x$, since this will eventually imply that $x_0 \leq x$.

Thus suppose that $f(x) + tx_0 = x$ for $||x|| = r$ and $0 < t < 1$ and $\sigma x_0 \leq x$ for some σ with $0 < \sigma < 1$. Using that f is increasing and sublinear, this gives

$$\sigma f(x_0) + tx_0 \leq x$$

Repeating the estimate we get

$$\sigma f^2(x_0) + tx_0 \leq x$$

After p repetitions we have

$$\sigma x_0 + tx_0 \leq \sigma f^p(x_0) + tx_0 \leq x.$$

which completes the proof. ∎

Our next proposition generalizes results in Lemma 2, Section 3, by proving that the local fixed point index of an isolated, nonzero fixed point of an increasing, sublinear map is usually one.

PROPOSITION 4. Let K be a normal cone in a Banach space X and $D \subset K$ a relatively open subset of K $(D = W \cap K,\ W$ open in $X)$ such that $y \in D$ and $0 \leq x \leq y$ imply $x \in D$. Let $f : D \to K$ be increasing and sublinear with $f(0) = 0$ and suppose f is a k-set-contraction, $k < 1$, with respect to a generalized measure of noncompactness μ. Define $f_s(x) = f(sx)$ for $0 \leq s \leq 1$. Assume that there exists an integer $p \geq 1$ and $e \in K - \{0\}$ such that f_s^p is e-increasing and e-sublinear on its domain for $0 < s \leq 1$. Then f has at most one nonzero fixed point in D; and if f has a nonzero fixed point x_0 and $U \subset D$ is a relatively open neighborhood of x_0 in K such that $0 \notin \overline{U} - U$, it follows that $i_K(f,U) = 1$.

PROOF. If x is in the domain of f_s^p it is easy to see that tx is also for $0 \leq t \leq 1$. Proposition 1 implies that f_s^p has at most one nonzero fixed point in D, and since every fixed point of f_s is a fixed point of f_s^p, f_s has at most one nonzero fixed point. Assume that f has a nonzero fixed point $x_0 \in D$. Then if V is a relatively open neighborhood of 0 in K such that $x_0 \notin \overline{V}$, Proposition 3 implies that $i_K(f,V) = 0$; and the additivity property of the fixed point index implies $i_K(f,U) = i_K(f,D)$. Thus to complete the proof it suffices to show that $A = \{x \in D : f_x(s) = x$ for some s with $0 \leq s \leq 1\}$ is a compact subset of D, because then the homotopy property of the index can be used to conclude that $i_K(f,D) = i_K(f_0,D) = 1$.

Suppose we can show that $A \subset B = \{y : 0 \leq y \leq x_0\}$. Define $B_1 = \{ty : y \in f(B),\ 0 \leq t \leq 1\}$ and generally define $B_n = \{ty : y \in f(B_{n-1} \cap B),\ 0 \leq t \leq 1\}$. It is not hard to see that $A \subset B_n$ for all n and that $\mu(B_n) \leq k^n \mu(B) \to 0$ (we use normality of K to ensure that B is bounded). It follows that A is compact.

Thus it only remains to show that if $f_s(x) = x$ for some s with $0 \leq s \leq 1$, then $x \leq x_0$. We can assume that $0 < s < 1$ and that $x \neq 0$. The assumption that f_s^p is e-increasing for $0 < s \leq 1$ shows that there exist positive constants α and β such that $x \leq \beta e$ and $\alpha e \leq x_0$, and this implies that for some σ with $0 < \sigma \leq 1$, $\sigma x \leq x_0$. If $s \in (0,1)$ is such that $f_s(x) = x$, we define $y_j = f_s^j(\sigma x)$. Notice that

$$y_0 = \sigma x = \sigma f_s(x) \leq f_s(\sigma x) = y_1$$

by the sublinearity of f_s. Generally, we see by induction that $y_j \leq y_{j+1}$ for $j \geq 0$. On the other hand, we have $y \leq x_0$; and if $y_j \leq x_0$, we find that

$$y_{j+1} = f_s(y_j) \leq f_s(x_0) \leq f(x_0) = x_0$$

This shows that $y_j \leq x_0$ for all j.

We claim that y_j approaches some element $y \in K$, and for an increasing sequence $\{y_j\}$ in a cone K, it is known that this assertion is equivalent to showing $\{y_j\}$ precompact. However, if $C = \{y_j : j \geq 0\}$, then

$$C = f(C) \cup \{y_0\}$$

This shows that $\mu(C) = \mu(f(C)) \leq k\mu(C)$, so $\mu(C) = 0$ and $y_j \to y$. By the continuity of f_s, it follows that $y = f_s(y)$, and the uniqueness of nonzero fixed points of f_s implies that $x = y \leq x_0$. ∎

If f in Proposition 4 is defined on all of K, the assumptions can be weakened. One can prove that all fixed points x of f_s satisfy $x \leq x_0$ without assuming uniqueness of such fixed points. Specifically, one has the following result, whose proof we leave to the reader.

PROPOSITION 5. Let K be a normal cone in a Banach space X and $f : K \to K$ an increasing, sublinear map such that $f(0) = 0$ and such that f is a k-set-contraction, $k < 1$, with respect to a generalized measure of noncompactness μ. Assume that f has exactly one nonzero fixed point $x_0 \in K$. Define $f_s(s) = f(sx)$ for $0 \leq s \leq 1$ and assume that there exists $e \in K - \{0\}$ such that if x is a nonzero fixed point of f_s for $s \in [0,1]$, then $\alpha e \leq x \leq \beta e$ for some constants $\alpha > 0$ and $\beta > 0$. Then if $U \subset K$ is open in the relative topology of K, $x_0 \in U$ and $0 \notin \overline{U} - U$, it follows that $i_K(f,U) = 1$.

REMARK 4. Proposition 4 is directly applicable to $f(x) = F_\tau(x)$ defined by equation (11), Section 3. The set D in that case is $\{x \in K : x(t) < M_0$ for all $t\}$, e is the function identically equal to one and p any integer such that $p\tau \geq \omega$.

We shall conclude this section by using fixed point index calculations to obtain a nonlinear version of the Krein-Rutman theorem [5] which contains a linear version as a special case.

PROPOSITION 6. Let K be a normal cone in a Banach space X. Let $D = \{x \in K : ||x|| < a,\ a > 0\}$ and let $f : \overline{D} \to K$ be an increasing function such that f is a k-set-contraction, $k < 1$, with respect to a generalized measure of noncompactness μ and $f(0) = 0$. Assume there exists a sequence of points $y_n \in K - \{0\}$ and a sequence of integers M_n such that $||y_n|| \to 0$, $f^{M_n}(y_n)$ is defined and $||f^{M_n}(y_n)|| > b > 0$, where b is independent of n. Then if $0 < r \leq min(b,a)$, there exist $x \in K$ with $||x|| = r$ and $t \geq 1$ such that $f(x) = tx$.

PROOF. If r is as above, let $V = \{x \in K : ||x|| < r\}$. If $x = f(x)$ for some x with $||x|| = r$, we are done, so suppose not. We claim that if $x \neq f(x)$ for $||x|| = r$, then $i_K(f,V) = 0$. Assuming this to be the case, consider the homotopy $f_s(x) = sf(x)$ for $0 \leq s \leq 1$. If $f_s(x) \neq x$ for $||x|| = r$ and

$0 \leq s \leq 1$, it follows that $i_K(f,V) = 1$, a contradiction. Thus we must have $f(x) = s^{-1}x$ for some x with $||x|| = r$ and $s \in (0,1]$.

Thus it only remains to prove that $i_K(f,V) = 0$. A simple compactness argument shows that there exists $\delta > 0$ such that $||x - f(x)|| \geq \delta$ for all $x \in K$ with $||x|| = r$. Select n so large that $||y_n|| < \delta$ and define $g(x) = f(x) + y_n$. If $f_t(x) = tf(x) + (1 - t)g(x)$ for $0 \leq t \leq 1$, $f_t(x) \neq x$ for $||x|| = r$ and $i_K(f,V) = i_K(g,V)$. If $g(x) = x$ for some $x \in \overline{V}$, it is not hard to see that $x \geq f^j(y_n)$ for all $j \geq 0$. Since $||f^{M_n}(y_n)|| > r$ and the cone is normal, we obtain that $||x|| > r$, a contradiction. It follows that g has no fixed points in $\overline{V}$, so that $i_K(g,V) = 0$. ∎

As a consequence of Proposition 6, we obtain a generalization of the usual linear Krein-Rutman theorem, at least for normal cones. In the statement of the theorem below recall that the number $\rho_K(L)$ defined by equation (23) is zero if L is a compact linear map.

PROPOSITION 7. Let K be a normal cone in a Banach space X and $L : X \to X$ a bounded linear map such that $L(K) \subset K$. Assume that $\rho_K(L) < r_K(L)$. Then there exists $x \in K - \{0\}$ such that $Lx = \lambda x$, where $\lambda = r_K(L)$.

PROOF. Define $\mu = \rho_K(L)$ and let s_n be a sequence of positive numbers with $\mu < s_n^{-1} < \lambda$ and $s_n \to \lambda^{-1}$. For notational convenience, fix n, write $s_n = s$ and define $g(x) = sL(x)$. Select N so large that $\gamma_K(g^N) = s^N\gamma_K(L^N) < 1$ and define a generalized measure of noncompactness μ by the formula

$$\mu(A) = \frac{1}{N} \sum_{j=0}^{N-1} \gamma(g^j(A)) \tag{25}$$

It is not hard to see that μ is a generalized measure of noncompactness and that there exists a constant $c < 1$ such that $\mu(g(A)) \leq c\mu(A)$ for every bounded $A \subset K$.

Define $r_K(g) = sr_K(L) = a > 1$ and select numbers a_1 and a_2 with $1 < a_1 < a_2 < a$. By definition there exists a sequence of unit vectors $u_m \in K$ (for m large enough) with $||g^m(u_m)|| \geq a_2^m$. If we define $y_m = a_1^{-m} u_m$, we have $y_m \to 0$ and $||g^m(u_m)|| > 1$. It follows from Proposition 6 that there exists $x \in K$ with $||x|| = 1$ and $t \geq 1$ such that $g(x) = tx$.

If we recall that g depends on n, as do x and t above, we have a sequence $s_n \to \lambda^{-1}$, $t_n \geq 1$ and $x_n \in K$ with $||x_n|| = 1$ such that

$$L(x_n) = s_n^{-1} t_n x_n$$

If $Lx = \alpha x$ for $x \in K - \{0\}$, a simple calculation gives (26) that $\alpha \leq \lambda$; so if we write $\lambda_n = s_n^{-1} t_n$, we have $\lambda_n \leq \lambda$ and $\lambda_n \to \lambda$.

To complete the proof it suffices to show that $A = \{x_n : n \geq 1\}$ has compact closure, because then by considering a subsequence we can assume that $x_n \to x$ and equation (26) implies that $Lx = \lambda x$.

Define $g(x) = \lambda^{-1} L(x)$ and notice that $x_n - g(x_n) = y_n \to 0$. If N is selected so that $\lambda^{-N} \gamma_K(L^N) < 1$ and μ is defined by equation (25), there exists a constant $c < 1$ such that $\mu(g(S)) \leq c\mu(S)$ for every bounded set $S \subset K$. As before μ is a generalized measure of noncompactness, so it suffices to prove $\mu(A) = 0$. Define $A_j = \{x_n : n \geq j\}$ and $B_j = \{y_n : n \geq j\}$. Since $x_n = g(x_n) + y_n$ we have for every $j \geq 1$

$$\mu(A) = \mu(A_j) \leq \mu(g(A_j) + B_j) \leq c\mu(A_j) + \mu(B_j) \tag{27}$$

Equation (27) implies that

$$\mu(A) \leq (1 - c)^{-1} \mu(B_j) \tag{28}$$

and since $\lim_{j \to \infty} \mu(B_j) = 0$, we are done. ∎

The normality of the cone in Proposition 7 is probably not necessary, but we have not been able to prove Proposition 7 (by an argument using the fixed point index) without the assumption of normality.

REFERENCES

[1] H. Amann, *On the number of solutions of nonlinear equations in ordered banach spaces*, J. Functional Analysis 11(1972), 346-384.

[2] K. L. Cooke and J. L. Kaplan, *A periodicity threshold theorem for epidemics and population growth*, to appear in Math. Biosciences.

[3] R. B. Grafton, *A periodicity theorem for autonomous functional differential equations*, J. Differential Equations 6(1969), 87-109.

[4] M. A. Krasnoselskii, *Positive solutions of operator equations*, P. Noordhoff Ltd., Groningen, The Netherlands, 1964.

[5] M. G. Krein and M. A. Rutman, *Linear operators leaving invariant a cone in a Banach space*, Amer. Math. Soc. Translation No. 26.

[6] Roger D. Nussbaum, *A periodicity threshold theorem for some nonlinear integral equations*, submitted for publication.

[7] ______, *Periodic solutions of some nonlinear integral equations*, to appear in the Proceedings of the Conference on Differential Equations held at the University of Florida, Gainesville, Florida, March, 1976.

[8] ______, *A global bifurcation theorem with applications to functional differential equations*, J. Functional Analysis 19(1975), 319-338.

[9] ______, *The fixed point index for local condensing maps*, Ann. Mat. Pura Appl. 89(1971), 217-258.

[10] ______, *The radius of the essential spectrum*, Duke Math. Journal 38(1970), 473-478.

[11] H. O. Peitgen and G. Fournier, *On some fixed point principles for cones in linear normed spaces*, preprint.

[12] H. Schaefer, *Topological Vector Spaces*, Springer-Verlag, New York, 1971.

[13] Hal Smith, *Periodic solutions for a class of epidemic equations*, preprint.

[14] ______, *On periodic solutions of delay integral equations modelling epidemics and population growth*, *Ph.D. dissertation*, University of Iowa, May, 1976.

[15] R. E. L. Turner and H. Schneider, *Positive eigenvectors of order preserving maps*, J. Math. Anal. Appl. 37(1972), 506-515.

SOME RECENT PROGRESS IN NEURO-MUSCULAR SYSTEMS[1)]

M. N. OĞUZTÖRELI AND R. B. STEIN
University of Alberta, Edmonton (Canada)

1. *Introduction*

The purpose of this talk is to present a short account of recent developments in neuromuscular systems. The motor systems of the brain do translate thought, sensation, and emotion into movement of the body. The initial steps in this process are not known clearly. The generation of voluntary movements is unknown. We do not know where the "orders" come from. Most of the information that is available concerns the circuits that execute these shadowy orders. The movement is the end product of a number of control systems that interact extensively. For a detailed study of the subject we refer to the recent edition of the classical book of V. B. Mountcastle (Ed.): Medical Physiology, Vol. II, T. C. Mosly Co., St. Louis, 1968.

Rhythmic oscillations are a prominent feature of all biological systems, and may have a periodicity ranging anywhere from years (e.g., the prey-predator cycles in ecology) to fractions of a millisecond (e.g., the oscillations in the uditory system to a tone of several *KHz*). A narrow band of these oscillations, which are observed in neuromuscular systems with a frequency of roughly 3-13 *Hz* (a period of 80-330 msec.), is referred as tremor or clonus. The study of these oscillations is a useful clinical value in the diagnosis of certain neuromuscular diseases such as Parkinson's and cerebellar diseases.

The Oscillatory movements one observes in living systems arise from a number of sources:

1) This work was supported by grants from the National Research Council of Canada to M. N. Oğuztöreli and the Medical Research Council and the Muscular Dystrophy Association of Canada to R. B. Stein.

1. oscillations intrinsic to the muscles;
2. oscillations arising from the interaction of a muscle with its load;
3. oscillations due to instabilities in the neural feedback pathways involved in the control of muscles;
4. oscillations or pattern generators contained within the central nervous system; and
5. oscillations imposed upon a system by another oscillatory system.

In a recent series of papers dealing with animal studies (Bawa, Mannard and Stein, 1975 a,b) and human studies (Bawa and Stein, 1975) a considerable amount of new data has been collected which relate to oscillations arising from the second, third, and fourth category. The present review will be concentrated on the oscillations from these three sources. The most recent references on the central, mechanical and reflex oscillations are summarized in Figure 1.

The basic model of a reflex system which we have discussed is shown in Figure 2, and the experimental arrangements for studying the properties of muscles in the cat and in man are described in Figure 3. A more detailed block diagram of the model of the mammalian neuromuscular system is shown in Figure 4 and Figure 5.

Data collected under a variety of loading conditions (Bawa et. al., 1975 a,b) are consistent with a simple second order model of skeletal muscle, which in many respects goes back to the work of A. V. Hill (1938). This muscle model contains five parameters:

- C: the contractile force produced internally by a single maximal shock to the muscle;
- B: the rate constant for the decay of this "active state";
- k_p: the stiffness of elastic elements in parallel with the active state element;
- k_i: the stiffness of elastic elements internal to the muscle, but in series with the active state ele-

Cybernetics, 1976.

[61] Tatton, M. G. Forner, S. D., Gerstein, G. L., Chambers, W.W., Lin, C. N., *The effect of postcentral cortical lesions on motor responses to sudden upper limb desplacements*, Brain Res. 96(1975), 108-113.

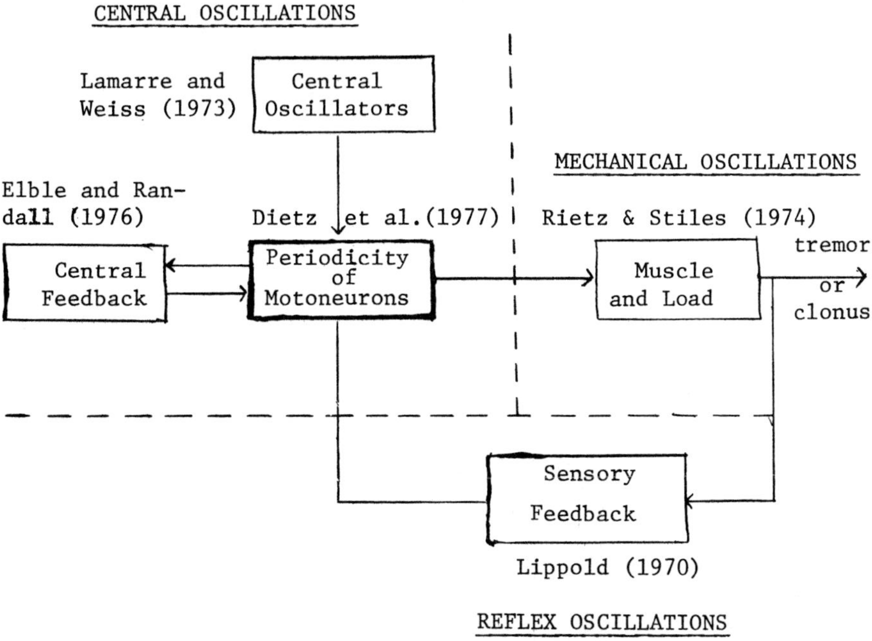

Figure 1. Summary of factors which may be involved in producing the normal 8-12 Hz tremor and the clonus, which follows a brief perturbation. These can be grouped into three major categories: central oscillations, mechanical oscillations and reflex oscillations. One reference in support of each factor has also been included. These factors are discussed in the text as well as some methods for distinguishing among them experimentally.

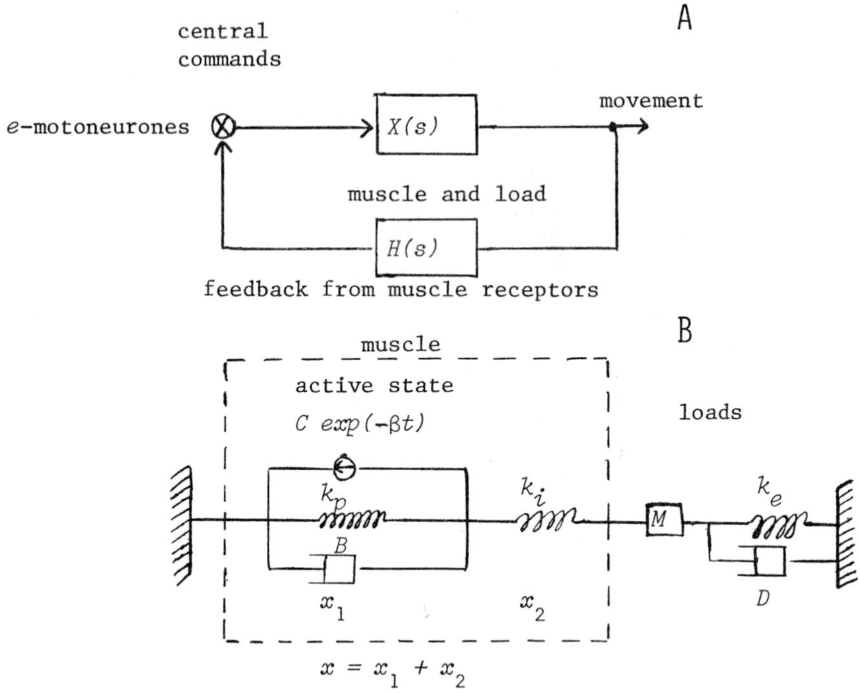

Figure 2. A block diagram of the model of the mammalian neuromuscular system analyzed in this study. The box labelled muscle and load in (A) is shown in more detail in (B). A nerve impulse produces an acitve state of magnitude C which decays exponentially with a rate constant β. The muscle contains a viscous element and parallel (k_p) and internal series (k_i) elastic elements which interact with external loads which may consist of a mass M, a spring k_e and a dashpot D. From Stein and Oguztoreli (1976b).

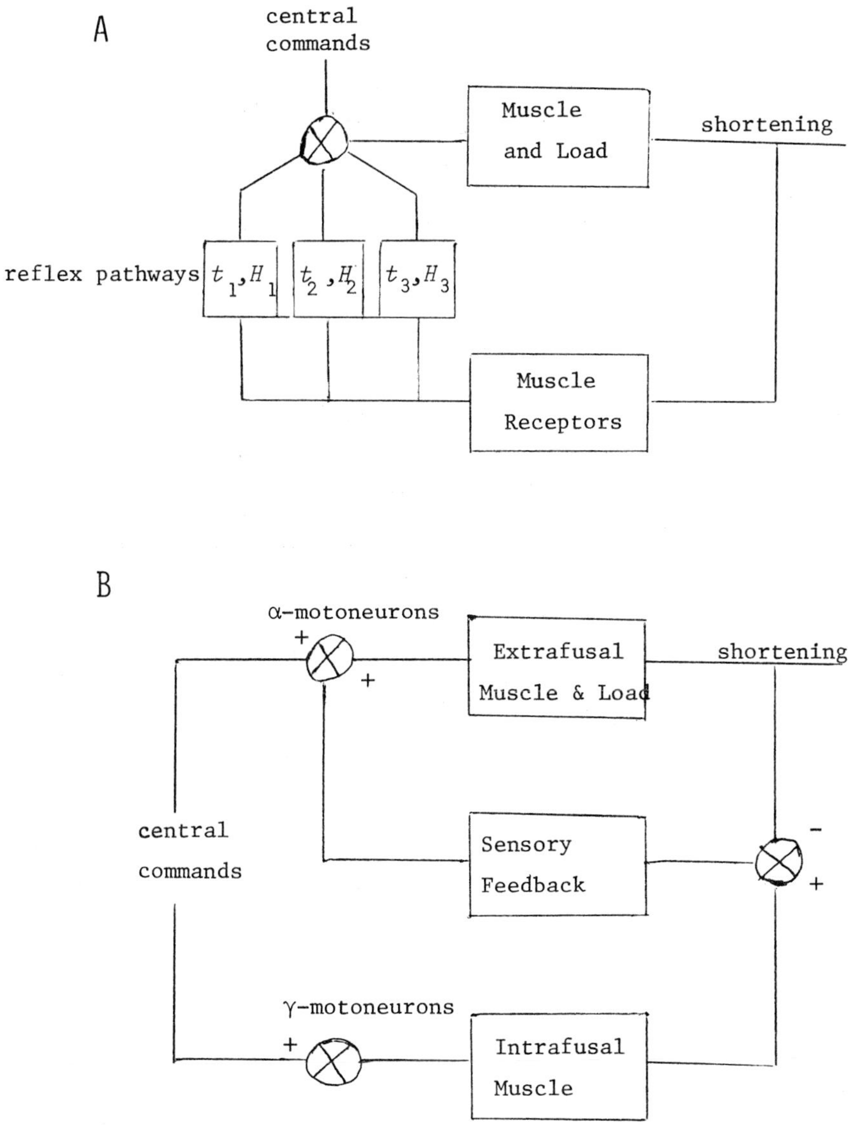

Figure 4. Modification of the basic model of Figure 1 to take into account A) multiple reflex pathways having different values of latency *(t)* and gain *(H)*, or B) coactivation of α- and γ-motoneurons by higher centres. Further discussion in the text. From Oguztoreli and Stein (1976b).

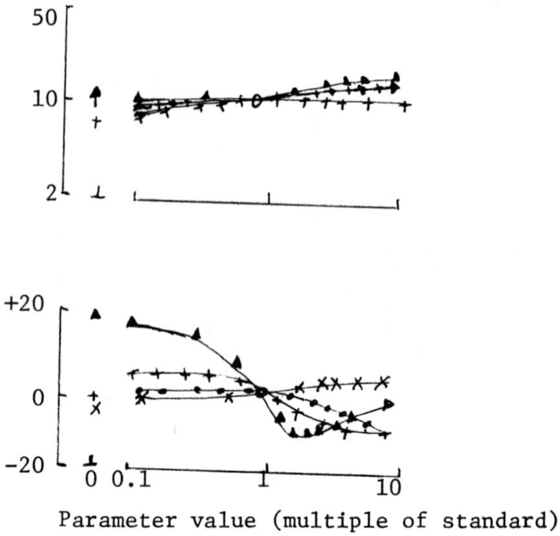

Figure 5. Dependence of the frequency of oscillation (q) and the rate constant (p) at which these oscillations grow $(p > 0)$ or decay $(p < 0)$ on the parameters of the muscle model: $\beta(\cdot)$, $k_p(\times)$, $k_i(+)$, $B(A)$. The scales for frequency and the parameter values are logarithmic. From Oguztoreli and Stein (1975).

ment; and

B: the viscosity of the active state element.

Each of these parameters depends on many factors such as the type of muscle (fast or slow), the state of the muscle (rested or fatigued), the length of the muscle, and the mean rate at which it is being stimulated. However, for a given set of conditions the parameters are effectively constant, even for quite wide fluctuations about the mean levels (Mannard and Stein, 1973). A linear second order model is then surprisingly good in predicting the responses of skeletal muscle. The relation of this simple model to the sliding flament model of muscle has been analyzed in Stein and Wong (1974).

In general, a muscle will be contracting against a load which can be specified in terms of three further parameters:

M: the mass or inertial load;

D: the viscosity or external damping of the load; and

k_e: the stiffness of the external elastic elements in the load.

A load containing a mass and spring itself constitutes a second order system, so that the interaction of a muscle with its load can produce a fourth order system. If the load contains only linear elements this system can be specified in terms of a linear transfer function between impulse (shock) input and length output

$$X(s) \quad \frac{-X_1}{(s + \beta)(s^3 + as^2 + bs + c)} \tag{1.1}$$

where

$$a = \frac{k_i + k_p}{B} + \frac{D}{M} \tag{1.2}$$

$$b = \frac{D(k_i + k_p) + B(k_i + k_e)}{MB} \tag{1.3}$$

$$c = \frac{k_p k_i + k_i k_e + k_e k_p}{MB} \tag{1.4}$$

$$X_1 = \frac{k_i C}{MB} \tag{1.5}$$

The sensory feedback from muscle depends on the length and velocity of the muscle and can be specified over a linear range by three further parameters (Matthews and Stein, 1969):

$$H(s) = H_1(s + \gamma)e^{-st_0} \tag{1.6}$$

where

H_1 = the sensitivity of the muscle receptors or in general the gain of the feedback pathway;

γ = the relative sensitivity of the receptor to length and velocity (γ has the dimensions of a rate constant and can be thought of as the frequency in radians/sec at which the length response and velocity response are equal); and

t_0 = pure time delays in the system which, in general, will arise in nervous conduction, sensory transmission and excitation-contraction coupling.

The response of a sensory receptor to changes in length and velocity from Eq. (1.6) will be simply

$$h(t) = H_1[\gamma\ell(t - t_0) + \dot{\ell}(t - t_0)] \tag{1.7}$$

where $\dot{\ell}(t)$ is the velocity of movement (the derivative of muscle length $\ell(t)$) at time t.

In general, the input to a muscle, $y(t)$, will be the sum of sensory feedback $h(t)$ and other inputs $i(t)$ from the central nervous system or under experimental conditions from shocks to the muscle nerve, i.e.

$$y(t) = h(t) + i(t). \tag{1.8}$$

We will be concerned here generally with brief perturbations, so that $i(t) = \delta(t)$, where $\delta(t)$ is the Dirac delta function.

Unfortunately, the linear range of muscle receptors is very

restricted and the sensitivity declines monotonically in either direction outside a linear region centered around the initial length. The hyperbolic tangent function has been chosen as a suitable function to describe this non-linearity in the initial formulation of the problem (Oğuztöreli and Stein, 1975), where a number of properties of the model have also been compared with experimental results. The good agreement between experiment and theory prompted these authors to examine more generally the properties of this model. Although in many respects it grossly simplifies the real system, even the linear version of the model contains 11 parameters, so they have also tried to examine the influence of each parameter on the characteristics of the model. Finally, the effects of the nonlinearity have been described and periodic oscillations discussed.

2. *Physiologically Significant Properties of the Neuro-Muscular Model*

The neuro-muscular system considered in section 1 can be described by an integro-differential difference equation of the form (cf. Oğuztöreli and Stein, 1975):

$$\ell(t) = x(t) + \int_0^t F\{H_1[\gamma\ell(u-t_0) + \dot{\ell}(u-t_0)]\}x(t-u)du \tag{2.1}$$

where, as mentioned before, t denotes time, $\ell(t)$ denotes the length of the muscle at time t, t_0 is the time-lag, γ is the rate constant which gives the frequency in radians/sec at which the length sensitivity and the velocity sensitivity of the muscle receptors are equal, $x(t)$ is the shortening generated by the muscle to a synchronous volley of nerve impulses (the twitch response) which is the inverse Laplace transform of the transfer function

$$X(s) = \frac{-X_1}{(s+\beta)(s^3+as^2+bs+c)} \tag{2.2}$$

where a, b, c and X are given by Eqs. (1.2)-(1.5), and

$$F\{h\} = \begin{cases} h & \text{in the linear regime,} \\ \frac{K}{2} \; tanh\ \{\frac{h}{2}\} & \text{in the nonlinear regime,} \end{cases} \tag{2.3}$$

where K is a positive constant which determines the saturation level in the muscle receptors (or in muscles themselves) limiting the magnitude of the oscillations.

By virtue of Eq. (2.2) the shortening $x(t)$ satisfies the homogeneous differential equation

$$x^{IV}(t) + c_3\dddot{x}(t) + c_2\ddot{x}(t) + c_1\dot{x}(t) + c_0x(t) = 0 \tag{2.4}$$

and the initial conditions

$$x(0) = \dot{x}(0) = \ddot{x}(0) = 0, \quad \dddot{x}(0) = -X_1 \tag{2.5}$$

where

$$c_0 = \beta c, \quad c_1 = c + \beta b, \quad c_2 = b + \beta a, \quad c_3 = a + \beta \tag{2.6}$$

and as before

$$X_1 = \frac{Ck_i}{MB} \tag{2.7}$$

It can be easily verified that the integral difference differential equation (2.1) is equivalent to the following nonlinear differential difference equation

$$\begin{aligned} \ell^{IV}(t) + c_3\dddot{\ }(t) + c_2\ddot{\ell}(t) + c_1\dot{\ell}(t) + c_0\ell(t) \\ = -X_1F\{H_1[\ell(t-t_0) + \dot{\ell}(t-t_0)]\} \end{aligned} \tag{2.8}$$

subjected to the conditions

$$\ell(0) = \dot{\ell}(0) = \ddot{\ell}(0) = 0, \quad \dddot{\ell}(0) = -X_1. \tag{2.9}$$

Furthermore, since the function $F\{h\}$ is analytic in h, and h depends on $\ell(t - t_0)$ and $\dot{\ell}(t - t_0)$ analytically, Eq. (2.1) admits a unique continuously differentiable solution for $t > 0$ if $\ell(t)$ and $\dot{\ell}(t)$ are given in the initial interval $-t_0 \leq t \leq 0$ (cf. Oğuztöreli, 1966):

$$\ell(t) = \phi_0(t), \quad \dot{\ell}(t) = \psi_0(t) \quad (-t_0 \leq t \leq 0) \tag{2.10}$$

where $\phi_0(t)$ and $\psi_0(t)$ are given continuous functions such that

$$\phi_0(0) = \psi_0(0) = 0. \tag{2.11}$$

This solution can be constructed by the method of successive continuations.

Eq. (2.1) incorporates explicitly the force, the muscle length and the time-lag feedback mechanism. In Eq. (2.8) the coefficients c_0, c_1, c_2 and c_3 inherit the role of the function $x(t)$. Obviously we have $\ell(t) \equiv x(t)$ if $H_1 = 0$, i.e., if there is no feedback gain influencing muscle.

Let us note that the solution $\ell(t)$ (as well as $\dot{\ell}(t)$, $\ddot{\ell}(t),\ldots$) of the Eq. (2.1) satisfying the conditions (2.10), (2.11) is always bounded for $t \geq 0$ in the nonlinear regime, but can generate unbounded solutions in the linear regime.

Further, $\ell(t) \equiv 0$ is the unique steady-state solution of the differential difference equation (2.8) where $F\{h\}$ is defined by the second equation in (2.3). The small oscillations around the steady-state solution in the nonlinear regime satisfy the following linear differential difference equation

$$\begin{aligned} v^{IV}(t) + c_3\dddot{v}(t) + c_2\ddot{v}(t) + c_1\dot{v}(t) + c_0v(t) \\ = -\lambda[\gamma v(t-t_0) + \dot{v}(t-t_0)] \end{aligned} \tag{2.13}$$

where

$$\lambda = \frac{H_1KX_1}{4}$$

Clearly, for $\lambda = H_1X_1$, Eq. (2.13) reduces to Eq. (2.8) where the feedback functional $F\{h\}$ is defined by the first equation in (2.3).

Consider the differential difference equation (2.13) and let us search for solutions of the form

$$v(t) = e^{rt} \tag{2.15}$$

where r is a constant. Substituting this expression into Eq. (2.13) we find the transcendental equation

$$\lambda(r + \gamma)e^{-t_0 r} + r^4 + c_3 r^3 + c_2 r^2 + c_1 r + c_0 = 0 \tag{2.16}$$

which is the characteristic equation of Eq. (2.13).

Now, employing the substitution

$$z = t_0 r$$

we transform Eq. (2.16) into the equation

$$e^z(z^4 + c_3 t_0 z^3 + c_2 t_0^2 z^2 + c_1 t_0^3 z + c_0 t_0^4) + \lambda t_0^3(z + \gamma t_0) = 0. \tag{2.17}$$

Clearly the conjugate of each complex root of Eq. (2.17) is also a root. We can show that Eq. (2.17) admits only finitely many pure imaginary roots. Further, by a theorem due to Pontryagin (Pontryagin, 1942) Eq. (2.17) has only finitely many roots on the right side of the imaginary axis in the complex z-plane, since $z\,e^z$ is the principal term in (2.17). On the other hand, in general, there are infinitely many roots lying on the left side of the imaginary axis. For further analysis of the distribution of the roots of Eq. (2.17) in the complex plane we refer to the book of Levin (Levin, 1964).

Let us note that there exist only finitely many multiple roots of Eq. (2.17) which can be easily verified by differentiating Eq. (2.17) and then eliminating e^z between Eq. (2.17) and the derived equation. In this way we obtain a polynomial equation in z which admits only finitely many roots.

Clearly, to each pair of conjugate complex characteristic roots $r_K = p_K \pm iq_K$ with multiplicity n_K correspond the particular solutions

$$t^\nu e^{p_K t} \cos q_K t,\ t^\nu e^{p_K t} \sin q_K t \quad (\nu = 0, 1, \ldots, n_K - 1). \tag{2.18}$$

These are the natural modes of the differential difference equation (2.13). The general solution of Eq. (2.13) is of the form

$$v(t) = \sum_{\kappa=1}^{\infty} e^{p_\kappa t}[V_\kappa(t) \cos q_\kappa t + W_\kappa(t) \sin q_\kappa t] \tag{2.19}$$

where $V_\kappa(t)$ and $W_\kappa(t)$ are polynomials in t of order $n_\kappa - 1$ with arbitrary coefficients such that the series is absolutely and uniformly convergent for $t \geq 0$ and differentiations are allowed four times termwise.

For a more systematic analysis of the differential difference equations (2.8) and (2.13) we refer to the books of Bellman and Cooke (1963), El'sgol'ts and Norkin (1973), Lakshmikantham and Ladas (1972), Lakshmikantham & Leela (1969), Myshkis (1955), Oguztoreli (1966) and Pinney (1958) where there can be found more general methods for qualitative and quantitative studies.

Let us note that periodic solutions and limit cycles exist in the nonlinear regime under suitable parametric conditions. The existence of periodic and/or sustained oscillations and their stability properties are of particular importance in the physiological studies. We remark that periodic oscillations in the neuromuscular model (2.1) are very rare in comparison with the set of all decaying oscillations. Generally, physiological tremors in this model are associated with decaying oscillations and pathological tremors are associated with periodic oscillations and limit cycles.

The neuro-muscular model (2.1) involves eleven structural parameters B, C, D, H_1, k_e, k_i, k_p, M, t_0, β, and γ, introduced in section 1. Denote a general parametric configuration by P:

$$P \equiv (B, C, D, H_1, k_e, k_i, k_p, M, t_0, \beta, \gamma). \tag{2.20}$$

In our study we restricted each parameter by the following relationships:

$$0 \le B \le 400N\text{-}sec/m,\ 0 \le C \le 4N,\ 0 \le D \le 10N\text{-}sec/m,$$
$$0 \le H_1 \le 10^6 m^{-1},\ 0 \le k_e \le 7 \times 10^4 N/m,\ 0 \le k_i \le 2 \times 10^4 N/m,$$
$$0 \le k_p \le 10^4 N/m,\ 0 \le M \le 10kg,\ 0 \le t_0 \le 0.5\ sec, \tag{2.21}$$
$$0 \le \beta \le 300\ sec^{-1},\ 0 \le \gamma \le 150\ sec^{-1},$$

in the respective units. Let R_{11} be the 11-dimensional parallelopiped defined by the inequalities (2.21).

We now denote by $x(t;P)$ the solution of Eq. (2.8) which satisfies the initial conditions (2.9)-(2.11) for a given $P \in R_{11}$. Then, observing the smoothness of coefficients and functions involved in Eq. (2.8) we immediately deduce that $x(t;P)$ depends on P analytically for $P \in R_{11}$ (cf. Oguztoreli, 1966, pp. 38-59). Hence, if μ is one of the coordinates of the point P, then

$$\frac{\partial^k \ \ell(t;P)}{\partial \ \mu^k} \tag{2.22}$$

exists and continuous for any k. Clearly, the same results are also true for the natural modes of the linear equation (2.13).

Although the partial derivatives (2.22) can be found by some complicated analytical techniques, the relevant equations are not suitable for an easy analysis to describe the vatiation of the function $\ell(t;P)$. To overcome these difficulties we used computer simulations.

Note that there exists only one independent oscillation in the case $H_1 = 0$, and, in addition to the infinitely many decaying natural modes, there are two independent oscillations of physiological significance in the case $H_1 > 0$, which will be distinguished as "low-frequency" and "high-frequency" oscillations.

We have chosen the following parametric configuration as "standard" aroung which we study the variations of $p(P)$ and $q(P)$ of the natural modes:

$$B = 40N\text{-}sec/m,\ C = 2.9N,\ D = 0.5N\text{-}sec/m,\ M = 0.3\ kg,$$
$$H_1 = 612m^{-1},\ k_e = 647N/m,\ k_i = 2200N/m,\ k_p = 880N/m, \tag{2.23}$$
$$t_0 = 0.025\ sec,\ \beta = 30\ sec^{-1},\ \gamma = 10\ sec^{-1}.$$

The M-, k_e- and t_0-variations of p_k and q_k $(k = 1,2)$ of low and high frequency oscillations about this standard set are considered in (Oğuztöreli and Stein, 1975) for comparison with experimental data. Here we have converted the parameters in this set to MKS units and have examined the dependence on each parameter systematically.

The results are shown in Figs. 6-8. Figure 6 shows the effect of varying the parameters of the muscle model: β, k_p, k_i and B. The effect of varying C is equivalent to varying H_1 which is considered in Fig. 8 with the other parameters of the feedback pathway: γ and t_0. The effects of the parameters of the muscle's load, M, D and k_e, are shown in Figure 6. Note in Fig. 6 that wide variation in the parameters of the muscle has very little effect on the frequency of oscillation, although the rate at which the oscillations grow or decay may be substantially increased or decreased (lower part of Fig. 6). Minima and maxima are also seen in the values of the rate constants. For example, increasing the viscosity of the muscle decreases the rate constant, and converts the oscillations from growing oscillations $(p > 0)$ to decaying oscillations $(p < 0)$. However, beyond a value about twice that of the standard set the rate constant again begins to increase.

Increasing the external mass (M) loading the muscle or the stiffness of the external spring (k_e) has a more marked effect on the natural frequency. In fact, on the double logarithmic plot of Fig. 7 it can be seen that the slope of the curves approaches $\pm\frac{1}{2}$ which is just the value expected for a mechanical mass-spring system in which the natural frequency ω is well known to be

$$\omega = \sqrt{k_e/M}\ . \tag{2.24}$$

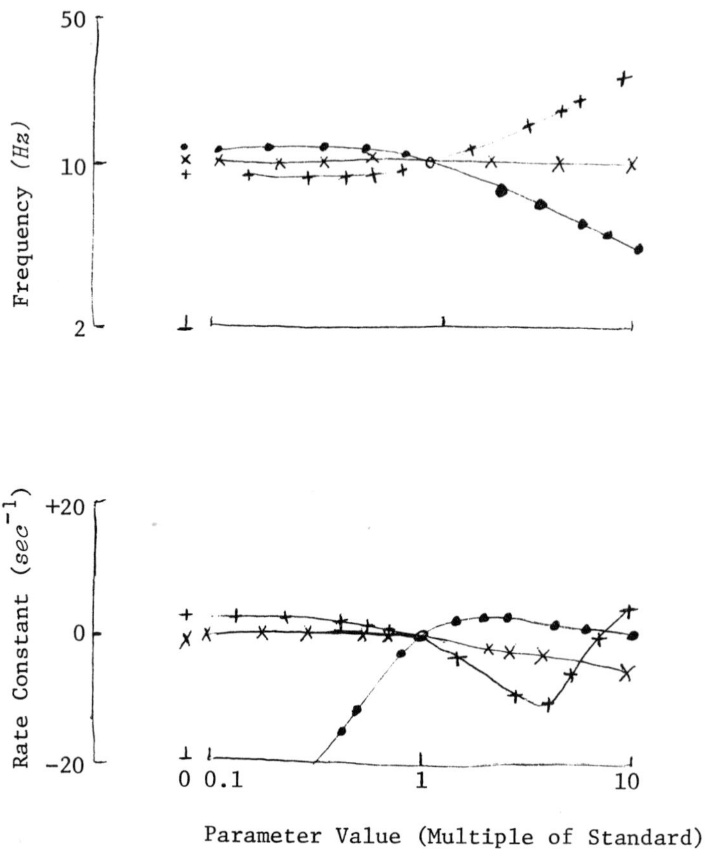

Figure 6. Dependence of the frequency of oscillation (q) and the rate constant (p) on the parameters of the load attached to a muscle: $M(\cdot)$, $D(\times)$, $k_e(+)$. The format is identical to Figure 5 for ease of comparison. From Oğuztöreli and Stein (1975).

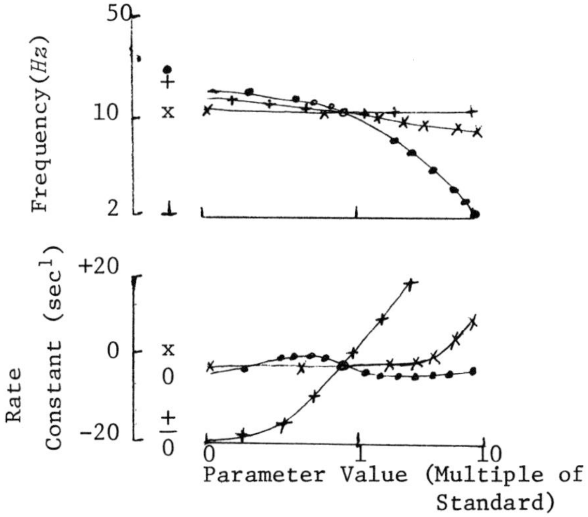

Figure 7. Dependence of the frequency of oscillation *(q)* and the rate constant *(p)* on the parameter of the neutral feedback system: $H_1(+)$, $\gamma(x)$, $t_0(\cdot)$. Format identical to Figure 5 and Figure 6. From Oguztoreli and Stein (1975).

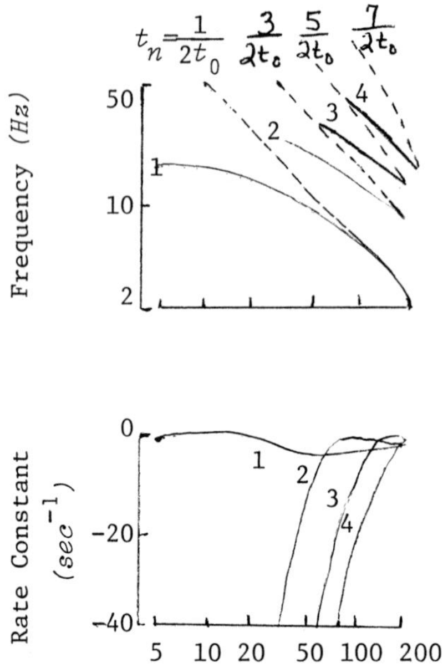

Figure 8. Dependence of the frequencies of oscillation and the rate constants of the first four oscillations (1-4) on the value of a pure time delay t_0 in the feedback pathway. Also shown by the interrupted lines are the predictions of a simple relationship (2.26). From Oguztoreli and Stein (1975).

The most marked tendency for growing oscillation (maximum value of rate constant) is observed with masses two to three times the standard value. The oscillations die away most quickly with values of the external spring four to five times the standard value.

The parameter with the most marked effect on the rate constant p is H_1 since this directly affects the loop gain of the feedback pathway. Increasing the output of the muscle C would have an identical effect and has therefore not been shown. The parameter γ also begins to have a marked effect on the tendency for oscillation if increased sufficiently. Note, however, that none of these parameters alters the frequency of the oscillations markedly (Figure 8). In contrast, increasing the feedback delay (t_0) has a marked effect on the frequency of oscillation without affecting the rate constant greatly. For large values of t_0, the frequency declines inversely with t_0 (a slope of -1 on the double logarithmic plot of Figure 8). With delays above about 30 msec secondary (and later tertiary) oscillations become prominent (Figure 9). Each oscillation has a maximum rate constant just greater than zero when its frequency is about 13 Hz, which corresponds to the frequency of the mechanical oscillations discussed above (due to the parameters k_e and M among others). The mechanical oscillations and those due to the feedback pathway then reinforce each other (see also Oğuztöreli and Stein, 1975). The time delay t_0 will produce a phase shift of 180° at a frequency f given by

$$f = \frac{1}{2t_0} \quad . \tag{2.25}$$

Similarly, a phase shift of $(2n - 1)$ (180)° will be produced at a frequency

$$f_n = \frac{2n - 1}{2t_0} \tag{2.26}$$

where n is a positive integer. In these instances the sign of

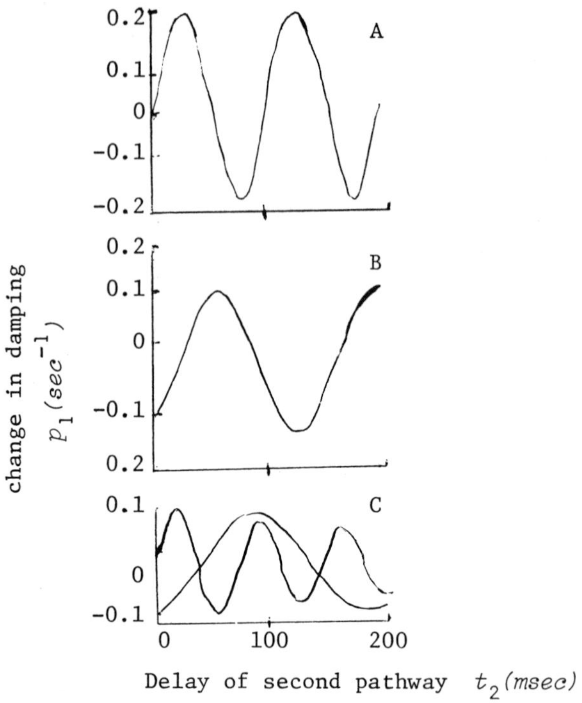

Figure 9. The effect on the damping p_1 (of an oscillation produced by one reflex pathway) of adding 1% extra gain in a second pathway with the delay t_2 indicated. The main pathway has a delay of A) 30 msec, B) 55 msec and C) 85 msec corresponding to spinal, cortical and transcerebellar pathways which have been studied in human subjects. A positive effect represents an enhancement of the existing oscillation while a negative effect represents a depression. From Oguztoreli and Stein (1976b).

the negative feedback pathway will tend to be reversed and so produce positive feedback. The interrupted lines of Figure 9 show the predicted frequencies from (2.26) for the first four integers, and it can be seen that the frequencies of the first four observed oscillations approach those predicted by this simple relation for large values of t_0.

Note that the model which we have developed for neuro-muscular oscillations is simple enough to permit many of its properties to be investigated analytically. Other properties can be conveniently studied by numerical methods. At the same time the model is complex enough to show interesting behavior which can be compared with experimental data (Joyce and Rack, 1974; Bawa and Stein, 1975). Such a comparison has been included elsewhere (Oğuztöreli and Stein, 1975).

3. *Effects of Multiple Pathways*

In the previous sections a model was presented for the study of the oscillations that occur in a neuro-muscular system. Two main sources were considered: the oscillations that arise from the interaction of a muscle with its load, and the oscillations produced by high gain in a sensory feedback pathway. However, in recent years it has become apparent that in a normal, functioning animal, even what was thought to be the simplest of reflex actions, namely the resistance of a muscle or group of muscles to stretch, is mediated through a number of reflex pathways. Some of these involve higher centers such as the motor cortex and possibly the cerebellum (Milner-Brown et. al., 1975). Extension of the neuro-muscular model to include multiple pathways is most desirable because of certain biological advantages. A block diagram of the extended model is given in Figure 4A.

The sensory receptors in muscle respond to the length ℓ and the velocity $\dot{\ell}$ of movement, and their response can be described by an equation of the form

$$h(t) = \sum_k H_k[\gamma_k \ell(t - t_k) + \dot{\ell}(t - t_k)]. \tag{3.1}$$

We will assume as a first approximation that the various reflex pathways from muscle receptors only differ in the latency of the response t_k and the magnitude of gain of the response H_k. We will consider three possible pathways $(k = 1,2,3)$ having different latencies. These three pathways correspond to the three pathways studied in the forearms of normal human subjects and will be referred to (Milner-Brown et. al., 1975) as the <u>spinal</u> $(t_1 = 30$ msec), <u>cortical</u> $(t_2 = 55$ msec) and <u>trans-cerebellar</u> $(t_3 = 85$ msec) pathways.

The improved neuro-muscular system is described by an integral difference differential equation of the form

$$\ell(t) = x(t) + \int_0^t F\{\sum_{k=1}^{3} H_k[\gamma_k \ell(u-t_k) + \dot{\ell}(u-t_k)]\}x(t-u)du \tag{3.2}$$

where t and $\ell(t)$ denote as before time and the length of the muscle at time t, respectively, and

t_k is the time-lag associated with the k-th pathway $(k = 1,2,3)$, with

$$0 < t_1 < t_2 < t_3 \tag{3.3}$$

γ_k is the rate constant in the k-th pathway which gives the frequency in rad/sec at which the length sensitivity and the velocity sensitivity of the muscle receptors are equal,

H_k is the gain of the k-th feedback pathway which indicates the sensitivity of the muscle receptors in the pathway,

$x(t)$ is the shortening generated by the muscle to a synchronous volley of nerve impulses (the twitch response) which is the inverse Laplace transform of the transfer function defined by Eqs. (1.1)-(1.5),

$F\{h\}$ is the feedback functional defined by Eq. (2.3).

It can be easily verified that the functional integral equation (3.2) is equivalent to the following nonlinear differential difference equation

$$\ell^{IV}(t) + c_3\dddot{\ell}(t) + c_2\ddot{\ell}(t) + c_1\dot{\ell}(t) + c_0\ell(t) = -X_1\, F\{\sum_{k=1}^{3} H_k[\gamma_k\ell(t - t_k) + \dot{\ell}(t - t_k)]\} \tag{3.4}$$

subjected to the conditions

$$\ell(0) = \dot{\ell}(0) = \ddot{\ell}(0) = 0, \quad \dddot{\ell}(0) = -X_1. \tag{3.5}$$

Note that Eq. (3.2) admits a unique continuously differentiable solution $\ell(t)$ for $t \geq 0$ if $\ell(t)$ and $\dot{\ell}(t)$ are given in the initial interval $-t_3 \leq t \leq 0$ (cf. Oğuztöreli, 1966):

$$\ell(t) = \phi_0(t), \; \dot{\ell}(t) = \psi_0(t) \quad (-t_3 \leq t \leq 0), \tag{3.6}$$

where $\phi_0(t)$ and $\psi_0(t)$ are given continuous functions such that

$$\phi_0(0) = \psi_0(0) = 0. \tag{3.7}$$

This solution can be constructed by successive continuations. Let us remark that the solutions are always bounded in the nonlinear regime. Small oscillations around the unique steady-state solution $\ell(t) \equiv 0$ satisfy the following linear differential difference equation:

$$v^{IV}(t) + c_3\dddot{v}(t) + c_2\ddot{v}(t) + c_1\dot{v}(t) + c_0v(t) = -\lambda \sum_{k=1}^{3} H_k[\gamma_k v(t - t_k) + \dot{v}(t - t_k)] \tag{3.8}$$

where λ is a constant such that

$$\lambda = \begin{cases} X_1 & \text{in the linear regime,} \\ -\dfrac{KX_1}{4} & \text{in the nonlinear regime.} \end{cases} \tag{3.9}$$

Now let us consider a general parametric configuration P:

$$P \equiv (B,C,D,H_1,H_2,H_3,k_e,k_i,k_p,K,M,t_1,t_2,t_3,\beta,\gamma_1,\gamma_2,\gamma_3), \qquad (3.10)$$

such that

$$\begin{gathered} 0 \leq B \leq 400N\text{-}sec/m,\ 0 \leq C \leq 4N,\ 0 \leq D \leq 10N\text{-}sec/m, \\ 0 \leq H_j \leq 10^6 m^{-1},\ 0 \leq k_e \leq 7 \times 10^4 N/m,\ 0 \leq k_i \leq 2 \times 10^4 N/m, \\ 0 \leq k_p \leq 10^5 N/m,\ 0 \leq M \leq 10\ kg,\ 0 \leq t_j \leq 0.5\ sec, \\ 0 \leq \beta \leq 300\ sec^{-1},\ 0 \leq \gamma_j \leq 150\ sec^{-1}, \\ 1 \leq K \leq 10 \\ (j = 1,2,3) \end{gathered} \qquad (3.11)$$

in the respective units. Let R_{18} be the 18-dimensional parallelepiped defined by the inequalities (3.11). We denote by $\ell(t;P)$ the solution of Eq. (3.2) satisfying the conditions (3.5)-(3.7). We have chosen again the parametric configuration (2.23) as "standard" around which we investigate the variations of $\ell(t;P)$.

It can be easily shown that there exists only one independent oscillation in the case $H_1 = H_2 = H_3 = 0$, and, in addition to the infinitely many decaying natural modes, there are two independent oscillations of physiological significance if at least one of H_1, H_2 and H_3 is positive, which will be distinguished as "low-frequency" and "high-frequency" oscillations.

Although the existence and smoothness of the partial derivatives $\dfrac{\partial^k \ell(t;P)}{\partial \mu^k}$ are assured where μ is one of the coordinates of the point P in R_{18}, a purely mathematical analysis to describe the variations of $\ell(t;P)$ with respect to μ is not easily available, since they are defined by rather complicated equations. We therefore made rather extensive computer simulations. Some of our computer findings are displayed in Figures 9, 10 and 11.

For the physiological background of the problem and some mathematical and computer results we refer to (Stein and Oğuztöreli, 1976).

4. *Acceleration Sensitivity*

In the previous sections models were developed to describe neuro-muscular systems in which muscle receptors respond to the length and velocity of the muscle. We now present a neuro-muscular model which is sensitive to the acceleration of the movement.

Let $\ell = \ell(t)$ be the length of the muscle at time t, and let τ_{1k} and τ_{2k} be the time constants indicating the velocity and acceleration sensitivity in the k-th pathway, respectively. In the models considered in sections 1 and 2, we ignored the acceleration sensitivity, i.e., we had $\tau_{2k} = 0$. It has been shown by Bawa et. al. (1976b) that the transfer function of the internal shortening is of the form

$$X_1(s) = \frac{-\frac{C}{MB}\{MS^2 + Ds + k_i + k_e\}}{(s+\beta)(s^3 + as^2 + bs + c)} \tag{4.1}$$

where all the symbols are as before.

The feedback from muscle spindles is supposed of the form

$$H(s) = \sum_{k=1}^{3} H_k\, e^{-st_k}(1 + s\tau_{1k})(1 + s\tau_{2k}) \tag{4.2}$$

where H_k and t_k are the gain and the time delay in the k-th pathway. The improved neuro-muscular system is described by the following functional integral equation

$$\ell(t) = x_1(t) + \int_0^t F\{\sum_{k=1}^{3} H_k[\ell(u - t_k) + (\tau_{1k} + \tau_{2k})\dot{\ell}(u - t_k) + \tau_{1k}\tau_{2k}\ddot{\ell}(u - t_k)]\}x_1(t - u)du \tag{4.3}$$

where $F\{h\}$ is defined by Eq. (2.3) and $x_1(t)$ is the interior

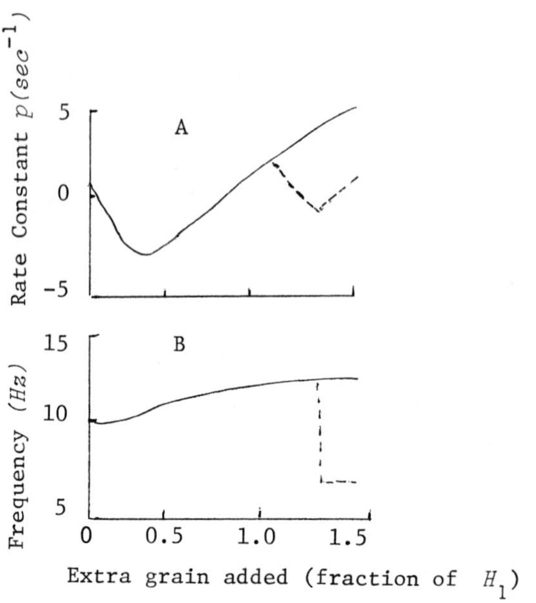

Figure 10. The effect of adding extra gain in the 85 msec pathway (solid lines) or in both the 55 and 85 msec pathways (interrupted lines). The extra gain added has been expressed as a fraction of the pre-existing gain in the 30 msec pathway ($H_1 = 765\ m^{-1}$). This gain was sufficient to produce slowly growing oscillations with a frequency of 10 Hz and a rate constant $p = 0.93$ sec^{-1}, but the oscillations could be converted into decaying oscillations ($p < 0$) by the addition of suitable amounts of extra gain in one or both pathways. From Oguztoreli and Stein (1976b).

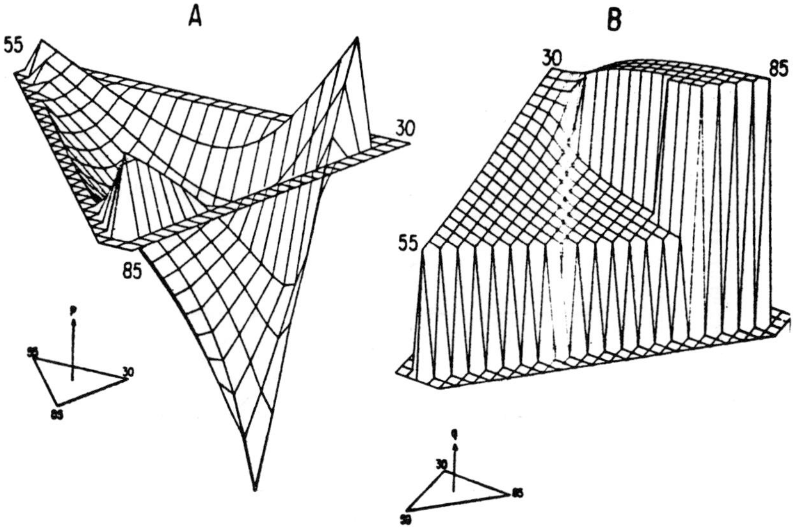

Figure 11. The effect of mixing a constant total gain $(H_1 + H_2 + H_3)$ between three pathways (with delays $t_1 = 30$ *msec*, $t_2 = 55$ *msec* and $t_3 = 85$ *msec*) on the damping (p) and frequency (q) of the most prominent reflex oscillations. A portion of the planes where *A*) $p = 0$ and *B*) $q = 0$ are shown for reference surrounding the figures. The accompanying figures on the left are intended for the purposes of orientation. The orientations of *A* and *B* are not the same. The percentage of the total gain in each pathway is indicated by the closeness of a point to each corner of the figure. From Oguztoreli and Stein (1976b).

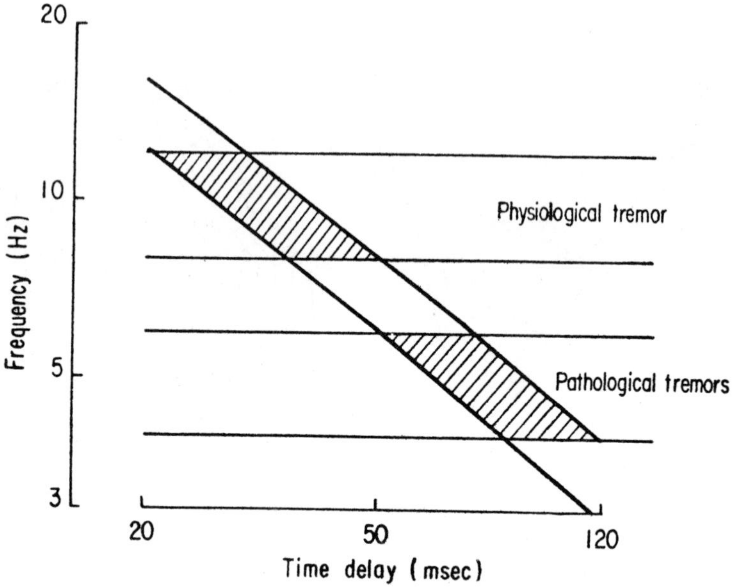

Figure 12. Frequency of reflex oscillations for different delays in the reflex pathway. The frequencies were calculated for two values of rate constants corresponding to a fast muscle (plantaris, upper diagonal line) and a slow muscle (soleus, lower diagonal line). Note that for delays (from sensory stimulus to the onset of contraction) in the range corresponding to spinal reflexes (20-50 msec) the frequencies are in the range of physiological tremor (8-12 HZ), while for the longer latencies of supraspinal reflexes (50-120 msec) the frequencies are in the range found in Parkinson's Disease and cerebellar disorders (4-6 Hz). Double logarithmic coordinates. From Stein and Oğuztöreli (1976a).

shortening of the muscle at time t whose Laplace transform is $X_1(s)$ defined by Eq. (4.1). Clearly $x_1(t)$ satisfies the differential equation

$$x_1^{IV}(T) + c_3\dddot{x}_1(t) + c_2\ddot{x}_1(t) + c_1\dot{x}_1(t) + c_0x_1(t) = 0 \qquad (4.4)$$

and is subjected to the initial conditions

$$x_1(0) = 0,\ \dot{x}_1(0) = -\frac{C}{MB}\xi_1,\ \ddot{x}_1(t) = -\frac{C}{MB}\xi_2,\ \dddot{x}_1(0) = -\frac{C}{MB}\xi_3 \qquad (4.5)$$

where

$$\xi_1 = M,\ \xi_2 = D - c_3M,\ \xi_3 = k_i + k_e + M(c_3^2 - c_2) - c_3D \qquad (4.6)$$

(cf. Stein and Oğuztöreli, 1976). It can be shown without any difficulty that all solutions of the nonlinear integral difference differential equation are bounded for $t \geq 0$. The solution of Eq. (4.3) which satisfies the initial conditions (2.9)-(2.11) can be constructed by successive continuations.

A rather comprehensive linear analysis together with computer simulations have been carried in (Stein and Oğuztöreli, 1976). Several questions pertaining to the stability of the model, periodic solutions, etc. in the nonlinear regime are still not answered.

We only note that the experimental values of the constants τ_{1k} and τ_{2k} are $\tau_{1k} \simeq 0.1s$ and $\tau_{2k} \simeq 0.023s$ (Poppele and Bowman (1970)).

Before closing this section we note that Eq. (3.4) reduces to the following difference differential equation of the neutral type in the linear regime.

$$\ell^{IV}(t) + c_3\dddot{\ell}(t) + c_2\ddot{\ell}(t) + c_1\dot{\ell}(t) + c_0\ell(t) + \lambda \sum_{k=1}^{3} H_k\{\sum_{j=0}^{4} P_{j,k}\,{}^{(j)}(t - t_k)\} = 0 \qquad (3.12)$$

where

$$\begin{aligned} P_{0,k} &= \xi_3 + c_3\xi_2 + c_2\xi_1 , \\ P_{1,k} &= (\tau_{1k} + \tau_{2k})P_{0,k} + (\xi_2 + c_3\xi_1), \\ P_{2,k} &= \tau_{1k}\tau_{2k} P_{0,k} + (\tau_{1k} + \tau_{2k})(\xi_2 + c_3\xi_1) + \xi_1, \\ P_{3,k} &= \tau_{1k}\tau_{2k}(\xi_2 + c_3\xi_1) + (\tau_{1k} + \tau_{2k})\xi_1, \\ P_{4,k} &= \tau_{1k}\tau_{2k}\xi_1, \end{aligned} \tag{3.13}$$

and the characteristic equation associated with Eq. (3.12) is

$$r^4 + c_3 r^3 + c_2 r^2 + c_1 r + c_0 + \lambda \sum_{k=1}^{3} H_k \left(\sum_{j=0}^{4} P_{j,k}\, r^j \right) e^{-t_k r} = 0. \tag{3.14}$$

5. *The Response of the Neuro-Muscular System to Centrally Generated Oscillations*

One of the most interesting problems in the neuro-muscular systems is the study of centrally generated oscillations. For the physiological evidences we refer to the work of Lamarre and Weiss (1973). Further references are also listed in the bibliography.

Consider the neuro-muscular model presented in section 3. Let

$$I(t) = \mu \sin (\Omega t + \phi) \tag{5.1}$$

be a harmonic oscillation generated in the central nervous system. The neuro-muscular model with the input

$$i(t) = \delta(t) + I(t) \tag{5.2}$$

is described by the integral difference differential equation

$$\ell(t) = x(t) + \int_0^t [I(u) + F\{h(u)\}]x(t - u)du \tag{5.3}$$

where $x(t)$ is defined by Eqs. (2.4)-(2.5), $F\{H\}$ is defined by Eq. (2.3), $h(t)$ is defined by Eq. (3.1), and $\ell(t)$ is the length of the muscle at time t, as before.

The functional integral equation (5.3) is equivalent to the following inhomogeneous difference differential equation:

$$\ell^{IV}(t) + c_3\dddot{\ell}(t) + c_2\ddot{\ell}(t) + c_1\dot{\ell}(t) + c_0\ell(t)$$

$$+ \lambda F\{\sum_{k=1}^{3} H_k[\gamma_k\ell(t-t_k) + \dot{\ell}(t-t_k)]\} = -X_i \quad sin(\Omega t + \phi) \quad (5.4)$$

where

$$\lambda = \frac{C}{MB}, \quad X_i = \frac{C\, k_i}{MB} \quad . \quad (5.5)$$

For the neuro-muscular model presented in section 4 we have the following functional integral equation

$$\ell(t) = x_1(t) + \int_0^t [I(u) + F\{h(u)\}]x_1(t - u)du \quad (5.6)$$

where $x_1(t)$ is defined by Eqs. (4.4)-(4.6). The corresponding inhomogeneous differential difference equation is complicated in the nonlinear regime, because of the fact that $\xi_1 \neq 0$ and $\xi_2 \neq 0$ in this case, and, in the linear regime we have the following inhomogeneous linear differential difference equation of the neutral type:

$$\ell^{IV}(t) + c_3\dddot{\ell}(t) + c_2\ddot{\ell}(t) + c_1\dot{\ell}(t) + c_0\ell(t)$$

$$+ \lambda \sum_{k=1}^{3} H_k\{\sum_{j=0}^{4} P_{i,k}\ell^{(j)}(t - t_k)\} = J(t) \quad (5.7)$$

where λ is as above, and

$$J(t) \equiv (\xi_3 + c_3\xi_2 + c_2\xi_1)I(t) + (\xi_2 + c_3\xi_1)\dot{I}(t) + \xi_1\ddot{I}(t)$$

$$= \mu\{[\xi_3 + c_3\xi_2 + (c_2 - \Omega^2)\xi_1]\}sin(\Omega t + \phi) \qquad (5.8)$$

$$+ \Omega(\xi_2 - c_3\xi_1)\ cos(\Omega t + \phi)$$

and ξ_1, ξ_2, ξ_3 and $P_{j,k}$'s are defined by Eqs. (4.6) and (4.8), respectively.

The dependence of the solutions of Eqs. (5.3) and (5.6) on Ω, ϕ and μ are of particular importance. We studied these problems mathematically and computationally. Comparison of our findings with experimental results is not yet complete, but there appears to be in good agreement. We omit the details here.

6. *Final Remarks*

Muscles are attached on opposite sides of a joint where two or more bones come together to form a nearly frictionless pivot. At least two muscles on opposite sides of a joint are required to provide a full range of movement in both directions, since individual muscles can pull but not push. At some joints several pairs of antagonistic msucles are attached to produce rotation, flexion and extension. At present we are working experimentally and mathematically on a system of a pair of antagonistic muscles. The describing equations are higher order and mathematical analysis is more complex, as expected. For the physiological background of these kind neuro-muscular systems and neural connections we refer to the books of V. B. Mountcastle: "Medical Physiology", and G.M. Shepherd: "The Synaptic Organization of the Brain", Oxford University Press (1974).

Further, we note that the effects of the γ-motor neurons (Figure 4) on the oscillations of the neuro-muscular system has been analysed in the linear regime. The results will soon appear elsewhere.

Finally, consider the expansion

$$th\ h = h - \frac{1}{3}h^3 + \frac{2}{15}h^5 - \frac{17}{315}h^7 + \dots + \frac{2^{2n}(2^{2n}-1)B_{2n}}{2n!}h^{2n+1} + \dots \tag{6.1}$$

which is valid for $|h| < \frac{\pi}{2}$, where B_n is the n-th Bernoulli number. Then, putting

$$\ell(t) = \sqrt{\varepsilon}\,u(t), \quad \lambda = \frac{KH_1X_1}{4}, \quad \nu = \frac{KH_1^3X_1}{48} \tag{6.2}$$

and omitting higher powers of ε, the difference differential equation (2.8) reduces to the following equation

$$u^{IV}(t) + c_3\dddot{u}(t) + c_2\ddot{u}(t) + c_1\dot{u}(t) + c_0u(t) + \lambda[\gamma u(t - t_0) + \dot{u}(t - t_0)] = \varepsilon\nu[\gamma u(t - t_0) + \dot{u}(t - t_0)]^3, \tag{6.3}$$

and, in the same approximation, centrally generated oscillations satisfy the equation

$$u^{IV}(t) + c_3\dddot{u}(t) + c_2\ddot{u}(t) + c_1\dot{u}(t) + c_0u(t) + \lambda[\gamma u(t - t_0) + \dot{u}(t - t_0)] = \mu\ sin(\Omega t + \phi) + \varepsilon\nu[\gamma u(t - t_0) + \dot{u}(t - t_0)]^3 . \tag{6.4}$$

Similar equations can be written for the neuro-muscular model described in section 3 involving multiple pathways. The situation is more complicated for the model considered in section 4.

A thorough nonlinear analysis of Eqs. (6.3) and (6.4) are of fundamental importance in the study of stability of physiological and pathological tremors. A partial analysis has been carried by Oguztoreli and Stein which will appear elsewhere.

REFERENCES

[1] Angel, A., and Lemon, R. B., *Sensorimotor cortical representation in the rat and the role of the cortex in the production of sensory myoclonic jerks*, J. Physiol., London 248(1975), 465-488.

[2] Armstrong, C. F., Huxley, A. F., and Julian, F. J., *Oscillatory responses in frog skeletal muscle fibres*, J. Physiol. London 186(1966), 26-27P.

[3] Bawa, P., Mannard, A., and Stein, R. B., *Effects of elastic loads on the contractions of cat muscles*, Biol. Cybern. 22(1976a), 129-137.

[4] Bawa, P., Mannard, A., and Stein, R. B., *Predictions and Experimental tests of a visco-elastic muscle model using elastic and inertial loads*, Biol. Cybern. 22(1976b), 139-145.

[5] Bawa, P., and Stein, R. B., *The frequency response of human soleus muscle*, J. Neurophysiol., (in press).

[6] Bellman, R., and Cooke, K. L., *Differential-Difference Equations*, New York-London: Academic Press, 1963.

[7] Bronk, D. W., and Ferguson, L.K., *The nervous control of intercostal respiration*, Am. J. Physiol. 110(1935), 700-707.

[8] Cody, F. W., Harrison, L.M., and Taylor, A., *Analysis of Activity of muscle spindles of the jaw-closing muscles during normal movements in the cat*, J. Physiol., London 253(1975), 565-582.

[9] Dietz, V., Freund, H. J., and Allum, J. H. J., *Physiological tremor and its relationship to motor unit activity in isometric voluntary contractions*.

[10] Elble, R. J., and Randall, J. E., *Motor-unit activity responsible for the 8 to 12 Hz component of human physiological finger tremor*, J. Neuro-phyiool. 39(1976), 370-383.

[11] El'sgol'ts, L. E., and Norkin, S. B., *Introduction to the Theory and Application of Differential Equations with Deviating Arguments*, Academic Press, 1973.

[12] Goodwin, G. M., Julliger, M., and Matthews, P. B. C., *The Effects of fusimotor stimulation during small amplitude stretching on the frequency-response of the primary ending of the mammalian muscle spindle*, J. Physiol., London 253 (1975), 175-206.

[13] Goodwin, G. M., and Luschei, E. S., *Discharge of spindle afferents from jaw-closing muscles during chewing in alert monkeys*, J. Neurophysiol. 38(1975), 560-571.

[14] Granit, R., *Receptors and Sensory Perception*, Yale University Press, New Haven, 1955.

[15] Granit, R., Kellerth, J. O., and Szumski, A. J., *Intracellular autogenetic effects of muscular contraction on extensor motoneurons; the silent period*, J. Physiol., London 182 (1966), 484-503.

[16] Grillner, S., and Udo, M., *Motor unit activity and stiffness of the contracting muscle fibres in the tonic stretch reflex*, Acta physiol. Scand. 81(1971), 422-424.

[17] Gurfinkel, V. S., and Osovets, S. M., *Mechanism of generation of oscillations in the tremor form of Parkinsonism*, Biophysics 18(1973), 781-790.

[18] Hagbarth, K. E., Wallin, G., Lofstedt, L., and Aquilonius, S. M., *Muscle spindle activity in alternating tremor of Parkinsonism and in clonus*, J. Neurol. Neurosurg. Psychiat. 38(1975), 636-641.

[19] Hill, A. V., *The heat of shortening and the dynamic constants of muscle*, Proc. Roy. Soc. (London) B 126(1938), 136-195.

[20] Houk, J. C., *The phylogeny of muscular control configurations*; in Drischel and Dettmar *Biocybernetics IV*, pp. 125-155 (VEB Gustav Fischer Verlag, Jena 1972).

[21] Joffroy, A. J., and Lamarre, Y., *Rhythmic unit firing in the precentral cortex in relation with postural tremor in a deafferented limb*, Brain Res. 27(1971), 386-389.

[22] Joyce, G. C., and Rack, P. M. H., *The effects of load and force on tremor at the normal human elbow joint*, J. Physiol., London 240(1974), 375-396.

[23] Julian, D. J., Sollins, K. R., Sollins, M. R., *A model for the transient and steady-state mechanical behaviour of contracting muscle*, Biophys. J. 14(1974), 546-561.

[24] Kernell, D., *The limits of firing frequency in cat lumbosacral motoneurones possessing different time course of afterhyperpolarization*, Acta physiol. Scand. 65(1965), 87-100.

[25] Lakshmikantham, V., Ladas, G. E., *Differential Equations in Abstract Spaces*, Academic Press (1972).

[26] Lakshmikantham, V., and Leela, S., *Differential and Integral Inequalities, Vol. 1,2.*, Academic Press (1969).

[27] Lamarre, Y., *Tremorgenic mechanisms in primates*; in Meldrum and Marsden *Advances in Neurology, Vol. 10, pp. 23-34* (Raven Press, New York, 1975).

[28] Lamarre, Y., and Weiss, M., *Harmaline-induced rhythmic activity of alpha and gamma motoneurons in the cat*, Brain Res. 63(1973), 430-434.

[29] Levin, B. Ja., *Distributions of Zeros of Entire Functions*, Providence, R. I., Amer. Math. Soc., 1964.

[30] Lippold, O. C. J., *Oscillation in the stretch reflex arc and the origin of the rhythmical 8-12 c/s component of physiological tremor*, J. Physiol., London 206(1970), 359-382.

[31] Llinas, R., and Volkind, R. A., *The olivo-cerebellar system: Functional properties as revealed by harmaline-induced tremor*, Exp. Brain Res. 18(1973), 69-87.

[32] Mannard, A., and Stein, R. B., *Determination of the frequency response of isometric soleus muscle in the cat using random nerve stimulation*, J. Physiol., London 229(1973), 275-296.

[33] Marsden, C. D., Merton, P. A., and Morton, H. B., *Servo Action in human voluntary movement*, Nature, London 238(1972), 140-143.

[34] Matthews, P. B. C., *Muscle spindles and their motor control*, Physiol. Rev. 44(1964), 219-288.

[35] Matthews, P. B. C., *Mammalian Muscle Receptors and Their Central Actions*, Arnold, London (1972).

[36] Matthews, P. B. C., and Stein, R. B., *The sensitivity of muscle spindle afferents to small sinusoidal changes in length*, J. Physiol., London 200(1969), 723-743.

[37] Mervill Jones, G., and Watt, D. G. D., *Observations on the control of stepping and hopping movements in man*, J. Physiol., London 219(1971), 709-727.

[38] Merton, P. A., *The silent period in a muscle of the human hand*, J. Physiol., London 114(1951), 183-198.

[39] Milner-Brown, H. A., Stein, R. B., and Yemm, R., *The contractile properties of human motor units during voluntary isometric contractions*, J. Physiol., London 228(1973), 285-306.

[40] Milner-Brown, H. A., Stein, R. B., and Yemm, R., *Synchronization of human motor units: possible roles of exercise and supra-spinal reflexes*, Electroenceph. clin. Neurophysiol. 28(1975), 245-254.

[41] Mountcastle, V. B., *Medical Physiology*, 12th ed., Mosby, 1974.

[42] Murphy, J. T., Wong, Y. C., and Wean, H. C., *Distributed feedback systems for muscle control*, Brain Res. 71(1974), 495-505.

[43] Myshkis, A. D.: *Lineare Differentialgleichungen mit nacheilendem Argument*, Berlin, Deutscher Verlag der Wiss, 1955.

[44] Oğuztöreli, M. N., *Time-Lag Control Systems*, New York-London, Academic Press, 1966.

[45] Oğuztöreli, M. N., and Stein, R. B., *An analysis of oscillations in neuro-muscular systems*, J. Math. Biol. 2(1975), 87-105.

[46] Oğuztöreli, M. N., and Stein, R. B., *The effects of multiple reflex pathways on the oscillations in neuro-muscular systems*, J. Math. Biol. 3(1976), 87-101.

[47] Padsha, S. M., Stein, R. B., *The bases of tremor during a maintained posture*, in *Control of Posture and Locomotion* (Stein, R. B., Pearson, K. G., Smith, R. S., Redford, J.B., eds.), pp. 415-419, New York, Plenum Press, 1973.

[48] Pinney, E., *Differential-Difference Equations*, Berkeley, Los Angeles, University of California Press, 1958.

[49] Podolsky, R. J., Nolan, A. C., Zaveler, S. A., *Cross-bridge properties derived from muscle isotonic velocity transients* Proc. Nat. Acad. Sci. U.S.A. 64(1969), 504-511.

[50] Pontryagin, L. S., *On the zeros of some elementary transcendental functions*, Izv. Akad. Nauk SSSR, Ser. Mat. 6(1942), 115-134.

[51] Poppele, R. E., and Bowman, R. J., *Quantitative description of linear behavior of mammalian muscle spindles*, J. Neurophysiol. 33(1970), 59-72.

[52] Poppele, R. E., and Kennedy, W. R., *Comparison between behavior of human and cat muscle spindles recorded in vitro*, Brain Res. 75(1974), 316-319.

[53] Pringle, J. W. S., *The contractil mechanism of insect fibrillar muscle*, Prog. Biophys. 17(1967), 1-60.

[54] Prochazka, A., Westerman, R. A., and Ziccone, S. P., *Discharge trains of single muscle spindle afferents in the conscious cat*, Third International Symposium on Motor Control, Varna, Bulgaria (1976).

[55] Rietz, R. R., and Stiles, J. N., *A viscoelastic-mass mechanism as a basis for normal postural tremor*, J. Appl. Physiol. 37(1974), 852-860.

[56] Rosenthal, N. P., McKean, I. A., Roberts, W. J., Terzuolo, C. A., *Frequency analysis of stretch reflex and its main subsystems in triceps surac muscle of the cat*, J. Neurophysiol. 33(1970), 713-749.

[57] Shepherd, G. M., *The Synaptic Organization of the Brain*, Oxford Univeristy Press, 1974.

[58] Stein, R. B., *The peripheral control of movement*, Physiol. Rev. 54(1974), 215-243.

[59] Stein, R. B. Oguztoreli, M. N., *Tremor and other oscillations in neuromuscular systems*, Biol. Cybernetics 22(1976), 147-157.

[60] Stein, R. B., Oguztoreli, M. N., *Does the velocity sensitivity of muscle spindles stabilize the stretch reflex?* Biol. Cybernetics 23(1976), 219-228.

APPLICATIONS OF THE SATURABILITY TECHNIQUE IN THE PROBLEM OF STABILITY OF NONLINEAR SYSTEMS*

V. M. Popov
University of Florida

In this paper, one studies the nonlinear Volterra equation

$$\text{(V)} \qquad \sigma(t) + \int_0^t g(t-\tau)\phi(\sigma(\tau))d\tau = f(t), \quad t \geqq 0$$

where $f : R_+ \to R$, $g : R_+ \to R$ and $\phi : R \to R$ are given. An important special case is given by the nonlinear system of ordinary differential equations

$$\text{(D)} \qquad \begin{aligned} \dot{x} &= A\,x - b\,\phi(\sigma), \quad x(0) = x_0, \quad t \geqq 0 \\ \sigma &= c^T x \,. \end{aligned}$$

One assumes that ϕ is nondecreasing, satisfies a global Lipschitz condition and $\phi(0) = 0$. The assumptions about g and f are very relaxed (they are automatically satisfied if equation (V) describes a system of the form (D), in which A is a Hurwitz matrix). The problem is to find a number $k > 0$ and conditions on g such that the solutions of (V) should be bounded and should have the property $\lim_{t\to\infty} \sigma(t) = 0$, for every function ϕ - as above - with the additional property: $\sup_{\sigma\neq 0} \phi(\sigma)/\sigma < k$. The result is expressed in terms of the Fourier transform $\hat{g}$ of the function g. One proves that, if $\hat{g}(0) < 0$, one can take $k = -1/\hat{g}(0)$, provided the following condition is satisfied: there exists $q > 0$ such that - if one denotes by $H(\omega)$ the expression $H(\omega) = (1 + \frac{q}{i\omega})\hat{g}(\omega)$ - one has the inequalities: 1^0 $Re\, H(\omega) > 0$, for every real $\omega \neq 0$, 2^0 $\lim_{\omega\to 0} Re\, H(\omega) > 0$ and $3^0 \lim_{\omega\to\infty} Re(\omega^2 H(\omega)) > 0$. The principal condition, 1^0, gives the counterpart (valid for nondecreasing functions ϕ) of a well-known frequency-domain criterion of absolute stability.

* Research supported in part by N. S. F. under Grant MCS 74-08184 A02.

In the last part of the paper, in order to treat a case in which the new criterion is obviously better than any criterion of absolute stability, one applies the above result in the case of the system (of the form (D)):

$$\dot{x} = -(1 + a)x - \phi(\sigma), \quad a > 0$$

$$\dot{y} = -z$$

$$\dot{z} = y - \phi(\sigma)$$

$$\sigma = x + y + z$$

which is well studied, since it represents one of the counterexamples of V. A. Pliss to the conjecture of M. A. Aizerman. In this particular case, the results of the paper imply that the asymptotic stability in the large of the trivial solution is secured for every nondecreasing ϕ whose graph lies in the Hurwitz sector - which is the best one could expect.

NONLINEAR EVOLUTION EQUATIONS AND NONLINEAR ERGODIC THEOREMS

Simeon Reich
The University of Chicago

In this paper we study certain aspects of the asymptotic behavior of generalized solutions of nonlinear evolution equations and of nonlinear nonexpansive semigroups in Banach spaces.

Let E be a real Banach space, $A \subset E \times E$ an accretive set, and $f \in L^1_{loc}(0,\infty;E)$. Suppose that $u : [0,\infty) \to E$ is the limit solution of the following quasi-autonomous Cauchy problem:

$$\begin{cases} u'(t) + Au(t) \ni f(t) & 0 < t < \infty \\ u(0) = x_0. \end{cases}$$

Motivated by a result of M. G. Crandall mentioned in [2, p. 166] we show that under certain conditions the weak $\lim_{t\to\infty} u(t)/t$ exists.

Now let $g : [0,\infty) \to [0,\infty)$ be a nonincreasing function of class C^1 such that $\lim_{t\to\infty} g(t) = 0$ and $\int_0^\infty g(t)dt = \infty$. Let x belong to $cl(D(A))$, and let u be the limit solution of the following initial value problem:

$$\begin{cases} u'(t) + Au(t) + g(t)u(t) \ni g(t)x & 0 < t < \infty \\ u(0) = x_0. \end{cases}$$

Define $\hat{A} = \{[z,w] \in cl(D(A)) \times E^{**}$: For each $[x,y] \in A$ there is $j \in J(x - z)$ such that $(y - w,j) \geq 0\}$. We show that if $x_0 \in D(\hat{A})$ and $0 \in R(A)$, then $\lim_{t\to\infty} ||\widehat{A}u(t)|| = 0$. Moreover, when certain other conditions are met, u is a strong solution and the strong $\lim_{t\to\infty} u(t) = Qx$ where Q is the sunny nonexpansive retraction of $cl(D(A))$ onto $A^{-1}(0)$. These results are inspired by [4, Theorems 10.11 and 10.12]. They can be used to derive iterative procedures for the construction of zeros of accretive sets.

The first ergodic theorems for nonlinear nonexpansive mappings and semigroups in Hilbert space were established by Baillon and by Baillon and Brézis [1]. Brézis and Browder [3] have recently simplified the original arguments in the discrete case and replaced the Cesàro method by more general summability methods. We obtain a similar extension in the continuous case.

For each positive s let $K(s,t) : [0,\infty) \to [0,\infty)$ be of bounded variation in $[0,\infty)$. Denote its total variation by $V(s)$. Suppose that the kernel K satisfies the following conditions: $\int_0^\infty K(s,t)dt = \lim_{T\to\infty} \int_0^T K(s,t)dt = 1$ for all s, $\lim_{s\to\infty} \int_0^T K(s,t)dt = 0$ for all finite T, and $\lim_{s\to\infty} V(s) = 0$. Let C be a bounded closed convex subset of a Hilbert space, $S : [0,\infty) \times C \to C$ a nonexpansive semigroup on C, $P : C \to F$ the nearest point map onto the nonempty fixed point set of S, and x a point in C. If $R(s,x) = \int_0^\infty K(s,t)S(t,x)dt$, then the weak $\lim_{s\to\infty} R(s,x)$ exists, and equals the strong $\lim_{t\to\infty} PS(t,x)$. We also point out those parts of the proof that are valid outside Hilbert space.

REFERENCES

[1] J. B. Baillon, and H. Brézis, *Une remarque sur le comportment asynptotique des semigroupes non linéaires*, to appear.

[2] H. Brézis, *Opérateurs Maximaux Monotones et Semigroupes de Contractions dans les Espaces de Hilbert*, North Holland, Amsterdam, 1973.

[3] H. Brézis, and F. E. Browder, *Nonlinear ergodic theorems*, to appear.

[4] F. E. Browder, *Nonlinear Operators and Nonlinear Equations of Evolution in Banach Spaces*, AMS, Providence, R.I., 1976.

Partially supported by NSF Grant MCS74 - 07495.

ON PURE STRUCTURE OF DYNAMIC SYSTEMS

D. D. Siljak
University of Santa Clara

Numerous natural, social, and technological systems which represent a dynamic interaction among a number of elements or subsystems, quite commonly either by design or fault, do not stay "in one piece" during operation. They are often subject to structural perturbations whereby groups of subsystems are disconnected from and again connected to each other in an unpredictable way. The effects of such perturbations on stability of various models in ecology, economics, and engineering were explored in a number of studies [1] where it was shown that a class of stable competitive systems is invulnerable to perturbations in the interconnection structure.

A general framework for studying the pure structural properties of dynamic systems was proposed in reference [2], where the notion of input and output reachability were introduced in terms of directed graphs (digraphs). A system is considered input reachable if each state can reach at least one output. Once a digraph is associated with a dynamic system and input (output) reachability is defined in digraph theoretic terms, we formulate the problem of vulnerability as follows: Does a removal of a line or a point from the digraph destroy input (output) reachability? In providing an answer to this question, it will be shown that input (output) decentralized systems are a structural counterpart to stable competitive systems in that they are invulnerable under the structural perturbations.

The plan of the paper is the following:

First, we fix the relationship between dynamic systems, directed graphs, and interconnection matrices. Inputs, states, and outputs of a system are represented by input, state, and output points of the corresponding digraph. The lines of the digraph and the entries of the related interconnection matrix are determined in the usual way by occurence of inputs, states, and outputs in the system equations.

Second, we introduce the notion of input and output truncated digraphs, and use the reachability and antecedent sets to define input and output reachability of dynamic systems. Then, using the path matrix we develop an algebraic criterion for input and output reachability which can be applied to dynamic systems using effective algorithms already available in digraph studies of computer systems [3].

Third, we introduce partitions and condensations of digraphs as related to dynamic systems, which provide an appropriate setting for analyzing the important reachability properties of large-scale systems composed of interconnected subsystems. In this context, we will formulate a pure cannonical structure of nonlinear systems with respect to input and output reachability.

Fourth, we formulate the notion of structural perturbations of composite dynamic systems as line removals and point removals from the corresponding condensations. Then, we define the notion of vulnerability of large-scale systems as loss of reachability due to the perturbations, and explore conditions under which dynamic systems are invulnerable. It will be shown that the class of input-output decentralized systems are invulnerable to arbitrary structural perturbations.

REFERENCES

[1] Siljak, D. D., *Connective Stability of Competitive Equilibrium*, Automatica 11(1975), 389-400.

[2] Siljak, D. D., *On Input-Output Reachability of Dynamic Systems*, International Journal of Systems Science, (To appear).

[3] Bowie, W. S., *Applications of Graph Theory in Computer Systems*, International Journal of Computer and Information Sciences 5(1976), 9-31.

OPTIMIZING AND EXTREMIZING NONLINEAR BOUNDARY VALUE PROBLEMS IN LENTICULAR ANTENNAS IN OCEANOGRAPHY, MEDICINE & COMMUNICATIONS: SOME SOLUTIONS AND SOME QUESTIONS

by

Robert L. Sternberg
Office of Naval Research
Boston, Massachusetts

1. *Introduction: The Scanning Problem at Low F-Numbers*

In the design of singlet microwave dielectric lenses in the electromagnetic spectrum and acoustic lenses for underwater sound applications in which wide angle scanning, or fixed off-axis operation, is required, it is frequently desired to achieve greater off-axis scan in one plane than in the other, wide angle azimuth performance usually being more necessary than elevation scan. Because of requirements for weight minimization and compact packaging for aircraft and submersible vehicle installations where both system volume and weight are at a premium, it is further generally desirable to extremize the lens in the sense of maximizing the lens aperture or minimizing the lens volume and sometimes simultaneously to achieve a minimum, or at any rate a near minimum, F-number.

On the other hand in both the microwave and acoustic applications for the most part -- excluding only those areas of medical acoustic radiography and related fields in which frequencies higher than say 10 megahertz are commonly used -- the wavelengths encountered are of the order of a millimeter, a centimeter, or a decimeter so that there is no need to restrict the lens design to spherical surfaces grindable by self-correcting motions as in optical lens manufacture, but rather, nonspherical lenses of quite complex shape may readily be used. With this relaxation of the design requirements, consideration in radar, and more recently in acoustics, was first given to nonspherical but rotationally invariant lenses, i.e., nonspherical lenses whose surfaces are surfaces of revolution about the lens axis such as the aplanatic, generalized aplanatic and bifocal or Doppler lenses of L. C. Martin [1], F. G. Friedlander [2], and R. L. Sternberg [3,4,5]. Such lenses, however, like any ordinary singlet lens, spherical or nonspherical, invariably suffer from astigmatism at off-axis points with the magnitude of that aberration increasing with increasing scan angle and decreasing F-number.

In order to circumvent this astigmatic defect of all single element rotationally invariant lenses at off-axis angles for the purpose of achieving increased scan in the azimuth plane at the anticipated sacrifice of a reduction in elevation scanning, we have been led to consider the possibility of designing a geometrically perfect nonspherical bifocal or Doppler lens in three-space, symmetric with respect to two orthogonal planes through the lens axis, one of which contains its design foci, and complete with minimal lens volume and a very low, or in some instances an even minimal, F-number. Such a nonspherical lens, if geometrically realizable, would then have axial symmetry, and also plane symmetry in both x and y, but no surfaces of revolution and would have balanced astigmatism; i.e., negative astigmatism on axis, zero astigmatism at its off-axis design points, and positive astigmatism at still further off-axis points and, consequently, would offer the opportunity for an optimized trade off between azimuth and elevation scanning capability by suitable optimal choice of the basic lens parameters.

Our objective in this paper is to solve the problem of the design of such geometrically perfect nonspherical low F-number extremized bifocal lenses in three-space and, subsequently, to outline computational procedures for optimization of the design for best azimuth versus elevation scanning trade off taking as our fundamental assumptions that the index of refraction n_0 is greater than unity and that Snell's law holds. The basic boundary value problem in nonmathematical form was first suggested to the writer many years ago by F. S. Holt as a result of successful related work by W. Ellis, E. Fine and G. Reynolds [6] on a simpler, microwave lens of waveguide, or constrained, type for which neither of the foregoing assumptions hold but for which the mathematical solution was vastly simplified by virtue of the waveguides forcing the rays within the lens to align themselves parallel to the lens axis. The solution we present for the present problem is an extension of Sternberg's solution [3] for the corresponding Doppler lens, ordinary differential equations problem, to three-dimensional partial differential equations.

2. *The Bifocal Boundary Value Problem*

Given the focal distance $|z_0|$, the lens diameter $2b_0$, the index of refraction of the lens material $n_0 > 1$, the off-axis design angle ψ_0 and the wavelength λ_0 of the incident electromagnetic or acoustic wave to be focused by the lens, the bifocal boundary value problem at very low or even minimal F-numbers, where $F = |z_0|/2b_0$, whose solutions, considered as functions of the lens parameters

$$(2.1) \qquad |z_0|, \quad 2b_0, \quad n_0 > 1, \quad \psi_0, \quad \lambda_0,$$

we will subsequently seek to optimize for azimuth versus elevation scanning trade off, may be formulated as follows.

With reference to Figure 1 let the equations of the lens surfaces S and S' be taken in the form

$$(2.2) \qquad S: \quad z = z(x,y), \qquad S': \quad z' = z'(x',y'),$$

and let a general ray R from the off-axis design focal point $F(x_0,y_0,z_0)$ pass through the lens surface S at the point x, y, z and through the surface S' at the point x', y', z', at each of which interfaces it is refracted in accordance with Snell's law, and let the ray emerge from the lens in the direction ψ_0, that is to say parallel to the y, z-plane and at the angle ψ_0 with respect to the z-axis. Let W denote the corresponding orthogonal wave front. In order to achieve a dual ray structure from a symmetrically placed dual focal point at $F'(x_0,-y_0,z_0)$ so that a general ray R' from the latter point emerges from the

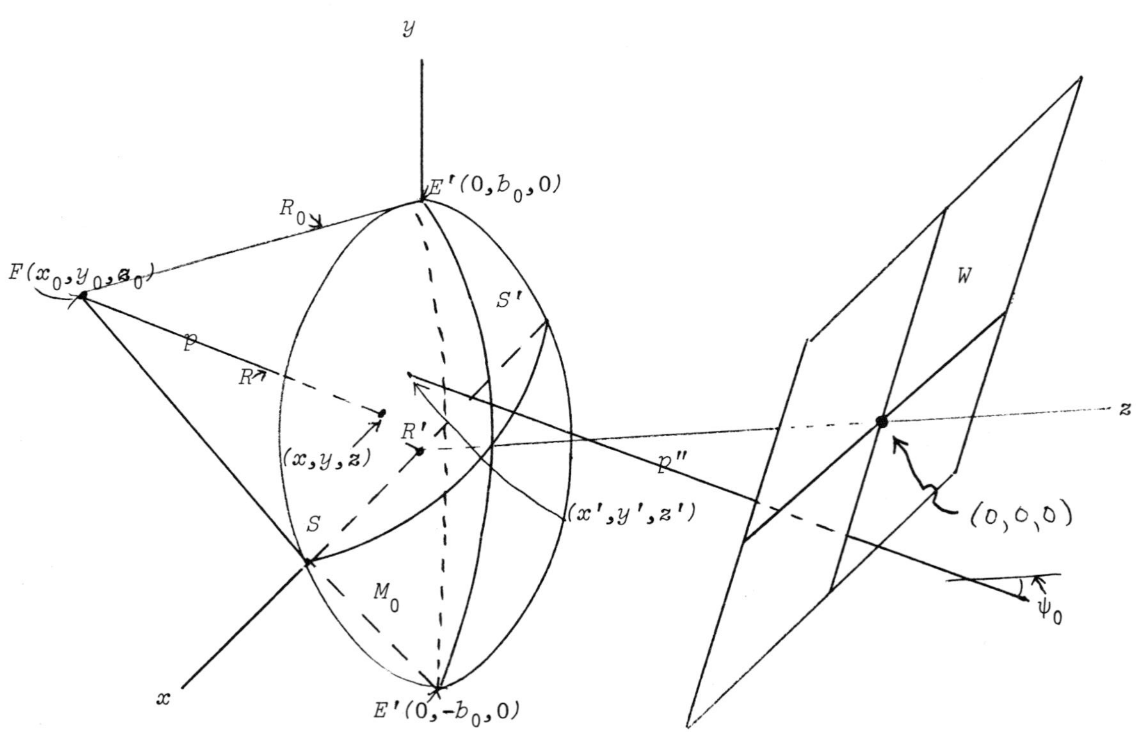

Figure 1. Bi-Focal Lens Problem.

lens in the corresponding dual direction $-\psi_0$ we require the lens surfaces S and S' to be symmetrical about the x, z-plane and, taking $x_0 = 0$, we can, without loss of generality, suppose the surfaces S and S' also to be symmetric about the y, z-plane. Finally, extremizing the lens design so as to obtain a maximal lens aperture, or what is the same thing, a minimal lens volume, by requiring that the lens surfaces S and S' intersect in a sharp boundary curve, or lens periphery, Γ, our problem then is to so determine the shape of the lens and the form of the functions $z = z(x,y)$ and $z' = z'(x',y')$ specifying the lens surfaces S and S' in (2.2) as to insure that all rays R from the focal point $F(x_0,y_0,z_0)$ incident on the lens within the lens periphery Γ are perfectly collimated by the lens into a parallel ray plane wave beam perfectly focused at infinity in the direction ψ_0 and, by symmetry, that all rays from the dual focal point $F'(x_0,-y_0,z_0)$ similarly incident on the lens within the periphery Γ are similarly perfectly focused by the lens at infinity in the dual direction $-\psi_0$.

Taking appropriate vector cross products between the incident ray R defined by appropriate angles ϕ and ψ and the normals to the lens surfaces S and S' in the illustration, and applying Snell's law at each interface, and using the Jacobian notation for appropriate two by two determinants of partial derivatives, we readily obtain for the lens surfaces S and S' the system of partial differential equations

$$\text{(2.3)}\quad \begin{aligned} S:&\quad \frac{\partial(z,y)}{\partial(\phi,\psi)} = F(\mathcal{A})\frac{\partial(x,y)}{\partial(\phi,\psi)}, \quad \frac{\partial(x,z)}{\partial(\phi,\psi)} = G(\mathcal{A})\frac{\partial(x,y)}{\partial(\phi,\psi)}, \\ S':&\quad \frac{\partial(z',y')}{\partial(\phi,\psi)} = F'(\mathcal{A})\frac{\partial(x',y')}{\partial(\phi,\psi)}, \quad \frac{\partial(x',z')}{\partial(\phi,\psi)} = G'(\mathcal{A})\frac{\partial(x',y')}{\partial(\phi,\psi)}, \end{aligned}$$

where $\mathcal{A}$ denotes the arguments $\mathcal{A} = (x,y,z,x',y',z')$ and the coefficient functions $F(\mathcal{A})$ to $G'(\mathcal{A})$ are of the forms

$$F(\mathcal{A}) = \frac{n_0(x'-x)p-xp'}{(z-z_0)p'-n_0(z'-z)p}, \quad G(\mathcal{A}) = \frac{n_0(y'-y)p-(y-y_0)p'}{(z-z_0)p'-n_0(z'-z)p},$$

(2.4)

$$F'(\mathcal{A}) = \frac{n_0(x'-x)}{p'\cos\psi_0-n_0(z'-z)}, \quad G'(\mathcal{A}) = \frac{n_0(y'-y)+p'\sin\psi_0}{p'\cos\psi_0-n_0(z'-z)},$$

and where p and p' denote the path length elements defined by the expressions

(2.5) $$p^2 = x^2 + (y - y_0)^2 + (z - z_0)^2,$$

$$p'^2 = (x' - x)^2 + (y' - y)^2 + (z' - z)^2,$$

and are thus the geometrical lengths of the segments of the ray indicated in the illustration.

In writing the partial differential equations (2.3) for the lens surfaces S and S' we have taken as independent variables the angles ϕ and ψ which the ray R makes with the z-axis and the x,z-plane at $F(x_0,y_0,z_0)$ and we have considered the unknown coordinates x, y, z and x', y', z' at which the ray R intersects the lens surfaces S and S' each to be functions of these angle variables. Thus, in this formulation in place of (2.2) we have assumed for the lens surfaces S and S' alternative parametric equations of the forms.

(2.6) $$S\colon\ z = z^*(\phi,\psi), \qquad S'\colon\ z' = z'^*(\phi,\psi),$$

or equivalently in (2.2) we suppose that $x = x^*(\phi,\psi)$, $y = y^*(\phi,\psi)$, $x' = x'^*(\phi,\psi)$ and $y' = y'^*(\phi,\psi)$ so that the functions $z = z(x,y)$ and $z' = z'(x',y')$ become functions $z = z^*(\phi,\psi)$ and $z' = z'^*(\phi,\psi)$ as in (2.6). Relating the angles ϕ and ψ to the unknown rectangular coordinates x, y, and z also are the equations

$$(2.7) \qquad \tan\phi = \frac{[x^2 + (y - y_0)^2]^{1/2}}{z - z_0}, \quad \cot\psi = \frac{[x^2 + (z - z_0)^2]^{1/2}}{y - y_0},$$

which equivalently may be taken as the definitions of the quantities ϕ and ψ.

It is to be noted that the partial differential equations (2.3) are nonlinear and, moreover, that the nonlinearity is inherent and cannot be removed by rewriting the equations in the apparently linear forms

$$(2.8) \qquad S: \frac{\partial z}{\partial x} = F(\mathcal{A}), \frac{\partial z}{\partial y} = G(\mathcal{A}), \; S': \frac{\partial z'}{\partial x'} = F'(\mathcal{A}), \frac{\partial z'}{\partial y'} = G'(\mathcal{A}),$$

which result from dividing out the Jacobians and taking x and y as the independent variables with $z = z(x,y)$ and $z' = z'^{**}(x,y)$ for this purpose. To see this we have only to note that with such a change of variables we have then also that

$$(2.9) \qquad \frac{\partial z'}{\partial x'} = \frac{\partial(z',y')}{\partial(x,y)} \Big/ \frac{\partial(x',y')}{\partial(x,y)}, \quad \frac{\partial z'}{\partial y'} = -\frac{\partial(z',x')}{\partial(x,y)} \Big/ \frac{\partial(x',y')}{\partial(x,y)},$$

since now $x' = x'^{**}(x,y)$ and $y' = y'^{**}(x,y)$ and there is no way to avoid the complications of this nonlinear relationship.

Essentially equivalent to either pair of the partial differential equations in (2.3) and capable of being substituted for one of those pairs of equations by Malus's theorem or, as may be shown by direct substitution, is the optical path length condition that

$$\begin{aligned} P &\equiv p + n_0 p' + p'' \\ &= [x^2 + (y - y_0)^2 + (z - z_0)^2]^{1/2} \\ &\quad + n_0[(x' - x)^2 + (y' - y)^2 + (z' - z)^2]^{1/2} \qquad (2.10) \\ &\quad + (a - z')\cos\psi_0 + y'\sin\psi_0 \\ &= P_0 = \text{Constant}, \end{aligned}$$

for all rays R from the focal point $F(x_0,y_0,z_0)$ to the wave front W propagating in the direction ψ_0 with instantaneous intercept a on the z-axis as illustrated and, applying this relation to the extremal rays R_0 and R_0' which pass through the lens extremities $E(0,b_0,0)$ and $E'(0,-b_0,0)$, we find for the focal point $F(x_0,y_0,z_0)$ the coordinates

$$(2.11) \qquad F: x_0 = 0,\ y_0 = [b_0^2 + z_0^2 sec^2\psi_0]^{1/2} sin\ \psi_0,\ z_0 = -|z_0|,$$

and for the constant P_0 in (2.10) the expression

$$(2.12) \qquad P_0 = [b_0^2 + z_0^2 sec^2\psi_0]^{1/2} + a\ cos\ \psi_0,$$

similarly as in Sternberg [3]. In terms of (2.11) the dual focal point is now $F'(x_0,-y_0,z_0)$.

The natural symmetry of the lens surfaces S and S' about the x, z-plane imposed by the requirement of providing for a dual ray structure from the dual focus $F'(x_0,-y_0,z_0)$ and the symmetry about the y, z-plane corresponding to the assumption that $x_0 = 0$ results in the four symmetry conditions.

$$(2.13) \qquad S: \begin{matrix} z(-x,y) = z(x,y), \\ z(x,-y) = z(x,y), \end{matrix} \qquad S': \begin{matrix} z'(-x',y') = z'(x',y'), \\ z'(x',-y') = z'(x',y'), \end{matrix}$$

as additional requirements on the functions defining the lens surfaces S and S' in (2.2) in terms of the variables x, y and x', y' which in terms of the angles ϕ and ψ as independent variables become of course much more complicated relations. The symmetry conditions will subsequently be seen to play a very strong role in the mathematics of the problem.

Applying the extremizing condition that the lens enjoy a maximal aperture or, equivalently, have a minimal volume in the sense described previously so that the lens surfaces S and S' intersect in a sharp periphery Γ passing through the lens extremities $E(0,b_0,0)$ and $E'(0,-b_0,0)$ as in the figure, we readily derive as the boundary conditions of the problem the requirement that the functions $z(x,y)$ and $z'(x',y')$ in (2.2) satisfy the relations

(2.14) $\qquad z(x,y) = z'(x,y) = 0,$

on the ellipse

(2.15) $\qquad \Gamma : (x^2/b_0^2 cos^2\psi_0) + (y^2/b_0^2) = 1.$

To derive (2.14) - (2.15) we use relations (2.10) to (2.13) all together and equate $z(x,y)$ to $z'(x,y)$.

Finally, similarly as in Sternberg [3,4,5] we have as the conditions for F-number minimization that the exiting segment of the ray R_0 from the focal point $F(x_0,y_0,z_0)$ be tangent to the lens surface S' at the lens extremity $E(0,b_0,0)$ and similarly for the corresponding ray from the dual focal point $F'(x_0,-y_0,z_0)$. Note that the lens in the illustration has an almost, but not quite, minimal F-number where $F = |z_0|/2b_0$.

Our bifocal lens boundary value problem now, mathematically speaking, is to solve the system of partial differential equations (2.3) - (2.4) subject to the conditions (2.10) to (2.15) with or without the optional additional condition of F-number minimization appended. We consider principally the nonminimal F-number version of the problem.

3. *Solution by Polynomial Approximations and Infinite Series*

A variety of analytic, approximate, and numerical approaches and solution algorithms appear viable for application to the solution of the bifocal lens boundary value problem with nearly minimal F-numbers and, with appropriate modifications, even to the problem with the F-number minimized. Three such solutions were given by the writer some time ago for the corresponding two-dimensional Doppler lens problem. Here we extend one of these, the writer's successive approximation method developed for the simpler ordinary differential equations problem to the present partial differential equations problem in three dimensions. As will be seen the method is exceedingly efficient. For background material and many details applicable to the present situation see Sternberg [3].

To the foregoing end, ellipticizing the two-dimensional successive approximation solution of the Doppler lens problem presented previously and appending appropriate Taylor series terms, we seek a solution to the bifocal boundary value problem (2.1) to (2.15) in the mixed forms

$$\text{(3.1)} \qquad S: \quad z = \lim_{N\to\infty} \sum_{n=0}^{2N+1} \beta_n(N)[(x^2/\cos^2\psi_0) + y^2]^n + \sum_{m,n=1}^{\infty} \beta_{mn} x^{2m}[(x^2/\cos^2\psi_0) + y^2 - b_0^2]^n ,$$

and

$$\text{(3.2)} \qquad S': z' = \lim_{N\to\infty} \sum_{n=0}^{2N+1} \beta_n'(N)[(x'^2/\cos^2\psi_0) + y'^2]^n + \sum_{m,n=1}^{\infty} \beta_{mn}' x'^{2m}[(x'^2/\cos^2\psi_0) + y'^2 - b_0^2]^n ,$$

assuming the F-number is nonminimal. Here the $\beta_n(N)$ and $\beta_n'(N)$ are the successive approximation coefficients obtained circa 1953 for the two-dimensional problem and the β_{mn} and β_{mn}' are correction, or completion, terms corresponding to the slight but finite deviations of the lens surfaces S and S' in the three-dimensional problem from a simple elliptic transformation of the corresponding, essentially two-dimensional, Doppler lens surfaces.

To obtain the coefficients $\beta_n(N)$ and $\beta_n'(N)$ in the polynomial approximations in (3.1) and (3.2) with the F-numbers nonminimal we begin exactly as in the two-dimensional Doppler problem by fitting $2N + 2$ rays $R_0, R_0', R_{N_1}, \ldots, R_{N_N}$ and R_{N_N}' from the focal **point** $F(x_0, y_0, z_0)$ in the y, z-plane cross section of the lens as in Figure 2 such that these rays are perfectly controlled in both direction and path length by the polynomial ap-

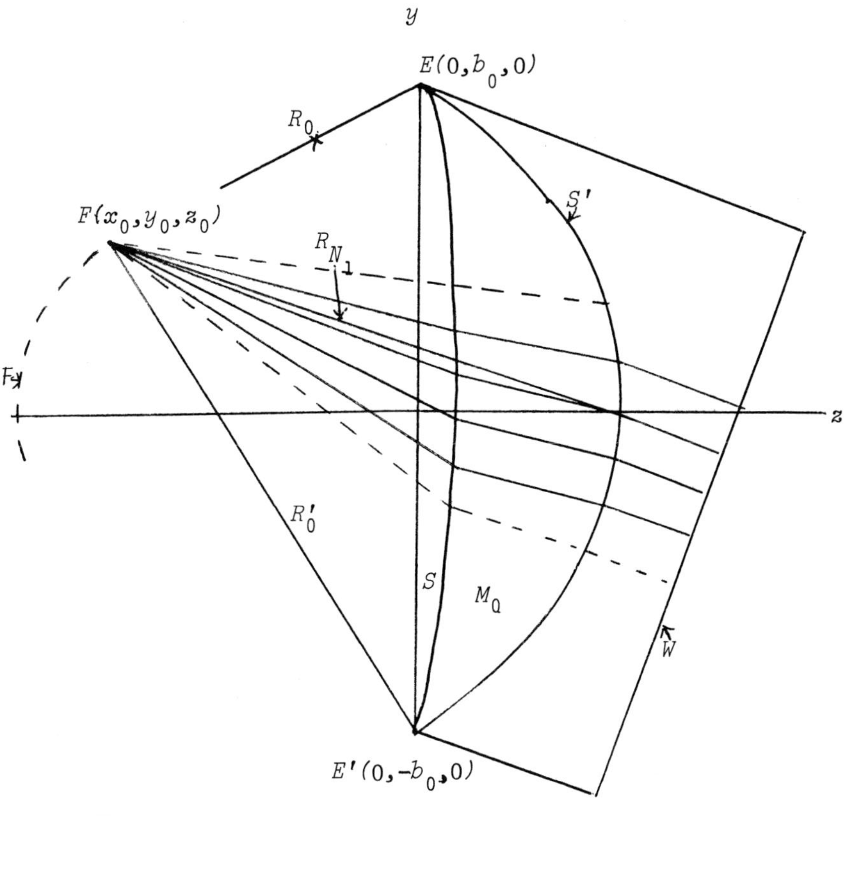

Figure 2. Rays R_0, R_0', R_{N_1}, R_{N_1}', ... R_{N_N}, R_{N_N}'

proximates to the lens surfaces S and S' of degree $4N+2$. As in the original work the technique hinges on solutions for approximate values of the eigenvalue like angles ϕ_1 and ϕ_1' in the illustration associated with the rays R_{N_1} and R'_{N_1} which pass through the lens vertices, or axial intercepts of the lens surfaces S and S', as shown. This solution in turn depends on solution of simultaneous determinantal equations of the form

$$\Delta(\phi_1,\phi_1') \equiv \begin{vmatrix} b_0^2 & b_0^4 & \cdots & b_0^{2n} & B_0 \\ 2b_0 & 4b_0^3 & \cdots & 2nb_0^{2n-1} & \tan\xi_0 \\ Y_j^2 & Y_j^4 & \cdots & Y_j^{2n} & B_0 - Z_j \\ 2Y_j & 4Y_j^3 & \cdots & 2nY_j^{2n-1} & \tan\Xi_j \end{vmatrix} = 0, \tag{3.3}$$

$$(j = 1,2,\ldots,N;\ n = 2N+1),$$

and

$$\Delta'(\phi_1,\phi_1') \equiv \begin{vmatrix} b_0^2 & b_0^4 & \cdots & b_0^{2n} & B_0' \\ 2b_0 & 4b_0^3 & \cdots & 2nb_0^{2n-1} & \tan\xi_0' \\ Y_j'^2 & Y_j'^4 & \cdots & Y_j'^{2n} & B_0' - Z_j' \\ 2Y_j' & 4Y_j'^3 & \cdots & 2nY_j'^{2n-1} & \tan\Xi_j' \end{vmatrix} = 0, \tag{3.4}$$

$$(j = 1,2,\ldots,N;\ n = 2N+1),$$

for the quantities ϕ_1 and ϕ_1'. In (3.3) and (3.4) the elements of the determinants $\Delta(\phi_1,\phi_1')$ and $\Delta'(\phi_1,\phi_1')$ and, hence the determinants themselves, depend only on ϕ_1 and ϕ_1'; i.e., each element of each determinant is either a constant or a function of ϕ_1 or ϕ_1' or both. Moreover, the functions defining the elements are simple elementary algebraic and trigonometric expressions so no great difficulty attends the solution of the system (3.3) - (3.4) for ϕ_1 and ϕ_1' for given N. We refer to Sternberg [3] for all details which appear there in full and need not be repeated.

Assuming the polynomial approximations in the solutions for

the lens surfaces S and S' in (3.1) converge to analytic limits and that the infinite series in (3.1) are suitably convergent it then follows that the coefficients β_{mn} and β'_{mn} in the latter parts of (3.1) can be obtained by a Cauchy-Kovalevsky type process, i.e., by the method of limits using known expressions for the derivatives $(d^k/dy^k)z(0,y)$ and $(d^k/dy'^k)z'(0,y')$ developed for the two-dimensional Doppler problem in Sternberg [3], together with the chain rule, to compute the quantities

$$(3.5) \qquad \frac{\partial^{i+j}}{\partial x^i \partial y^j} \lim_{N\to\infty} \sum_{n=0}^{2N+1} \beta_n(N)[(x^2/\cos^2\psi_0) + y^2]^n , \quad (i,j = 0,1,2,\ldots\infty),$$

and

$$(3.6) \qquad \frac{\partial^{i+j}}{\partial x'^i \partial y'^j} \lim_{N\to\infty} \sum_{n=0}^{2N+1} \beta'_n(N)[(x'^2/\cos^2\psi_0) + y'^2]^n , \quad (i,j = 0,1,2,\ldots\infty),$$

required in the process. For practical purposes we can sometimes even determine a few of the coefficients by trial and error using ray tracing methods to check the results. We use the latter method to obtain an approximate value of a single coefficient β'_{11} in the example worked out below.

Computational experience with the mixed forms of solution (3.1) - (3.2) indicates that the convergence of both the limit expressions and the infinite series in (3.1) and (3.2) is exceedingly rapid even when the F-numbers are nearly minimal. Thus, for example, for the case of a bifocal lens with F-number one-half at 20 degrees off-axis having as basic parameter values the numbers

$$(3.7) \qquad |z_0| = 9, \quad 2b_0 = 18, \quad n_0 = 1.594, \quad \psi_0 = 20°,$$

and λ_0 arbitrary, and taking the simplest case $N = 1$ in (3.1)-(3.2), as in Figure 3 we obtain for the coefficients $\beta_n(N)$ and $\beta'_n(N)$ the values

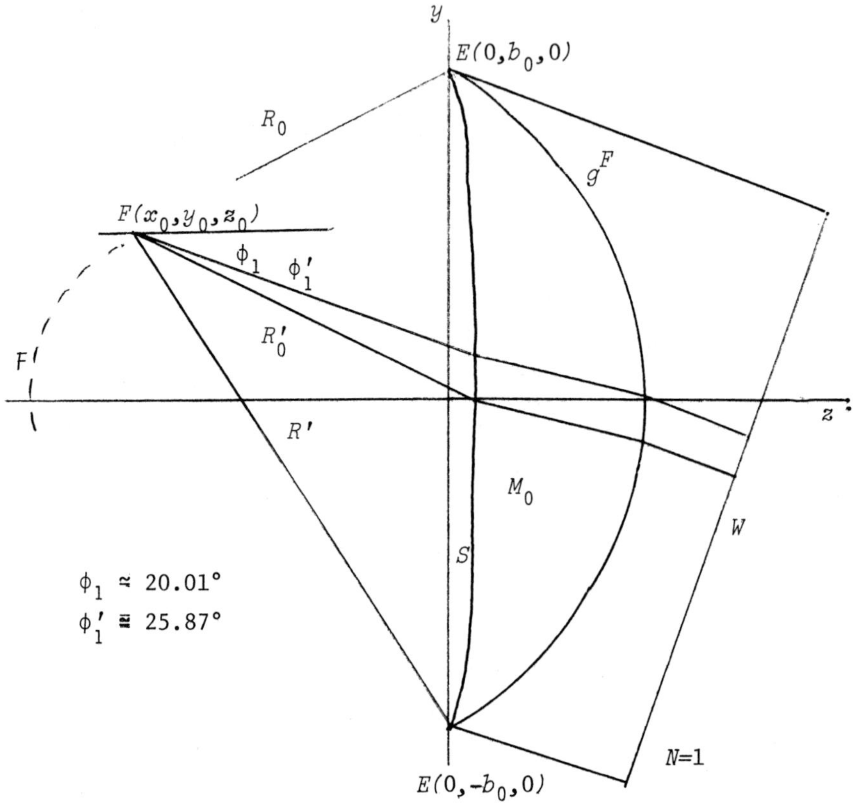

Figure 3. Example B -Focal Lens

$$(3.8)\qquad S: \begin{aligned} \beta_0(1) &= 2.68\text{x}10^{-1}, \\ \beta_1(1) &= 9.20\text{x}10^{-5}, \\ \beta_2(1) &= -6.65\text{x}10^{-5}, \\ \beta_3(1) &= 3.02\text{x}10^{-7}, \end{aligned} \qquad S': \begin{aligned} \beta_0'(1) &= 5.369\text{x}10^{0}, \\ \beta_1'(1) &= -5.569\text{x}10^{-2}, \\ \beta_2'(1) &= -7.266\text{x}10^{-5}, \\ \beta_3'(1) &= -7.166\text{x}10^{-7}, \end{aligned}$$

which, with all β_{mn} and β_{mn}' values zero, yields the ray tracing results shown in Figure 4(a) and, which, with all of the latter coefficients zero except for β_{11}' which we find by trial and error has approximately the value $\beta_{11}' = -1.5\text{x}10^{-5}$, yields the ray tracing data shown in Figure 4(b). In the first case the wave front is seen to be flat to within 0.01 units over a circular region of diameter 18 $\cos\,\psi_0$ units, or to be flat to within one part in 1690, and in the second case it is seen to be flat to within 0.005 units over the same circular region, or to be flat to within one part in 3380.

By the quarter wavelength Rayleigh criterion for diffraction limited beam forming, the latter case of the foregoing example lens with $F = 1/2$ at 20 degrees would be capable of forming a collimated beam of half-power width of approximately one-fifteenth of a degree, or four minutes of arc, or twice this with a one-eighth wavelength Rayleigh criterion, a very sharply focused beam in either case for a bifocal lens with so very low an F-number.

Finally, if the F-number is actually to be a minimum, the foregoing treatment may need modification along the lines of the related treatment of the generalized aplanatic lens design problem. In particular if such is to be the case, it may very well be best to give up the analytic expressions for the lens surface S' in (3.1)-(3.2) and determine S' from S by the path length condition after previously determining S as before. See Sternberg [4] for treatment of the comparable problem for generalized aplanatic lenses.

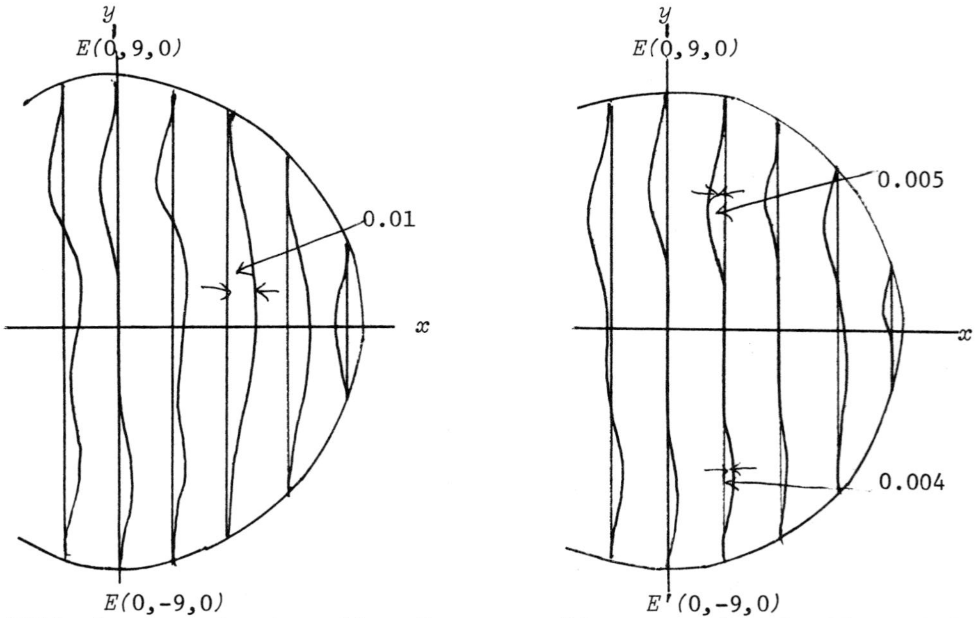

(a) Elliptic approximation $\beta'_{11} = 0$

(b) Corrected with $\beta'_{11} = -1.5 \times 10^{-5}$

Figure 4. Ray Tracing Results on Example Lens.

4. *Optimization for Azimuth vs. Elevation Scanning Trade Off*

After solving the bifocal lens boundary value problem we then seek to adjust the basic lens parameters $|z_0|$, n_0, ψ_0 in (2.1), particularly the index of refraction n_0 and the off-axis design angle ψ_0, so as to optimize the lens for best balance of azimuth versus elevation scanning performance. Generally we seek a maximum azimuth scanning capability for given acceptable elevation performance. In carrying out this azimuth versus elevation scanning trade off, any of several varied but more or less equivalent optimization criteria may be applied. We outline one below.

Thus, with reference to Figure 5, given a solution algorithm for the bifocal lens surfaces S: $z = z(x,y)$ and S': $z' = z'(x',y')$ as in (2.2) or in the parametric forms (2.6) satisfying all of the boundary, symmetry and side conditions of the problem (2.1) to (2.15) for a given set of lens parameters (2.1), we may take as a typical optimization condition the requirement of maximization of say $\alpha\psi_0$ over the parameter space $n_0 > 1$ and $|z_0|$ greater than the value of $|z_0|$ for which the F-number is a minimum at the given n_0, subject to the condition

$$\max_{x,y} |P(x,y,z,x',y',z',x_0^*,y_0^*,z_0^*,n_0,\psi_0^*) - P_0^*| < \Delta\lambda_0 \tag{4.1}$$

as x_0^*,y_0^*,z_0^* varies over a suitable focal surface F back of the lens and ψ_0^* scans the sector $|\psi_0^*| < \alpha\psi_0$ for say $\alpha = 3/2$ and $\Delta\lambda_0$ satisfies a more or less stringent Rayleigh criterion, say, $\Delta\lambda_0 < (1/8)\lambda_0$ or $(1/4)\lambda_0$. In (4.1) x, y, z and x', y', z' are of course the usual points on the lens surfaces S and S' in (2.3) at which a ray R intersects the surfaces except that here we take R from the variable focal point $F^*(x_0^*,y_0^*,z_0^*)$ rather than from the design focal point $F(x_0,y_0,z_0)$ and $P(x,y,z,x',y',z',x_0^*,y_0^*,z_0^*,n_0,\psi_0^*)$ and P_0^* are variable and constant path lengths, the former being given by

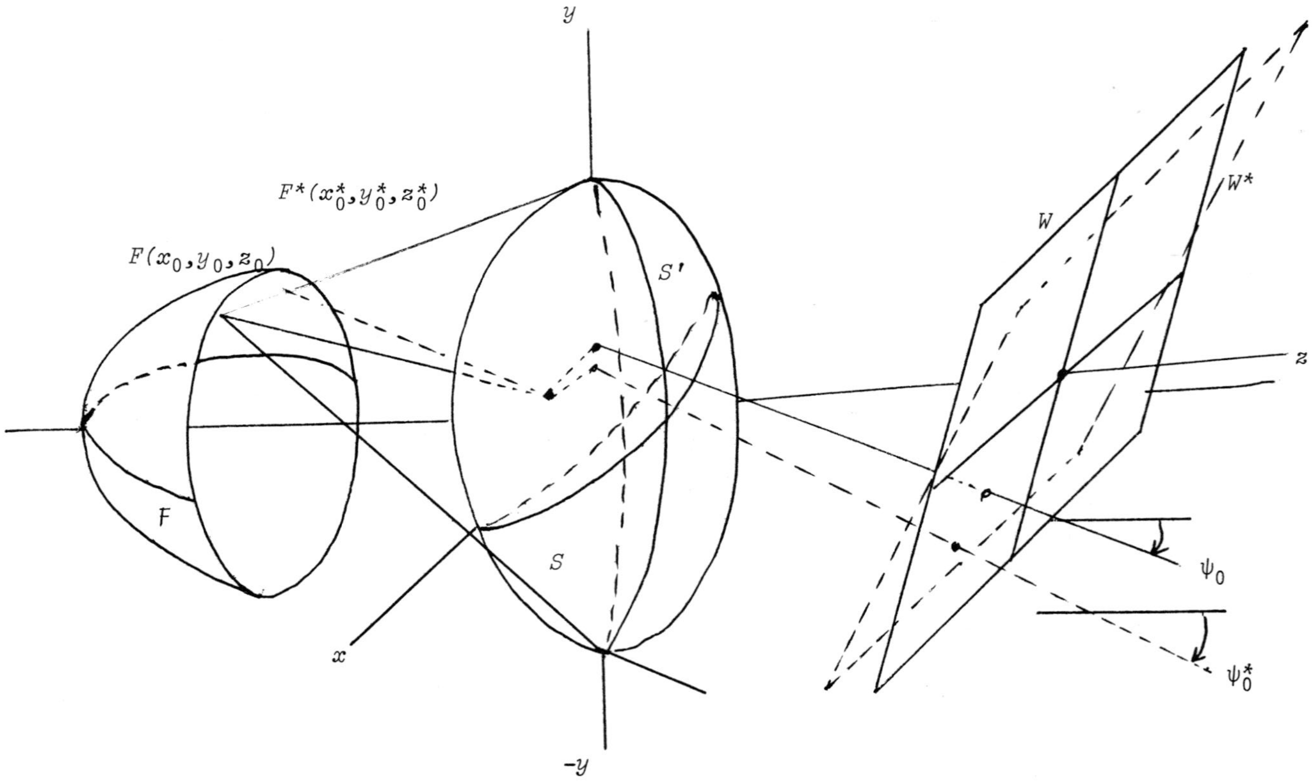

Figure 5. Optimization for Azimuth vs. Elevation Scanning Trade Off.

$$(4.2) \qquad P \equiv p^* + n_0 p^{*\prime} + p^{*\prime\prime}$$
$$= [x^2 + (y - y_0^*)^2 + (z - z_0^*)^2]^{1/2}$$
$$+ n_0[(x' - x)^2 + (y' - y)^2 + (z' - z)^2]^{1/2}$$
$$+ y' \sin \psi_0^* + (a - z') \cos \psi_0^* ,$$

and, the latter being given approximately by,

$$(4.3) \qquad P_0^* \cong [b_0^2 + z_0^{*2} \sec^2 \psi_0^*]^{1/2} + a \cos \psi_0^* ,$$

similarly, except for the approximations, as in (2.10) and (2.12). The focal surface F in the illustration on which the variable focal point $F^*(x_0^*, y_0^*, z_0^*)$ is located may be approximately specified, as ψ_0^* varies, as the ellipsoid of revolution formed by rotating around the z-axis the curve of intersection of rays R_0^* and $R_0^{*\prime}$ traced backwards through the lens extremities $E(0, b_0, 0)$ and $E'(0, -b_0, 0)$ from the wave front W^* propagating in the direction ψ_0^* until they meet. More particularly, the focal surface F is to be determined empirically during the maximization process on $\alpha\psi_0$ and thus, the maximization in (4.1) is really over x, y and also F.

There are no analytic methods known to the writer to carry out this optimization procedure but it appears to present no great difficulties on an approximate and admittedly somewhat rough basis on a computer using a suitably designed numerical ray tracing program. Final computational results have not yet been attained but are anticipated to be available before long.

The parameters $\alpha = 3/2$ and $\Delta\lambda_0$ used in connection with the maximization process in (4.1) are somewhat arbitrary. The first, α, serves to limit the azimuth scan to some specified multiple of ψ_0 while the quantity $\Delta\lambda_0$ usually taken as $\lambda_0/8$ or $\lambda_0/4$ or somewhere in between, is as before the Rayleigh criterion for diffraction limited focusing of plane waves, satisfaction of which in (4.1) ensures excellent beam forming capabilities of the lens throughout the azimuth and elevation scan; i.e., for $F^*(x_0^*, y_0^*, z_0^*)$ anywhere on the surface F within the limits prescribed by $\psi_0^* \leq \alpha\psi_0$ and the criterion (4.1).

5. *General Theory, Applications and Some Open Questions*

In concluding this interim discussion of the problem of the design of bifocal azimuth versus elevation optimized and extremized scanning lenses at low F-numbers a few remarks are in order on existence theory, applications, and unanswered open questions for future study.

As must be apparent in the foregoing sections we are not able at this time to give mathematical proofs of either the existence of solutions to the boundary value and optimization problems or of the convergence of the limiting processes and infinite series appearing in the formal solutions presented and outlined.

Nevertheless a few simple things can be proved. First, by Malus's theorem which asserts that the rays and the wave fronts are always orthogonal, or more or less equivalently, by the path length condition, it is readily seen that if we drop either one of the two pairs of symmetry conditions, i.e., either the symmetry requirements on lens surface S or those on lens surface S' in (2.13), then a continuum of solutions to the boundary value problem (2.1) to (2.15) exists corresponding to any reasonably simple choice for the remaining symmetric lens surface. That is to say, we can always solve the problem if we merely require symmetry properties on one of the lens surfaces S and S', the F-number of course in any case being greater than or equal to its minimum.

Moreover by considering sequences of such solutions to the boundary value problem and the obvious consequence of Snell's law and the path length condition that the lens thickness tends to zero as the index of refraction n_0 tends to infinity, and by applying the path length condition once again, it is readily verified that the bifocal boundary value problem (2.1) to (2.15) has the unique limiting solution

(5.1) $$S_\infty, S'_\infty: \quad \frac{x^2}{a^2} + \frac{y^2}{b^2} + \frac{(z - z_0)^2}{c^2} = 1$$

where

(5.2) $$a^2 = c^2 = b_0^2 \cos^2\psi_0 + z_0^2, \qquad b^2 = b_0^2 + z_0^2 \sec^2\psi_0,$$

when $n_0 = \infty$. Here the F-number is a minimum when $|z_0| = b_0 \sin\psi_0 \cos\psi_0$ incidentally. Direct calculation with the path length condition and the symmetry conditions yields (5.1) and (5.2) and the uniqueness follows from the failure of any other candidate symmetrical zero thickness lenses to provide for a symmetrical dual ray structure from the dual focal point $F'(x_0, -y_0, z_0)$.

While the foregoing is admittedly not much in the way of existence theory, we are led by it -- and more especially, by the very strong computational evidence regarding both the boundary value problem and, on a more hazardous basis, the optimization process for solving the azimuth versus elevation scanning trade off -- to formulate the following:

Existence Conjecture: Given a set of lens parameters $|z_0|$, b_0, n_0, ψ_0 and λ_0 as in (2.1) with $n_0 > 1$ it appears that if the F-number, $F = |z_0|/2b_0$, is sufficiently large and in any event greater than $b_0 \sin\psi_0 \cos\psi_0$, and if ψ_0 is sufficiently small and in any event less than $\pi/2$, then there exists one and only one solution of the bifocal lens boundary value problem (2.1) to (2.15). Moreover, for given $n_0 > 1$ and $\psi_0 < \pi/2$ it appears that there exists a least value of the F-number, $F = |z_0|/2b_0$, for which there exists a solution of the problem. Finally, for given b_0 it appears that there exists one or more optimal sets of choices for $|z_0|$, n_0 and ψ_0 yielding one or more optimal solutions of the azimuth versus elevation scanning trade off problem for given λ_0 and Rayleigh criterion.

That is to say we believe these are the facts. We cannot prove them but hope someone may sometime be able to do so while in the interim the computational work progresses.

Applications of these lenses are mainly anticipated at present to be in the field of underwater acoustic imaging and beam forming for use in a variety of technical problem areas. In research submersibles in oceanography and in working submersibles in ocean floor seabed construction work, acoustics offers an alternative to optical imaging systems and wide angle scanning is of signal importance. Applications in medicine for both diagnostics and also for acoustic therapy and perhaps also for acoustics enhanced--by heat generation--X-ray therapy are other potential applications of significance. The F-number in the latter of these plays a unique roll--the lower it is the greater the relative concentration of sound energy and, hence, also heat generation, at the cite to be treated with X-rays versus that present at intervening tissue layers and, therefore, the greater the enhancement of the theraputic effects of the X-rays and the greater the usefulness of the treatment. In communications by line-of-sight microwave links, lenses of the waveguide type have recently been investigated for satellite to satellite and satellite to ground communications where multiple channels in widely differing directions are needed. This latter may also be an eventual area of applications of bifocal lenses of the types considered here provided sufficiently light weight dielectrics are eventually developed for microwave lens construction. Typical configurations for acoustic imaging in oceanography and medical diagnostics and for satellite communications applications are illustrated in Figure 6.

Finally, in closing this paper it seems worthwhile to note

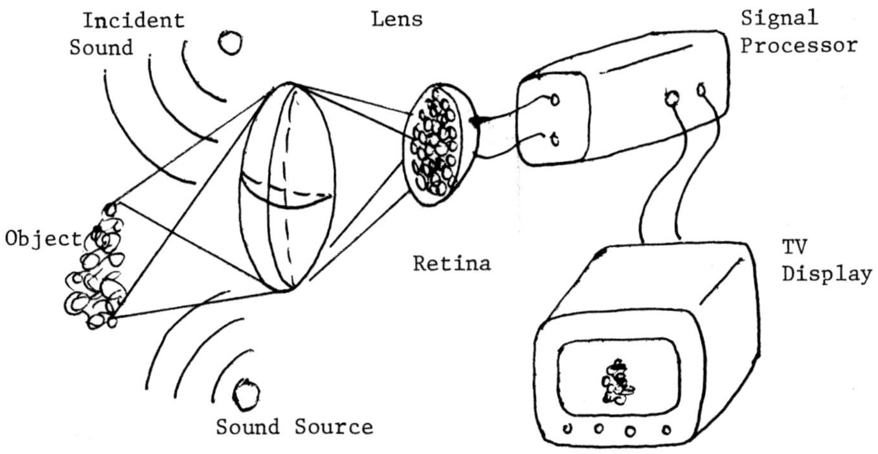

(a) Configuration for Acoustic Imaging

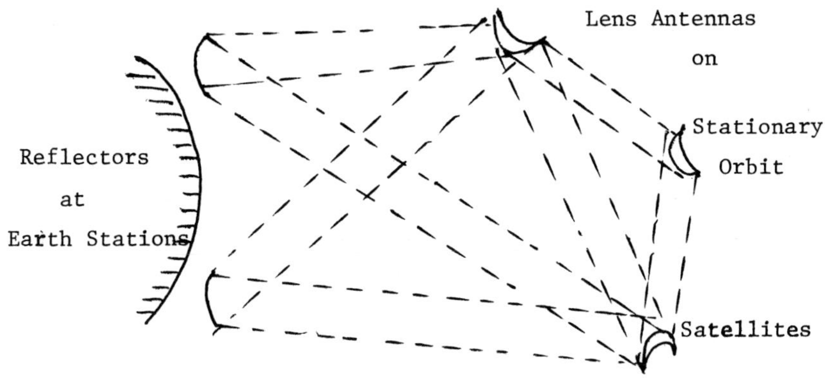

(b) Configuration for Satellite Microwave Link

Figure 6. Applications to Underwater Acoustic Imaging in Oceanography and Medicine and to Satellite Microwave Communications Problems.

some open questions for future research. Included here are not only the already indicated needs for a firm mathematical existence theory and perhaps also further developments of solution techniques and algorithm improvements but also such matters as investigation of various new lens materials and materials properties such as in acoustics, the matter of the energy partition in solid lens materials between transverse and longitudinal waves, and also various questions of environmental effects of temperature and ambient pressure on lens materials. Fabrication methods and materials for very light weight large satellite and communications antennas in the microwave region of the electromagnetic spectrum and development of high quality practical quarter-wave matching layers for reflection loss control and reduction in both microwaves, and especially in acoustics, are other fields in which additional research would be useful. Spot diameter measurements in finite focal length doublet lens systems would add impetus to potential applications of nonspherical lenses of the type we have considered to problems in medicine and related acoustic inspection systems. Lastly the field is open for development of a general mathematical theory of multiple element nonspherical lenses and a thorough going computer modeling and experimental quantitative comparison of the advantages and disadvantages of nonspherical versus spherical lenses in all the different applications including the azimuth versus elevation scanning trade off problem studied here needs to be made; for additional background material on these latter questions and particularly on experimental matters see also Sternberg and Goldberg [7], Corbett [8], Brown [9] and Corbett, Middleton and Sternberg [10].

ACKNOWLEDGMENTS

During the period of the project which intermittently spanned more than two decades the writer was associated with many persons and several institutions. Thanks are given to all those at Laboratory for Electronics, Inc. who made contributions to the work during the period 1955 to 1963 and to Mr. Gerry Stevens and Mr. William Greene of the Newport Laboratory of the Naval Underwater Systems Center, and to Mr. Kevin Corbett and Dr. Foster Middleton of the Ocean Engineering Department of the Univeristy of Rhode Island, for supporting experimental and numerical work related to the project done during the years 1974 to 1976. Finally, as noted in the introduction Dr. F. S. Holt, then at the Air Force Cambridge Research Laboratory, originally suggested the basic boundary value problem to the writer in nonmathematical form circa 1953.

REFERENCES

[1] L. C. Martin, *Wide-Aperture Aplanatic Single Lenses*, Proceedings of the Physical Society, (London) Vol. 56(1944), 104-113.

[2] F. G. Friedlander, *A Dielectric-Lens Aerial for Wide-Angle Beam Scanning*, Journal of the Institute of Electrical Engineers 93(1946), Pt-III-A, 658-662.

[3] R. L. Sternberg, *Successive Approximation and Expansion Methods in the Numerical Design of Microwave Dielectric Lenses*, Journal of Mathematics and Physics 34(1955), 209-235.

[4] R. L. Sternberg, *On a Multiple Boundary Value Problem in Generalized Aplanatic Lens Design and Its Solution in Series*, Proceedings of the Physical Society (London) 81(1963), Pt. 5, 902-924.

[5] R. L. Sternberg, *Surface Series Expansion Coefficients for Numerical Design of Scannable Aspherical Microwave and Acoustic Lenticular Antennas*, Proceedings of the IEE (England) 121(1974), 1351-1354.

[6] W. Ellis, E. Fine, and G. Reynolds, *A Point Source Binormal Lens*, Air Force Cambridge Research Laboratory Report, 1961.

[7] R. L. Sternberg and H. B. Goldberg, *Off-Axis Short Focal Length Dielectric Lens Antennas*, Proceedings National Aerospace Electronics Conference, Dayton, Ohio, 1961.

[8] K. T. Corbett, *Design, Construction and Evaluation of a Nonspherical Acoustic Lens*, Ocean Engineering Report, University of Rhode Island, October, 1974.

[9] J. Brown, *Microwave Lenses*, Methuen & Co Ltd.(London) 1953.

[10] K. T. Corbett, F. H. Middleton, and R. L. Sternberg, *Nonspherical Acoustic Lens Study*, Journal of the Acoustical Society of America 59(1976), 1104-1109.

A MODEL FOR THE STATISTICAL ANALYSIS OF CELL RADIOSENSITIVITY DURING THE CELL CYCLE*

Janice O. Tsokos and Chris P. Tsokos
University of South Florida

1. *Introduction*

There is abundant evidence for variation in cell sensitivity to radiation during the cell cycle, Barendsen, L. H. (1967), Brown, B. W., Thompson, J. R., Barkley, T., Suit, H., and Withers, H. R. (1974), Burch, P. R. J. (1957), Hall, E. J., and Bedford, J. S. (1964), Mauro, F., and Madoc-Jones (1969), Sinclair, U. K. (1970), Terasima, J., and Tolmack, L. J. (1963), Thompson, L. H., and Humphrey, R. M. (1969), and Whitmore, G. F., Gulyas, S., and Botond, J. (1965). A compartmental model to approximate the varying radiosensitivity of cells with position in the cell cycle has been developed and applied to the problem of choosing optimal strategies in radiotherapy by Brown, B. W., and Thompson, J. R. (1972, 1974). The variation is attributed to changes in the target number, Sinclair, U. K. (1961), where, using the assumption of multitarget theory, Barendsen, G. W. (1967), it is assumed that each cell has n targets susceptible to radiation damage and when all n targets have been damaged, the cell is destroyed. Targets are repaired concurrently, Elkind, M. M., and Sinclair, U. K. (1965), Elkind, M. M. (1967), and Withers, H. R. (1970). A dose of D_0 rads is required to obtain as many hits as targets and the rate of target damage is thus equal to the dose rate in D_0's. The target number ranges from a value of one during mitosis to a maximum of perhaps twenty in S phase. The model, shown in Figure 1.1 below, approximates these changes by dividing the cycle into just two phases, one a sensitive phase where damage of a single target kills the cell, and the other a resistant phase in which the cells have a moderately high target number.

*The authors wish to acknowledge the USF Research Council Release Time Award for pursuing this study and United State Air Force Office of Scientific Research, Under Grant No. AFOSR-74-2711.

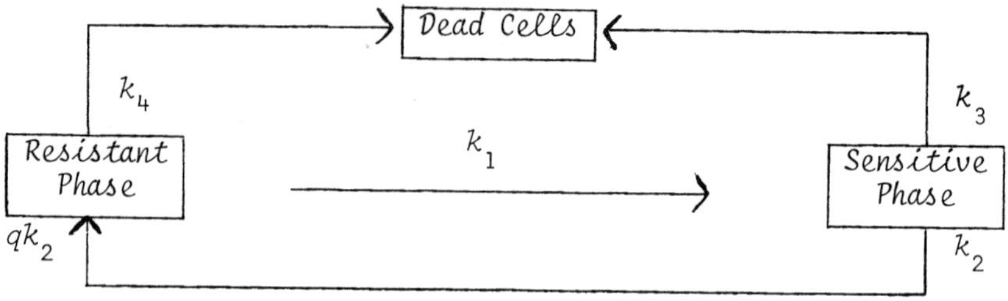

Figure 1.1

We summarize the parameters of the model as follows:

a) the proportions of the cell population which occupy the sensitive and resistant compartments when treatment begins, S_0 and R_0,

b) the rates at which cells cycle between resistant and sensitive phases, k_1 and k_2,

c) the rates of cell death from the two phases, k_3 and k_4,

d) the cell cycle time, T,

e) the factor q which expresses the increase in cell number due to mitosis (after which cells are assumed to enter the resistant phase immediately),

and

f) the target number n of cells in the resistant phase.

There being only one target in the sensitive phase, k_3 is equal to the dose rate (in units of D_0). Since it is known that a fraction of the dose delivered does single hit killing (due to irreparable damage) in the resistant phase, that fraction is also a parameter of the model, designated w, and k_4 is equal to w times the dose rate, for low dose rate therapy. For simplicity the small amount of multi-hit killing from the resistant phase is neglected. For high dose rates k_4 becomes a function of the proportions of single hit (w) and multiple-hit $(1 - w)$ killing.

It is clear that several of these parameters are subject to individual variations or uncertainties, for example, those parameters which are related to the time spent in the sensitive and resistant phases and the rates of cycling between them. Radiosensitivity is known to vary with cell age, and the cell cycle phase distribution is perturbed by the radiation itself. The proportion of the cell cycle which is radiosensitive most probably ranges between 5 and 30%, Brown, B. W., Thompson, J. R., Barkley, T., Suit, H., and Withers, H. R. (1974). The dispersion of these values implies that it is appropriate to consider S_0 and R_0 as random variables with some prescribed probability distribution.

In this paper we shall consider the solutions of the system of differential equations which characterize the above model when the rate constants k_1 and k_2, together with initial proportions of sensitive and resistant cells S_0 and R_0, are treated as random variables with specified probability distributions. The analysis is performed under the assumption that the above random variables may be characterized by the uniform and beta probability distributions. In view of the range of variation of these variables and their inherent characteristics these probability distributions seem quite relevant.

In Section 2 we shall describe the radiosensitivity model, and the statistical characterization of this system will be presented in Section 3. The numerical results and conclusions of the study are given in Section 4.

2. *Radiosensitivity Model*

The basic system which describes the progression of cells from the sensitive to the resistant phase of the cycle and the destruction of cells in both compartments is schematically shown in Figure 1.1. In order to discuss the surviving proportions of cells in the compartmental model of Figure 1.1, we normalize $S^*(t)$ and $R^*(t)$ (number of cells in the sensitive or resistant phase at time t):

$$S(t) = \frac{S^*(t)}{S^*(0) + R^*(0)}$$

and

$$R(t) = \frac{R^*(t)}{S^*(0) + R^*(0)} .$$

The kinetic equations which describe the above radiosensitivity model for continuous radiation are given by

$$(2.1)\quad \begin{aligned} \frac{dS(t)}{dt} &= -(k_1 + k_4)S(t) + qk_2R(t) \\ \frac{dR(t)}{dt} &= k_1S(t) - (k_2 + k_3)R(t). \end{aligned}$$

The solution to the above system of deterministic equations is given by

$$S(t) = C_1 exp\ \{(-A + W)t\} + C_2 exp\ \{-(A + W)t\}$$

$$R(t) = d_1 exp\ \{(-A + W)t\} + d_2 exp\ \{-(A + W)t\}$$

where

$$d_1 = \frac{1}{k_2T} - d_2$$

$$d_2 = \left\{\frac{k_1 + k_4 - A_1 - W}{-2qk_2W}\right\}\left\{\frac{q}{T} - \frac{k_1 + k_4 - A + W}{k_1T}\right\}$$

$$A = \frac{k_1 + k_2 + k_3 + k_4}{2}$$

$$B = k_1k_3 + k_1k_2(1 - q) + k_4(k_2 + k_3)$$

$$W = \sqrt{A^2 - B}$$

$$C_1 = \frac{qk_2d_1}{k_1 + k_4 - A + W}$$

$$C_2 = \frac{1}{k_1T} - C_1 .$$

For specific values of the parameters and initial conditions $S^*(0)$ and $R^*(0)$, one can plot the solutions (trajectories),

$S(t)$ and $R(t)$, of the differential system (2.1) and obtain the behavior of the proportions of cells in the sensitive and resistant phases as a function of time. In what follows we shall give some of the basic theory in studying the mean behavior of $S(t)$ and $R(t)$. That is, we would like to study the above differential system when the proportion of cells, S_0 (R_0) in the sensitive (resistant) phase at time zero is considered as a random variable.

3. *Statistical Characterization of the System*

A general form of a differential system can be written in vector-matrix notation as

$$\dot{\underline{U}}(t) = \underline{h}(\underline{U}(t),t) \tag{3.1}$$

with initial conditions $\underline{U}(0) = \underline{C}$. We shall assume that the initial conditions $\underline{U}(0)$ behave as random variables and are characterized by the joint probability density function $g_0(\underline{C})$. Our aim is to determine the joint probability density function of $\underline{U}(t)$ at a given time t. To obtain this probability distribution we shall make use of the Liouville theorem whose proof can be found in Soong (1973).

<u>THEOREM</u> (Liouville). Suppose that a random solution (in the mean square sense of (3.1) exists. Then the joint probability density function of $\underline{U}(t)$, $g(\underline{u},t)$ satisfies the equation

$$\frac{\partial g(u,t)}{\partial t} + \sum_j \frac{\partial (gh_j)}{\partial u_j} = 0, \quad g(\underline{u},0) = g_0(\underline{c}) \tag{3.2}$$

where h_j and u_j are the jth component of the vectors $\underline{h}$ and $\underline{u}$. The equation (3.2) can be solved by examining its associated Lagrange system. We can write equation (3.2) as

$$\frac{\partial g(u,t)}{\partial t} + g(\underline{u},t)\nabla\cdot\underline{h}(\underline{u}(t),t) + \sum_j h_j \frac{\partial g(u(t),t)}{\partial u_j} = 0 \tag{3.3}$$

where $\nabla\cdot\underline{h}$ is the divergence of the vector $\underline{h}$. The Lagrange system of equation (3.3) can be written as

$$\frac{dt}{1} = \frac{-dg(\underline{u},t)}{g(\underline{u},t)\nabla\cdot\underline{h}(\underline{u}(t),t)} = \frac{du_1}{h_1} = \cdots = \frac{du_n}{h_n} . \tag{3.4}$$

Let us assume that the general solution of system (3.1) is in a general form as

$$\underline{u}(t) = f(\underline{c},t) \tag{3.5}$$

and its inverse form is given by

$$\underline{c} = f^{-1}(\underline{u},t). \tag{3.6}$$

Thus, the first equality of equation (3.4) yields the desired solution, namely, the joint probability density function of $\underline{u}(t)$;

$$g(\underline{u},t) = g_0(\underline{c})\; exp\left\{- \int_0^t \nabla\cdot\underline{h}[\underline{u} = f(\underline{c},\tau),\tau]d\tau\right\}\Bigg|_{\underline{c}=f^{-1}(\underline{u},t)}. \tag{3.7}$$

Now, for the radiosensitivity model we proceed as follows. The system of differential equations (2.1) can be written in matrix-vector form as

$$\dot{\underline{U}}(t) = \begin{bmatrix} \dot{S}(t) \\ \dot{R}(t) \end{bmatrix} = \begin{bmatrix} -(k_1 + k_4) & k_2 q \\ k_1 & -(k_2 + k_3) \end{bmatrix} \underline{U}(t) \tag{3.8}$$

where

$$\underline{U}(t) = \begin{bmatrix} S(t) \\ R(t) \end{bmatrix} \text{ and initial conditions } \underline{C} = \begin{bmatrix} S_0 \\ R_0 \end{bmatrix} .$$

Thus, we can write

$$\dot{\underline{U}}(t) = \begin{bmatrix} -(k_1 + k_2)S(t) + k_2 qR(t) \\ k_1 S(t) - (k_2 + k_3)R(t) \end{bmatrix}$$

$$= h(\underline{U}(t),t). \tag{3.9}$$

The particular solution to equation (3.9) subject to the initial condition $\underline{C} = [S_0, R_0]^T$ is

$$(3.10) \qquad \underline{U}(t) = \begin{bmatrix} S(t) \\ R(t) \end{bmatrix} =$$

$$= \frac{1}{(\alpha_1 - \alpha_2)} \begin{bmatrix} \alpha_1 e^{\lambda_1 t} - \alpha_2 e^{\lambda_2 t} & \alpha_1\alpha_2 e^{\lambda_2 t} - \alpha_1\alpha_2 e^{\lambda_1 t} \\ e^{\lambda_1 t} - e^{\lambda_2 t} & \alpha_1 e^{\lambda_2 t} - \alpha_2 e^{\lambda_1 t} \end{bmatrix} \begin{bmatrix} S_0 \\ R_0 \end{bmatrix},$$

where $\lambda_1 = -A + W$, $\lambda_2 = -(A + W)$

$$\alpha_1 = \frac{qk_2}{\lambda_1 + (k_1 + k_2)} \quad \text{and} \quad \alpha_2 = \frac{qk_2}{\lambda_2 + (k_1 + k_4)} .$$

If we let

$$D = \begin{bmatrix} \alpha_1 e^{\lambda_1 t} - \alpha_2 e^{\lambda_2 t} & \alpha_1\alpha_2 e^{\lambda_2 t} - \alpha_1\alpha_2 e^{\lambda_1 t} \\ e^{\lambda_1 t} - e^{\lambda_2 t} & \alpha_1 e^{\lambda_2 t} - \alpha_2 e^{\lambda_1 t} \end{bmatrix},$$

then equation (3.10) can be written as

$$(3.11) \qquad \underline{U}(t) = f(\underline{c}, t) = D\underline{C}$$

or

$$\underline{C} = (\alpha_1 - \alpha_2) D^{-1} \underline{U}(t)$$

where

$$D^{-1} = \frac{1}{(\alpha_1 - \alpha_2)^2 \exp(\lambda_1 + \lambda_2)t} \begin{bmatrix} \alpha_1 e^{\lambda_2 t} - \alpha_2 e^{\lambda_1 t} & \alpha_1\alpha_2 \left(e^{\lambda_1 t} - e^{\lambda_2 t}\right) \\ e^{\lambda_2 t} - e^{\lambda_1 t} & \alpha_1 e^{\lambda_1 t} - \alpha_2 e^{\lambda_2 t} \end{bmatrix}.$$

Thus,

$$\underline{C} = \frac{1}{(\alpha_1-\alpha_2)} \begin{bmatrix} \alpha_1 e^{-\lambda_1 t} & \alpha_2 e^{-\lambda_2 t} & \alpha_1\alpha_2(e^{-\lambda_2 t}-e^{-\lambda_1 t}) \\ e^{-\lambda_1 t} - e^{-\lambda_2 t} & & \alpha_1 e^{-\lambda_2 t} - \alpha_2 e^{-\lambda_1 t} \end{bmatrix} \begin{bmatrix} S(t) \\ R(t) \end{bmatrix}$$

Also,

$$\nabla\cdot\underline{h} = \begin{bmatrix} \frac{\partial}{\partial s} \\ \frac{\partial}{\partial R} \end{bmatrix} \cdot\underline{h} = -(k_1 + k_2 + k_3 + k_4).$$

The joint probability density function of the number of cells in the sensitive and resistant phases is given by

$$g(S,R,t) = g_0(S_0,R_0)\ exp\left\{-\int_0^t -(k_1+k_2+k_3+k_4)d\tau\right\}\Big|_{\underline{c}=f^{-1}(\underline{u},t)}$$

$$= exp(k_1+k_2+k_3+k_4)t\ g_0\left\{\frac{1}{(\alpha_1-\alpha_2)}\left[(\alpha_1 e^{-\lambda_1 t} - \alpha_2 e^{-\lambda_2 t})S(t) + \alpha_1\alpha_2(e^{-\lambda_2 t} - e^{-\lambda_1 t})R(t)\right],\right.$$

$$\left.\frac{1}{(\alpha_1-\alpha_2)}\left[(e^{-\lambda_1 t} - e^{\lambda_2 t})S(t) + (\alpha_1 e^{-\lambda_2 t} - \alpha_2 e^{-\lambda_1 t})R(t)\right]\right\}.$$

From the above equation one can obtain the marginal densities of $S(t)$ and $R(t)$ and thus one can find the mean behavior of the proportion of cells in the sensitive and resistant phases, respectively.

For the present study we shall investigate the mean behavior of $S(t)$ and $R(t)$ subject to the random variables S_0 and R_0 being characterized by i) the uniform probability density function and ii) the beta probability distribution. These types of probabilistic functions should be quite realistic characterizations of the behavior of the cell population.

Case (i). In this case we shall obtain the expected behavior of $S(t)$ and $R(t)$ when randomness is introduced into the system with S_0 and R_0 being stochastic variables uniformly distributed over the desired interval. Using the information given by the probability distribution of the proportion of cells, S_0, in the sensitive phase at time zero, is uniformly distributed over the interval 0.05 to 0.30. That is,

$$g(S_0) = \begin{cases} \frac{1}{0.25}, & 0.05 \leq S_0 \leq 0.30 \\ 0, & \text{elsewhere,} \end{cases}$$

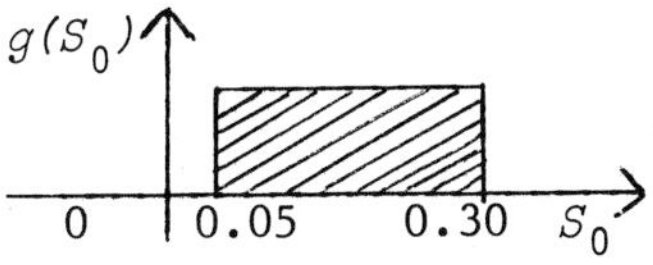

Similarly, the proportion of cells, R_0, in the resistant phase at time zero is considered to be uniformly distributed over the interval 0.70 to 0.95 as given below:

$$g(R_0) = \begin{cases} \frac{1}{0.25}, & 0.70 \leq R_0 \leq 0.95 \\ 0, & \text{elsewhere,} \end{cases}$$

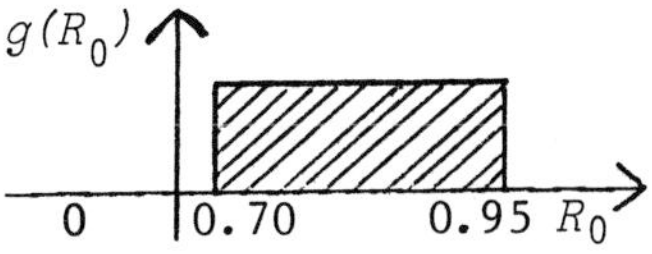

Note that the relationship between the behavior of S_0 and R_0 must be preserved, namely, their sum must equal unity.

Case (ii). In this case we investigate the mean behavior of $S(t)$ and $R(t)$ subject to S_0, (R_0), the proportion of cells in the sensitive (resistant) phase at time zero being probabilistically characterized by the beta distribution.

The beta distribution of S_0 is defined over the interval 0.05 to 0.30 given by

$$g(S_0) = \begin{cases} \frac{k_1(\alpha+\beta-1)!}{(\alpha-1)!(\alpha-2)!} S_0^{\alpha-1}(1-S_0)^{\beta-1}, & 0.05 \leq S_0 \leq 0.30, \alpha, \beta > 0 \\ 0, & \text{elsewhere} \end{cases}$$

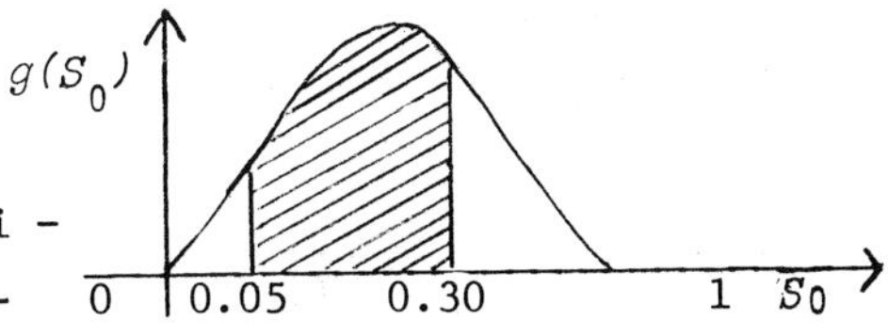

where α and β are parameters which determine the actual shape of the distribution and k_1 is the truncation factor which is calculated so that the total mass of the probability function is concentrated at the desired interval.

Similarly, R_0, the proportion of cells in the resistant phase at time zero is beta distributed,

$$g(R_0) = \begin{cases} \dfrac{k_2(\alpha+\beta-1)!}{(\alpha-1)!(\alpha-1)!} R_0^{\alpha-1}(1-R_0)^{\beta-1}; & 0.70 \le R_0 \le 0.95, \alpha, \beta > 0 \\ 0, \text{ elsewhere} & \end{cases}$$

where α and β are as above and k_2 is the truncation factor for $g(R_0)$. In the present case the parameters of the beta distribution were estimated using the maximum likelihood method.

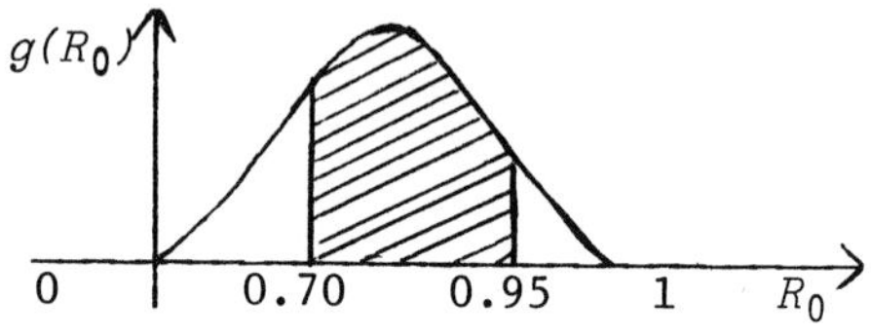

4. *Results and Conclusions*

Given such a mathematical model, the parameters may be varied one at a time, while holding the remaining parameters constant at chosen conditions, and the effect of the varied parameters on cell survival during radiation may be computed.

In the present study we have chosen to look at cell survival as a function of time through the duration of a single cell cycle at radiation doses of 5-25 D_0's / cycle. Values of D_0 for aerobic mammalian cells vary, but 150 rads is common. Assuming a 48 hour cell cycle, continuous radiation of 5-25 D_0's / cycle would be about 15-75 rads / hour. This is in the radio-therapeutic range.

Cell survival patterns as functions of time and radiation dose are seen in Figures 4.1 to 4.4. The proportion of cells in the sensitive phase has been estimated on the basis of experimental evidence to range between 5% and 30%. We have chosen to look at the pattern with initial conditions at the bottom extreme of the range, that is, 5% sensitive, 95% resistant, and at a convenient value near its mean, 20% sensitive to 80% resistant. Figure 4.1 shows $R(t)$ at 5, 15 and 25 D_0's, and Figure 4.2 $S(t)$

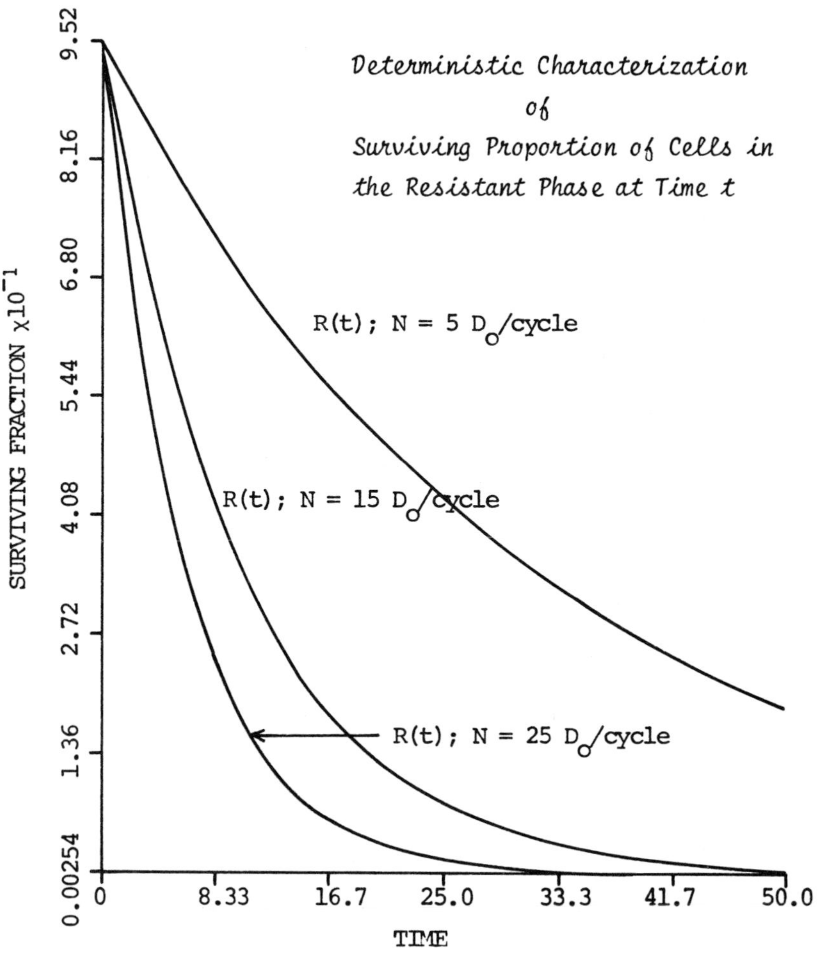
Deterministic Characterization
of
Surviving Proportion of Cells in
the Resistant Phase at Time t
R(t); N = 5 D_o/cycle
R(t); N = 15 D_o/cycle
R(t); N = 25 D_o/cycle
SURVIVING FRACTION $\chi 10^{-1}$
9.52
8.16
6.80
5.44
4.08
2.72
1.36
0.00254
0
8.33
16.7
25.0
33.3
41.7
50.0
TIME

Figure 4.1

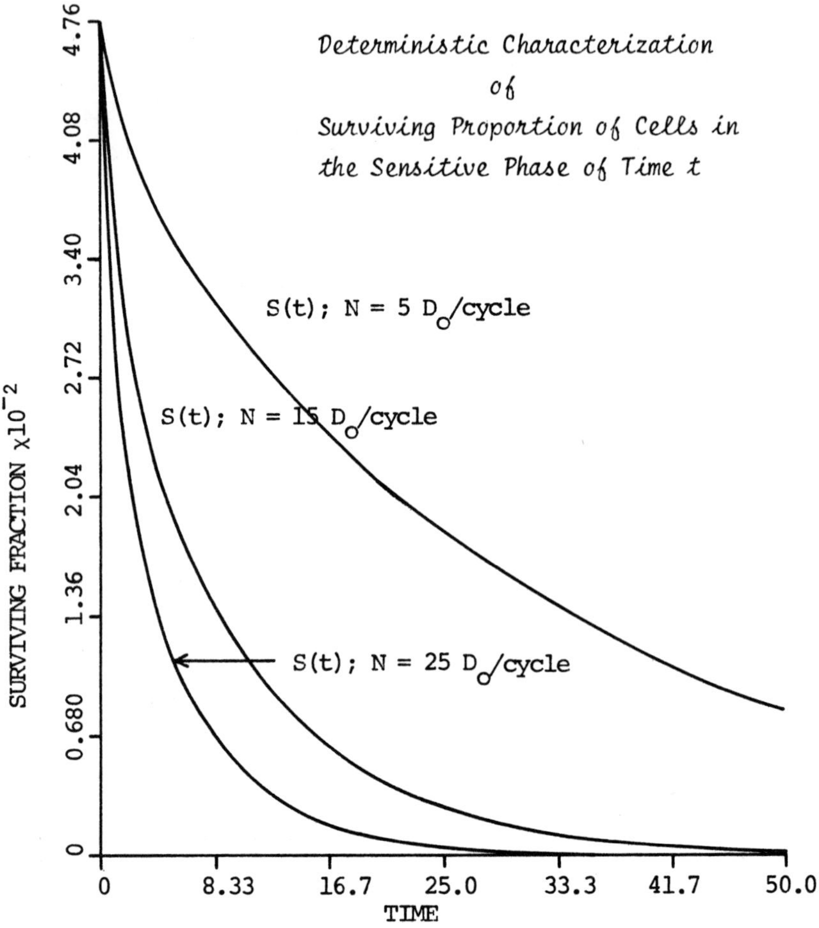
Deterministic Characterization
of
Surviving Proportion of Cells in
the Sensitive Phase of Time t
S(t); N = 5 D₀/cycle
S(t); N = 15 D₀/cycle
S(t); N = 25 D₀/cycle
SURVIVING FRACTION x10⁻²
4.76
4.08
3.40
2.72
2.04
1.36
0.680
0
0
8.33
16.7
25.0
33.3
41.7
50.0
TIME

Figure 4.2

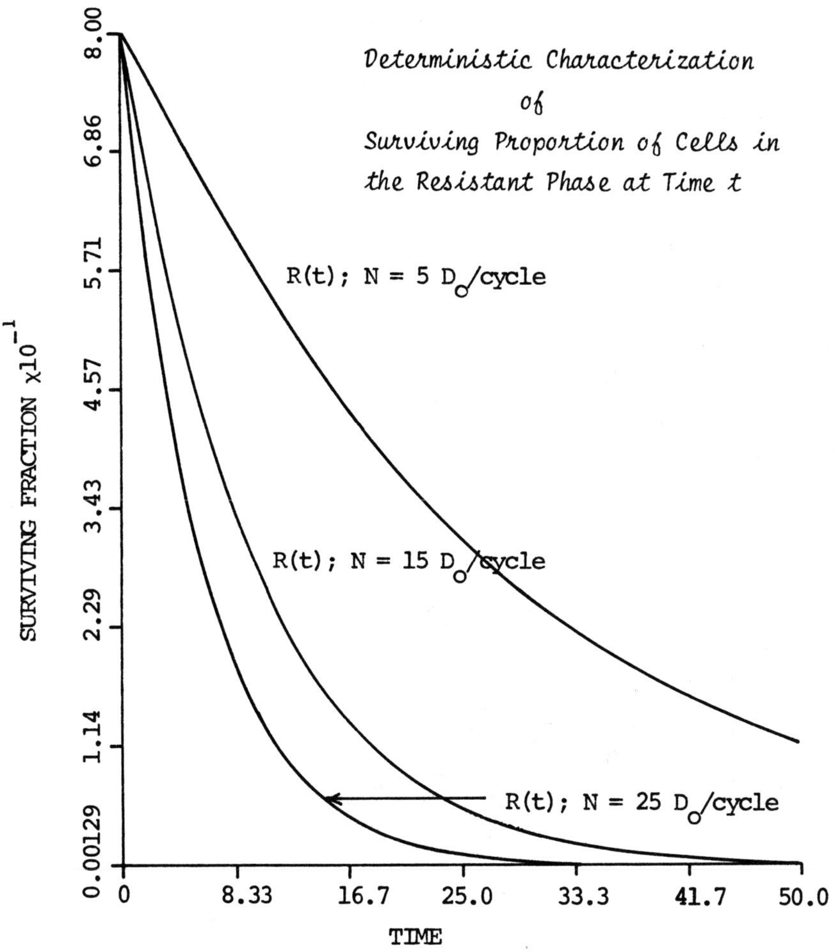
Deterministic Characterization
of
Surviving Proportion of Cells in
the Resistant Phase at Time t
R(t); N = 5 Do/cycle
R(t); N = 15 Do/cycle
R(t); N = 25 Do/cycle
SURVIVING FRACTION x10^-1
8.00
6.86
5.71
4.57
3.43
2.29
1.14
0.00129
0
8.33
16.7
25.0
33.3
41.7
50.0
TIME

Figure 4.3

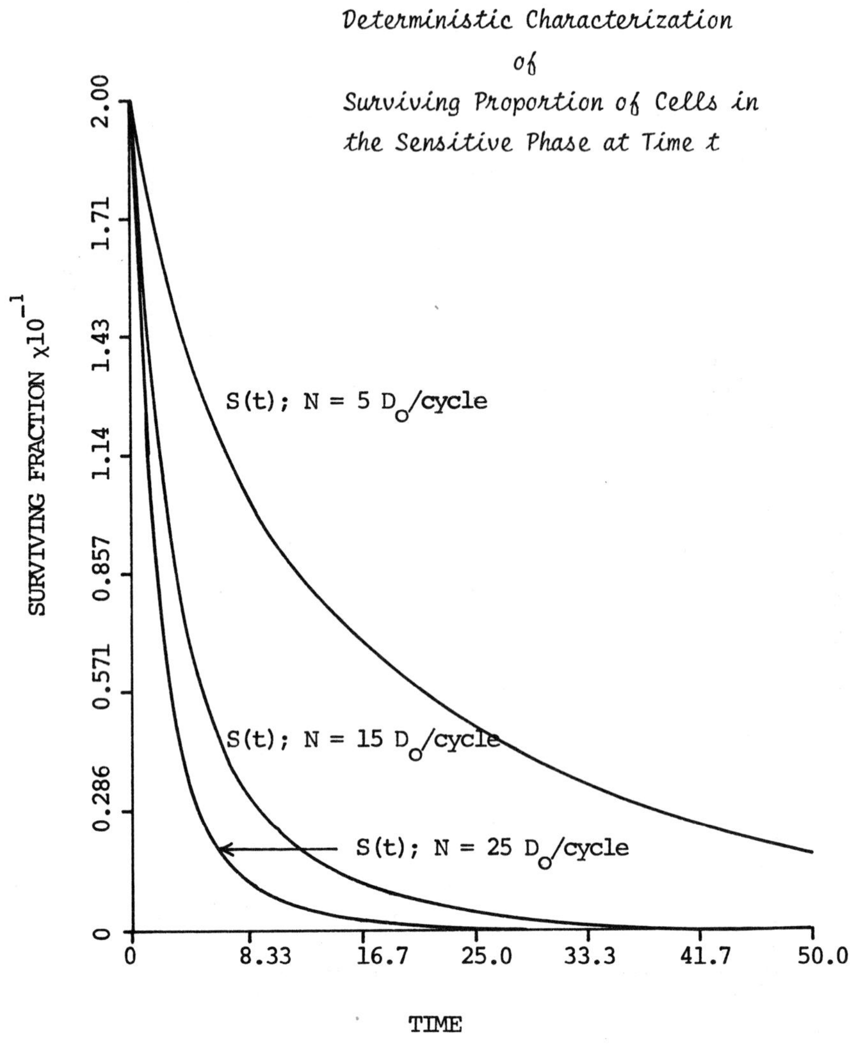

Deterministic Characterization
of
Surviving Proportion of Cells in
the Sensitive Phase at Time t
SURVIVING FRACTION χ10^-1
2.00
1.71
1.43
1.14
0.857
0.571
0.286
0
S(t); N = 5 D_o/cycle
S(t); N = 15 D_o/cycle
S(t); N = 25 D_o/cycle
0
8.33
16.7
25.0
33.3
41.7
50.0
TIME

Figure 4.4

at 5, 15 and 25 D_0's with initial conditions of 95% resistant and 5% sensitive cells. Note the more rapid fall in surviving sensitive fraction with time.

Figure 4.3 gives $R(t)$ and Figure 4.4 $S(t)$ at 5, 15 and 25 D_0's/cycle with initial conditions of 80% resistant, 20% sensitive cells.

Such information could be utilized to estimate the differential effects of a given continuous radiation dose rate or total dosage on, say, a rapidly growing tumor surrounded by slowly growing normal tissue, such as connective tissue, or similarly for the case of a slow growing tumor located within a population of cells which must continue to divide to maintain life, such as the intestinal epithelium.

Since most of the parameters of the model may be said to be subject to random variation it was also of interest to look at results obtained if one or more of the parameters was treated as a random variable with some specified probability distribution. We chose to do a statistical characterization in which the proportions of cells in the resistant and sensitive phases at time zero are taken to be random variables. This undoubtedly is nearer to the truth than the assumption that they have constant values.

The uniform and beta probability distributions were selected to characterize the stochastic variables. Cell survival patterns obtained using the uniform probability distribution for R_0 and S_0, Figure 4.5, are not appreciably different from the deterministic solutions except at low dose rate, where the deterministic characterization tends to slightly under-estimate the surviving fraction of resistant cells and over-estimate the surviving fraction of sensitive cells. The solution tends to converge at higher dosages.

Since in radiotherapy it is desirable to use the lowest dose rate consistent with obtaining the desired result -- an acceptable level of tumor tissue destruction -- this could be useful information.

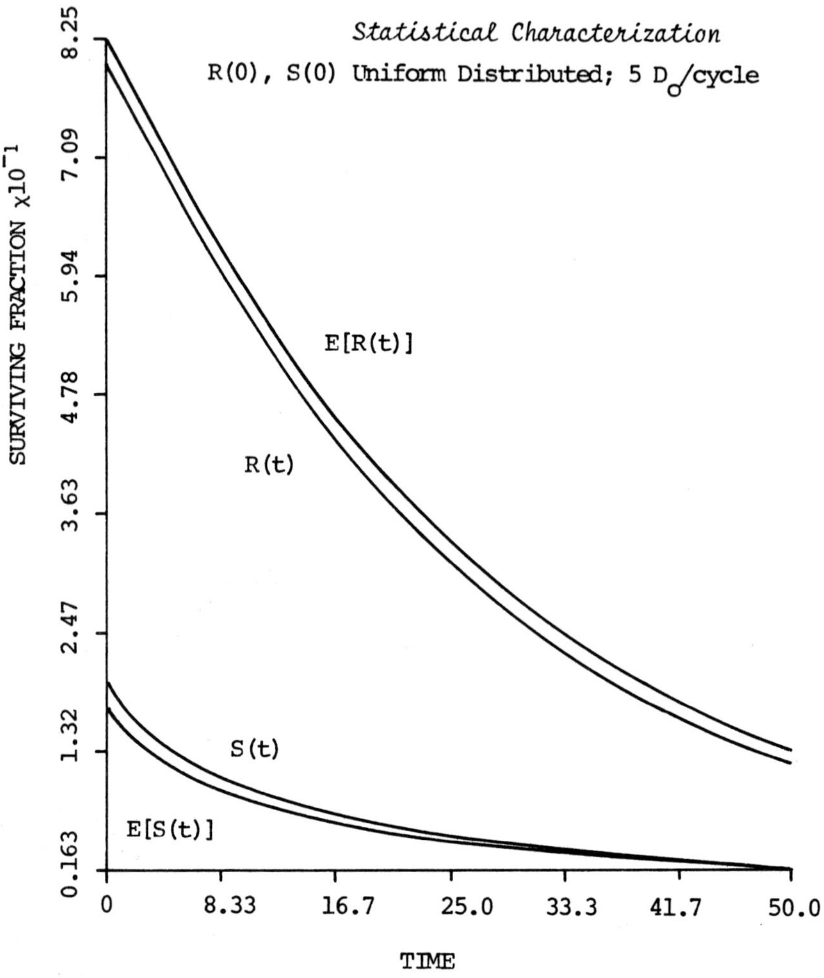

Statistical Characterization
R(0), S(0) Uniform Distributed; 5 D_o/cycle
SURVIVING FRACTION x10^-1
8.25
7.09
5.94
4.78
3.63
2.47
1.32
0.163
E[R(t)]
R(t)
S(t)
E[S(t)]
0
8.33
16.7
25.0
33.3
41.7
50.0
TIME

Figure 4.5

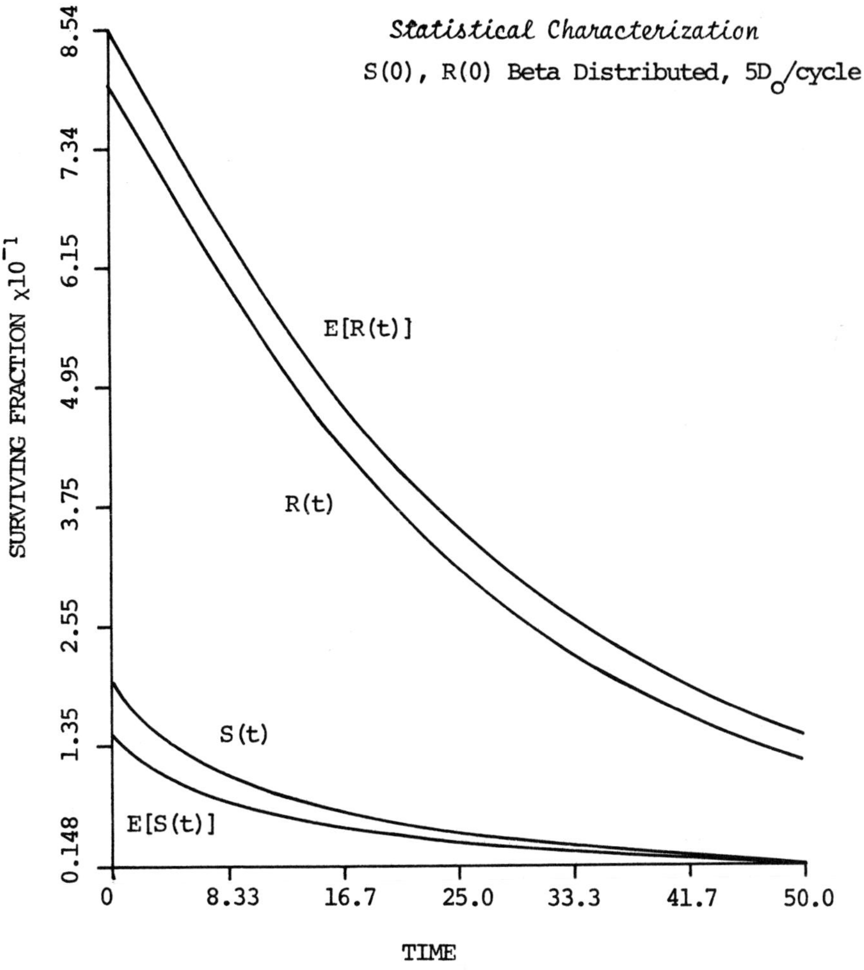
Statistical Characterization
S(0), R(0) Beta Distributed, 5D$_o$/cycle
SURVIVING FRACTION $\chi 10^{-1}$
8.54
7.34
6.15
4.95
3.75
2.55
1.35
0.148
E[R(t)]
R(t)
S(t)
E[S(t)]
0
8.33
16.7
25.0
33.3
41.7
50.0
TIME

Figure 4.6

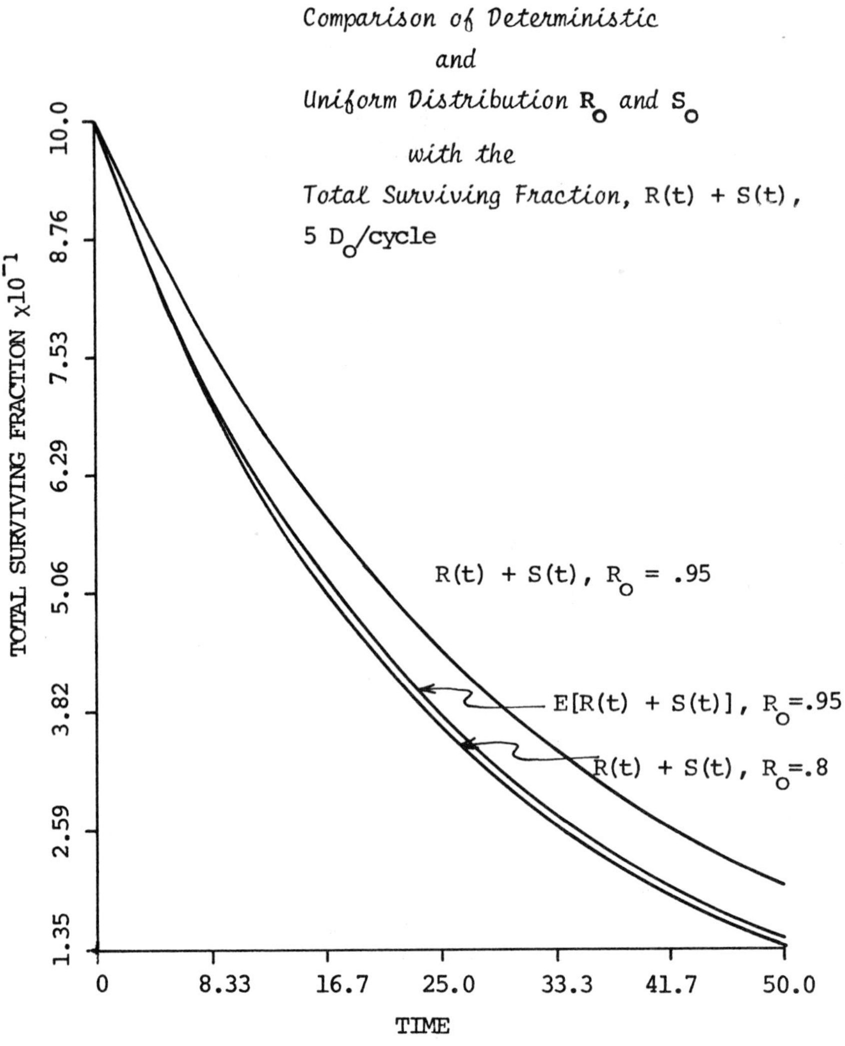
Comparison of Deterministic
and
Uniform Distribution R_o and S_o
with the
Total Surviving Fraction, R(t) + S(t),
5 D_o/cycle
TOTAL SURVIVING FRACTION x10^-1
10.0
8.76
7.53
6.29
5.06
3.82
2.59
1.35
R(t) + S(t), R_o = .95
E[R(t) + S(t)], R_o=.95
R(t) + S(t), R_o=.8
0
8.33
16.7
25.0
33.3
41.7
50.0
TIME

Figure 4.7

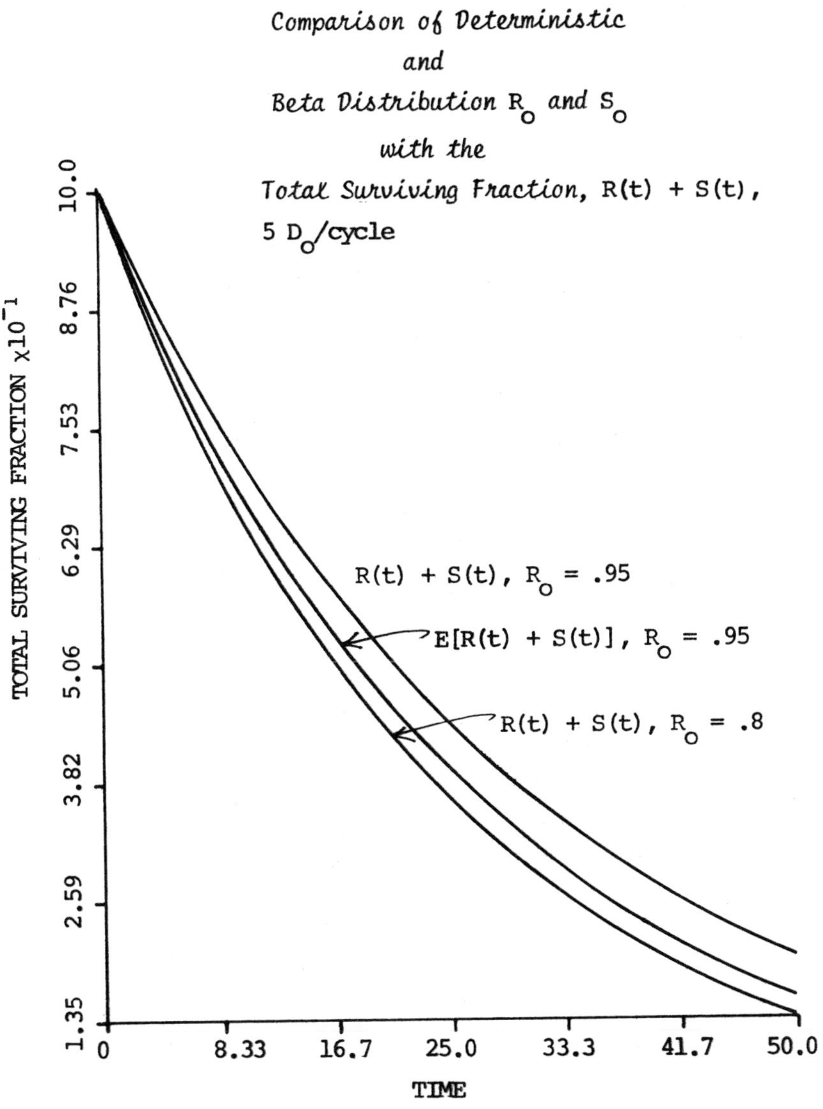
Comparison of Deterministic
and
Beta Distribution R_o and S_o
with the
Total Surviving Fraction, R(t) + S(t),
5 D_o/cycle
TOTAL SURVIVING FRACTION $\chi 10^{-1}$
10.0
8.76
7.53
6.29
5.06
3.82
2.59
1.35
R(t) + S(t), R_o = .95
E[R(t) + S(t)], R_o = .95
R(t) + S(t), R_o = .8
0
8.33
16.7
25.0
33.3
41.7
50.0
TIME

Figure 4.8

Use of the beta probability distribution, which probably better characterizes the variation in the parameters than does the uniform distribution, leads to greater differences between expected values and the deterministic solutions, Figure 4.6. The same pattern recurs -- overestimation of $S(t)$ and under-estimation of $R(t)$, with gradual convergence at higher dose rates.

Figures 4.7 and 4.8 represent a comparison of the results obtained at 5 D_0/cycle for all the deterministic and statistical cases considered, in terms of total surviving fraction of the population of irradiated cells. It allows comparison of the rates of cell population depletion for the deterministic solutions and both statistical modes. At the lowest dose rate considered -- 5 D_0/cycle -- there is some real difference between the predictions obtained using the various conditions. This is especially clear when we consider that the curve for 0.8 resistant to 0.2 sensitive, near the mean of the two extreme conditions, still underestimates the total surviving fraction, that is, over-estimates the amount of cell killing, as predicted by the better statistical estimate. At still lower doses the divergence would be greater yet. It is generally true that when significant differences between the deterministic and statistical behavior of a model exist, researchers using such models will obtain better estimates with the use of a statistical formulation. Furthermore, it is well known that the stochastic results should converge to the deterministic behavior if the phenomenon was in fact deterministic.

Work to extend this study to statistical treatment of some of the other parameters inherent in the model are in progress.

Acknowledgments

The authors wish to express their gratitudes to Dr. James Thompson of the M. D. Anderson Hospital and Tumor Institute for helpful discussions and providing the data utilized.

REFERENCES

[1] Barendsen, G. W., *Theoretical and Experimental Biophysics*, ed. Arthur Cole, Marcel Dekkor, Inc., New York, New York, 1967, 167-231.

[2] Brown, B. W., Thompson, J. R., Barkley, T., Suit, H., and Withers, H. R., *Theoretical considerations of dose rate factors influencing radiation strategy*, Radiobiology 110 (1974), 197-202.

[3] Brown, B. W., and Thompson, J. R., *Combat models as applied to radiotherapy*, Proc. 17th Conference on the Design of Experiments in Army Research, Development and Testing, Part I, 1972, 33-60.

[4] Burch, P. R. J., *Some aspects of relative biological efficiency*, Brit. J. Radio., 30(1957), 524-529.

[5] Elkind, M. M., and Sinclair, U. K., *Recovery in X-irradiation mammalian cells*, in Current Topics in Radiation Research, V. 1, ed. M. Ebert and A. Howard, North Holland Pub. Co., Amsterdam, 1965, 165-220.

[6] Elkind, N. M., *Sublethal X-ray damage and its repair, mammalian cells*, in Current Topics in Radiation Research, ed. G. Silini, North Holland Pub. Co., Amsterdam, 1967, 558-586.

[7] Fletcher, G. H., *Textbook of Radiotherapy*, Lea and Febiger, Philadelphia, Pennsylvania, 1966, 110, 161-8, 179-80, 439, and others.

[8] Gillette, E. L., Withers, H. R., and Tannock, I. F., *The age-sensitivity of epithelial cells of mouse small intestine*, Radiology 96(1970), 639-663.

[9] Hall, E. J., Bedford, J. S., *Dose Rate: Its effects on the survival of He La cells irradiated with gamma rays*, Rad. Res. 22(1964), 305-315.

[10] Mauro, F., and Madoc-Jones, *Age response to X-radiation of murine lymphomia cells synchronized in vivo*, Proc. Nat. Acad. Sci. U.S. 63(1969), 686-91.

[11] Pierquin, B., and Dutreix, A., *Toward a new system in Curie-therapy*, Brit. J. Radiol. 40(1967), 184-86.

[12] Regand, C., *Radium therapy of cancer at the Radium Institute of Paris, technique, biological principles and results*, Am. J. Roentgeol. Rad. Therapy and Nuclear Med. 21(1929), 1-24.

[13] Rossi, H. H., Biavati, M. H., and Gross, W., *Local energy density in irradiated tissues, radiobiological significance*, Rad. Res. 15(1961), 431-39.

[14] Sinclair, U. K., *Dependence of radiosensitivity upon cell*

age in time vs. dose relationships in radiation biology as applied to radiotherapy, Brookhaven National Laboratory, Pub. 50203 (C-57), 1970, 97-107.

[15] Soong, T. T., *Random Differential Equations in Science and Engineering*, Academic Press, New York, New York, 1973, 327.

[16] Terasima, J., and Tolmack, L. J., *Variations in several responses of He La cells to X-irradiation during the division cycle*, Biophys. J., 3(1963) 11-33.

[17] Thompson, L. H., and Humphrey, R. M., *Response of mouse L-P 59 cells to X-irradiation in the G_2 phase*, Int. J. Radiation Biol. 15(1969), 181-84.

[18] Whitmore, G. F., Gulyas, S., and Botond, J., *Radiation sensitivity through the cell cycle and its relationship to recovery in cellular radiation biology*, Proc. 18th Am. Symp. Fundamental Cancer Research, Houston, Williams and Wilkins, Baltimore, Maryland, 1965, 423-441.

[19] Wideroe, R., *Radiobiology and radiotherapy*, Ann. N. Y. Acad. Sci. 161(1969), 357-67.

[20] Withers, H. R., *Capacity for repair in cells of normal and malignant tissues in time vs. dose relationships in radiation biology as applied to radiotherapy*, Brookhaven National Laboratory, pub. 50203. (C-57), 1970, 54-69.

VOLTERRA INTEGRAL EQUATIONS AND NONLINEAR SEMIGROUPS

G. F. Webb
Vanderbilt University

We are concerned with the nonlinear Volterra integral equation

$$(1) \qquad x(t) = y(t) + \int_0^t g(t - s, x(s))ds, \quad t \geq 0,$$

where H is a Hilbert space, $y : [0,\infty) \to H$ is given, $g : [0,\infty) \times H \to H$ satisfies a Lipschitz condition in its second place, and $x : [0,\infty) \to H$ is the unknown function. Our approach is to associate with the solutions of (1) a semigroup of nonlinear operators acting in a Banach space X, so that (1) is then converted to an abstract ordinary differential equation in X. Our theory applies to equations of a more general form than (1), and so we reformulate (1) as a special case of the nonlinear functional equation

$$(2) \quad x(t) = h - G(\phi) + G(x_t), \; t \geq 0, \; x_0 = \phi \in L^p, \; x(0) = h \in H.$$

In (2) we suppose that $1 \leq p < \infty$, $r > 0$ or $r = \infty$, H is a Hilbert space, $L^p = L^p(-r,0;H)$, $X \overset{def}{=\!=} L^p \times H$, $G : L^p \to H$ is Lipschitz continuous, and $x_t \in L^p$ is defined for each $t \geq 0$ by $x_t(\theta) = x(t + \theta)$ for a.e. $\theta \in (-r,0)$. The following theorem is proved:

<u>THEOREM</u>. For each $\{\phi,h\} \in X$ there exists a unique solution $x(\phi,h)(t)$ to (2). Define the family of nonlinear operators $T(t)$, $t \geq 0$ in X by $T(t)\{\phi,h\} \overset{def}{=\!=} \{x_t(\phi,h), x(\phi,h)(t)\}$. Then,

(i) $T(t + s) = T(t)T(s)$, s, $t \geq 0$;

(ii) $||T(t)\{\phi,h\} - T(t)\{\psi,k\}||_X \leq Me^{\omega t}||\{\phi,h\} - \{\psi,k\}||_X$ for $t \geq 0$, $\{\phi,h\}$, $\{\psi,k\} \in X$, where M,ω are constants and, in general, $M > 1$;

(iii) $T(t)\{\phi,h\}$ is continuous as a function from $[0,\infty)$ to X for each fixed $\{\phi,h\} \in X$;

(iv) the infinitesimal generator A of $T(t)$, $t \geq 0$ (that is, the mapping $A\{\phi,h\} \overset{def}{=} \lim_{t\to 0} t^{-1} (T(t)\{\phi,h\} - \{\phi,h\}))$ is $A\{\phi,h\} = \{\phi', \lim_{t\to 0} t^{-1} (G(x_t(\phi,h)) - G(\phi))$ with $D(A) = \{\{\phi,h\} :$ ϕ is absolutely continuous, $\phi' \in L^p$, $h = \phi(0)$, and $\lim_{t\to 0} t^{-1} (G(x_t(\phi,h)) - G(\phi))$ exists$\}$.

If in addition, G is Frechet differentiable at each $\phi \in X$ and there exists a constant C such that for all $\phi_1, \phi_2, \psi_1, \psi_2 \in X$

$$|G'(\phi_1)\psi_1 - G'(\phi_2)\psi_2| \leq C(||\phi_1 - \phi_2|| + ||\psi_1 - \psi_2||),$$

$$|G'(\phi_1)\psi_1| \leq C(1 + ||\phi_1||)||\psi_1|| ,$$

then

(v) $A\{\phi,h\} = \{\phi', G'(\phi)\phi'\}$ for all $\{\phi,h\} \in D(A)$;

(vi) $D(A)$ is dense in X;

(vii) A is closed in the sense that if $\{\phi_n,h_n\}$ converges to $\{\phi,h\}$, $\{\phi_n,h_n\} \in D(A)$, and $A\{\phi_n,h_n\}$ converges to $\{\psi,k\}$ then $\{\phi,h\} \in D(A)$ and $A\{\phi,h\} = \{\psi,k\}$

(viii) if $\{\phi,h\} \in D(A)$, then $T(t)\{\phi,h\} \in D(A)$ for $t \geq 0$, and $d/dt\ T(t)\{\phi,h\} = AT(t)\{\phi,h\}$ for $t \geq 0$, $T(0)\{\phi,h\} = \{\phi,h\}$.

Invited Addresses and Research Reports

THE MATHEMATICAL ANALYSIS OF A FOUR-COMPARTMENT STOCHASTIC MODEL OF ROSE BENGAL TRANSPORT THROUGH THE HEPATIC SYSTEM

D. H. Anderson[1,3*], J. Eisenfeld[2,3],
S. I. Saffer[3], J. S. Reisch[3], & C. E. Mize[4]

Introduction

Clinical differentiation among the major categories of liver disease is quite difficult. To hopefully aid in this diagnostic differentiation, a test based on measuring the disappearance rate of radioactive Rose Bengal leads to a four-compartment mathematical model of tracer transport. It is the purpose of this paper to present the findings of the mathematical analysis of this model.

Development of the Mathematical Model

The physiological basis for the model centers on the transfer of bilirubin, the details of which have been presented elsewhere [7,8]. Bilirubin transport, which forms the basis for many tests used in determining the nature of certain blood and liver disorders, is especially important in this model because it is the excess of bilirubin in the blood and body tissues that causes jaundice, the yellow discoloration associated with hepatic disease. Due to recent studies and normal physiology, the idealized pathway of radioactive Rose Bengal transport through the biliary system suggested is illustrated in Figure 1, in which the first, second, and fourth compartments are sampled in time.

[1] Department of Mathematics, Southern Methodist University, Dallas Texas 75275.

[2] Department of Mathematics, University of Texas at Arlington, Arlington, Texas 76019.

[3] Department of Medical Computer Science, University of Texas Health Science Center, Dallas, Texas 75235.

[4] Department of Pediatrics, University of Texas Health Science Center, Dallas, Texas 75235.

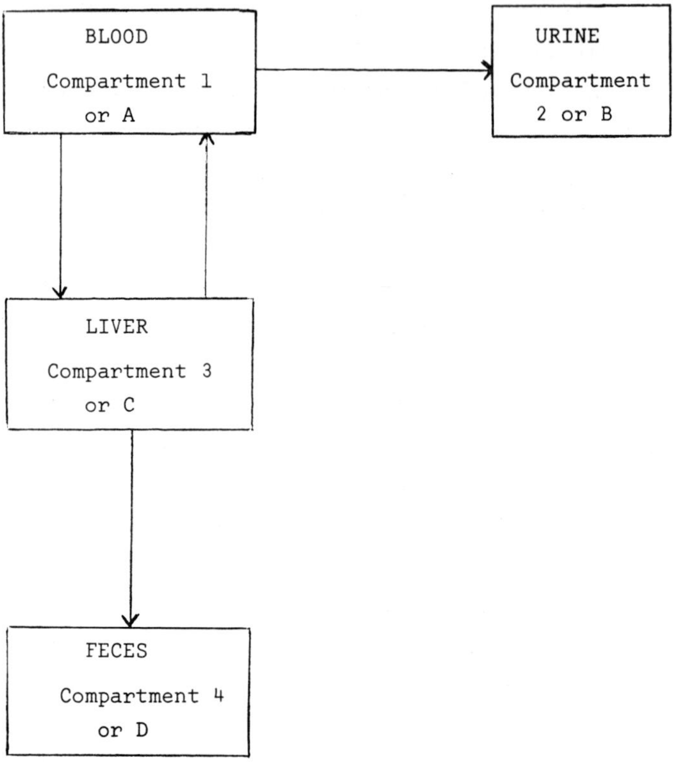

Figure 1. Pathway of Rose Bengal

When a radioactive tracer is introduced into a system, it will be distributed into the various parts of the system. The mathematical study of these processes comes under the general heading of compartmental analysis [5]. If the simplest visualization of the processes is assumed, that is, that the time rate of change of the amount $X_i(t)$ of tracer in any compartment i is given by the linear "input minus output" to that compartment, then the basic mathematical model consists of a system of first order linear differential equations with constant fractional transfer coefficients k_{ij} from compartment j to i. This system can be written in matrix form as

$$\dot{X}(t) = KX(t), \quad X(0) = X_0,$$

for time $t \geq 0$.

One of the reasons that classical compartmental analysis has limited success in its attempt to reflect hepatic function is due to the sensitivity of curve fitting of linear combinations of exponential functions to a small number of data points, and in a reasonable period of time (say, 72 hours) it may be possible to obtain only three or four urine or fecal specimens. Another related shortcoming of classical compartmental analysis techniques is the short interval of time required between blood samples, and thus the large number of samples that must be taken and analyzed--making the technique not very appealing as a clinical procedure.

The alternative method that is proposed herein is a discretized model that will utilize the small number of urine and fecal data points and allows blood samples to be fewer in number and thus spaced further apart.

A finite difference approach is taken for the differential equation. Let h be a small positive number. For all nonnegative integers m, let $t_m \equiv mh$ and $X_m \equiv X(t_m)$. Thus $\dot{X} = KX$ implies

$$(X_{m+1} - X_m)/h \approx KX_m$$

or

(1) $$X_{m+1} \approx (I + hK)X_m.$$

Note that

$$X_m \approx (I + hK)^m X_0$$

and

$$X(t) = e^{tK} X_0 = \lim_{h \to 0^+} (I + hK)^{t/h} X_0.$$

Let

(2) $$P \equiv [P_{ij}] \equiv (I + hK)^*$$

where * denotes transpose of the matrix. From (1), the basic equation for the discrete case should be

$$X_{m+1} = P^* X_m$$

for all $m \geq 0$. The physiological model illustrated in Figure 1 and the sampling procedures strongly suggest a "closed" compartmental system [5]; hence $K \equiv [K_{ij}]$ must satisfy the fundamental condition

(3) $$-k_{ii} = \sum_{j \neq i} k_{ji}$$

for all i, where $k_{ji} \geq 0$ if $j \neq i$, $k_{ii} \leq 0$ for all i, and the summation is over all j except for $j = i$. By equations (2) and (3),

$$P_{ii} + \sum_{j \neq i} P_{ij} = (hk_{ii} + 1) + \sum_{j \neq i} hk_{ji}$$

$$= 1 + h(k_{ii} + \sum_{j \neq i} k_{ji}) = 1,$$

so that each row sum of P is unity. Also if $i \neq j$, then always $p_{ij} = hk_{ji} \geq 0$; provided $i = j$, then $p_{ii} = hk_{ii} + 1$ for all i, and for those integers i such that $k_{ii} \neq 0$, then $p_{ii} \geq 0$ under the condition that h is chosen no larger than $-k_{ii}^{-1}$. Hence it appears that, to represent the activity of this hepatic system, the matrix P could be stochastic. To support this notion, suppose $p_{ij} (i \neq j)$ is interpreted as the portion (percentage) of radioactive Rose Bengal in the ith compartment transferred to the jth compartment during, say, a one hour time period, and assume that these p_{ij}'s remain constant over the time range in which patient data is collected. The Markov chain interpretation of the system is that $\{p_{ij}\}$ is the set of transition probabilities and the four states are blood (b), urine (u), liver (l), and feces (f). The information given in Figure 2 says that the transition matrix is

$$P = \begin{array}{c} \\ b \\ u \\ l \\ f \end{array} \begin{array}{c} \begin{array}{cccc} b & u & l & f \end{array} \\ \begin{bmatrix} p_{11} & p_{12} & p_{13} & 0 \\ 0 & 1 & 0 & 0 \\ p_{31} & 0 & p_{33} & p_{34} \\ 0 & 0 & 0 & 1 \end{bmatrix} \end{array}$$

where

$$p_{11} \equiv 1 - p_{12} - p_{13}, \quad p_{33} \equiv 1 - p_{31} - p_{34};$$

the stochastic condition on P requires that

$$0 < p_{12} + p_{13} < 1, \quad 0 < p_{31} + p_{34} < 1.$$

The system of difference equations which relates the amounts of tracer in the different compartments over time is

$$X(t_{m+1}) = P^*X(t_m), \quad X(0) = X_0,$$

by (1) and (2), or in component form,

$$A(t + 1) = p_{11}A(t) + p_{31}C(t),$$

$$B(t + 1) = p_{12}A(t) + B(t),$$

$$C(t + 1) = p_{13}A(t) + p_{33}C(t),$$

$$D(t + 1) = p_{34}C(t) + D(t),$$

$$A(0) = 1, \quad B(0) = C(0) = D(0) = 0,$$

where $A(t)$ is the amount of radioactive Rose Bengal present in the blood compartment at time t, and $t = 0,1,2,\ldots,72$ (typical sampling is over 72 hours with blood samples at hours 4, 12, 24, 48 and 72; fecal and urine samples at hours 24, 48, and 72).

The use of numbers p_{ij} between zero and one (percentages) rather than quantitative rates gives certain advantages of this model over the differential equation model: (1) the number of parameters to be estimated is lowered, thus the present model is more suitable for clinical studies in which the number of data points is small; (2) since each patient's biliary system may be functionally different, a percentage transferred may better reflect liver function than actual quantitative amounts; (3) convenient linear constraints and upper and lower bounds can be placed on the p_{ij}'s

Analysis of the Mathematical Model

Observe that there are two absorbing states in the model and that the Markov chain is absorbing. Rewritten in the usual canonical form [1,6], the transition matrix is

$$C \equiv \begin{array}{c} \\ u \\ f \\ b \\ l \end{array} \begin{array}{c} \begin{array}{cccc} u & f & b & l \end{array} \\ \left[\begin{array}{cc|cc} 1 & 0 & 0 & 0 \\ 0 & 1 & 0 & 0 \\ \hline p_{12} & 0 & p_{11} & p_{13} \\ 0 & p_{34} & p_{31} & p_{33} \end{array}\right] \end{array} \equiv \left[\begin{array}{c|c} I & O \\ \hline R & Q \end{array}\right].$$

This canonical form yields

$$C^m = \left[\begin{array}{c|c} I & O \\ \hline (I + Q + \ldots + Q^{m-1})R & Q^m \end{array}\right] \tag{4}$$

for any positive integer m.

To consider the asymptotic behavior of the Markov chain, which is reflected by $\lim_{m\to\infty} C^m$, the limit of Q^m as $m \to \infty$ must be considered. From the standard theory of absorbing Markov chains [1,6],

(5) $$\lim_{m\to\infty} Q^m = 0;$$

a simpler proof of this fact can be given for this particular model. For, let λ be any eigenvalue of the matrix Q, and let $||\cdot||_\infty$ be the matrix (row) norm induced by the vector supremum norm [2, p.67]. Since the magnitude of any eigenvalue of a matrix cannot exceed a matrix norm of that matrix, and since each row of R has a positive entry, then

$$|\lambda| \leq ||Q||_\infty = max\{p_{11} + p_{13}, p_{31} + p_{33}\} < 1.$$

Hence, by a well-known theorem of matrices, equation (5) holds.

Now consider a closed form for the limit as $m \to \infty$ of the lower left matrix in the partitioned form of C^m. That corner matrix is equal to

(6) $$\left(\sum_{i=0}^{m-1} Q^i\right) (I - Q)(I - Q)^{-1} R = (I - Q^m)(I - Q)^{-1} R$$

provided $(I - Q)^{-1}$ exists. Again, by the standard theory of absorbing Markov chains, this inverse matrix does exist. However, an alternative verification can be given for this fact. For, it has been noted that $||Q||_\infty < 1$. Now assume to the contrary that $I - Q$ is singular. Then there is a 2×1 vector $x \neq 0$ such that $(I - Q)x = 0$ or $x = Qx$. As a consequence,

$$||x||_\infty = ||Qx||_\infty \leq ||Q||_\infty \, ||x||_\infty < ||x||_\infty,$$

a contradiction.

This inverse matrix

(7) $$M \equiv \begin{array}{c} \\ b \\ l \end{array} \begin{array}{c} \begin{array}{cc} b & l \end{array} \\ \begin{bmatrix} m_{11} & m_{12} \\ m_{21} & m_{22} \end{bmatrix} \end{array} \equiv (I - Q)^{-1}$$

is called the fundamental matrix for the Markov chain, and regarding b as the first nonabsorbing state and l as the second nonabsorbing state, then m_{ij} has the interpretation of giving the expected number of times the Markov process is in state j, having started initially in state i, before the process enters an absorbing state [1,6].

Equations (4) - (6) yield

$$C^{\infty} \equiv \lim_{m\to\infty} C^{m} = \left[\begin{array}{c|c} I & 0 \\ \hline (I - Q)^{-1}R & 0 \end{array}\right],$$

which is a stochastic matrix - an inherited property through the limit from C^m. Hence

(8) $$F \equiv (I - Q)^{-1}R \equiv MR$$

likewise is stochastic and thus can be represented by

(9) $$F \equiv \begin{array}{c} \\ b \\ l \end{array}\begin{array}{c} \begin{array}{cc} u & f \end{array} \\ \left[\begin{array}{cc} F_1 & 1 - F_1 \\ F_2 & 1 - F_2 \end{array}\right] \end{array}.$$

This reduces by two the number of entries in the F-matrix that must be estimated. The proper interpretation is that the i, j-entry of F gives the probability that the system ends in absorbing state j, starting from the nonabsorbing state i.

Let $t = 0,1,2,\ldots,72$. As an immediate consequence of the definition of $A(t)$, $B(t)$, $C(t)$, and $D(t)$,

(10) $$A(t) + B(t) + C(t) + D(t) = 1$$

for all t, so that for any hour t on which each of the blood, urine, and feces is sampled, $C(t)$, the amount of radioactive Rose Bengal present in the liver at time t, is given by

$$1 - A(t) - B(t) - D(t).$$

Since the Rose Bengal is initially only in the blood, then

(11) $$[(B(t)\ D(t))\ (A(t)\ C(t))] = [(0\ 0)\ (1\ 0)]C^{t}$$
$$= [(1\ 0)(I - Q^{t})F\ (1\ 0)Q^{t}],$$

where equations (4), (6), and (8) have been utilized. These partitioned matrices in turn imply that

(12) $$(B(t)\ D(t)) = (1\ 0)(I - Q^t)F$$

and

(13) $$(A(t)\ C(t)) = (1\ 0)Q^t.$$

From the second and fourth difference equations, it follows that

$$B(t+1) > B(t), \quad D(t+1) > D(t).$$

Thus the amounts of tracer in the urine and feces increase as time passes. Moreover if time was allowed to increase without bound, and since

$$\lim_{t\to\infty} Q^t = 0,$$

then an appeal to equation (12) yields that

$$\lim_{t\to\infty} B(t) = F_1, \quad \lim_{t\to\infty} D(t) = 1 - F_1.$$

Similarly, (13) kicks out the result that ultimately $A(t)$ and $C(t)$ decline, and in fact approach zero as $t \to \infty$.

Equation (13) can also be substituted in (12):

$$\begin{aligned}(B(t)\ D(t)) &= (1\ 0)F - ((1\ 0)Q^t)F\\ &= (1\ 0)F - (A(t)\ C(t))F.\end{aligned}$$

In component form this reduces to the two equations

(14) $$B(t) = (1 - A(t))F_1 - C(t)F_2,$$

(15) $$D(t) = 1 - F_1 - A(t)(1 - F_1) - C(t)(1 - F_2),$$

for all t. Equation (15) can be rewritten as

$$(F_1 - A(t)F_1 - C(t)F_2) + A(t) + C(t) + D(t) = 1.$$

If the left hand side of equation (14) is substituted into the last equation, then (10) results. Thus, of the two equations (14) and (15), only (14) is retained since (15) reduces to (10) and as a result indicates no new information.

Statement (14) relates the entries of the matrix F directly to the patient data. Also a pair of equations can be written which connects F directly to Q. The natural starting point is

$$R = (I - Q)F. \tag{16}$$

Since the elements of R can be expressed in terms of the elements of Q,

$$p_{12} = 1 - p_{11} - p_{13},$$

$$p_{34} = 1 - p_{31} - p_{33},$$

equation (16) becomes

$$\begin{bmatrix} 1-p_{11}-p_{13} & 0 \\ 0 & 1-p_{31}-p_{33} \end{bmatrix} =$$

$$\begin{bmatrix} (1 - p_{11})F_1 - p_{13}F_2 & (1- p)(1- F_1) - p_{13}(1- F_2) \\ -p_{31}F_1 + (1 - p_{33})F_2 & -p_{31}(1 - F_1) + (1 - p_{33})(1 -F_2) \end{bmatrix}$$

The resulting four component equations have two which are redundant; the other two can be expressed as

$$F_2 = F_1p_{31} + F_2p_{33}, \tag{17}$$

$$1 - F_1 = (1 - F_1)p_{11} + (1 - F_2)p_{13}, \tag{18}$$

which gives a straightforward relationship between the elements of F and Q.

Equations (10), (14), (17), and (18) constitute basic relationships in the model. From them much information can be gleaned. For instance, (17) and (18) allow one to find the entries of F and M in terms of the p_{ij}'s, and to get orderings among these entries. For, the last two equations can be viewed as linear equations in F_1 and F_2. Let

$$s \equiv p_{12}(p_{31} + p_{34}),$$

$$d \equiv s + p_{13}p_{34}.$$

The constraints on the p_{ij}'s show that $0 < d < 1$. The determinant of coefficients of the system (17) - (18) is $d \neq 0$; thus there always exists a unique solution for F_1 and F_2 in terms of the p_{ij}'s. These are calculated to be

$$F_1 = s/d, \quad F_2 = p_{12}p_{31}/d.$$

Clearly, $F_1 > F_2$. Moreover, the matrix

$$M = FR^{-1} = \begin{bmatrix} F_1 & 1 - F_1 \\ F_2 & 1 - F_2 \end{bmatrix} \begin{bmatrix} p_{12}^{-1} & 0 \\ 0 & p_{34}^{-1} \end{bmatrix}$$

$$= \begin{bmatrix} F_1 p_{12}^{-1} & (1 - F_1)p_{34}^{-1} \\ F_2 p_{12}^{-1} & (1 - F_2)p_{34}^{-1} \end{bmatrix} .$$

Thus

$$m_{11} = F_1 p_{12}^{-1} > F_2 p_{12}^{-1} = m_{21}$$

$$m_{22} = (1 - F_2)p_{34}^{-1} > (1 - F_1)p_{34}^{-1} = m_{12} .$$

The inequality $F_1 > F_2$ says that the probability that Rose Bengal will eventually end up in the urine (absorbing state) having stated from the blood is greater than the probability that the Rose Bengal will end up in the urine having started from the liver. Another inequality, $m_{11} > m_{21}$, indicates that the expected number of hours in the blood before eventually ending in the urine having started from the blood is greater than the expected number of hours in the blood before ending in the urine having started from the liver.

Estimation of the Model's Parameters

Probably the most important problem and certainly the most

difficult task in the analysis of the model is that of obtaining estimates of the model's parameters. As has already been shown, every parameter of interest can be expressed in terms of the four transfer paramenters p_{12}, p_{13}, p_{31}, and p_{34}, so it is sufficient to devise a method of estimating just these four parameters.

A crude upper estimate on each of the four p_{ij}'s can be found by a simple least-squares fit of data to equation (14) re-written as

$$(19)\quad (A(t) + B(t) + D(t) - 1)/B(t) = (F_1/F_2)((A(t) - 1)/B(t)) + 1/F_2 .$$

The data used can be only for those hours at which the blood, urine, and feces are all sampled (typically, $t = 24$, 48, and 72, so that three data points are used). Equation (19) is interpreted in the form of an affine function

$$y = mx + b.$$

Let $\hat{m}$ and $\hat{b}$ be the optimal linear regression parameter estimates associated with (19). Define $\hat{F}_1 \equiv \hat{m}/\hat{b}$ and $\hat{F}_2 \equiv 1/\hat{b}$. Suppose it turns out that $\hat{F}_1 > \hat{F}_2$ (as it should be since in the model $F_1 > F_2$). Then from equations (17) and (18), the following estimates of p_{ij} - ratios are obtained:

$$p_{12}/p_{13} = (\hat{F}_1 - \hat{F}_2)/(1 - \hat{F}_1),$$

$$p_{31}/p_{34} = \hat{F}_2/(\hat{F}_1 - \hat{F}_2).$$

Thus by the linear constraints on the p_{ij}'s,

$$((1 - \hat{F}_1)/(\hat{F}_1 - \hat{F}_2)) + 1 = (p_{13} + p_{12})/p_{12} < 1/p_{12},$$

and hence

$$p_{12} < \frac{\hat{F}_1 - \hat{F}_2}{1 - \hat{F}_2} .$$

In like manner,

$$p_{13} < \frac{1 - \hat{F}_1}{1 - \hat{F}_2} ,$$

$$p_{31} < \hat{F}_2/\hat{F}_1,$$

$$p_{34} < (\hat{F}_1 - \hat{F}_2)/\hat{F}_1.$$

To visualize these bounds, consider the percentages of injected Rose Bengal remaining in measured body compartments for Patient D.

Hour	Blood	Urine	Feces
4	34.19		
12	21.50		
24	16.60	20.63	4.24
48	11.21	32.78	33.08
72	6.65	43.46	45.42

For this patient, $\hat{F}_1 = .4668$, $\hat{F}_2 = .3171$, and the bounds are

$$p_{12} < .2191 \qquad p_{13} < .7809$$

$$p_{31} < .6793 \qquad p_{34} < .3207$$

There are no cases among the patients tested where $\hat{F}_1$ is not less than unity. Unfortunately, there exist instances when $\hat{F}_2 \geq \hat{F}_1$. This is the price one pays for not using all the data and/or for using a simplified approximation method. In such cases, of course, the values of $\hat{F}_1$ and $\hat{F}_2$ are not used.

A nonlinear programming (NLP) method which calculates these transfer parameters has been developed by Saffer [8,7] and consists of a combination of Monte Carlo trials, Gauss least squares, and direct pattern search techniques. An alternative method has now been developed which seems to give better accuracy and is much faster. A great deal of time and effort was spent in matching algorithms and functions to be analyzed. Different combinations precipitated a variety of difficulties such as ill-conditioning of matrices, matrices with small enough determinants as to be considered computationally singular, computed stepsizes so small as to render unattainable the convergence of the algorithm, and con-

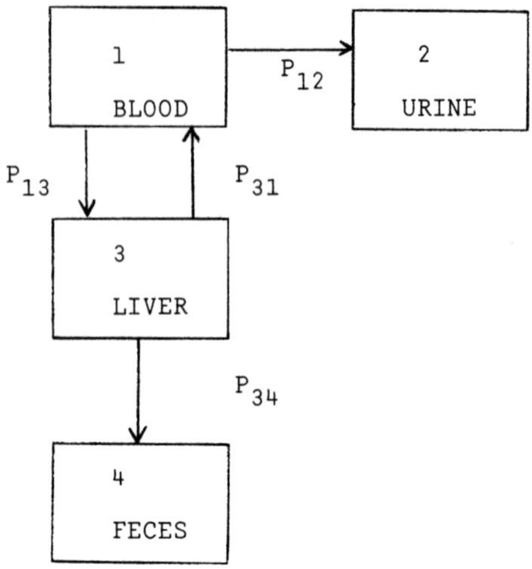

Figure 2

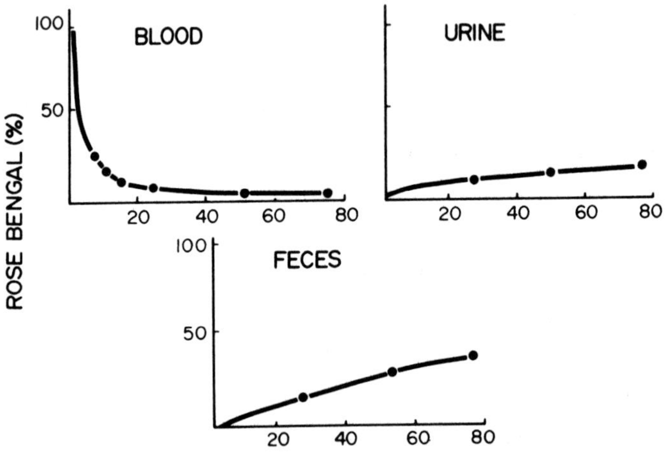

Figure 3

vergence of routines to points outside the feasible region of solution. Finally, the following program proved successful. This particular program is executed to compute the p_{ij} -values for Patient D.

```
      PROGRAM MINFZX (INPUT,OUTPUT)
      EXTERNAL FUNCT
      DIMENSION X(4),H(10),G(4),W(12)
      N=4
      NSIG=5
      MAXFN=1000
      IOPT=0
      DATA X/.06,.38,.09,.007/
      CALL ZXMIN(FUNCT,N,NSIG,MAXFN,IOPT,X,H,G,F,W,IER)
      PRINT 9,X(1),X(2),X(3),X(4),F,W(2),W(3),IER
    9 FORMAT(1H0,F13.5)
      END
      SUBROUTINE FUNCT(N,X,F)
      DIMENSION X(N),A(73),B(73),C(73),D(73)
      A(1)=1.
      B(1)=0.
      C(1)=0.
      D(1)=0.
      DO 5 I=1,72
      A(I+1)=A(I)*(1.-X(1)-X(2))+X(3)*C(I)
      B(I+1)=B(I)+X(1)*A(I)
      C(I+1)=C(I)*(1.-X(3)-X(4))+X(2)*A(I)
      D(I+1)=D(I)+X(4)*C(I)
    5 CONTINUE
      FAA=(.3419-A(5))**2.+(.2150-A(13))**2.
      FA=(.1660-A(25))**2.+(.1121-A(49))**2.+(.0665-A(73))**2.
      FB=(.2063-B(25))**2.+(.3278-B(49))**2.+(.4346-B(73))**2.
      FD=(.0424-D(25))**2.+(.3308-D(49))**2.+(.4542-D(73))**2.
      F=FAA+FA+FB+FD
      RETURN
      END
```

In the algorithm, *X(1),..., X(4)* replace p_{12}, p_{13}, p_{31}, and p_{34} respectively. The point of the program is to select that set of *X(J)* - values, $J = 1,2,3,4,$ in the feasible region so as to "best" reconstruct via the difference equations the measured patient data. The criterion for "best" is to select the four *X(J)* - values which minimize the sum-of-squares function F. The routine used to carry out the minimization is the ZXMIN subroutine in the IMSL (International Mathematical and Statistical Library) Version 5 Package, which uses a quasi-Newton method for finding the minimum of a real function of N variables. It is based on the Harwell library routine VA10A, Harwell, England [4]. NSIG is an input convergence criterion which is the number of digits of accuracy required in the parameter estimates; the criterion is satisfied if on two successive iterations, the parameter estimates $X(J)$, $J = 1,\ldots,4$, agree to NSIG digits. MAXFN is an input which is the maximum number of calls to the subroutine FUNCT allowed. In the execution of program MINFZX for various patients, it is found that MAXFN = 300 is normally sufficient with NSIG = 5. The initial parameter estimates of $X(1) = .06$, $X(2) = .38$, $X(3) = .09$, and $X(4) = .007$ are selected on the basis of average p_{ij} - values as computed for previous patients. The results of running the program suggest that the function F has a unique global minimum; to test this many random (but reasonable) starting values of $X(J)$ are used. The output of MINFZX for Patient D is

$$p_{12} = .0321 \qquad p_{13} = .2619$$

$$p_{31} = .1071 \qquad p_{34} = .0153,$$

and the value of the function at the final parameter estimates is $F = .0266$. Also printed is $W(2) = 115$, the number of function evaluations (calls to subroutine FUNCT) performed, $W(3) = 5.9$, an estimate of the number of significant digits in the final parameter estimates, and IER = 0, an error parameter which when set to zero on output implies that convergence was achieved and no errors occurred. The total CPU execution time for this program is 1.05 seconds on a CDC CYBER 70, model 72.

The following table gives a comparison of the results of the two methods for Patient D

Method	p_{12}	p_{13}	p_{31}	p_{34}	F_1	$1 - F_1$	m_{11}	m_{12}
MINFZX	.0321	.2619	.1071	.0153	.4950	.5049	15.42	33.00

NLP	.032	.26	.12	.015	.49	.51	15	32

Once the p_{ij}'s have been computed by the MINFZX program, there is a method of double checking their values. Certain inequality relationships can be established directly between the data and the p_{ij}'s; these inequalities are derived from the difference equations. For instance, the second difference equations implies that

$$B(t + 1) \geq p_{12}A(t)$$

for all $t = 0,1,\ldots,72$. Since $A(0) = 1$ and $B(1) = p_{12}$, then inductively it follows that

$$p_{12} < \min_{t\geq 2} B(t). \tag{20}$$

Likewise the first difference equation implies that

$$A(t + 1) > (1 - p_{12} - p_{13})A(t)$$

for all $t = 0,1,\ldots,72$. This along with $A(0) = 1$ allows the conclusion by an inductive argument that

$$A(t) \geq (1 - p_{12} - p_{13})^t$$

for all $t \geq 1$, with equality only when $t = 1$. This last statement suggests some feedback to the clinician, for if the blood can be sampled at hour one, then the estimate $1 - A(1)$ is obtained for $p_{12} + p_{13}$. If this blood sample is not taken, the inequality still allows the linear constraint on $p_{12} + p_{13}$ to be tightened to

$$1 - \min_{t\geq 1} A(t)^{1/t} \leq p_{12} + p_{13} < 1. \tag{21}$$

The fourth difference equation says that $D(2) = p_{13}p_{34}$. Also $D(1) = D(0) = 0$. Since $D(t)$ is an increasing function, the inequality

$$p_{13}p_{34} \leq \min_{t\geq 2} D(t) \tag{22}$$

obtains, with equality only when $t = 2$. Finally, a combination of the first and third equations yields

$$A(t+1) > p_{31}C(t) > p_{13}p_{31}A(t-1)$$

for all $t \geq 1$. This observation allows that

$$A(2t) > (p_{13}p_{31})^t$$

or that

$$p_{13}p_{31} < \min_{t \geq 1} A(2t)^{1/t} \tag{23}$$

obtains. Thus inequalities (20) - (23) provide simple checks of the computed p_{ij}'s against the patient data.

Observations and Conclusions

An attempt was made to discover any patterns among the estimated parameter values that would hint at obtaining a more conclusive differentiation among physiological abnormalities of the hepatic system. It is observed that information of significance possibly may be obtained from the various ratios p_{ij}/p_{mn}. All ratios of this form are calculated, and the three that seem to have promise of indicating some differentiation are p_{12}/p_{34}, p_{13}/p_{34}, and p_{31}/p_{34}. For example, the ratio p_{31}/p_{34} gives the percentage of radioactive Rose Bengal recirculating to the blood compared to the percentage going to the feces. A large value of this ratio may be an indication of an obstruction within the patient.

A standard statistical discriminant analysis program, BMD07M [3], was run to see if the values of the three ratios along with the other data could be used to properly classify a control group of fourteen patients in their known diagnosis of atresia or hepatitis. Upon the selection by the program of the first discriminator (p_{31}/p_{34}), all patients but one were correctly classified; an atresia patient was incorrectly classified in the hepatitis group. On the next step, the initial variable along with the second selected (p_{12}/p_{34}) succeeded in classifying all patients correctly.

Finally, it is clear that the validation of the model will come about only through a comparison, for future patients, of the differential diagnosis prediction of the model with that diagnosis found clinically.

Acknowledgment

We would like to thank Patricia Daniel, Sonja Sandberg, and Sarah Stead for doing some of the programming and computing necessary in this research.

References

[1] Bhat, U.N., *Elements of Applied Stochastic Processes*, New York, John Wiley and Sons, 1972.

[2] Blum, E.K., *Numerical Analysis and Computation: Theory and Practice*, Reading, Massachusetts, Addison-Wesley Publishing Co., 1972.

[3] Dixon, W.J., ed., *BMD Biomedical Computer Programs*, Second edition, University of California Press, 1968.

[4] Fletcher, R., *Fortran subroutines for minimization by quasi-Newton methods, Report R7125 AERE*, Harwell, England.

[5] Jacquez, J. A., *Compartmental Analysis in Biology and Medicine*, Amsterdam, Elsevier Publishing Co., 1972.

[6] Maki, D. P., and M. Thompson, *Mathematical Models and Applications*, Englewood Cliffs, New Jersey, Prentice-Hall, Inc., 1973.

[7] Saffer, S. I., et. al., *The comparison of a four-compartment and a five-compartment model of Rose Bengal transport through the hepatic system*, Proceedings of the International Conference on Nonlinear Systems and Applications, Academic Press, 1977.

[8] Saffer, S. I., et. al., *Use of nonlinear programming and stochastic modeling in the medical evaluation of normal-abnormal liver function*, IEEE Transactions on Biomedical Engineering, Vol. BME-23, No. 3, May, 1976.

A MORE EFFICIENT ALGORITHM FOR AN OPTIMAL TOUR OF A HEALTH CARE CONSUMER WITH MULTIPLE HEALTH CARE FACILITIES IN EACH COUNTY

A. J. G. Babu and Wei Shen Hsia
The University of Alabama

1. *Introduction*

An exposition of shortage [3] and distribution [4] of Alabama's supply of physicians was published recently. The detailed data and discussion [3] give a reasonable estimate of the shortage of physicians speciality-wise and county-wise in Alabama. W. D. Bottom [2] also gives an outline of the national health planning measures proposed by the government under PL 93-641. In Bottom's paper, it is pointed out that it will not be possible to have all types of health care facilities in every county.

Therefore, irrespective of how efficient the planning may be, it would not be possible to have all the health care facilities in every county or city. In [4], Bridgers commented "systems of care always seem to be based upon the premise that the consumers of health services are mobile but the providers are immobile". Hence the process of a single person, family, or a collection of people traveling to visit the health care facilities at various neighboring counties or cities is inevitable.

The problem concerned then is to find an optimal tour with a view to completing M specified jobs (or M specified health facilities needed by a patient) by visiting some or all N stations (or N health care centers). In contrast with the traveling salesman problem, it is possible here to complete one or more specified jobs at each station.

This problem was considered by Bansal and Kumar in their paper [1], in which a recurrence relation was developed to find the optimal tour. While the method provides the true optimal tour and the order in which stations are visited, the computational complexity would increase enormously as the number of the stations increases. It is the purpose of this paper to develop a more efficient algorithm by modifying theirs.

Bansal and Kumar's method is reviewed briefly in section 2. In section 3, the modified algorithm is presented, along with some discussion. An example to illustrate this algorithm is given in section 4.

2. *Bansal and Kumar's Algorithm* [1]

Suppose there are N stations, A_0, $A_1, \ldots, A_{n-1}$, where A_0 denotes the starting station. Assume it is possible to complete more than one job at each visited station A_i. The problem considered is that of finding an optimal tour to complete M given jobs. Therefore, the problem concerned is completing M jobs rather than visiting N stations.

Bansal and Kumar [1] give a recurrence relation to find the optimal tour:

$$f(A_0;n;M) = \min_i[(A_0,A_i) + f(A_i;n - 1;M)], \tag{1}$$

where

$n \leq N - 1$,

$[A_0,A_i]$ = shortest distance from A_0 to A_i, $f(A_i;n - 1;M)$ can be found from the recurrence relation:

$$f(A_i;r;m) = \min\{\min_{j\varepsilon J_{1,i}}[(A_i,A_j) + f(A_j;r - 1;p)], \quad \min_{j\varepsilon J_{2,i}}[(A_i,A_j) + f_1(A_j;r - 1;m')]\} \tag{2}$$

where

$M \geq m > p$ or $m';i = 1,\ldots,N - 1$; $r = 0,1,\ldots,N - 2$;

(A_i,A_j) = shortest distance between A_i and A_j.

$f(A_i;r;m)$ = the minimum length of the path from station A_i to A_0 after completing m distinct jobs by visiting r stations,

$f\ (A_i;r;m')$ = the second minimum length of the path from station i to home station after completing m' distinct jobs by visiting r stations.

$J_{1,i}$ = set of all j's such that the minimum distance path from A_j to A_0 does not include A_i.

$J_{2,i}$ = set of all j's such that the minimum distance path from A_j to A_0 includes A_i.

While this method provides a true optimal path and eliminates the possibility of errors, it gives no criterion for stopping the operation of the recurrence relation. Therefore, the computational effort might increase enormously as the size of the problem increases.

3. *Modified Algorithm*

Let $K = min\{M,N - 1\}$. For $k = 1,\ldots,K$, define

$S_1(k)$ = set of all possible routes for visiting k stations,

$S_2(k)$ = set of all possible routes for completing M jobs after visiting k stations,

and

$U(k)$ = shortest commuting distance among all routes in $S_2(k)$.

Note that $S_2(k)$ is a subset of $S_1(k)$ and $S_2(k)$ might be empty. Define $U(k) = +\infty$, if $S_2(k) = \phi$.

Assume the problem is solvable, i.e., there is at least one collection of stations via which all M jobs can be done. Let $T(0)$ be a fairly large number, say, 10^{99}. For $k = 1,\ldots,K$, the kth iteration consists

<u>STEP 1</u>: Is there a route in $S_1(k)$ such that the commuting distance of that route is less than $T(k-1)$? If the answer is "Yes", go to STEP 2. If the answer is "No", then $T(k - 1)$ is the optimal distance, stop.

<u>STEP 2</u>: Let $T(k) = min\{T(k - 1),U(k)\}$. Set $k = k + 1$, go to STEP 1. Note that the maximum number of stations required to be visited (excluding the starting station) to complete all M jobs is bounded by M and $N - 1$. Therefore, at most $K = min\{M, N - 1\}$ iterations are required in this algorithm.

Let S be an arbitrary collection of stations, denote the shortest commuting distance around all the stations in S by T_S.

A collection of stations S^* is called a refinement of S, if $S \subset S^*$. It is obvious that $T_S \leq T_{S^*}$, if S^* is a refinement of S.

LEMMA 1: If there is no set with i stations $(i = 1,\ldots,K)$ for which the commuting distance is less than $T(i - 1)$, then there is no set with j stations $(i < j \leq K)$, for which, the commuting distance is less than $T(i - 1)$.

PROOF. This is a result of the above observation, since each set with j stations $(j > i)$ is a refinement of some set with i stations.

Q.E.D.

COROLLARY 1.1: If the answer is "No" in STEP 1 of the algorithm, $T(k - 1)$ is the required optimal distance.

This corollary assures the stopping criterion set up in STEP 1.

LEMMA 2: $T(K - 1) = T(0)$ if and only if $S_2(k) = \phi$ for $k = 1, \ldots, K - 1$.

PROOF. This follows immediately from STEP 2.

Q.E.D.

From the assumption that the problem is solvable, $T(K)$ must be less than $T(0)$. Therefore, the algorithm will stop after at most K iterations. In fact, the algorithm generates a non-increasing sequence of real numbers and the algorithm will be stopped if it is impossible to continue to generate this non-increasing sequence.

Because the total computing time needed depends on K, it would save considerable computing time, if some pre-analysis is done before the algorithm starts. Some basic rules are stated below.

If station A_i in a tour adds no new facility to other stations of the tour, station A_i can be dropped from the tour. For instance, in the example given in [1], since station A_1 adds no new facility to those available at the rest of the stations A_2, A_3, A_4 in the tour $A_0A_1A_2A_3A_4A_0$, station A_1 can be dropped.

On the other hand, if station A_i contains a certain needed facility which is not available at any other station, then A_i must be a member of the optimal tour. Again, consider the same example given in [1], stations A_2, A_3, A_4 must be on the optimal tour. Since the tour $A_0A_2A_3A_4A_0$ also provides all required facilities, it is the optimal tour.

4. *Numerical Example*

Consider an example of 13 stations $(A_0, A_1, \ldots, A_{12})$ and 9 jobs $(J_1, \ldots, J_9)$ where an arbitrarily chosen distance matrix satisfying the triangle inequality with job facilities at each station is given below in Table 1.

JOB FACILITIES AVAILABLE		0	1	2	3	4	5	6	7	8	9	10	11	12
·	0	∞	21	59	95	100	82	61	93	103	96	66	44	38
$J_1J_2J_4$	1	21	∞	50	86	91	74	69	94	111	107	74	52	46
$J_3J_6J_9$	2	59	50	∞	45	51	44	39	64	82	81	75	59	53
$J_1J_2J_5$	3	95	86	45	∞	20	41	63	56	79	103	100	92	87
J_1J_5	4	100	91	51	20	∞	37	58	45	68	92	97	87	82
$J_3J_6J_8$	5	82	74	44	41	37	∞	31	36	68	72	68	70	55
$J_2J_5J_8$	6	61	69	39	63	58	31	∞	41	53	52	46	40	37
$J_4J_6J_7$	7	93	94	64	56	45	36	41	∞	36	55	78	72	69
$J_1J_2J_7$	8	103	111	82	79	68	68	53	36	∞	32	60	80	79
J_2J_6	9	96	107	81	103	92	72	52	55	32	∞	40	59	65
J_1J_7	10	66	74	75	100	97	68	46	78	60	40	∞	31	37
J_5J_6	11	44	52	59	92	87	70	40	72	80	59	31	∞	15
$J_6J_7J_8$	12	38	46	53	87	82	55	37	69	79	65	37	15	∞

TABLE 1

where it is assumed that no facility is available in A_0.

Table 2 to Table 5 exhibit all information of traveling distance involving 2 stations to 5 stations (not including A_0), respectively, and terminating with A_0. The following example explains how to use those tables. Suppose we want to find the minimum tour involving 4 stations (not including A_0) starting from A_2 and ending up with A_0: check the minimum entry of the 2nd row of table 4, the value is 132 at position (2,6); next check the minimum value in the 6th row of table 3, the value is 93 at position (6,11); then check the minimum value in the 11th row of table 2, the value is 53 at position (11,12). This method not only gives that the traveling distance is 132 (from table 4) but also that the associated tour is $A_2A_6A_{11}A_{12}A_0$. Hence, the total distance of the complete tour $A_0A_2A_6A_{11}A_{12}A_0$ is 59 + 132 = 191.

Since no tour involving only three stations (not including A_0) has all job facilities, by the definition of $T(i)$ defined in section 3, $T(0) = T(1) = T(2) = T(3) = 10^{99}$.

Table 4 shows that one can complete all jobs by visiting 4 stations say, $A_1A_2A_{11}A_{12}$. Therefore $S_2(4) \neq \phi$ and $T(4) =$ $21 + 162$ which is the distance from A_0 to A_1 plus the traveling distance of the tour $A_1A_2A_{11}A_{12}A_0$. In fact, there are several other tours which can complete all jobs and have traveling distances 183, e.g., $A_0A_{11}A_{12}A_2A_1A_0$, $A_0A_1A_2A_{12}A_{11}A_0$, etc.

Table 5 shows that there cannot be a route containing five stations (not including A_0) for which the commuting distance is less than or equal to 183. By the stopping criterion set up in STEP 1, $T(4) = 183$ is the optimal solution.

	1	2	3	4	5	6	7	8	9	10	11	12
1	∞	109	181	191	156	130	187	214	203	140	96	84
2	71	∞	140	151	126	100	157	185	177	141	103	91
3	107	104	∞	120	123	124	149	182	199	166	136	125
4	112	110	115	∞	119	119	138	171	188	163	131	120
5	95	103	136	137	∞	92	129	171	168	134	114	93
6	90	98	158	158	113	∞	134	156	148	112	84	75
7	115	123	151	145	118	102	∞	139	151	144	116	107
8	132	141	174	168	150	114	129	∞	128	126	124	117
9	128	140	198	192	154	113	148	135	∞	106	103	103
10	95	134	195	197	150	107	171	163	136	∞	75	75
11	73	118	187	187	152	101	165	183	155	97	∞	53
12	67	112	182	182	137	98	162	182	161	103	59	∞

TABLE 2

	1	2	3	4	5	6	7	8	9	10	11	12
1	∞	141	190	201	166	144	196	225	210	149	105	105
2	134	∞	152	163	136	114	166	196	184	150	112	112
3	170	116	∞	130	133	138	158	193	206	175	145	146
4	175	122	124	∞	129	133	147	182	195	172	140	141
5	158	115	145	147	∞	106	138	182	175	143	123	114
6	153	110	167	168	124	∞	148	170	155	121	93	96
7	178	135	160	155	128	116	∞	150	158	153	125	128
8	195	153	183	178	160	128	138	∞	135	135	133	138
9	191	152	207	202	164	127	157	146	∞	115	112	124
10	158	146	204	207	160	121	180	174	143	∞	84	96
11	136	130	196	197	162	115	174	194	165	126	∞	82
12	142	124	191	192	147	121	171	193	168	112	88	∞

TABLE 3

	1	2	3	4	5	6	7	8	9	10	11	12
1	∞	162	216	215	180	162	210	239	219	158	167	158
2	155	∞	178	180	150	132	180	210	193	159	141	141
3	191	157	∞	142	147	156	172	207	215	184	174	175
4	196	163	136	∞	143	151	161	196	204	181	169	170
5	179	156	157	159	∞	124	152	196	184	152	152	143
6	174	151	179	180	145	∞	166	186	164	130	122	125
7	199	176	172	167	142	134	∞	164	167	162	154	157
8	216	194	195	190	174	146	152	∞	144	144	162	167
9	212	193	219	214	178	145	171	160	∞	124	141	153
10	179	187	216	219	174	139	194	188	152	∞	113	125
11	193	173	208	209	176	150	188	208	186	152	∞	139
12	195	189	203	204	170	147	197	214	180	180	141	∞

TABLE 4

	1	2	3	4	5	6	7	8	9	10	11	12
1	∞	182	233	234	198	199	228	255	231	213	204	226
2	208	∞	192	194	168	161	198	226	205	188	211	194
3	244	177	∞	163	165	185	190	223	227	213	231	228
4	249	183	167	∞	161	180	179	212	216	210	226	223
5	232	176	183	173	∞	153	170	212	196	181	209	196
6	227	180	205	194	174	∞	195	197	176	159	179	178
7	252	196	198	181	160	163	∞	180	179	191	211	210
8	269	214	221	204	192	175	170	∞	156	173	219	220
9	265	213	245	228	196	174	189	178	∞	153	198	206
10	236	207	242	233	192	168	212	204	181	∞	170	184
11	232	209	234	223	222	219	214	232	219	205	∞	162
12	261	231	229	218	212	216	236	269	279	253	165	∞

TABLE 5

It is worth noting that if the recurrence relations (1), (2) defined by Bansal and Kumar are used, nine tables are required in order to find the optimal solution.

REFERENCES

[1] S. P. Bansal, and S. Kumar, *Optimal tour with multiple job facilities at each station*, Indian Journal of Mathematics 13(1971), 45-49.

[2] W. D. Bottom, *National health planning analysis of a new approach*, The Alabama Journal of Medical Sciences 13(1976), 67-73.

[3] W. F. Bridgers, *Alabama's physician shortage - an estimate of its size and distribution by county and by specicality groups*, The Alabama Journal of Medical Sciences 12(1975), 280-294.

[4] W. F. Bridgers, *A study of turnover of Alabama's physician force - trends in geographic distribution*, The Alabama Journal of Medical Sciences 13(1976), 78-89.

STABILITY OF NON-COMPACT SETS

Prem N. Bajaj
Wichita State University

1. *Introduction*

In this paper stability of non-compact sets, in Semi-dynamical Systems (s.d.s.), will be examined. S.d.s. are continuous flows defined for all future time (non-negative t). Natural examples of s.d.s. are provided by functional differential equations for which existence and uniqueness conditions hold. S.d.s. offer new and interesting notions (e.g., a start point, a singular point) which are not possible in dynamical systems.

Stability of non-compact sets has been studied by numerous authors and continues to be a topic of interest, (e.g., Hajek [5], [6]). We examine the stability of a closed set with compact boundary. On the phase space, we take the hypothesis of rim-compactness which is much weaker than local compactness.

This paper is divided into sections. After stating the basic notions, we pass on to limit sets, prolongations and their limit sets. For a rim-compact space, we prove that each of positive limit set and prolongational limit is weakly negatively invariant, and contains no start points; moreover if the prolongational limit set is nonempty and compact, so is the positive limit set. The last section is devoted to stability and asymptotic stability.

2. *Basic Notions*

2.1 Definitions. A *semi-dynamical system* (s.d.s.) is a pair (X,π) where X is a topological space and π is a continuous map from $X \times R^+$ into X satisfying the conditions:

$$\pi(x,0) = x, \quad x \in X, \quad \text{(identity axiom)}$$

$$\pi(\pi(x,t),s) = \pi(x,t+s), \quad x \in X; \quad t, s \in R^+ \quad \text{(semi-group axiom)}$$

(R^+ is the set of nonnegative reals with usual topology). For brevity $\pi(x,t)$ will be denoted by xt, the set $\{xt\colon x \in M \subset X,$

$t \in K \subset R^+\}$ by MK etc.

For any t in R^+, the map $\pi^t : X \to X$ is defined by $\pi^t(x) = xt$. *Negative funnel* $F(x)$ from x is the set $\{y \in X : yt = x$ for some $t \in R^+\}$. For any t in R^+, the function π^{-t} defined on X with values in the set of subsets of X is given by $\pi^{-t}(x) = \{y \in X : yt = x\}$. Clearly $F(x) = \cup\{\pi^{-t}(x) : t \in R^+\}$. Positive trajectory $\gamma^+(x)$, positive invariance are defined as in dynamical systems. *Negative trajectory* $\gamma^-(x)$ from any point x is a maximal non-empty subset of $F(x)$ such that for any y, z in $\gamma^-(x)$, if we let $t(y) = inf\ \{s \geq 0 : ys = x\}$ and $t(z) = inf\ \{s \geq 0 : zs = x\}$, then $y \in z[0,t(z)]$ or $z \in y[0,t(y)]$. In general there will be more than one negative trajectory from x. A negative trajectory $\gamma^-(x)$ will be called a *principal negative trajectory* if the set $\{t \in R^+ : yt = x$ for some y in $\gamma^-(x)\}$ is unbounded.

A subset K of X is said to be negatively invariant (weakly negatively invariant) if the negative funnel (at least one negative trajectory) from each point of K lies in K.

For any x in X, let $E(x) = \{t \geq 0 : yt = x$ for some y in $X\}$. Escape time of x is said to be infinite or $sup\ E(x)$ according as $E(x)$ as unbounded or bounded. A point with zero escape time is said to be a *start point*. See [1], [2] for some results on start points.

2.2 Notation. Through out (X,π) denotes a semi-dynamical system where X is taken to be Hausdorff. $N(x)$ denotes the neighborhood filter [8, p. 78] of x. A net in X will be referred to as x_i where i is in the directed set and x_i is its image.

3. *Limit Sets, Prolongations and Prolongational Limit Sets*

3.1 Definitions. Positive limit set $\Lambda(x)$, positive prolongation $D(x)$, and positive prolongational limit set $J(x)$ of a point x in X are defined by

$\Lambda(x) = \{y \in X :$ there exists a net t_i in R^+, $t_i \to \infty$ such that $xt_i \to y\}$.

$D(x) = \{y \in X :$ there exists a net x_i in X, $x_i \to x$, a net t_i in R^+ such that $x_i t_i \to y\}$.

$J(x) = \{y \in X :$ there exists a net x_i in X, $x_i \to x$, a net t_i in R^+, $t_i \to \infty$ such that $x_i t_i \to y\}$.

3.2 Proposition. Let $x \in X$. Then

$$\Lambda(x) = \bigcap\{Cl(\gamma^+(xt)) : t \in R^+\}.$$

3.3 Remarks. The fact that the sets $D(x)$ and $J(x)$ are defined as in dynamical systems and that these are defined for 'future time' only tempts one to conclude that their properties as displayed in dynamical systems will also hold in s.d.s. But this is not true. The past affects the future. This leads us to define another system of prolongations and their limit sets by

$$B(x) = \bigcup\{D(xt) : t \in R^+\} \quad \text{and} \quad L(x) = \bigcup\{J(xt) : t \in R^+\}.$$

Results on these are announced elsewhere.

3.4 Definitions. [7, p. 111] A topological space is said to be rim-compact (or semi-compact) if it has a base of open sets with compact boundaries.

3.5 Theorem. Let X be rim-compact. Let $x \in X$. Let $\Lambda(x)$ be nonempty and compact. Then

(a) $Cl(xR^+)$ is compact

(b) $\Lambda(x)$ is connected.

3.6 Theorem. Let X be rim-compact. Let $x \in X$. Then each of $\Lambda(x)$ and $J(x)$ is weakly negatively invariant and contains no start points.

Proof. We outline the proof for $J(x)$ only. The proof for $\Lambda(x)$ is similar.

Let $y \in J(x)$. Then there exists a net x_i in X, $x_i \to x$, a net t_i in R^+, $t_i \to +\infty$ such that $x_i t_i \to y$. Let $t > 0$ be arbitrary but fixed. We may suppose $t_i > t$ for every i. Now consider the net $x_i(t_i - t)$ in X. If $x_i(t_i - t)$ converges to y, then $x_i(t_i - t)t \to yt$, moreover $x_i(t_i - t)t = x_it_i \to y$, so that $yt = y$.

If $x_i(t_i - t)$ does not converge to zero, there exists a nbd V of y such that a subnet of $x_i(t_i - t)$, is in $X - V$; we may take V open, with its boundary $F_r(V)$ compact. For simplicity of notation, let the subnet be $x_i(t_i - t)$ itself. Since $x_it_i \to y$, there exists an I in the directed set such that $x_it_i \in V$ whenever $i \geq I$. Now for each $i \geq I$, there exists an s_i, $t_i - t \leq s_i < t_i$ such that $x_is_i \in F_r(V)$. By compactness of $F_r(V)$, x_is_i has a cogt subnet. Let $x_is_i \to z \in F_r(V)$. Since $0 < t_i - s_i \leq t$, the net $t_i - s_i$ has a cogt subnet; let $t_i - s_i \to s$, where $0 \leq s \leq t$. Then $x_is_i(t_i - s_i) = x_it_i \to y$ and $x_is_i(t_i - s_i) \to zs$, so that $zs = y$. But $z \in F_r(V)$, $y \in V$ and V is open, therefore $s \neq 0$.

In either case y is not a start point, and there exists a point (yt or z) in $J(x)$ which is in the negative funnel of y. Existence of negative trajectory from y, now, follows from Hausdorff's maximality principle.

The following theorems, interesting in their own right, will be needed in the sequel.

3.7. Theorem. Let $x \in X$. Let $\omega \in \Lambda(x)$. Then $J(x) \subset J(\omega)$.

3.8. Theorem. Let X be rim-compact. Let $x \in X$. Let $J(x)$ be non-empty and compact. Then $\Lambda(x)$ is non-empty (and compact).

Proof. If $\gamma^+(x) \cap J(x) \neq \phi$, then $xs \in J(x)$ for some $s \in R^+$. Since $J(x)$ is closed and positively invariant, it follows, easily, from the compactness of $J(x)$, that $\Lambda(xs)$, and so $\Lambda(x)$, is non-empty.

If $\gamma^+(x) \cap J(x) = \phi$, let, if possible, $\Lambda(x)$ be empty. In this case we can choose a nbd U of $J(x)$ such that $U \cap \gamma^+(x) = \phi$. Since X is rim-compact and $J(x)$ compact, we may take U closed and $F_r(U)$ compact. Let $y \in J(x)$. Then there exists a net x_i in $X - U$, $x_i \to x$, a net t_i in R^+, $t_i \to +\infty$ such that $x_i t_i \varepsilon U$ and $x_i t_i \to y$. For each i, pick $s_i \varepsilon R^+$ such that $0 < s_i \leq t_i$ and $x_i s_i \varepsilon F_r(U)$. Since $x_i s_i$ has a cogt subnet, let the net $x_i s_i$, itself, converge to $z \varepsilon F_r(U)$. If the net s_i is bounded, it has a subnet converging to s for some $s \varepsilon R^+$, so that $xs = z$ contradicting that $\gamma^+(x) \cap U = \phi$. Consequently, net s_i is unbounded, and, so, $z \varepsilon J(x)$ which contradicts that $J(x) \subset Int\ U$. Hence $\Lambda(x)$ is non-empty.

4. *Stability and Asymptotic Stability*

4.1. Definition. Let M be a subset of X. Then M is said to be *stable* if given a neighborhood U of M, there exists a nbd V of M such that $VR^+ \subset U$.

4.2. Proposition. Let M be a subset of X.

(a) If M is stable, it is positively invariant.

(b) M is stable if and only if given a nbd U of M, there exists a nbd V of M such that $F(x) \cap V = \phi$ for every $x \varepsilon F_r(U)$, $F(x)$ being the negative funnel from x.

4.3. Theorem. Let M be a closed subset of X with compact boundary.

(a) If M is stable, $D(M) = M$

(b) If X is rim-compact, and $D(M) = M$, then M is stable.

For a closed set M with compact boundary, $D(M)$ is characterized by the following:

4.4. Theorem. Let M be a closed subset of X with compact boundary. Then

(a) $D(M) = \cap \{Cl(UR^+) : U \text{ is a nbd of } M\}$.

(b) $D(M) = \cap \{U : U \text{ is a closed, positively invariant nbd of } M\}$.

In particular it follows that $D(M)$ is closed.

Proof. (b): Let K be the intersection of all closed, positively invariant nbds of M. Let $x \varepsilon M$. Then $D(x) = \cap \{Cl(UR^+) :$

$U \in N(x)\}$. For every closed, positively invariant nbd W of M, $Cl(WR^+) = W$ holds. Since a nbd of M is a nbd of x, it follows that $D(x) \subset K$. As $x \in M$ is arbitrary, $D(M) \subset K$.

Next let $y \notin D(M)$. In particular $y \notin D(x)$ for every $x \in F_r(M)$. For each $x \in F_r(M)$, there exists a nbd U_x of x, a nbd V_x of y such that $V_x \cap U_x R^+ = \phi$. Let $\{U_1,\ldots,U_n\}$ be a finite subcover of the open cover $\{U_x : x \in F_r(M)\}$ of the compact set $F_r(M)$. Let $V = \cap \{V_i : i = 1,2,\ldots,n\}$ and $W = \cup \{U_i : i = 1,2,\ldots,n\}$. Clearly $V \cap (M \cup W)R^+ = \phi$ and, so, $y \notin Cl((M \cup W)R^+)$. But $Cl((M \cup W)R^+)$ is a closed, positively invariant nbd of M. Hence $y \notin K$.

The following theorem, used in stability problems, shifts the emphasis from the set to its boundary.

<u>*4.5. Theorem.*</u> Let M be a closed subset of X with compact boundary. Let $y \in D(x) - M$ for some x in M. Then $y \in D(z)$ and $z \in D(x)$ for some $z \in F_r(M)$.

<u>*Proof.*</u> If $x \in F_r(M)$, let $z = x$. If $x \in int\, M$ there exists a net x_i in $int\, M$, $x_i \to x$, a net t_i in R^+ such that $x_i t_i \notin M$ and $x_i t_i \to y$. For each i, pick s_i, $0 < s_i < t_i$ such that $x_i s_i \in F_r(M)$. By compactness of $F_r(M)$, $x_i s_i$ has a cogt subnet. For clarity of notation, let $x_i s_i \to z \in F_r(M)$. Clearly $y \in D(z)$ and $z \in D(x)$.

<u>*4.6. Definition.*</u> A subset M of X is said to be an *attractor* if its region of attraction $A(M) = \{y \in X : \phi \neq \Lambda(x) \subset M\}$ is a nbd of M. An attractor M is said to be a *uniform attractor* if given a compact set K in $A(M)$, and a nbd U of M, there exists $T > 0$ such $xt \in U$ for every $t \geq T$ and every $x \in K$. A set M which is a stable attractor is said to be *asymptotically stable*.

<u>*4.7. Theorem.*</u> Let X be rim-compact. Let M be a closed subset of X with compact boundary. Let M be positively invariant. If M is a uniform attractor, it is asumptotically stable.

Proof. We need show that M is stable. Let V be a neighborhood of M. We may take V to be open, $F_r(V)$ compact and $Cl\ V \subset A(M)$.

For the compact set $F_r(V)$, let $T > 0$ be of the definition of uniform attraction. For $x \in F_r(V)$, let $\tau(x) = inf\ \{t \geq 0 : x[t,\infty) \subset V\}$. Since M is a uniform attractor, $\tau(x)$ is well defined and $\tau(x) \leq T$. Clearly $x\tau(x) \in F_r(V)$. Moreover, $x[0,\tau(x)] \cap M = \phi$. (This follows from the observation that M is positively invariant). Consider $F = \bigcup\{x[0,\tau(x)] : x \in F_r(V)\}$. We assert that F is compact. Let $x_i t_i$, $x_i \in F_r(V)$, be any net in F, $0 \leq t_i \leq \tau(x_i) \leq T$. t_i, being bounded, has a cogt subnet. W.l.g. let $t_i \to t$. Similarly let $\tau(x_i) \to \tau$, so that

$$0 \leq t \leq \tau \leq T. \qquad (*)$$

Also, x_i being in a compact set $F_r(V)$, has a cogt subnet. As before, let $x_i \to x \in F_r(V)$. Now $x_i t_i \to xt$. We have to show that $xt \in F$, i.e., $t \leq \tau(x)$. Now $x_i\tau(x_i) \to x\tau \in F_r(V)$ (notice that $F_r(V)$ is closed). Therefore $\tau \leq \tau(x)$. Hence, by (*), $0 \leq t(\leq\tau) \leq \tau(x)$, and so $xt \in x[0,\tau(x)] \subset F$. This proves the compactness of F. X being T_2, F is closed, Next $F \cap M = \phi$ as $F \cap x[0,\tau(x)] = \phi$ for each $x \in F_r(V)$. Let $U = V - F$. Then U is a nbd of M and $UR^+ \subset V$. Hence M is stable.

4.8. Remarks. The proof is significant for many reasons. We have avoided the "local compactness" condition for space X. $\tau(x)$ is not necessarily continuous, as can easily be seen. Moreover, the proof is constructive in the sense that not only stability of M is established, but also the neighborhood U corresponding to given neighborhood V is actually found such that $UR^+ \subset V$. Moreover, neighborhood U found out is the LARGEST such neighborhood.

REFERENCES

[1] Prem N. Bajaj, *Start points in semi-dynamical systems*, Funkcialaj Ekvacioj, 13(1971), 171-177.

[2] Prem N. Bajaj, *Connectedness properties of start points in semi-dynamical systems*, Funkcialaj Ekvacioj, 14(1971), 171-175.

[3] N. P. Bhatia, and G. P. Szegö, *Dynamical Systems, Stability Theory, and Applications*, Springer-Verlag, New York, 1967.

[4] N. P. Bhatia, *Weak attractors in dynamical systems*, Bol. Soc. Mat. Mex., 11(1966), 56-64.

[5] Otomar Hajek, *Absolute stability of noncompact sets*, J. Differential Equations 9(1971), 496-508.

[6] Otomar Hajek, *Ordinary and asymptotic stability of noncompact sets*, J. Differential Equations 11(1972), 49-65.

[7] J. R. Isbell, *Uniform Spaces, Math. Surveys, No. 12*, American Math. Society, 1964.

[8] S. Willard, *General Topology*, Addison-Wesley, Reading, Ma.

PARAMETRIC EXCITATION AS THE MEANS OF ENERGY TRANSFER IN QUANTAL SYSTEMS WITH REFERENCE TO CARCINOGENESIS AND BIOENERGETICS

T. W. Barrett
University of Tennessee Center for the Health Sciences

When quantal systems are expressed in four parameter form, the local potential defining a quantal system results in a singularity in topological mapping known as a Riemann-Hugoniot catastrophe with the equation of state: Weber's equation. On the other hand, by the application of the same form of analysis, the local potential for two interacting systems is a mapping singularity known as a butterfly catastrophe with the equation of state: Mathieu's equation. The main result obtained by this analysis is that if the radial quantum number of the first system is twice that of the second, parametric excitation of the second system occurs due to the availability of a less entropic configuration for the total joined system, compared with the two systems considered separately. In the light of this observation, carcinogenic potency may be an example of parametric excitation.

The equations of motion of the bioenergetic theory of oxidation-reduction coupling of electron flow, the theory of superconductivity, and that of the conformon are in the same form as the butterfly catastrophe potential function, indicating the generality of the result obtained by this analysis.

The original paper of this abstract will be published in the Journal of Nonlinear Analysis: Theory, Methods, and Applications.

This research was supported by American Cancer Society Institute Research Grant IN-85-J-11, University of Tennessee.

QUANTUM STATISTICAL FOUNDATIONS FOR STRUCTURAL INFORMATION THEORY AND COMMUNICATION THEORY

T. W. Barrett
University of Tennessee Center for the Health Sciences

A confusion exists concerning the relation between the Shannon theory of information (1), hereafter called *CT* (communication theory), and the information theory developed initially by Gabor (2) and which, more recently, has been subsumed under quantum theory (3). The latter theory will hereafter be called *SIT* (structural information theory). The two theories can be simply related within a context of quantum statistics according to the following analysis.

The treatment of acoustical quanta and photons, i.e., elementary signals, in terms of the theoretical structure of mainstream quantum mechanics is possible if the following definitions are made:

(i) A system is a mechanical device which generates signals by sampling from white noise or filtering white noise.

(ii) Whereas in the quantum mechanical formulation of atomic events a wave function Ψ defines a ***state***, in acoustics and optics, this wave function defines a waveshape amplitude modulating a sinusoidal *signal*.

(iii) Whereas in the quantum mechanical formulation of atomic events $|\Psi|^2$ defines a *probability distribution* - of locating an electron - in acoustics and optics this wave function completely describes a *signal* generated by sampling from white noise.

By this translation of the definition of atomic state into signal amplitude modulation terms, which is the equivalent of causing a wave event to conform to a wave mechanical state, the physics of elementary signals follows well known laws (4).

Consider now a macroscopic system which has s degrees of freedom; then a signal generated from that system will be specified by the values of the s coordinates: Δf_i or signal bandwidths, and Δt_i or signal durations. Every system has its own phase space with twice as many dimensions as the system has degrees of freedom. Different signals generated from the system can be mathematically described by points in phase space.

The signals generated from the system will change with time (τ) and the corresponding point in phase space, called the phase point of the system, describes a line in phase space called a phase trajectory.

In defining a "subsystem" we follow a known definition: "Consider a macroscopic body or system of bodies. We assume that the system is closed, i.e., that it does not interact with other bodies. Now consider a small part of this system, very small in comparison with the whole system but nevertheless macroscopic; clearly, if the whole system contains a sufficiently large number of particles, the number contained even in a small part of it may still be very large. Such comparatively small, yet macroscopic systems we shall call 'subsystems' " (4, p. 2).

We shall assume "statistical equilibrium" defined: If a closed macroscopic system generates a signal that for any macroscopic subsystem all "macroscopic" frequencies approximate very closely to their mean value, then the closed system is said to have generated a signal in statistical equilibrium. Statistical independence will also be assumed, which means that a signal generated from one subsystem has no effect on a signal or signals generated from another subsystem.

The concept of "density matrix" can then be applied to signal generation. The knowledge of the density matrix permits the calculation of the mean value of any frequency of the system, and also the generation of the different values of these frequencies.

Consider now some subsystem, and define its "stationary signals" as those signals resulting if its interaction with neighboring parts of the closed system is neglected. Let $\psi_n(\Delta f)$ be the normalized wave functions [without the time (τ) factor] of these signals, where Δf stands in this instance for the coordinates of the subsystem and the index n for the set of all quantum numbers labelling the stationary signals, the energy of these signals being denoted by E_n. Assume that at a given instant of time the subsystem generates a signal completely described by the wave function Ψ. This function can then be expanded in terms of the complete set of functions $\psi_n(\Delta f)$:

$$\Psi = \sum_n c_n \psi_n \tag{1}$$

The mean value of an arbitrary frequency f of this signal can be calculated in terms of the coefficients c_n by:

$$f_0 = \sum_n \sum_m c_n{}^* c_m f_{nm} \tag{2}$$

where

$$f_{nm} = \int \psi_n{}^* \hat{f}\, \psi_m \, d(\Delta f) \tag{3}$$

are the matrix elements of the frequency f ($\hat{f}$ being the corresponding operator).

The transition from a complete to an incomplete quantum mechanical description of the subsystem can, in a certain sense, be regarded as averaging over its different Ψ signals. As a result of such an averaging process, the products $c_n{}^* c_m$ will give rise to coefficients which we shall denote by w_{mn} which form a double sequence (as both indices vary) and which cannot be expressed as the products of coefficients forming a single sequence. The mid-frequency, f_0, of a signal will be given by an expression of the type:

$$f_0 = \sum_m \sum_n w_{mn} f_{nm} \tag{4}$$

The set of coefficients w_{mn} [in general functions of time (τ)] forms the density matrix in the energy representation. In statistics it is called the statistical matrix. This statistical matrix plays, in quantum statistics, the part which the distribution function plays in classical statistics. The coefficients w_n enable the generation of one quantum signal or another of certain midfrequency f_0, but only indirectly indicate the bandwidth Δf and duration Δt of the signals. A signal generated by sampling from white noise is completely described by the function $|\Psi|^2$:

$$|\Psi|^2 = \sum_n \sum_m w_{mn} \psi_n{}^* \psi_n \tag{5}$$

From the definition of matrix elements we may write:

$$\sum_m w_{mn} \psi_m = \hat{w} \psi_n \quad \text{where } \hat{w} \text{ is a statistical operator,} \tag{6}$$

and hence:

$$\sum_n \sum_m w_{mn} \psi_n{}^* \psi_m = \sum_n \psi_n{}^* \hat{w} \psi_n \tag{7}$$

permitting the following definition for the generation of bandwidths Δf_i and durations Δt_i:

$$dw_{\Delta f} = \Sigma \psi_n{}^* \hat{w} \psi_n \cdot d(\Delta f); \tag{8}$$

$$dw_{\Delta t} = d(\Delta t) \int \psi_{\Delta t}{}^* \hat{w} \psi_{\Delta t} d(\Delta f), \tag{9}$$

where $d(\Delta t) = d(\Delta t_1) d(\Delta t_2) \ldots d(\Delta t_s)$.

In order to define the time (τ) derivative of the statistical matrix it is necessary to define the coefficients w_{mn} as functions of time:

$$\dot{w}_{mn} = \frac{i}{C}(E_n - E_m) w_{mn} \tag{10}$$

where $C = \hbar$ for wave optics
$= 1$ for wave acoustics.

As $$(E_n - E_m) w_{mn} = \sum_l (w_{ml} H_{ln} - H_{ml} w_{ln}) \tag{11}$$

where H_{mn} are matrix elements of the Hamiltonian $\hat{H}$ of the system, we may write:

$$\dot{\hat{w}} = \frac{i}{C}(\hat{w}\hat{H} - \hat{H}\hat{w}) \tag{12}$$

where $C = \hbar$ for wave optics
$= 1$ for wave acoustics.

As the statistical matrices of the subsystems must be stationary, from the definition of statistical equilibrium it follows that the matrices w_{mn} of all the subsystems must be diagonal. The problem of defining the statistical matrix is hence reduced to that of calculating the coefficients $w_n = w_{nn}$ and the midfrequency f_0 for any signal may be defined:

$$f_0 = \sum_n w_n f_{nn} \tag{13}$$

As the coefficient of signal generation, w, is a quantum mechanical integral of motion, the logarithm of this coefficient is:

$$\log w_n^{(a)} = \alpha^{(a)} + \beta E_n^{(a)} \tag{14}$$

where the index a distinguishes the different subsystems and α and β are constant coefficients.

In analogy with the quantum statistics of atomic events (4, p. 21), we may now introduce the concept of the number of quanta or elementary signals of a closed system "belonging to" a certain infinitesimal energy level and denote this number by $d\Gamma$. Each signal generated by the system as a whole can be specified by giving the signal of each subsystem and the number $d\Gamma$ will be represented by the product:

$$d\Gamma = \prod_a d\Gamma_a \tag{15}$$

of the numbers $d\Gamma_a$ of elementary quantum signals of the subsystems (such that the sum of the energies of the subsystems lies in the given interval of the energy of the whole system).

Consider now one subsystem and let w_n be the coefficient of signal generation of this subsystem. In order to obtain an expression for the generation of a signal the energy of which lies between E and $E + dE$, we define $\Gamma(E)$ as the number of quantal signals with energies less than or equal to E and the required number of signals with energies between E and dE is:

$$\frac{d\Gamma(E)}{dE}\, dE \tag{16}$$

The resulting distribution of signal energies is then:

$$W(E) = \frac{d\Gamma(E)}{dE}\, w(E) \tag{17}$$

The "width" ΔE of the curve $W = W(E)$ may be defined as the width of a rectangle whose height is equal to the value of the function $W(E)$ at its maximum and whose area is equal to unity:

$$W(\bar{E})\Delta E = 1 \quad \text{or} \quad w(\bar{E})\Delta\Gamma = 1, \tag{18}$$

where

$$\Delta\Gamma = \frac{d\Gamma(\bar{E})}{dE}\, \Delta E \tag{19}$$

is the number of quantum signals corresponding to the range ΔE of energy values.

The quantity $\Delta\Gamma$ is also called the statistical weight of the subsystem and its logarithm

$$S = log\ \Delta\Gamma \tag{20}$$

is the entropy, S, of the subsystem. Entropy is a measure of disorder and is also defined:

$$S = k\ log\ W, \tag{21}$$

where k is Boltzmann's constant.

Whereas in thermodynamics, W, the disorder parameter, is defined as the probability that the system will exist in the state it is in relative to all possible states it could be in, we shall define it here as the number of quantal signals sampled or generated by the system relative to all the possible signals the system could have sampled or generated.

SIT acoustical quanta are defined (3):

$$f_0 \cdot t_0 = \Delta f \cdot \Delta t = 1/2(2\eta + 1) = \xi, \quad \eta = 0,1,2,\ldots \quad (22)$$

Light quanta may be similarly defined if the right hand side of equation (22) is multiplied by $\hbar$ or Planck's constant divided by 2π. Confining the discussion to acoustical quanta (with $\eta = 0$, equation (22) defines the "logon" of Gabor (2), figure 1): the total energy of the subsystem may be obtained from equation (14) as a function of n which is related to η by the limitation:

$$\eta \leq n \quad (23)$$

For $\eta < n$, a series of signals might be generated from a subsystem such that:

$$\sum_i \xi_{\eta_i} = \xi_n \quad (24)$$

Furthermore, it is clear that the system will produce more quanta for $\xi = 1/2$ $(\eta = 0)$, than for $\xi = 3/2$ $(\eta = 1)$, or $\xi = 5/2$ $(\eta = 2)$, etc. ... with n constant. Therefore, the number of quanta $\Delta\Gamma$ in the system is a measure of the number of quantum elementary signals sampled or generated by the system relative to all the possible signals the system could have sampled or generated, i.e., it is a measure of disorder and equivalent to W in equation (21), with $W_{\eta=0} > W_{\eta=1} > W_{\eta=2}$ ... for constant ΔE or constant n.

Suppose now that a Maxwell demon or a biological membrane changes the entropy of the total system with no change in system energy, all energy for the change being provided by the demon or the membrane. Then other states of the system are produced with the coefficients of signal generation given by equation (10). If the entropy of a second later system state is less than a first state we may define this difference to be:

$I = S_1 - S_2$, where $S_1 = k \, log \, W_1$, or the entropy of the first state, (25)

and $S_2 = k \, log \, W_2$, or the entropy of the second state.

Now, I is CT information and exists only in a system of at least two states, being a measure or difference, or, equivalently: CT information only exists in the case of a system which changes as a function of time (τ). This kind of information has been called negentropy (5).

The relation between SIT and CT is, therefore, the relation between W_i or $\Delta\Gamma$ and I. The difference between SIT and CT concerns the difference between disorder and negentropy. Whereas equation (22) defines quanta of order, equation (25) defines a difference in disorder. The relation of the two concepts is provided by the entropy equation (20) or (21) but it is a mistake to take them as identical. They are sometimes confused by making the following errors.

The first error is to assume that equation (25) is the same as equation (20). Because equation (25) can be expressed: $I = k \, log \, (W_1/W_2)$, one may mistakenly assume the existence of a one-state system (independent of τ), when in fact, equation (25) which defines CT information, refers to a two-state system (which is a function of τ). This error produces a Picasso-like τ-dimensionally collapsed picture of the true state of affairs with no distinction between CT and SIT.

The second error is to assume that because the energy of the system is constant (which means n is constant), then the informational content of the system is also constant (which means η is also constant). One then incorrectly assumes that the system's energy level, which is a function of n, completely determines the size of the quantal signals generated by the system and hence their number, when, in fact, η, which may be less than n [see (23)] determines each signal's size, the number of which determines the system entropy. The number n does not determine the system entropy but merely limits it.

Statistical mechanics and quantum mechanics are parent to *CT* and *SIT* respectively, and they are related by quantum statistics. In definition (iii) above, a system was defined as sampling or filtering white noise. If the capability of a system is generalized to include the sampling of random events, in general, then one may offer the following cautionary remark. Statistical mechanical concepts are inapplicable until limitations set by quantum mechanical considerations determine the size of, or amount of disorder in, the quanta into which the physical data is divided. Until the size of the repertoire of signals for a set energy amount is determined, statistical mechanics, and "ipso facto", *CT*, is inapplicable. One cannot make statistical mechanical predictions concerning an energy amount before a division into quanta is made, because quanta vary in size and thus the statistical data base, i.e., W or $\Delta\Gamma$, (even with a set energy amount) may vary. It is this energy division, i.e., imposition of order, which quantum mechanics, or, if signals are considered, *SIT*, describes.

We now turn to a description of system linearity. The question: what is a system's transfer function, is related to the question: what is the elementary or minimal quantum signal [(2), (3)] for that system. When the elementary signal has been established for a system, then the application of such a signal will result in a linear transduction of that elementary signal - whether the system is characterized as linear or non-linear. I now turn to the reasoning behind this contention.

The conception of a system's linearity or non-linearity in transducing a signal depends on whether the duration of a signal is permitted to vary or not. A system which responds not only to a signal's frequency and bandwidth but also to that signal's midperiod and duration will be classified as non-linear when responding to a sinusoidal signal - if non-linearity is defined, as it usually is, with respect to the system's response to sinusoidal signals of arbitrary duration and midperiod. If a system responds also to the secular form of a signal, the principle of superposition does not hold for that system, and it is characterized as non-linear. Thus, a system can be, and usually is, defined as non-linear with respect to signals whose definition does not include time as a dynamic variable. In essence, what is presently regarded as a linear system is an a-historical system which performs according to a Fourier theorem description of function analysis. A Fourier theorem description has been previously criticized (2). Thus, the terms: linear or non-linear, are not absolute descriptions but are relative to a-historical systems which treat temporal variables in a very cavalier manner.

Suppose, on the other hand, a system analyses a signal's temporal dimensions as well as its frequency and bandwidth. If we are considering a system which might be called isentropic, i.e., in which the entropy is everywhere and always the same, there will be dynamic relations between the temporal, frequency and bandwidth variables.

One can pursue further the conception of non-linearity: according to Wiener [(6), p. 98], the average response of a non-linear system may be characterized by

$$r(t,\alpha) = \sum_m \sum_n G_m[K_m(t+\tau_1,\ldots t+\tau_m),\alpha] \; X \; G_n[H_n(t+\tau_1,\ldots t+\tau_n),\alpha], \tag{26}$$

where K_m and K_n are kernel expressions, G_m and G_n are orthogonal functionals of α, and α is a stochastic variable. This characterization is due to the following definitions of an unknown system's response,

$$f_{un}(t,\alpha) = \sum_n G_n[K_n(t+\tau_i,\ldots t+\tau_n),\alpha], \tag{27}$$

and a known system's response,

$$f_k(t,\alpha) = \sum_n G_n[H_n(t+\tau_1,\ldots t+\tau_n),\alpha]. \tag{28}$$

Both system responses (27) and (28) are elicited by white noise and together give equation (26). Thus equation (26) is the multiple of the responses of a known and an unknown system in parallel, both responding to white noise.

The K_n are kernels in the G_n orthogonal functional expansion. Thus, the H_n, which are also kernels in another G_n functional expansion, will multiply with the K_n only if they are of the same order: i.e., if $m = n$ in equation (26).

Lee and Schetzen (7) substituted for this crucial observation, concerning the possible multiplication of certain orthogonal functionals obtained by the product of orthogonal polynomials, the further observation that signals delayed by different times are also orthogonal and thus

$$\overline{y(t)y_1(t)} = \overline{\sum_{n=1}^{\infty} G_n[h_n, x(t)]x(t - \sigma)}, \tag{29}$$

where σ = an adjustable delay time, $x(t)$ = signal imput, h_n = set of kernels of the unknown system, $y(t)$ = the output of the unknown system in parallel with the known system and $y_1(t)$ = the output of the delay circuit or known system. The bar indicates an average value.

By the use of a series of delayed signals one may, they suggest, avail oneself of the properties of an orthogonal expansion of series but obtain those properties in a technically simpler way. Another advantage is that the expansion by delayed signals can be truncated in any part of the series with no approximation error. Truncation of a polynomial series, of course, leads to an approximation error. The Lee and Schetzen method, however, which is now of general use in non-linear analysis, has a drawback which we shall point out below.

We now consider the result of such an analysis of a so-called non-linear system - whether by using orthogonal polynomials or orthogonal delayed signals. Suppose white noise is applied to a system and that system is demonstrated to be non-linear by using these methods. Let us suppose that the third kernel, K_3, turns out to be significant. Now the third kernel is, according to the original Wiener analysis (but not, unfortunately, according to the Lee and Schetzen analysis), the third term in an orthogonal expansion of a series of functionals obtained by the multiplication of orthogonal polynomials. According to Wiener, exactly what series of orthogonal polynomials one uses is according to one's predilection and he chose a Laguerre polynomial expansion [see (3)]. However, for very cogent physical reasons (3). I have suggested that a Weber-Hermite polynomial expansion is required. Of even more consequence, a Weber-Hermite expansion also defines the modulating envelopes for a series of elementary signals of increasing size (3).

Now the point to be made is that if, after the aforementioned system analysis in which, e.g., the third Weber-Hermite polynomial kernel is found to be significant, one then applied to that system an elementary signal of the third kind, i.e., an elementary signal whose modulating envelope is that Weber-Hermite polynomial, <u>the response of that system will be linear</u>, and, furthermore, linear in the sense of signal input-output transduction in the case of a well defined elementary signal, rather than with respect to a sinusoidal signal.

This conclusion is based on the observation that if

$$r(t,\alpha) = G_3[K_3(t+\tau_1,\ldots t+\tau_3,\alpha] \; X \; G_3[H_3(t+\tau_1,\ldots t+\tau_3)] \tag{30}$$

in the case of a system responding to white noise in parallel with a system of known response, it follows that the response of this dual system elicited by a signal whose envelope is defined by an expression also defining that kernel (K_3), is

$$r(t,\alpha) = [G_3(K_3(t+\tau_1,\ldots t+\tau_3),\alpha]^2, \tag{31}$$

and thus the response of the unknown system is

$$f_{un}(t,\alpha) = G_3(K_3(t+\tau_1,\ldots t+\tau_3)). \tag{32}$$

We may define an elementary signal, therefore, as <u>that signal which produces a linear impulse response in a system</u>. It is understood, of course, that we are now using the terms linear and non-linear as relative to the input-output characteristics of a system - whether the applied signal is a sinusoid or not. By the application of structural information theory (3), therefore, a "non-linear" system may be "linearized". Indeed, any system is non-linear or linear with respect to certain signals. One may, therefore, question the anchoring of these terms to a-historical systems of the Fourier analysis type and propose that the terms "linear" and "non-linear" are relative. This suggestion has an implication of some consequence for characterizing linearity in secular systems to which we now turn.

The smallest signal which is transduced linearly by any system, is an elementary signal for that system. Thus, there is more than one elementary signal and the series of elementary signals are defined by the boundary conditions (3):

$$f_0 t_0 = (1/2)(2n + 1), \quad n = 0,1,2,\ldots,$$
$$\Delta f \Delta t = (1/2)(2n + 1), \quad n = 0,1,2,\ldots, \tag{33}$$

where f_0 is signal midfrequency in Hertz, t_0 is signal midperiod in seconds/cycle, Δf is signal bandwidth in cycles and Δt is signal duration in seconds. These conditions define a series of signals, the modulating envelopes of which are modified Weber-Hermite polynomials:

$$D_n(x) = e^{-(1/2)x^2} H_n(x), \quad n = 0,1,2,\ldots, \tag{34}$$

where the $H_n(x) = (-1)^n e^{x^2} \partial^n e^{-x^2}/\partial x^n$ are Hermite polynomials. The smallest elementary signal, $(n = 0)$, has a modulating envelope which is the Weber-Hermite polynomial of the zeroth order, i.e., it is Gaussian, and is defined by the expressions

$$e^{-(t-t_0)^2} e^{i2\pi f_0 t}, \text{ in the time domain, } e^{-\pi^2 (f-f_0)^2} e^{-i2\pi t_0 f}, \text{ in the frequency domain,} \tag{35}$$

The series of elementary signals of increasing size defined in equations (33) and (34) are thus a four signal parameter series development (3) of the original single two parameter formulation of an elementary signal (2). The point I wish to make concerns the importance of the smallest elementary signal defined by expressions (35).

Suppose, using the Wiener method of non-linear analysis (6), one applies white noise to a known and unknown system in parallel. The response possibilities of the known system may be characterized by the Weber-Hermite polynomial series, because all that Wiener requires in his method is that the series used be orthogonal. Both the response of the known and unknown system are multiplied together, and due to the orthogonal nature of the polynomial series used, only those polynomials will remain in the final output of this parallel system which characterize the response of the unknown system.

To take a particular case, suppose the output of an unknown system is characterized only by the zeroth polynomial: i.e., the system is linear in the usually accepted sense of the term. This is because both

$$\sum_{i=1}^{\infty} f_{0_i} = \int \psi_n(t)\, dx(t,\alpha), \tag{36}$$

(where α is a random variable, $x(t,\alpha)$ describes a Brownian motion, the $\psi_n(t)$ function exists in L_2 space and $\int \psi_n(t)\,dx(t,\alpha)$ is a definition for white noise (6)),

and $\sum_{i=1}^{\infty} e^{-(t-t_{0_i})^2} e^{i2\pi f_{0_i} t} = \sum_{i=1}^{\infty} D_{0_i}(x) cox(360x)^0 = \int \psi_n(t)\,dx(t,\alpha).$

However,

$$\sum_{k=m}^{\infty} D_k(x) \cos \left(\frac{180x}{1/2(2k+1)} \right)^0 \neq \int \psi_n(t)\,dx(t,\alpha), \quad m > 0, \tag{37}$$

because there is a lower bound on the permissible frequencies. Yet

$$\sum_{i=1}^{\infty} D_{k_i}(x) \cos \left(\frac{180x}{1/2(2k+1)} \right)^0, \quad k = 0,1,2,3,\ldots, \tag{38}$$

defines a linear addition of variables giving superposition for a bounded set.

Thus, a particular system, if it can linearly transduce sinusoids, i.e., functions defined over two signal parameters, must also linearly transduce the functions $D_0(x)cos(360x)^0$ defined over four signal parameters. On the other hand, other systems which can linearly transduce only elementary signals other than $D_0(x)cos(360x)^0$ - elementary signals defined for $k = s$, $s > 0$, in expression (38) - transduce all other elementary signals, for $k \neq s$, non-linearly. $D_s(x)cos(180x/(\frac{1}{2})(2s + 1))^0$ is thus a "non-linear" system's elementary signal and is transduced linearly.

It follows, therefore, that the $D_0(x)cos(360x)^0$ elementary signal is unique in that for this signal, and only for this signal, a system which is linear in two dimensional space is also linear in four dimensional space. We now turn to applications of these methods in biology.

In physiology, the Lee and Schetzen adaptation of the Wiener method has been used in studies of the human pupil light reflex (9), the insect retina receptor system (10), and the fish retina receptor system (11). On the other hand, we have used elementary signals in studies of the auditory cortex (12), inferior colliculus (13), medial superior olive (14) and cochlea (15).

One may ask whether the cochlea is a linear or non-linear transducer of signals defined over two dimensions (i.e., sinusoids). The answer is that there are non-linearities at certain stimulus conditions [(16), chapter 6], but in the optimum range for stimulus transduction, de Boer (7), using a process similar to the Lee and Schetzen (7) modification of the Wiener white noise analysis (6), and studying the response of single auditory nerve fibers, has demonstrated a linear filtering process with a response remarkably similar to the elementary signal $D_0(x)cos(360x)^0$ [(13), see Figures 6 and 8]. The implication of this study is that for optimum signals, the cochlea transduces linearly.

Finally, one may observe that in the case of a transducer like the cochlea, the fluid mechanics of which is designed not merely to transduce a sinusoid or a four parameter signal linearly, but must also retain separation of those four parameters for later separate signal processing in the central nervous system, emergent transfer properties must occur in the cochlear fluid mechanics so that fluid movements elicited by the Δf signal parameter, although coupled to fluid movement elicited by the f_0 signal parameter, are yet qualitatively different. It is, perhaps, for this reason that the cochlea is a spiral in most mammals.

REFERENCES

[1] C. E. Shannon, *Bell System Tech. J.*, 27(1948), p. 379 & 623.

[2] D. Gabor, *J. Inst. Electr. Engrs. (London)*, 93(1946), p. 429; 94(1947), p.369; *Nature* 159(1947), p. 591.

[3] T. W. Barrett, *J. Sound & Vibration* 20(1972), p.407; 25(1972), p. 638; 39(1975), p. 265; 41(1975), p. 259; *Acustica* 27 (1972), p. 44; 27(1972), p. 90; 29(1973), p. 65; 33(1975), p. 149; 35(1976), p. 80; *J. Acoust. Soc. Am.* 54(1973), p. 1092; *TIT J. Life Sci.* 1(1971), p. 129.

[4] L. D. Landau and E. M. Lifshitz, *Statistical Physics*, Addison-Wesley, Reading, Massachusetts, 1958.

[5] L. Brillouin, *Science and Information Theory, 2nd Edition*, Academic Press, New York, 1962.

[6] N. Wiener, *Nonlinear Problems in Random Theory*, MIT Press, Cambridge, Massachusetts, 1958.

[7] Y. W. Lee and M. Schetzen, *Research Laboratory of Electronics*, Massachusetts Institute of Technology Quarterly Progress Report No. 60, 1961.

[8] J. F. Barrett, *J. Electronic Control* 15(1963), p. 567; A.G. Bose, *Research Laboratory of Electronics*, Massachusetts Institute of Technology Technical Report No. 309, 1956; M. B. Brilliant, *Research Laboratory of Electronics*, MIT Technical Report No. 345, 1958; D. A. George, *Research Laboratory of Electronics*, MIT Technical Report No. 355, 1959.

[9] L. Stark, *Neurological Control Systems: Studies in Bioengineering*, Plenum, New York, 1968.

[10] P. Z. Marmarelis, and G. D. McCann, *Kybernetik* 12(1973), p.74.

[11] P. Z. Marmarelis and K. I. Naka, *Science* 175(1972), p. 1276; *J. Neurophysiol.* 36(1973), p. 605; 36(1973), p. 619; 36 (1973), p. 634; *IEEE Trans. Biomed. Eng.* BME 21(1974), p.88.

[12] T. W. Barrett, *Brain Research* 28(1971), p. 579; *Exp. Neurology* 34(1972), p. 1; 34(1972), p. 484; *Beh. Biology* 8(1973), p. 299.

[13] T. W. Barrett, *Beh. Biology* 9(1973), p. 189.

[14] T. W. Barrett, *Brain Res. Bull.* 1(1976), p. 209.

[15] T. W. Barrett, *Experientia* 30(1974), p. 1287; *Acustica* 33 (1975), p. 102; *Math. Biosciences* 29(1976), p. 203.

[16] P. Dallos, *The Auditory Periphery: Biophysics and Physiology*, Academic Press, New York, 1973.

[17] E. deBoer, in *Regulation and Control in Physiological Systems* (editors, A. S. Iberall and A. C. Guyton), Instrument Society of America, Pittsburgh, 1973, pp. 187–94.

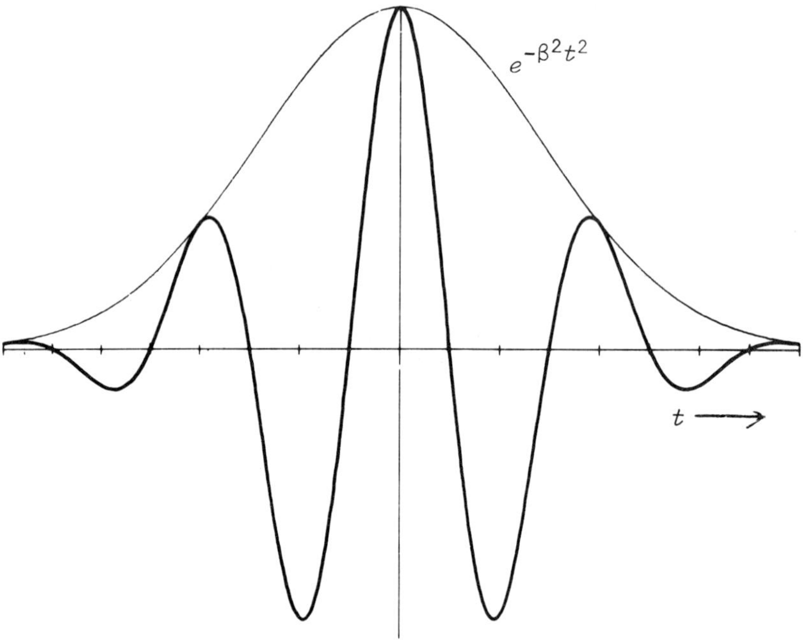

FIGURE 1

The amplitude modulated signal shown is the elementary signal in the time domain defined by the bounding conditions: $\Delta f \Delta t = f_0 t_0 = 1/2$, where Δf is signal band-width, Δt is signal duration, f_0 is signal midfrequency and t_0 is signal midperiod. The definition of this signal in the time domain is:

$$s(t) = exp[-(t - t_0)^2] exp\ (i 2\pi f_0 t)$$

In the frequency domain the definition is:

$$S(f) = exp[-\pi^2 (f - f_0)^2] exp(-2\pi t_0 f).$$

This is the smallest elementary signal of a series of increasing size defined by the bounding conditions:

$$\Delta f \Delta t = f_0 t_0 = 1/2\,(2n + 1), \quad n = 0,1,2,\ldots,$$

The amplitude modulation of the signal shown is the Weber-Hermite polynomial of the zeroth order, i.e., a Gaussian envelope. The amplitude modulations for the series of elementary signals are the

Weber-Hermite polynomials:

$$D_n(x) = N_n H_n(\xi) exp[-\xi^2], \quad n = 0,1,2,...$$

where N_n is a constant such that:

$$\int_{-\infty}^{+\infty} D_n^*(x) D_n(x) dx = 0$$

and:

$$H_n(\xi) = (-1)^n exp[\xi^2] \frac{\partial^n}{\partial \xi^n} exp[-\xi^2]$$

are Hermite polynomials.

The Weber-Hermite functions or parabolic cylinder functions are solutions to Weber's equation:

$$\frac{d^2D}{dx^2} + (\lambda - \xi^2)D = 0$$

where ξ is a dimensionless independent variable, and λ is a dimensionless eigenvalue. Schrodinger's equation (the amplitude equation for matter waves) is in the form of Weber's equation. The relation of the amplitude equation to the energy equation or Hamiltonian, H, is given by writing Weber's equation in the form:

$\mathsf{H}D = \mathsf{E}D$, where E is energy.

BOUNDARY VALUE PROBLEMS FOR NONLINEAR DIFFERENTIAL EQUATIONS

T. T. Bowman
University of Florida

In [1], the solutions for small ε of the differential system

(1) $$x' = Ax + f(t,x,x',\beta,\varepsilon)$$

with boundary conditions

(2) $$B_1x(a) + B_2x(b) = 0$$

were considered. Here $x = col\ (x_1,\ldots,x_n)$, A is an $n \times n$ matrix, βa parameter from a compact subset of the Euclidean space E^v, ε a small real parameter, and f a continuous differentiable function from an open subset of E^{2n+v+2} into E^n with the condition that $f(t,x,x',\beta,0) = 0$. On the boundary condition it is assumed that B_1 and B_2 are $m \times n$ matrices such that the $m \times 2n$ matrix (B_1,B_2) has rank m.

Let $U(\cdot)$ be an $n \times p$ matrix whose columns are independent solutions of the linear part $(\varepsilon = 0)$ of (1) and (2). Using the alternative method of Cesari [2], equations (1) and (2) are replaced by equivalent equations of the form

(3) $$x = U\alpha + K(I - Q)f(\cdot,x,x',\beta,\varepsilon)$$

(4) $$Qf(\cdot,x,x',\beta,\varepsilon) = 0$$

where α is a $p \times 1$ matrix, K and Q are operators on appropriate function spaces. A unique solution $x^*(\alpha,\beta,\varepsilon)$ of (3) for sufficiently small ε is obtained by application of the Banach fixed point theorem. To approximate the solution $x^*(\alpha,\beta,\varepsilon)$ to higher orders than one an explicit formula for K is given in [1].

Let $V(\cdot)$ be a $q \times n$ matrix whose rows are independent solutions of the adjoint linear problem

$$\tilde{Y}' = -\tilde{Y}\tilde{A}$$

with boundary conditions

$$\tilde{Y}(a) = \tilde{\sigma}B_1 \quad \text{and} \quad \tilde{Y}(b) = -\tilde{\tilde{\sigma}}\tilde{B}_1$$

where σ is any vector in E^n. Define

$$H(\alpha,\beta,\varepsilon) = \int_a^b V(s)f(s,X^*(s),X^{*\prime}(s),\beta,\varepsilon)ds.$$

It is shown in [3], that (3) and (4) have a solution $X^*(\alpha,\beta,\varepsilon)$ if and only if $H(\alpha,\beta,\varepsilon) = 0$. An application of the implicit function theorem gives a limited version of a theorem found in [1].

THEOREM. Suppose α_0 is E^p and β_0 in E^v are such that $H(\alpha_0,\beta_0,0) = 0$ and the $q \times (p + v)$ matrix $|\partial H/\partial\alpha \ \partial H/\partial\beta|$ evaluated at $(\alpha_0,\beta_0,0)$ has rank q. Then equations (1) and (2) have a solution $X(\varepsilon)$ for sufficiently small ε.

In [3], Nagle gives a similar development for the nonautonomous boundary value problem. In [1], the main concern is with the autonomous case, but the introduction of the parameter β gives a unified treatment of both cases.

REFERENCES

[1] T. T. Bowman, *Boundary value problems for weekly nonlinear differential equations*, submitted.

[2] L. Cesari, *Alternative methods in nonlinear analysis*, International Conference on Differential Equations, H. A. Antosiewicz, editor, Academic Press, New York, 1975, pp. 95-148.

[3] R. K. Nagle, *Boundary value problems for nonlinear ordinary differential equations*, dissertation, University of Michigan, 1975.

GLOBAL SOLUTION FOR A PROBLEM OF NEUTRON TRANSPORT WITH TEMPERATURE FEEDBACK [1]

G. Busoni[2], V. Capasso[3], and A. Belleni-Morante[2]

1. *Introduction*

The behaviour of a nuclear reactor with temperature-dependent feedback was recently studied by using the diffusion approximation of neutron transport and by assuming that feedback was acting only through the multiplication factor [1-4].

In [5] the same problem was briefly examined by one of the authors, under the assumption that the macroscopic cross sections were linearly dependent on temperature. The nonlinear neutron transport problem under consideration was first transformed into a nonlinear abstract problem; the theory of evolution equations in Banach spaces was then used to prove existence and uniqueness of a solution, defined on a suitably small time interval $[0,\bar{t}]$.

In this paper we consider the same problem, referring for simplicity to a slab of thickness $2L$ with cross sections depending on temperature in a sufficient "smooth" way (See Section 2).

In this way, by introducing two auxiliary linear neutron transport problems we can prove that the nonlinear problem under consideration admits a unique solution belonging to a suitable Banach space and defined on any arbitrarily fixed finite time interval $[0,t_0]$. We shall also indicate how the solutions of the two auxiliary linear problems provide lower and upper bounds for the solution of the nonlinear problem.

We finally remark that, with some formal complications, most of the results that follow can be generalized to the case of ener-

[1] Work performed under the auspices of CNR (Gruppo Nazionale per la Fisica Matematica).

[2] Istituto di Matematica Applicata, Universita di Firenze, Firenze, Italy.

[3] Istituto di Analisi Matematica, Universita di Bari, Bari, Italy.

gy dependent neutron transport in a nonhomogeneous convex body with anisotropic scattering, and with delayed neutrons.

2. The Nonlinear Initial-Value Problem

If we refer for simplicity to a slab of thickness $2L$, the problem examined in [5] can be written in the following form, (see also [6] and [7]):

(1) $$\frac{\partial N}{\partial t} = -\mu v \frac{\partial N}{\partial x} - v\Sigma N(x,\mu;t) + \frac{v\gamma}{2}\int_{-1}^{1} N(x,\mu';t)d\mu'$$

(2) $$\frac{dT}{dt} = -h(T - T_c) + K\int_{-L}^{L} dx' \int_{-1}^{1} N(x',\mu';t)d\mu'$$

with $|x| < L$, $|\mu| \leq 1$, $t > 0$; where $N(x,\mu;t)$ is the neutron density; γ, v, Σ, h, K are all positive functions of T whose definition is given in [5]; and T_c is the coolant temperature.

System (1)-(2) must be supplemented with an initial condition and with the usual nonre-entry boundary conditions:

(3) $$N(x,\mu;0) = N_0(x,\mu); \quad T(0) = T_0$$

(4) $$N(-L,\mu;t) = 0 \quad \forall\mu\in]0,1] \; ; \quad N(+L,\mu;t) = 0 \quad \forall\mu\in[-1,0[.$$

where $N_0(x,\mu)$ is an assigned nonnegative function and T_0 is a given real constant.

Here we shall modify the hypotheses made in [5], by assuming that Σ, γ, and K depend on the temperature as follows:

(5) $$\Sigma = \Sigma_0(1 + a_1(\theta)); \quad \gamma = \gamma_0(1 + a_2(\theta)); \quad K = K_0(1 + a_3(\theta))$$

where

(6) $$\theta = \theta(t) = T(t) - T_c; \quad \theta(0) = \theta_0 = T_0 - T_c$$

and where Σ_0, γ_0, K_0 are positive constants.

Moreover, each of the a_i's $(i = 1,2,3)$ is assumed to be a real valued function which satisifes the assumptions:

(a1) $-1 \leq a_i(y) \leq \bar{a}_i < +\infty \qquad \forall\, y \in \underline{\underline{R}}$

(a2) $a_i(y)$ is a continuous function of $y \in \underline{\underline{R}}$, with $a_i(0) = 0$

(a3) the derivative $b_i(y) = da_i/dy$ exists and is such that $|b_i(y)| \leq \bar{b}_i \qquad \forall\, y \in \underline{\underline{R}}$

(a4) $b_i(y)$ is a continuous function of $y \in \underline{\underline{R}}$.

We remark that relations (5) and assumption (a1) lead to the inequalities

$$
\begin{aligned}
&0 \leq \Sigma' \leq \Sigma_0(1 + a_1(y)) \leq \Sigma'' = \Sigma_0(1 + \bar{a}_1) \qquad \forall\, y \in \underline{\underline{R}} \\
&0 \leq \gamma' \leq \gamma_0(1 + a_2(y)) \leq \gamma'' = \gamma_0(1 + \bar{a}_2) \qquad \forall\, y \in \underline{\underline{R}} \qquad (7) \\
&0 \leq K' \leq K_0(1 + a_3(y)) \leq K'' = K_0(1 + \bar{a}_3) \qquad \forall\, y \in \underline{\underline{R}}
\end{aligned}
$$

If we now put

$$u_1(t) = N(x,\mu;t) \quad \text{and} \quad u_2(t) = \theta(t), \tag{8}$$

where $u_i(t)$ $(i = 1,2)$ must be interpreted as a map from $[0,+\infty[$ into the real Banach space x_i (see Appendix A), system (1)+(4) then becomes:

$$\frac{d}{dt}u_1 = (B_1+v\gamma_0 J_1)u_1(t) - v\Sigma_0 a_1(u_2(t))u_1(t) + v\gamma_0 a_2(u_2(t))J_1 u_1(t) \qquad t > 0 \tag{9}$$

$$\frac{d}{dt}u_2 = -hu_2(t)K_0 J_2 u_1(t) + K_0 a_3(u_2(t))J_2 u_1(t) \qquad t > 0 \tag{10}$$

$$\lim_{t\to 0^+} u_1(t) = N_0(x,\mu); \qquad \lim_{t\to 0^+} u_2(t) = \theta_0 \tag{11}$$

or in a more compact form (see Appendix B):

$$
\begin{cases}
\dfrac{d}{dt}u(t) = (B + J)u(t) + F(u(t)) \qquad t > 0 \\[2ex]
\lim\limits_{t\to 0^+} u(t) = u_0
\end{cases} \tag{12}
$$

Where we have essentially used the same definitions as in [5] (see Appendix A),

$$u(t) = \begin{pmatrix} u_1(t) \\ \\ u_2(t) \end{pmatrix} \qquad u_0 = \begin{pmatrix} N_0(x,\mu) \\ \\ \theta_0 \end{pmatrix}$$

System (12) is now a nonlinear initial-value problem in the space $X = X_1 \times X_2$ (Appendix B).

In the sequel, we shall say that $u = u(t)$ is a strong solution in X of the nonlinear problem (12) over the interval $[0,\bar{t}]$ if (i) $u(t)$ is strongly continuous at any $t \in [0,\bar{t}]$, (ii) $u(t)$ is strongly differentiable at any $t \in [0,\bar{t}]$, (iii) $u(t)$ satisfies (12). ([13-16]).

3. *The Auxiliary Linear Problems*

Together with (9), we consider the following "lower" problem

(9') $$\frac{d}{dt} w_1 = (B_0 - v\Sigma'' I + v\gamma' J_1) w_1(t) \qquad t > 0$$

(11') $$\lim_{t \to 0^+} w_1(t) = w_{10} \equiv N_0(x,\mu)$$

where we now assume $N_0(x,\mu) \in \mathcal{D}(B_0) \cap X_1^+$.

The lower problem (9')+(11') has the form of an initial-value problem in the Banach space X_1. Due to Lemma C1, the unique strong solution $w_1 = w_1(t)$ of (9')+(11') can be put in the form

(13) $$w_1(t) = exp\ (t(B' + J'))w_{10}, \qquad t \geq 0$$

with

(14) $$||w_1(t)|| \leq exp\ ((v\gamma' - v\Sigma'')t)||w_{10}||_1, \qquad t \geq 0$$

provided that $w_{10} \in \mathcal{D}(B_0)$ ([8], Chap. 9). Moreover, since $w_{10} \in \mathcal{D}(B_0) \cap X_1^+$, we obtain from (13) and (C4)

(15) $$w_1(t) \in X_1^+ \qquad \forall\, t \geq 0$$

We also consider the following "upper" problem

(9") $$\frac{d}{dt} W_1 = (B_0 - v\Sigma' I + v\gamma'' J_1) W_1(t) \qquad , t > 0$$

(11") $$\lim_{t \to 0^+} W_1(t) = W_{10} \equiv N_0(x,\mu)$$

which can be dealt with in an analogous way. Hence

(16) $$||W_1(t)||_1 \leq \exp\,((v\gamma'' - v\Sigma')t)\,||W_{10}||_1, \qquad t \geq 0$$

(17) $$W_1(t) \in X_1^+ \qquad \forall\, t \geq 0$$

provided that $N\,(x,\mu) \in \mathcal{D}(B_0) \cap X_1^+$. In (16) $W_{10} = w_{10}$.

We finally remark that both the "lower" and the "upper" system are standard linear transport problems [10,11].

4. *A Priori Properties of* $u(t)$.

Let $[0,t_0]$ be a finite time interval with $t_0 > 0$ arbitrarily fixed and let us assume that system (12) has a strong solution $u = u(t)$ defined on some subinterval $[0,t] \subset [0,t_0]$

If we subtract (9') from (9) and (11') from (11) and if we put $\delta = u_1 - w_1$ we then obtain

(18) $$\frac{d}{dt}\delta = B_0\delta(t) - v\Sigma(t)\delta(t) + v\gamma(t)J_1\delta(t) + \phi(t)$$

(19) $$\lim_{t \to 0^+} \delta(t) = 0$$

where

(20) $$\phi(t) = v(\Sigma'' - \Sigma(t))w_1(t) + v(\gamma(t) - \gamma')J_1 w_1(t)$$

and where (see (5), (7) and (8)):

(21) $$\Sigma(t) = \Sigma_0(1 + a_1(u_2(t))) \leq \Sigma''; \quad \gamma(t) = \gamma_0(1+a_2(u_2(t)))\leq\gamma''$$

Since $u(t) = \begin{bmatrix} u_1(t) \\ u_2(t) \end{bmatrix}$ is a strong solution of system (12) over $[0,\bar{t}]$, it follows from (21) and from (a2) of Section 2 that $\Sigma(t)$ and $\gamma(t)$ are continuous at any $t \in [0,\bar{t}]$. Relation (18) then shows that $\phi(t)$ is also continuous over $[0,\bar{t}]$, because $w(t)$ is given by (13); moreover $\phi(t) \in X^+$, $\forall\, t \in [0,\bar{t}]$ provided that $N_0(x,\mu) \in \mathcal{D}(B_0) \cap X_1^+$. (See (15)). By a standard procedure, [12], problem (18)+(19) can then be transformed into a Volterra integral equation, which is linear in $\delta(t)$ and which can be solved by the usual method of successive approximations. It is then easy to show that $\delta(t) \in X_1^+$ $\forall\, t \in [0,\bar{t}]$ since $\phi(t) \in X_1^+$ $\forall\, t \in [0,\bar{t}]$. Hence $u_1(t) = w_1(t) + \delta(t) \in X_1^+$ $\forall\, t \in [0,\bar{t}]$, and we have also

$$(22) \qquad 0 \le w_1(t) \le u_1(t) = N(x,\mu;t) \qquad \forall\, t \in [0,\bar{t}]$$

In order to find an upper bound for $N(x,\mu;t)$ we subtract (9) from (9") and (11) from (11"). If we proceed as for the lower bound we can finally prove the following theorem.

THEOREM 1: If system (12) has a strong solution $u = u(t)$, $t \in [0,\bar{t}]$ and if $N_0 \in \mathcal{D}(B_0) \cap X_1^+$, then the first component of $u(t)$ is such that

$$0 \le w_1(t) = w_1(x,\mu;t) \le u_1(t) = N(x,\mu;t) \le W_1(t) = W_1(x,\mu;t)$$

We finally evaluate an upper bound for $|u_2(t)|$ and for $||u(t)||$. We have from (10) and the second of (11)

$$u_2(t) = exp\,(-ht)\,\theta_0 + \int_0^t exp\,(-h(t-s))(K(s)J_2u_1(s))ds$$

where $t \in [0,\bar{t}]$ and where $0 \le K(s) = K_0(1 + a_4(u(s))) \le K''$. Hence

$$(23) \quad |u_2(t)| \le exp(-ht)|\theta_0| + K'' \int_0^t exp(-h(t-s))||W_1(s)||_1 ds.$$

On the other hand, we have from (16)

$$||W_1(t)||_1 \le ||W_{10}||_1 \max_{t\in[0,t_0]} exp(\langle v\gamma'' - v\Sigma'\rangle t) \quad \forall\, t \in [0,t_0]$$

or

(24) $$||W_1(t)||_1 \leq m||N_0||_1 \qquad \forall\, t \,\varepsilon [0,t_0]$$

It then follows from (23)

(25) $$|u_2(t)| \leq exp(-ht)|\Theta_0| + mK''(||N_0||_1)(1 - exp(-ht))/h$$
$$\leq |\Theta_0| + \frac{mK''}{h}||N_0||_1 \qquad \forall\, t \,\varepsilon [0,\bar{t}]$$

Finally, since $||u_1(t)||_1 \leq ||W_1(t)||_1$ due to Theorem 1 we obtain

$$||u(t)|| = K_0||u_1(t)|| + h|u_2(t)| \leq m||N_0||_1(K_0 + K'') + h|\Theta_0| \equiv M$$

Hence

(26) $$||u(t)|| \leq M \qquad \forall\, t \,\varepsilon [0,\bar{t}]$$

provided that the assumptions listed in Theorem 1 are satisfied.

5. *The Nonlinear Operator* F.

Due to assumptions (a1)-(a4), from the definition of F (Appendix B), it can be shown that

(27) $$||F(f) - F(g)|| \leq \alpha(||f|| + ||g||)||f - g|| \quad \forall\, f,g \,\varepsilon\, X$$

where

(28) $$\alpha = \frac{1}{2}\left(\frac{v}{h}(\Sigma_0\bar{b}_1 + \gamma_0\bar{b}_2) + \bar{b}_3\right)$$

It can also be shown that the Frechet derivative F_f of F exists at any $f \,\varepsilon\, X$. By using assumptions (a1) and (a3) and Lemma A1 it is easy to verify that $F_f \,\varepsilon\, \mathcal{B}(X)$, $\forall\, f \,\varepsilon\, X$.

6. *Global Strong Solution of System (12)*.

We first formally transform system (12) into the following integral equation:

$$u(t) = Z(t)u_0 + \int_0^t Z(t - s)(Ju(s) + F(u(s)))ds$$

where $Z(t)$ is the semigroup generated by B.

It was shown in [5] that equation (29) has a unique solution $u \in \mathcal{C}([0,\bar{t}],X)$ provided that $\bar{t}$ is small enough.

This solution is now also a strong solution of (12) since satisfies the conditions listed in [13] provided that $u_0 \in \mathcal{D}(B)$. Hence

THEOREM 2: If assumptions (a1)-(a4) of Section 2 are satisfied and if $u_0 \in \mathcal{D}(B)$, a time $\bar{t} > 0$ exists such that the nonlinear problem (12) admits a unique strong solution $u = u(t)$, $t \in [0,\bar{t}]$.

Moreover, if u_0 is such that $N_0(x,\mu) \in \mathcal{D}(B_0) \cap X_1^+$, the first component of $u(t)$ satisfies the inequalities of Theorem 1 and (26).

Also, owing to inequality (26) it is possible to show that a unique solution exists for (12) starting from time $\bar{t}$ up to time $2\bar{t}$. By induction it is possible to show that the following theorem holds:

MAIN THEOREM: If assumptions (a1)-(a4) of Section 2 are satisfied and if u_0 is such that $N_0(x,\mu) \in \mathcal{D}(B_0) \cap X_1^+$, the nonlinear problem (12) admits a unique strong solution $u = u(t)$ $\forall\, t \in [0,\bar{t}_0]$, where t_0 is an arbitrarily fixed finite value of t. Moreover the first component of $u(t)$ satisfies the inequalities of Theorem 1.

7. *Concluding Remarks*

We note that assumptions (a1)-(a4) require that the cross sections are nonnegative and continuously differentiable functions of the temperature, with bounded derivatives. Hence, (a1)-(a4) are at least quite plausible from a physical point of view. Furthermore, $N_0(x,\mu)$ is a density of particles and consequently, $N_0(x,\mu) \geq 0$. Thus the condition $N_0(x,\mu) \in X_1^+$ has a precise physical meaning. On the other hand $N_0 \in \mathcal{D}(B)$ is a standard condition ([9], Chap. 8) and it implies also that the initial neutron density has to satisfy the nonre-entry condition (4). Under these assumptions, the main theorem states that the abstract version of system (1)-(4) admits a unique strong solution $u = u(t)$ over any finite time interval (and not over a suitably small one).

Moreover, the first component of $u(t)$ (See (8)) is nonnegative and bounded as shown by Theorem 1 whereas the second component satisfies inequality (25).

Finally we remark that the bounds for the components of $u(t)$ can be found by solving two suitable linear problems.

Detailed proofs of the theorems listed above will appear elsewhere.

Appendix A

In order to write system (1)-(4) as an abstract initial value problem, we introduce the following real Banach spaces [5]:

$$X_1 = L^1([-L,L] \times [-1,1]); \qquad ||f_1||_1 = \int_{-1}^{1} d\mu \int_{-L}^{L} dx\, f_1(x,\mu)$$

$$X_2 = \underline{\underline{R}} \qquad ; \qquad ||f_2||_2 = |f_2|$$

We also define the operators

(A1) $$J_1 f_1 = \frac{1}{2}\int_{-1}^{1} f_1(x,\mu)d\mu \qquad \mathcal{D}(J_1) = X_1,\ \mathcal{R}(J_1) \subset X_1$$

(A2) $$J_2 f_1 = \int_{-L}^{L} dx' \int_{-1}^{1} d\mu' f_1(x',\mu'),\ \mathcal{D}(J_2) = X_1,\ \mathcal{R}(J_2) \subset X_2$$

(A3) $$B_0 f_1 = -v\mu \frac{\partial f_1}{\partial x}\ ;\ \mathcal{D}(B_0) = \Big\{ f_1 \in X_1 \,|\, \mu \frac{\partial f_1}{\partial x} \in X_1;$$

$$f_1(-L,\mu) = 0,\ \forall\, \mu \in]0\ ,1];\ f_1(L,\mu) = 0,$$

$$\forall\, \mu \in [-1,0[\Big\}\ \ \mathcal{R}(B_0) \subset X_1.$$

(A4) $$B_1 f_1 = B_0 f_1 - v\Sigma_0 f_1,\ \mathcal{D}(B_1) = \mathcal{D}(B_0),\ \mathcal{R}(B_1) \subset X_1$$

where $\partial f_1/\partial X$ is a distributional derivative.

From the mathematical theory of neutron transport (see [9], Chap. 8) the following lemmas are known:

LEMMA A1: (a) $J_1 \in \mathcal{B}(X_1)$, $||J_1f_1||_1 \leq ||f_1||_1$; (b) $J_2 \in \mathcal{B}(X_1, X_2)$, $||J_2f_1||_2 \leq ||f_1||_1$; (c) $B_0 \in \mathcal{G}(1,0)$, $B_1 \in \mathcal{G}(1,-v\Sigma_0)$ ([8]); (d) $||exp(tB_0)f_1||_1 \leq ||f_1||_1$, $\forall\ t \geq 0$, $||exp(tB_1)f_1||_1 \leq exp(v\Sigma_0 t)||f_1||_1$ $\forall\ t \geq 0$, if we denote by $exp(tB)$ the semigroup generated by B.

LEMMA A2: (a) $exp(tB_0)[X_1^+] \subset X_1^+$; $exp(tB_1)[X_1^+] \subset X_1^+$; (b) $J_1[X_1^+] \subset X_1^+$, $J_2[X_1^+] \subset X_2^+$, where we denote by X_i^+ the closed positive cone of X_i $(i = 1,2)$.

Appendix B

Let X be the real Banach space $X_1 \times X_2$ (see Appendix A) with norm

$$||f|| = K_0||f_1||_1 + h||f_2||, \quad f = \begin{pmatrix} f_1(x,\mu) \\ f_2 \end{pmatrix}, \quad f_i \in X_i, \quad i = 1,2$$

where K_0 and h are used in the preceding definition to adjust dimensions.

We define the following operators:

$$\text{(B1)} \qquad B = \begin{pmatrix} B_1 & 0 \\ 0 & -hI \end{pmatrix}, \quad \mathcal{D}(B) = \mathcal{D}(B_0) \times X_2, \quad \mathcal{R}(B) \subset X$$

$$\text{(B2)} \qquad J = \begin{pmatrix} v\gamma_0 J_1 & 0 \\ K_0 J_2 & 0 \end{pmatrix}, \quad \mathcal{D}(J) = X, \quad \mathcal{R}(J) \subset X$$

$$\text{(B3)} \qquad F(f) = \begin{pmatrix} -v\Sigma_0 a_1(f_2)f_1 + v\gamma_0 a_2(f_2)J_1 f_1 \\ K_0 a_3(f_2)J_2 f_1 \end{pmatrix} \quad \mathcal{D}(F) = X, \quad \mathcal{R}(F) \subset X$$

From Lemma A1, it follows

LEMMA B1: (a) $B \in \mathcal{G}(1,-z_0)$, $z_0 = min\{h, v\Sigma_0\}$; (b) $J \in \mathcal{B}(X)$, $||J|| \leq z_1$, $z_1 = h + v\gamma_0$.

We observe that (a) implies that B generates a strongly continuous semigroup $\{Z(t) = exp(tB),\ t \geq 0\}$ such that

$$||Z(t)|| \leq exp(-z_0 t) \qquad \forall\, t \geq 0.$$

Appendix C

If we define

(C1) $$B' = B_0 - v\Sigma'' I, \quad \mathcal{D}(B') = \mathcal{D}(B_0)$$

(C2) $$J' = v\gamma' J_1 \quad , \quad \mathcal{D}(J') = X_1$$

it follows from Lemmas A1 and A2:

LEMMA C1: (a) $B' \in \mathcal{G}(1, -v\Sigma'')$; (b) $J' \in \mathcal{B}(X_1)$; $||J'|| \leq v\gamma'$; (c) the semigroup $\{exp(tB'),\ t \geq 0\}$ and the operator J' both map the positive cone X^+ into itself.

It follows from Lemma C1 that $(B' + J')$ generates the strongly continuous semigroup $\{exp(t(B' + J')),\ t \geq 0\}$ such that

(C3) $$||exp(t(B' + J'))||_1 \leq exp((v\gamma' - v\Sigma'')t), \quad \forall\, t \geq 0$$

(C4) $$exp(t(B' + J'))[X_1^+] \subset X_1^+ \qquad \forall\, t \geq 0$$

REFERENCES

[1] D. H. Nguyen, *Nucl. Sci. Eng.* 50(1973), 370.

[2] D. H. Nguyen, *Nucl. Sci. Eng.* 52(1973), 292.

[3] D. H. Nguyen, *Nucl. Sci. Eng.* 55(1974), 307.

[4] D. B. Reister, and P. L. Chambre', *Nucl. Sci. Eng.* 48(1972), 211.

[5] A. Belleni-Morante, *Nuc. Sci. Eng.* 59(1976), 56.

[6] G. I. Bell, and S. Glasstone, *Nuclear Reactor Theory*, Van Nostrand Reinhold Co., New York (1970).

[7] V. C. Boffi, *Fisica del Reattore Nucleare*, Patron, Bologna (1974).

[8] T. Kato, *Perturbation Theory for Linear Operators*, Springer Pub. Co., Inc., New York (1966).

[9] M. G. Wing, *An Introduction to Transport Theory*, John Wiley & Sons, Inc., New York (1962).

[10] J. Mika, *Nukleonik*, 9(1967), 46.

[11] A. Belleni-Morante, *Le Matematiche* 25(1970).

[12] R. S. Phillips, *Trans. A. M. S.* 74(1953), 199.

[13] I. Segal, *Ann. of Math.* 78(1963), 339.

[14] G. E. Ladas, and V. Lakshmikantham, *Differential Equations in Abstract Spaces*, Academic Press, New York (1972).

[15] T. Kato, *Proc. Symp. Appl. Math.* 17(1965), 50.

[16] T. Kato, *Lecture Notes in Mathematics, Vol. 448*, p. 25-70, Springer Pub. Co., Inc., New York (1975).

OPTIMAL HARVESTING FOR THE LOGISTIC AND GOMPERTZ GROWTH CURVE

J. E. Chance
Pan American University

Introduction

Life scientists in many fields have observed a close correspondence between the growth of living systems and the logistic and Gompertz curves, developed to mathematically describe the size of a living system as a function of time. In horticulture, the Gompertz curve has been used successfully to describe the height of an avocado graft [1] in an example of what researchers in this area call Mitscherlich's law of growth. Logistic growth has been found to be a viable model of reproduction in fish species [2] as well as the growth of certain cancerous cells [3]. Good correspondence between the growth of certain cancerous human cells and the Gompertz curve is found in [4].

Recent interest in optimum scheduling of fractionated radiation dosage for cancer tumor treatment has encouraged questions relating to optimal harvest of life systems whose growth curves are either logistic or Gompertz. That is, if a species of fish (or the cancerous cells of a tumor) reproduce according to these laws, can judicious harvesting of portions of this population increase the yield? If so, what is the maximum yield that can be obtained -- how viable is a species that reproduces according to these laws? This paper will consider the problem of optimal harvesting. The idea for this paper came from material presented by H. T. Banks in the Chautauqua Course for Mathematical Modeling of Biomedical Systems in November, 1975.

Suppose that a population with an initial size of N_0 of a unit monotonically increases to a limit population of 1 unit with a population size at time t of $N(N_0,t)$ units. A yield of .95 unit can be obtained by a single harvest at time, t_{95}, or k-harvests can be performed at times $0 = t_0, t_1, \ldots, t_{k-1}$, $t_k = t_{95}$, harvesting fractions H_i $i = 0,1,2,\ldots,k$ of the population at each of these times to give a yield of $\sum_{i=0}^{k} H_i$. Can this scheme of fractional harvesting give a better yield than .95 units, and if so, is there an optimal strategy for harvesting that maximizes the yield?

Yield Formula

Suppose one considers the function $N(N_0,t)$ on the interval $[0,r]$ with $N(N_0,0) = N_0$. Select numbers $t_0,t_1,\ldots,t_k$ in this interval such that

$$0 = t_0 < t_1 < t_2 < \ldots < t_k = r$$

and at each of these points t_i, select a number H_i called a fractional harvest such that

$$\begin{aligned}
H_k &= N(N_{k-1},\Delta t_{k-1}) \qquad (\Delta t_k = t_{k+1} - t_k) \\
N_{k-1} &= N(N_{k-2},\Delta t_{k-2}) - H_{k-1} \\
&\vdots \\
N_{k-i} &= N(N_{k-i-1},\Delta t_{k-i-1}) - H_{k-i}
\end{aligned} \tag{1}$$

and $H_0 = 0$.

H_0, the initial fractional harvest will be assumed to be zero, but in many cases it may be necessarily non-zero to optimize the yield.

The numbers $N_1,N_2,\ldots,N_{k-1}$ are the "new" initial populations that remain after each fractional harvest.

The yield for k-fractional harvests, $Y(k)$, is given by

$$\begin{aligned}
Y(k) &= \sum_{i=0}^{k} H_i = \sum_{i=1}^{k-1} H_i + H_k \\
&= \sum_{i=1}^{k-1} [N(N_{k-i-1},\Delta t_{k-i-1})-N_{k-i}] + N(N_{k-1},\Delta t_{k-1}) \\
&= \sum_{i=0}^{k-1} [N(N_{k-i-1},\Delta t_{k-i-1}) - N_{k-i}] + N_0 \\
&= \sum_{i=0}^{k-1} [N(N_i,\Delta t_i) - N_i] + N_0
\end{aligned} \tag{2}$$

$$= \sum_{i=0}^{k-1} \frac{N(N_i,\Delta t_i) - N_i}{\Delta t_i} \Delta t_i + N_0$$

DEFINITION: For $N(N_0,t)$ continuous and having continuous derivation on $[0,r]$, let

$$G : [0,1] \times [0,r], \rightarrow R \text{ with}$$

$$G(x,y) = \frac{N(x,y) - N(x,0)}{y} \quad \text{if } y > 0$$

$$= N'(x,0) \quad \text{if } y = 0.$$

G is continuous throughout this closed rectangle, and (2) can be written as

$$Y(k) = \sum_{i=0}^{k-1} G(N_i,\Delta t_i)\Delta t_i + N_0$$

For fixed numbers $t_i, G(N_i,\Delta t_i)$ attains its maximum value (since it is continuous) at points, say, $(\hat{N}_i,\Delta t_i)$ $i = 0,1,2,\ldots, k-1$, and

$$Y(k) = \sum_{i=0}^{k-1} G(N_i,\Delta t_i)\Delta t_i + N_0 \leq \sum_{i=0}^{k-1} G(\hat{N}_i,\Delta t_i)\Delta t_i + N_0. \tag{3}$$

$Y(k)$ will attain this upper bound if positive H_i can be chosen from (1) in such a manner that $N_i = \hat{N}_i$. If so, $Y(k)$ will be maximized for a given partition $t_0, t_i, \ldots, t_k$. Once this strategy is applied to maximize Y for a given partition, the method of LaGrange Multipliers can be used to optimize the partition choice.

The Logistic Curve

The function

$$N(N_0,t) = \left[\left(\frac{1 - N_0}{N_0}\right)e^{-Bt} + 1\right]^{-1} \tag{4}$$

is known as the Logistic Curve with initial population N_0, asymptotic population of 1 unit and growth parameter B. $(0 < N_0 < 1)$

The following facts about the Logistic Curve will be useful for the ensuing discussion

(i) If t_I is the inflection point for (4), then $N(t_I) = .5$.

(ii) (4) satisfies the differential equation $\frac{dN}{dt} = BN(1-N)$, $N(0) = N_0$.

(iii) For $0 \leq x < t_I$

$$N(t_I - x) + N(t_I + x) = 1.$$

One can simplify the notation used in the definition for an interval size, a, by writing

$$G(t,a) = \frac{N(t) - N(t - a)}{a}$$

$$\frac{dG}{dt} = \frac{1}{a}\left[N'(t)-N'(t-a)\right] = \frac{1}{a}\left[BN(t)(1-N(t)) - BN(t-a)(1 - N(t-a))\right]$$

$$= \frac{B}{a}\left[N(t)-N(t-a)\right]\left[1 - (N(t)+N(t-a))\right]$$

Using (iii) it can be seen that

$\frac{dG}{dt} = 0$ if $t = t_I + \frac{a}{2}$, and since N is an increasing function

$$\frac{dG}{dt} > 0 \text{ if } t < t_I + \frac{a}{2}$$

$$\frac{dG}{dt} < 0 \text{ if } t > t_I + \frac{a}{2}.$$

Thus G is maximized on the interval $[t_I - \frac{a}{2}, t_I + \frac{a}{2}]$.

One can observe by (iii) that if $N\left(t_I + \frac{t_i}{2}\right) = .5 + x_i$, then $N\left(t_I - \frac{\Delta t_i}{2}\right) = .5 - x_i$. For partition $t_0, t_i, \ldots, t_k$, a fractional harvest exists that satisfies (1). In this case

$$N_i = 5 - X_i, \quad N(N_{i-1}, \Delta t_{i-1}) = .5 + X_{i-1}$$

$H_i = X_i + X_{i-1}$, so that the maximum yield given in (3) can be attained, and is given by

$$Y_{(k)} = \sum_{i-1}^{k-1} \left[N(t_I + \frac{\Delta t_i}{2}) - N(t_I - \frac{\Delta t_i}{2}) \right] + N(\Delta t_0). \tag{5}$$

(The harvest on the first interval Δt_0 cannot be maximized, since N_0 cannot be controlled.) (5) must be maximized as a function of the variables $\Delta t_0, \Delta t_1, \ldots, \Delta t_{k-1}$ subject to the constraint that

$$\sum_{i=0}^{k-1} \Delta t_i = r.$$

Using LaGrange Multipliers, one locates the critical points for

$$F(\Delta t_0, \Delta t_i, \ldots, \Delta t_{k-1}) = \sum_{i=1}^{k-1} \left[N\left(t_I + \frac{\Delta t_i}{2}\right) - N\left(t_I - \frac{\Delta t_i}{2}\right) \right] + N(\Delta t_0)$$

$$+ \lambda \left[r - \sum_{i=0}^{k-1} \Delta t_i \right]$$

$$0 = \frac{\partial F}{\partial \Delta t_0} = N'(\Delta t_0) - \lambda$$

$$0 = \frac{\partial F}{\partial \Delta t_i} = \frac{1}{2} N'\left(t_I + \frac{\Delta t_i}{2}\right) + \frac{1}{2} N'\left(t_I - \frac{\Delta t_i}{2}\right) - \lambda$$

$$i = 1, 2, \ldots, k-1. \tag{6}$$

Since $N\left(t_I + \frac{\Delta t_i}{2}\right) + N\left(t_I - \frac{\Delta t_i}{2}\right) = 1$, $N'\left(t_I + \frac{\Delta t_i}{2}\right) = N'\left(t_I - \frac{\Delta t_i}{2}\right)$, so that (6) becomes

$$\lambda = N'\left(t_I + \frac{\Delta t_i}{2}\right) \qquad i = 1, 2, \ldots, k-1$$

or

$$N'\left(t_I + \frac{\Delta t_i}{2}\right) = N'\left(t_I + \frac{\Delta t_j}{2}\right) \qquad i, j \neq 0.$$

Since these points are to the right of the inflection point, one can see by using (ii) that $t_I + \frac{\Delta t_i}{2} = t_I + \frac{\Delta t_j}{2}$, or $\Delta t_i = \Delta t_j$. Denote this common spacing as Δt, then for $i = 0$,

$$N'(\Delta t_0) = \lambda = N'\left(t_I + \frac{\Delta t}{2}\right) \quad \text{and}$$

$$\Delta t_0 = t_I + \frac{\Delta t}{2} .$$

An examination of the second partial derivatives indicates that F is maximized at this critical point.

Yield for the Logistic Curve

In what follows, it is assumed that $N_0 < .5$. If not, an initial fractional harvest H_0 should be made. This initial harvest would raise the yield, but we shall assume a minimal initial condition.

With an initial population size N_0, the time required to reach a population size of .95 unit, t_{95}, is

$$.95 = \frac{1}{\left(\frac{1 - N_0}{N_0}\right) e^{-Bt_{95}} + 1}$$

or
$$t_{95} = \frac{1}{B} \, log\left[19(1 - N_0)/N_0\right]$$

Using the conditions found by the LaGrange Multiplier, the population increases until it reaches the size $.5 + x$. The harvest $H_1 = 2x$ is made leaving a new initial population of $.5 - x$. Again the population is allowed to increase to the size $.5 + x$, and a harvest $H_2 = 2x$ is made. This process is continued a total of $k - 1$ times until the time t_{95} is reached and the total remaining population of $.5 + x$ is harvested. The time required for the population to increase from $.5 - x$ to

$.5 + x$ is t_1 where

$$.5 + x = \frac{1}{\left(\frac{.5 + x}{.5 - x}\right)e^{-Bt_1} + 1}$$

or
$$t_1 = \frac{2}{B}\, log\left(\frac{.5 + x}{.5 - x}\right) .$$

The time t_2 required for the population size to increase from N_0 to $.5 + x$ is

$$.5 + x = \frac{1}{\left(\frac{1 - N_0}{N_0}\right)e^{-Bt_2} + 1}$$

$$t_2 = \frac{1}{B}\, log\left[\left(\frac{.5 + x}{.5 - x}\right)\left(\frac{1 - N_0}{N_0}\right)\right] .$$

For a total of k harvests

$$(k - 1)t_1 = t_{95} - t_2$$

$$\frac{2(k - 1)}{B}\, log\left(\frac{.5 + x}{.5 - x}\right) = \frac{1}{B}\, log\left[19\left(\frac{1 - N_0}{N_0}\right)\right] - \frac{1}{B}\, log\left[\left(\frac{.5+x}{.5-x}\right)\left(\frac{1-N_0}{N_0}\right)\right]$$

$$(2k - 1)\, log\left(\frac{.5 + x}{.5 - x}\right) = log\ 19$$

from which it is seen that

$$x = .5\left[\frac{19^{\frac{1}{2k - 1}} - 1}{19^{\frac{1}{2k - 1}} + 1}\right] .$$

The total yield

$$Y(k) = (k - 1)2X + X + .5$$

$$= (2k - 1)X + .5$$

$$= \frac{(2k - 1)}{2} \left[\frac{19^{\frac{1}{2k-1}} - 1}{19^{\frac{1}{2k-1}} + 1} \right] + .5 \, .$$

One can observe that the yield is independent of N_0 and B and

$$\lim_{K \to \infty} Y(k) = \frac{\log 19}{4} + .5 = 1.236109, \text{ an upper limit on yield.}$$

Table 1 summarizes harvest schedules and yields using $B = .5$. It should be noted that in Table 1, if the initial population is larger than the N_0 given, an initial harvest H_0 will be required to achieve the indicated yield.

TABLE 1

k Number of Harvests	t_{95} $N(t_{95}) = .95$	Δt Harvest Spacing	H_i Harvest	N_0 Initial Population	$Y(k)$ Yield for k Harvests
2	7.85184	3.92592	.454803	.272598	1.18221
3	7.06665	2.35555	.28622	.35689	1.21555
4	6.73015	1.68254	.20727	.396365	1.22544
5	6.5432	1.30864	.162136	.418932	1.22961
6	6.42423	1.07071	.133045	.433478	1.23175
7	6.34187	.905981	.112766	.443617	1.23298
8	6.28147	.785184	.097834	.451083	1.23376

The Gompertz Curve

The Gompertz curve is of the form

$$N(t) = KA^{(b^t)}$$

a double exponential with S-shaped graph for $0 < b < 1$. With initial value $N_0 < 1$ and final value 1 the curve becomes

$$N(N_0, t) = N_0^{(b^t)}$$

By calculations analogous to those done in the previous section, it can be shown that

(i) For fixed interval size Δt, $G(t,\Delta t)$ attains its maximum value for $t = t_I + \alpha\Delta t$, where $0 < \alpha < 1$.

(ii) Using LaGrange Multipliers, the optimal partition is $\Delta t_1 = \Delta t_2 = \cdots = \Delta t_{k-1} = \Delta t$, and $\Delta t_0 = t_I + \alpha\Delta t$.

(iii) Letting $\beta = 1 - \alpha$, $N_0 = N(t_I - \beta\Delta t)$

$$Y(k) = \sum_{i=0}^{k-1}\left[N(t_I + \alpha\Delta t) - N(t_I - \beta\Delta t)\right] + N(t_I - \beta\Delta t)$$

$$= k\left[N(t_I + \alpha\Delta t) - N(t_I - \beta\Delta t\right] + N(t_I - \beta\Delta t).$$

But $k\Delta t = t_{95}$ so that

$$Y(k) = t_{95}\,\frac{N(t_I + \alpha\Delta t) - N(t_I - \beta\Delta t)}{\Delta t} + N(t_I - \beta\Delta t)$$

and

$$\lim_{K\to\infty} Y(k) = \hat{t}_{95}N'(t_I) + N(t_I)$$

where $\hat{t}_{95}$ is the time required for the Gompertz Curve to reach .95 with an initial value of $N_0 = N(t_I)$. This limit can be calculated by use of the following:

(a) $$t_I = -\frac{\log\left(\log N_0^{-1}\right)}{\log b}$$

(b) $$N(t_I) = \frac{1}{N_0^{\log N_0^{-1}}}$$

(c) to find $\hat{t}_{95}$, $N_0 = N_0^{\frac{1}{\log N_0^{-1}}}$, or $N_0 = e^{-1}$.

This gives $\hat{t}_{95} = \dfrac{\log\left[\log \frac{20}{19}\right]}{\log b}$, and $N'(t_I) = (-e^{-1})\log b$.

Thus $$\lim_{K\to\infty} Y(k) = -\log\left[\log \frac{20}{19}\right]e^{-1} + e^{-1} = 1.46055.$$

Thus, no yield will exceed this number. As in the case of the Logistic Curve the yield is independent of N_0 and b. To calculate a harvest schedule for a fixed b and a fixed positive integer k, an iteration technique is used. To maximize $G(t,a)$ for a fixed a, t must satisfy the equation

$$N_0^{b^t} - b^{-a}[N_0]^{b^t \cdot b^{-a}} = 0.$$

The iteration technique is thus: *(fix k,b)*

(a) Choose N_0

(b) Calculate $a = \frac{1}{k} t_{95} = \dfrac{\log\left(\frac{\log .95}{\log N_0}\right)}{k \log b}$.

(c) Let $r = b^{-a}$ and solve the equation

$$N_0^{b^t} - r\left[N_0^{b^t}\right] = 0, \quad \text{for } t \text{ which gives}$$

$$t = \frac{1}{\log b}\left[\log \log r - \log\left((1 - r) \log N_0\right)\right] .$$

(d) If $t = a$, $N(t - a) = N_0$ and the correct N_0 has been chosen. If not, then the sign of $t - a$ indicates whether or not to choose a new N_0 that is larger or smaller than the previous choice. A new choice for N_0 is substituted into (b) and the result is repeated until $t = a$.

Table 2 summarizes these results to give harvest schedules for $B = .5$. Again, if N_0 is larger than the value indicated in the table, an initial harvest H_0 must be made to achieve the indicated yield.

TABLE 2. Yield for the Gompertz Curve ($B = .5$)

H_i Harvest	$Y(k)$ Yield	t_{95} $N(t_{95}) = .95$	N_0 Initial Population	k Number of Harvests
.605085	1.31858	2.71849	.1084	2
.406032	1.40193	1.6826	.1837	3
.300317	1.42929	1.21239	.2280	4
.237106	1.44126	.9464	.2557	5
.195522	1.44759	.77626	.2745	6

REFERENCES

[1] J. E. Chance and C. O. Foerster, Jr., *The growth curve for an avocado graft*, Journal of the Rio Grande Valley Horticultural Society 27(1973), 71-73.

[2] R. T. H. Beverton, and S. T. Holt, *On the Dynamics of Exploited Fish Populations*, Fish. Invest. Lond. 2(1957), 19.

[3] B. Jansson, and L. Revesz, *Analysis of the growth of tumor cell populations*, Math. Biosciences 19(1974), 131-154.

[4] J. Aroesty, T. Lincoln, N. Shapiro, and G. Boccia, *Tumor growth and chemotherapy: mathematical methods, computer simulations, and experimental foundations*, Math Biosciences 17(1973), 243-300.

[5] H. Hethcote and P. Waltman, *Theoretical determination of Optimal treatment schedules for radiation therapy*, Radiat. Res. 56(1973), 150-161.

MAXIMUM AND MINIMUM DEGENERACY SET OF LINEAR TIME-INVARIANT DELAY-DIFFERENTIAL SYSTEMS OF THE NEUTRAL TYPE

A. K. Choudhury*
Howard University

1. *Introduction*

Since the paper of Popov [11] much attention has been paid to the study of pointwise degeneracy of delay and neutral type differential equations from the view point of control theory. The main point in this connection is related to the fact that a pointwise degenerate system may be considered as a result of using a so-called linear delay-feedback (Popov 1971). If we look at the problem from this side a question that naturally arises is the one of the characterization of the point t_d where degeneracy starts, since this point may be viewed as being the one where the output of the system is driven to zero by delayed state-feedback. First results in this direction belong to Popov, 1972 and important extensions has been made by Kappel [7,8] and Lang [9]. The aim of this paper is to obtain similar results for neutral delay-differential equations and more precisely to prove the following theorem.

THEOREM. Consider the equation

$$\dot{x}(t) = \sum_{i=0}^{m} B_i x(t - ih) + \sum_{i=1}^{m} c_i \dot{x}(t - ih).$$

Suppose that there exist $d \neq 0$ and $t_1 > 0$ such that all solutions defined for $t \geq -mh$ satisfy $d^T x(t) = 0$ for $t \geq t_1$; then there exists an integer $k_0 \geq 2$ such that all solutions defined for $t \geq -mh$ satisfy $d^T x(t) = 0$ for $t \geq (k_0 - 1)h$ and for any $\hat{t} < (k_0 - 1)h$ there exists a solution such that $d^T x(t)$ is not identically zero on $\hat{t} < t < (k_0 - 1)h$. Moreover $k_0 - 1 \leq (n - 1)m$.

* This research work was partially supported by NASA Grant No. 9010.

REMARK. Examples show that $k_0 = 2$ is possible. The theorem will be proved after some preparations.

2. Preliminary Computations

Define $G_i(\sigma) = (\sigma I - B_0)^{-1}(B_i + \sigma C_i)$; $i = 1,2,\ldots,m$.

Denote $\Delta(\sigma) = \det(\sigma I - B_0)$.

Construct $Q_0(\sigma) = I$, $Q_1(\sigma) = G_1(\sigma)\Delta(\sigma)$

$$Q_k(\sigma) = \sum_{i=1}^{m} Q_{k-i}(\sigma)G_i(\sigma)\Delta^i(\sigma).$$

LEMMA 1. $Q_k(\sigma) = \sum_{i=1}^{m} G_i(\sigma)Q_{k-i}(\sigma)\Delta^i(\sigma)$.

PROOF. By using definition and changing the order of summation.

Define $P_{r,j}(\sigma) = \sum_{i=1}^{m} Q_{r-i}(\sigma)G_{j+i-1}(\sigma)\Delta^i(\sigma)$.

LEMMA 2.

$$P_{r+1,j}(\sigma) = \sum_{i=1}^{r} G_i(\sigma)\Delta^i(\sigma)P_{r-i+1,j}(\sigma) + \Delta^{r+1}(\sigma)G_{r+j}(\sigma)$$

with the conventions $G_i(\sigma) = 0$, $i > m$, $P_{r,j}(\sigma) = 0$ $r < 0$

PROOF. Use the definition and compute.

Define the block matrix

$$H(\sigma) = \begin{pmatrix} G_1(\sigma) & \cdots & G_{m-1}(\sigma) & G_m(\sigma) \\ I & \cdots & 0 & 0 \\ \cdots & \cdots & \cdots & \cdots \\ 0 & \cdots & I & 0 \end{pmatrix}$$

Then direct computation shows that

$$H^r(\sigma)\cdot\Delta^r(\sigma) = \begin{pmatrix} P_{r,1}(\sigma) & \ldots & P_{r,m-1}(\sigma) & P_{r,m}(\sigma) \\ P_{r-1,1}(\sigma)\Delta(\sigma) & \ldots & P_{r-1,m-1}(\sigma)\Delta(\sigma) & P_{r-1,m}(\sigma)\Delta(\sigma) \\ \ldots & \ldots & & \ldots \\ P_{1,1}(\sigma)\Delta^{r-1}(\sigma) & \ldots & P_{1,m-1}(\sigma)\Delta^{r-1}(\sigma) & P_{1,m}(\sigma)\Delta^{r-1}(\sigma) \\ \ldots & \ldots & & \ldots \\ 0 & & \Delta^r(\sigma)I & 0 \end{pmatrix}$$

for $r < m$, and

$$H^r(\sigma)\Delta^r(\sigma) = \begin{pmatrix} P_{1,1}(\sigma) & \ldots & \ldots & P_{r,m}(\sigma) \\ \vdots & \ldots & \ldots & \vdots \\ P_{r-m+1,1}(\sigma)\Delta^{r-1}(\sigma) & \ldots & & P_{r-m+1,m}(\sigma)\Delta^{r-1}(\sigma) \end{pmatrix}$$

for $r \geq m$.

Introduce the notation

$$H^k(\sigma) = \begin{pmatrix} H_1^k(\sigma) & \ldots & \ldots & H_m^k(\sigma) \\ \vdots & & & \vdots \\ H_1^{k-m+1}(\sigma) & \ldots & \ldots & H_m^{k-m+1}(\sigma) \end{pmatrix}$$

<u>LEMMA 3</u>. $H_i^k(\sigma) = \sum_{\ell=1}^{\min\{k,m\}} H_1^{k-\ell}(\sigma)G_{i+\ell-1}(\sigma), \quad i = 1,\ldots,m.$

<u>PROOF</u>. Direct computation.

<u>LEMMA 4.</u> $\sigma H_i^k(\sigma) = \sum_{j=0}^{k-1} B_{k,j}H_i^{k-j}(\sigma) + B_{k,k+i-1} + \sigma C_{k,k+i-1}$

where

$$B_{k,i} = B_{k-1,i} + C_{k-1,k-1}B_{i-k+1}, \quad i = 0,1,\ldots,k+m-1, \quad k \geq 2$$

$$B_{1,i} = B_i$$

$$C_{k,i} = C_{k-i,i} + C_{k-1,k-1}C_{i-k+1}$$

$$C_{1,i} = C_i$$

and we adopt the conventions

$$B_i = C_i = 0, \quad i < 0, \quad C_{k,i} = 0, \quad i < k$$

$$B_i = C_i = 0, \quad i > m, \quad B_{k,i} = C_{k,i} = 0 \quad \text{for} \quad i \geq k + m$$

PROOF. By induction on k; easy to check for $k = 1$. Assume it is true for powers $1 < \ell \leq k$; use lemma 2, use induction for a representation of $H_1^{k-\ell+1}(\sigma)$; use also $\sigma G_j(\sigma) = B_j + \sigma C_j + B_0 G_j(\sigma)$; use then the recurrence formulae for $B_{k,j}$, $C_{k,j}$. Change the summation order and make carefully all computations. It is better to consider separately the cases $k < m$ and $k \geq m$

LEMMA 5. $\sigma^r H_i^k(\sigma) = \sum_{j=0}^{k-1} B_{k,j}^r H_i^{k-j}(\sigma) + B_{k,k+i-1}^r + \sum_{j=0}^{k-1} \sigma^{j+1} C_{k,k+i-1}^{r-j}$

$$B_{k,\lambda}^{r+1} = \sum_{j=0}^{\lambda} B_{k,j}^r \, B_{k-j,\lambda-j} \, ,$$

$$C_{k,\lambda}^{r+1} = \sum_{j=0}^{k-1} B_{k,j}^r C_{k-j,\lambda-j} + B_{k,\lambda}^r; \quad B_{k,j}^1 = B_{k,j}$$

REMARK. In the formula for $B_{k,\lambda}^{r+1}$ if $\lambda > k - 1$ then summation ends at $k - 1$, since $B_{k-\lambda,\lambda-j} = 0$ if $j \geq k$.

3. *Representation Formula and Necessary Conditions for Degeneracy*

PROPOSITION. Consider solutions of the equation defined by initial functions of the form $[\Delta(D)]^k \phi$ with ϕ analytic and $(D^s \phi)(-ih) = 0$, $s = 0,1,\ldots,nk$, $i = 0,1,\ldots,m$. Then for $(r - 1) \leq t \leq rh$,

$$x(t) = \sum_{i=1}^{m} [P_{r,i}(D)\Delta^{k-r}(D)\phi](t - (j + r - 1)h), \quad D = \frac{d}{dt}$$

where $P_{r,i}$ are the polynomials with matrix co-efficients defined in section 2.

PROOF. By induction using lemma 2.

COROLLARY 1. If for some $(k-1)h \leq t_1 < kh$ all solutions satisfy $d^T x(t) = 0$ for some $t \geq t_1$, then $d^T P_{k,i}(\sigma) = 0$.

PROOF. If $d^T x(t) = 0$ for $t \geq t$ then by properly choosing ϕ, we get $d^T P_{k,i}(\sigma) = 0$.

COROLLARY 2. If for all solutions $d^T x(t) = 0$ for $t \geq t_1$, then $d^T H_i^{(n-1)m+1}(\sigma) = 0$.

PROOF. $d^T P_{k,i}(\sigma) = 0$ implies by the definition of $H_i^k(\sigma)$ that $DH^{k+m-1}(\sigma) = 0$, where,

$$D = \begin{pmatrix} d^T & 0 & \dots & 0 \\ 0 & d^T & \dots & 0 \\ \dots & \dots & \dots & \dots \\ 0 & 0 & \dots & d^T \end{pmatrix}$$

D is a nm x nm and if $k - 1 > (n-1)m$ one may use Cayley-Hamilton theorem to deduce that $DH^{nm}(\sigma) = 0$ and considering the structure of $\overset{nm}{H}(\sigma)$ it follows that $d^T H_i^{(n-1)m+1}(\sigma) = 0$.

COROLLARY 3. If for all solutions $d^T x(t) = 0$ for $t \geq t_1$, $(k-1)h \leq t_1 < kh$, then $d^T C^r_{k,k+i-1} = 0$. $r \geq 1$, $i = 1,2,..,m$.

PROOF. The assumption will imply by corollary 1 that $d^T H_i^k(\sigma) = 0$, hence $d^T \sigma^r H_i^k(\sigma) = 0$ for all $r \geq 1$; use lemma 5, divide by σ^r and let $\sigma \to \infty$.

4. *Proof of the Theorem*

STEP 1. Let x be a solution of the given equation; then for $t > (k-1)h$ it will satisfy

$$\dot{x}(t) = \sum_{i=0}^{k+m-1} B_{k,i} x(t - ih) + \sum_{i=k}^{k+m-1} C_{k,i} \dot{x}(t - ih)$$

where $B_{k,i}$, $C_{k,i}$ were defined in lemma 4.

PROOF. Replace derivative terms in the equation successively.

STEP 2. Let x be as above; then for $t > (k-1)h$

$$x^{(r)}(t) = \sum_{i=0}^{k+m-1} B^{r}_{k,i} x(t - ih) + \sum_{\ell=0}^{r-1} \sum_{i=k}^{k+m-1} C^{r-\ell}_{k,i}\ x^{(\ell+1)}(t - ih)$$

where $B^{r}_{k,i}$, $C^{r-\lambda}_{k,i}$ were defined in lemma 5.

PROOF. By induction using step 1.

STEP 3. Assume $d^T x(t) = 0$ for $t \geq t_1$, $t_1 \varepsilon [(k-1)h, kh]$ (for all solutions); then by corollary 2 $d^T H_i^{(n-1)m+1}(\sigma) = 0$ and as in corollary 3 $d^T C^{r-j}_{(n-1)m+1,\lambda} = 0$ for $j = 0,1,\ldots, r-1$, $r \geq 1$, $(n-1)m+1 \leq \lambda \leq nm$. Use step 2 for $t \geq (n-1)mh$ and the above remark to get

$$d^T x^r(t) = \sum_{i=0}^{nm} d^T B^r_{(n-1)m+1,i}\ x(t - ih) \quad \text{for all } r \geq 1.$$

For large r enough that allows elimination of $x(t - ih)$ and for $y(t) = d^T x(t)$ one gets $y^{(\rho)}(t) + \sum_{\sigma=0}^{\rho-1} a_\sigma y^{(\sigma)}(t) = 0$; since $y(t) = 0$ for $(k-1) \leq t_1 \leq t$ and $y(t)$ is analytic it follows that $y(t) = 0$ for $t \geq (n-1)mh$.

STEP 4. Let k be such that $d^T x(t) = 0$ for $(k-1)h \leq t_1 < t < kh$ for all solutions and assume $d^T P_{k-1,i}(\sigma) = 0$, then $d^T x(t) = 0$ for $t \geq (k-2)h$.

PROOF. $d^T P_{k-1,i}(\sigma) = 0$ implies $d^T H_i^{k-1}(\sigma) = 0$ and as in corollary 3, it implies that $d^T C^{r-j}_{k-1,\lambda} = 0$, $j = 0,1,\ldots, r-1$; $r > 1$, $k-1 \leq \lambda \leq k+m-2$. Use step 2 for $t \geq (k-2)h$ to get

$$d^T x^{(r)}(t) = \sum_{i=0}^{k+m-2} d^T B^r_{k-1,i} x(t - ih);$$ use the same argument as in step 3 to deduce a differential equation for y and then $y(t) = 0$ for $t \geq t_1$ will imply that $y(t) = 0$ for $t \geq (k-2)h$. This completely proves the theorem.

REMARK. The result is general; it contains in particular the case of delay equations $(C_i = 0)$. This special case can also be obtained from the results of Kappel [8] and Lang [9].

REFERENCES

[1] Asner Jr., B. A., and Halanay, A., *Algebraic theory of pointwise degenerate delay-differential systems*, J. Differential Equations 14(1973), 293-306.

[2] Asner Jr., B. A., and Halanay, A., *Delay-feedback using derivatives for minimum time linear control problems*, J. Math. Anal. Appl. 48(1974), 257-63.

[3] Brooks, R. M., and Schmitt, K., *Pointwise completeness of differential difference equations*, J. Math. Rocky Mountain 3(1972), 11-14.

[4] Charrier, P., and Haugazeau, Y., *Sur la degenerescence des equations differentielles lineaires autonomes avec retards*, J. Math. Anal. Appl 52(1975), 42-55.

[5] Choudhury, A. K., *Necessary and sufficient conditions of pointwise completeness of linear time-invariant delay-differential systems*, International J. of Control 16(1972), 1083-1100.

[6] ______, *On the pointwise completeness of delay-differential systems of the neutral type*, SIAM J. Math. Anal. (1976).

[7] Kappel, F., *On the degeneracy of functional differential equations*, Proceedings of International Conference on Differential Equations, edited by H. Antosiewicz, Academic Press, Los Angeles, pp. 434-448.

[8] ______, *Degenerate difference-differential equations*, Algebraic theory, Private Communication.

[9] Lang, H., *Some comments on pointwise degeneracy of delay-differential systems*, Matematiska Institutionen Stockholm, Sweden, 1975, No. 1.

[10] Popov, V. M., *Pointwise degeneracy of linear time invariant delay-differential equations*, J. Differential Equations 11(1972), 541-561.

[11] ______, *Delay-feedback, time-optimal, linear time invariant control systems*, 1971 NRL-MRC Conference, edited by L. Weiss, Academic Press, Washington, D.C., pp. 545-552.

ON THE CONTROLLABILITY TO CLOSED SETS OF NONLINEAR AND RELATED LINEAR SYSTEMS

E. N. Chukwu
Cleveland State University

Consider the nonlinear control process

$$\dot{x} = f(t,x,u) \quad \text{in} \quad C^1[E \times E^n \times E^m] \tag{1}$$

and its related linear system

$$\dot{x} = A(t)x + B(t)u \tag{2}$$

where $A(t) = \frac{\partial f}{\partial x}(t,0,0)$, $B(t) = \frac{\partial f}{\partial u}(t,0,0)$. Let G be a closed target set function. Let $L^-(t_0) = \bigcup_{t>t_0} L(t,t_0)$ where $L(t,t_0)$ is the controllable set of (2) at time t_0; that is, the set of initial states $x_0 \in E^n$ such that there exists some admissible control u for which the solution $x(t) = x(t;t_0,x_0,u)$ of

$$\dot{x}(t) = f(t,x(t),u(t)), \; x(t_0) = x_0 \tag{3}$$

satisfies $x(t) \in G(t)$. Then

$$L^-(t_0) = \bigcup_{t>t_0} (X^{-1}(t,t_0)G(t) + X(t_0)\mathcal{C}(t,t_0)) \tag{4}$$

where $X(t,t_0)$ is the fundamental matrix of $\dot{x} = A(t)x$ with $X(t_0,t_0) = I$ the identity matrix, and $X(t_0) = X(t_0,0)$. Here $\mathcal{C}(t,t_0)$ is the controllable set of (2) on $[t_0,t]$.

<u>THEOREM 1</u>. If $L^-(t_0) = E^n$, then (1) is locally controllable to G in the following sense: Given any trajectory $x^0(\;;t_0,x_0,u_0) \equiv x^0(\;)$ of (3) with u_0 admissible and $x^0(t_0) = x_0$ such that for some $t > t_0$, $x^0(t) \in G(t)$ then there is a neighborhood $N(x_0)$ of x_0 such that for each $\bar{x}_0 \in N(x_0)$ there exists a control u^* on $[t_0,t]$ such that $x(t;t_0,\bar{x}_0,u^*) \in G(t)$ for some $t > t_0$.

<u>THEOREM 2</u>. Consider the system (1) with a fixed closed target G. Suppose for $\varepsilon > 0$ there is a Lyapunov function V_ε with the following property:

H(1) For each $t_0 \geq 0$ and each $\varepsilon > 0$ there exists $\delta = \delta(t_0,\varepsilon) > 0$ such that for all $t \geq t_0$, $V(t_0,x) < \varepsilon$ whenever $0 < d(x,G) < \delta$;

H(2) There exists some $\delta_1 > 0$ such that $V^1(t,x) \leq -\delta_1 |f(t,x,u(x))|$ for all $x \in S^c(G,\varepsilon) = \{y : d(t,S) \geq \varepsilon\}$.

H(3) $f(t,x,u)$ satisfies the following condition: for each x_0 in G^c, complement of G, there exists $\eta > 0$ and a continuous $g : [0,\infty) \to [0,\infty)$ with the property that $\int_0^\infty g(t)dt = \infty$.

H(4) $x \in G^c$ and $|x - x_0| < \eta$ then

$$|f(t,x,u)| \geq g(t).$$

Then G is globally asymptotically stable.

The two results now yield sufficient conditions for the controllability to G of (1).

A DISCRETE-TIME NONLINEAR $(m + n)$-PERSON LABOR MARKET SYSTEM WITH UNCERTAINTY

Peter Coughlin
Middlebury College

1. *Introduction*

This paper is concerned with two important questions in economics. First, given a labor market with n labor applicants and m employers with incomplete information on the behavior of directly competing individuals and wage setting through a wage bidding process in which bids and offers are based on expectations, will each participant be able to select an optimal solution to each of the three resulting decisions? Second, will there exist such labor markets which have possible equilibria which imply the existence of persistent unemployment and no adjustment of the wages being paid? Both questions are answered in the affirmative by this paper.

Models of interactive labor markets and the existence of optimal decisions and equilibria have been considered in the recent literature. Spence considered labor characteristics and signaling decisions and equilibria. Mirman and Porter and Porter have been concerned with decisions under uncertainty and stochastic equilibria. Persistent equilibria with involuntary unemployment in labor markets with a continuum of agents have been considered in Futia.

Disequilibrium decisions have been considered by Barro and Grossman and in the Benassy papers. Benassy has considered fix-price equilibria with disequilibrium decisions. Dynamic adjustment and persistent unemployment has been analyzed in Tobin.

Sequential equilibria under price uncertainty have been developed and analyzed by Radner.

Keynes and Leijonhufvud were concerned with the possibility of the existence of labor markets with behavioral equilibria with unemployment.

Section 2 of this paper presents a mathematical model of an

interactive labor market with n labor applicants and m employers and deduces the relevant structure for the triple decision process for each participant. Each participant has incomplete information on the behavior of all the rival economic agents which is summarized in a subjective probability distribution. Wages are determined by a bidding process in which the labor applicants decide on bids and the employers decide on their offers on the basis of their expectations on the behavior of the other labor applicants and employers, respectively, and the functional relationships which they perceive for the remaining structure of the labor market. Wages may vary from labor applicant to labor applicant and the wage schedule that emerges from the bidding may be a disequilibrium wage schedule (in the sense of Walras) which has corresponding unemployment.

The labor applicants select a job from their job offers which then determines the level of employment for each employer. Each firm then decides what level of production will maximize profits. This process results in an allocation of jobs, wages, labor services, and output in every time interval. Labor applicants and employers are then assumed to revise their expectations on the basis of observations on the wage bids of the other labor applicants and the wage offers of the other employers, respectively. Agents may thus adjust their expectations to incorporate new information by adapting their subjective distributions and behavior.

Section 3 presents the main results, and Section 4 provides the detailed proofs of these results and some directions for further research. This paper proves that for each labor applicant in the defined model there is an optimal solution to the decision problems of selecting wage and job offers under uncertainty (if the firms are competitive) or without uncertainty (if there is a monopoly firm), selecting the level of output which will maximize profits at the market determined level of employment, and selecting revised expectations.

This paper then proves that in any such labor market there is a set of expectations which imply that the actions based on these expectations will confirm the expectations so that the actions will be repeated for the remaining time intervals. Since actions become fixed, all the variables in the labor market become fixed, and the system will be in a behavioral equilibrium. An implication of the existence of an equilibrium in every such labor market is the existence of equilibria with unemployment and a direct example is provided. The unemployment is persistent since the structure of the market which each participant perceives is reinforced and accepted on the basis of the observations on the economic behavior of the other individuals. Labor applicants may desire jobs and receive none but since their expectations imply the optimal strategy for acquiring the desired employment if the expectations are fixed the strategy will be repeated and unemployment will recur. Involuntary unemployment may thus exist in a labor market equilibrium.

2. *The Model*

This section defines a mathematical model corresponding to a labor market with incomplete information on the actions of competing participants, disequilibrium wages, and a wage bidding process. The model is defined by introducing the relevant sets, topological spaces, mappings, and assumptions.

$\exists$ a set of time intervals,

(2.1) $T = \{1,2,\ldots,t_n\}$ where $t \in T$.

$\forall\, t\ \exists$ a finite set of labor applicants,

(2.2) $I = \{1,2,\ldots,n\}$ where $i \in I$.

$\forall\, i\ \exists$ a finite set of possible jobs,

(I.1) $E_i = \{1,\ldots,k_i\}$ where $e_i \in E_i$

$\forall\ j_i\ \exists$ a space of possible wage bids,

(I.2) $(W_{j_i}, \tau(W_{j_i}))$ with $W_{j_i} \subset R_+$

& $\tau(W_{j_i})$ is the usual top. on a subset of R_+ where $w_{j_i} \varepsilon W_{j_i}$. Assume W_{j_i} is compact.

(I.1) & (I.2) → ∃ a space of possible wage bids for E_i,

(I.3) $(W_i, \tau(W_i))$ with $W_i \subset R_+^{k_i}$ & $\tau(W_i)$ is the usual top. on a subset of $R_+^{k_i}$ $w_i \varepsilon W_i$. W_i is compact.

(I.3) → ∃ a space of possible wage bids by the other labor

(I.4) appl., $(B_i, \tau(B_i))$ with $B_i \subset R_+^{\tau}$, where $\tau = \sum_{\substack{h=1 \\ h \neq i}}^{n} k_h$, where $b_i \varepsilon B_i$. B_i is compact.

∃ a space of possible wage offers received,

(I.5) $(0_i, \tau(0_i))$ with $0_i \subset B_i$ & $\tau(0_i)$ is the usual top. on a subset of $R_+^{k_i}$, where $0_i \varepsilon 0_i$.

∃ a set of possible labor applicant allocations,

(I.6) $A_i = \{(j_i, w_{j_i}) : j_i \varepsilon J_i \ \& \ w_{j_i} \varepsilon W_{j_i}\}$,

∃ a preference pre-ordering over A_i,

(I.7) $(A_i, \underset{\sim}{>}_i)$,

Assume that the pre-ordering is complete over A_i, and bounded below and continuous over each W_{j_i}.

∃ a utility function defined over A_i,

(I.8) $U_i : A_i \to R_+$

Assume that U_i is continuous with respect to each $(W_{j_i}, \tau(W_{j_i}))$. ∃ a subjective probability space expressing expectations, $(B_i, \beta(B_i), \mu_i)$

(I.9)

where $\beta(B_i)$ is the Borel sets & $\mu_i \varepsilon M(B_i)$, the set of all possible probability measures on $(B_i, \beta(B_i))$. (I.5), (I.6), & (I.8) → ∃ an acceptance mapping, ψ_i: $0_i \to 2^{A_i}$

(I.10)

where $\forall\, 0_i$

$$\psi_i(0_i) = \{a_i' \in A_i : j_i \in J_i,\ w_{j_i} = 0_{j_i},\ U_i(a_i') \geq U_i(a_i)\}$$

(I.10.a)

$\exists$ an acceptance choice function,

$$\phi_i : \{\psi_i(0_i) : 0_i \in 0_i\} \to A_i$$

(I.11)

where $\phi_i(\psi_i(0_i)) \in \psi_i(0_i) \quad \forall\, 0_i \in 0_i$

$\exists$ an expected offers (labor applicant demand) function,

$$\theta_i : B_i \times W_i \to 0_i$$

(I.12)

Assume that the function from $((B_i \times W_i),\ \tau(B_i) \times \tau(W_i))$ into $(0_i,\ \tau(0_i))$ is continuous. (I.1)-(I.12) $\to \exists$ an optimal wage bid mapping, $\chi_i : M_i \to 2^{W_i}$

(I.13)

where $\chi_i(\mu_i) = \{w_i' \in W_i : EU_i(w_i') \geq EU_i(w_i) \quad \forall\, w_i \in W_i\} \qquad \forall\, \mu_i$

(I.13.a)

$\exists$ a wage bid choice function $\xi_i : \{\chi_i(\mu_i) : \mu_i \in M_i\} \to W_i$

(I.14)

where $\xi_i(\chi_i(\mu_i)) \in \chi(\mu_i) \qquad \forall\, \mu_i \in M_i$

$\exists$ a space of possible observations on B_i, $(V_i,\ \tau(V_i))$ where $v_i \in V_i$

(I.15)

where $V_i \subset B_i$ & $\tau(V_i)$ is the standard top. on a subset of Euclid. space. $\exists$ an observation function, $\Lambda_i : B_i \to V_i$

(I.15.a)

Assume Λ_i is a continuous function. $\exists$ a loss function, $L_i : V_i \times M_i \times M_i \to R_+$.

(I.16)

Assume that L_i is a continuous function from $(V_i \times M_i \times M_i,\ \tau(V_i) \times \tau(M_i) \times \tau(M_i))$ into $(R_+,\ \tau(R_+))$, where $\tau(M_i)$ is the topology of weak convergence. (I.15) & (I.16) $\to \exists$ an optimal expectations map, $\lambda_i : M_i \times V_i \to 2^{M_i}$

(I.17)

where $\lambda_i(\mu_{it},v_{it}) = \{\mu_i' \in M_i : L_i(v_{it},\mu_{it},\mu_i') \leq L_i(v_{it},\mu_{it},\mu_i),$

$\forall \mu_i \in M_i\}$

(I.17.a)

$\exists$ a measure choice function,

$$\delta_i : \{\lambda_i(\mu_i,v_i) : \mu_i \in M_i, \quad v_i \in V_i\} \to M_i$$

(I.18)

where $\delta_i(\lambda_i(\mu_i,v_i)) \in \lambda_i(\mu_i,v_i)\ \forall\ (\mu_i,v_i) \in M_i \times V_i$, and $\mu_i \in \lambda_i(\mu_i, v_i) \to \delta_i(\lambda_i(\mu_i,v_i)) = \mu_i$. Assume δ_i is a cont. fcn. from $(2^{M_i}, \tau(2^{M_i}))$ into $(M_i, \tau(M_i))$. $\forall\ t\ \exists$ a finite set of employers (firms), $J = \{1,\ldots,m\}$ where $j \in J$

(2.3)

$\forall\ j\ \exists$ a set of jobs, $E_j = \{1,2,\ldots,k_j\}$ where $e_j \in E_j$

(E.1)

$\exists$ a space of possible wage offers, for each e_j,

$(W_{e_j}, \tau(W_{e_j}))$ where $w_{e_j} \in W_{e_j}$

(E.2)

where $W_{e_j} \subset R_+$ and $\tau(W_{e_j})$ is the standrard top. on a subset of Euc. space. Assume W_{e_j} is compact $\forall e_j$. (E.1) & (E.2) $\to \exists$ a space of possible wage offers, $(W_j, \tau(W_j))$ where $w_j \in W_j$

(E.3)

where $w_j = \Pi_{e_j=1}^{k_j} W_{e_j}$. W_j is compact. (E.3) $\to \exists$ a space of possible wage offers by the other employers, $(0_j, \tau(0_j))$ where $0_j \in 0_j$

(E.4)

where $0_j = \Pi_{\substack{h=1\\h\neq j}}^{m} W_h$ and $\tau(0_j) = \Pi_{\substack{h=1\\h\neq j}}^{m} \tau(W_h)$

$\exists$ a space of possible bids received, $(B_j, \tau(B_j))$ where $b_j \in B_j$

(E.5)

where $B_j = \Pi_{i=1}^{n} (W_j \cap W_i)$ and $\tau(B_j)$ is the standard top. on a subset of Euc. space.

$\exists$ a space of possible expected numbers of acceptances,

$(N_j, \tau(N_j))$ where $n_j \in N_j$

(E.6)

where $N_j = \{x \in R_+^{k_j} : 0 \leq (n_j)_h \leq n; \sum_{h=1}^{k_j} (n_j)_h\}$, $(n_j)_h$ is the hth component of n_j, and $\tau(N_j)$ is the usual top. on a subset of Euc. space. $\exists$ a space of possible outputs, $(Q_j, \tau(Q_j))$ where $q_j \in Q_j$

(E.7)

where $Q_j \subset R_+$ and $\tau(Q_j)$ is the standard top. on a subset of Euc. space. $\exists$ a production possibilies function, $F_j : N_j \to 2^{Q_j}$

(E.8)

where F_j is continuous from $(N_j, \tau(N_j))$ into $(2^{Q_j}, \tau(2^{Q_j}))$ and $F_j(n_j)$ is compact $n_j \in N_j$. $\exists$ a space of possible output prices, $(P_j, \tau(P_j))$ where $p_j \in P_j$

(E.9)

where $P_i \subset R_{++}$. Assume that P_j is compact. $\exists$ a profit function, $\Pi_i : N_j X W_j X Q_j X P_j \to R$

(E.10)

Assume Π_j is a continuous fcn. from $((N_j X W_j X Q_j X P_j), (\tau(N_j) X \tau(W_j) X \tau(Q_j) X \tau(P_j))$ into $(R, \tau(R))$.

$\exists$ an expected price (demand) function, $D_j : Q_j \to P_j$

(E.11)

Assume D_j is a continuous fcn. $\exists$ an expected labor supply function, $S_j : B_j X 0_j X W_j \to N_j$

(E.12)

Assume S_j is a continuous function from $((B_j X 0_j X W_j), \tau(B_j) X \tau(0_j) X \tau(W_j))$ into $(N_j, \tau(N_j))$. $\exists$ a subjective probability space expressing expectations,

$(0_j, \beta(0_j), \mu_j)$

(E.13)

where $\beta(0_j)$ is the Borel sets on 0_j and $\mu_j \in M_j(0_j)$, the set of all possible probability measures on $(0_j, \beta(0_j))$. (E.1) & (E.6)-(E.11) $\to \forall w_j$ $\exists$ a profit maximization map, $\eta_{n_j} : N_j \to 2^{Q_j}$

(E.14)

where $\forall\, n_j \in N_j$

$$\eta_{w_j}(n_j) = \{q_j' \in F_j(n_j) : \Pi_j(n_j, w_j, q_j')) \geq \Pi_j(n_j, w_j, q_j, p_j(q_j)) \quad \forall\, q_j \in Q_j\}$$

(E.14.a)

$\exists$ an output choice function $\forall\, w_j$,

$$\alpha_{w_j} : \{\eta_{w_j}(n_j)) \in \eta_{w_j}(n_j)\} \to n_j \in N_j$$

(E.15)

where $\alpha_{w_j}(\eta_{w_j}(n_j)) \phi\, \eta_{w_j}(n_j) \quad \forall\, n_j \in N_j$. Assume α_{w_j} is a continuous fcn. from $(\{\eta_{w_j}(n_j) : n_j \in N_j\}, \tau(\eta_{w_j}))$ into $(Q_j, \tau(Q_j))$, where $\tau(\eta w_j)$ is the standard top. on a subset of the space of subsets of Q_j.

(E.1)-(E.15) $\to \exists$ a wage offer mapping,

$$\Delta_j : M_j \to 2^{W_j}$$

(E.16)

where $\forall\, \mu_j \in M_j$

$$\Delta_j(\mu_j) = \{w_j' \in W_j : E\Pi_j(w_j') \geq E\Pi_j(w_j) \quad \forall\, w_j \in W_j\}.$$

(E.16.a)

$\exists$ a wage offer choice function,

$$\beta_j : \{\Delta_j(\mu_j) : \mu_j \in M_j\} \to w_j \in W_j$$

(E.17)

where $\beta_j(\Delta_j(\mu_j)) \in \Delta_j(\mu_j) \quad \forall\, \mu_j \in M_j$.

$\exists$ a space of possible observations on 0_j,

$(V_j, \tau(V_j))$ where $v_j \in V_j$

(E.18)

where $V_j \subset R_+^k$ and $\tau(V_j)$ is the standard top. on a subset of R^k. $\exists$ an observation function,

$$\Lambda_j : 0_j \to V_j$$

(E.18.a)

Assume that Λ_j is a continuous function from $(0_j, \tau(0_j))$ into $(V_j, \tau(V_j))$. $\exists$ a loss function,

$$L_j : V_j \times M_j \times M_j \to R_+$$

(E.19)

Assume that L_j is a continuous function from $(V_j \times M_j \times M_j, \tau(V_j) \times \tau(M_j) \times \tau(M_j))$ into $(R_+, \tau(R_+))$. (E.18) & (E.19) $\to \exists$

an optimal expectations map,

$$\lambda_j : V_j \times M_j \to 2^{M_j}$$

(E.20)

where $\forall (v_j, \mu_j) \in V_j \times M_j$

$$\lambda_j(v_{jt}, \mu_{jt}) = \{\mu_j' \in M_j : L_j(v_{jt}, \mu_{jt}, \mu_j') \leq L_j(v_{jt}, \mu_{jt}, \mu_j), \forall \mu_j \in M_j\}$$

(E.20.a)

$\exists$ a measure choice function,

$$\delta_j : \{\lambda_j(v_j, \mu_j) : (v_j, \mu_j) \in V_j \times M_j\} \to M_j$$

(E.21)

where $\delta_j(\lambda_j(v_j, \mu_j)) \in \lambda_j(v_j, \mu_j)$

and $\mu_j \in \lambda_j(v_j, \mu_j) \to \delta_j(\lambda_j(v_j, \mu_j)) = \mu_j$.

Assume δ_j is a continuous function from $(\{\lambda_j(v_j, \mu_j) :$

$(v_j, \mu_j) \in V_j \times M_j\}$ $\tau(\lambda_j))$ into $(M_j, \tau(M_j))$.

(2.1)-(E.21) the system relation,

$$(M_I \times M_J, T, R_h)$$

(2.4)

where $M_I = \Pi_{i=1}^{n} M_i$,

$M_J = \Pi_{j=1}^{m} M_j$,

and $R_h : M_I \times M_J \to M_I \times M_J$.

(2.4.a)

Let $\mathcal{L}$ denote the above class of labor markets, with $\mathcal{L}_h \in \mathcal{L}$.

Variable Relations

Decision subsystems:

Labor applicants:

$$\mu_{it} \to b_{it}$$

(i:1) wage bid selection

$$0_{it} \to a_{it}$$

(i:2) allocation selection

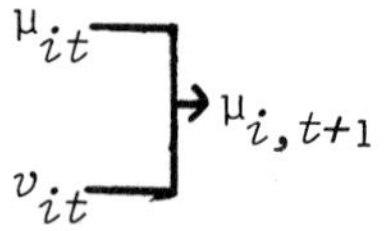

(i:3) revised expectations selection

Employers:

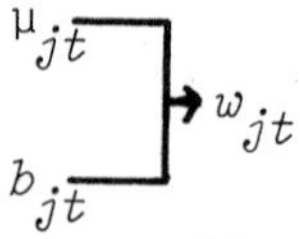

(j:1) wage offer selection

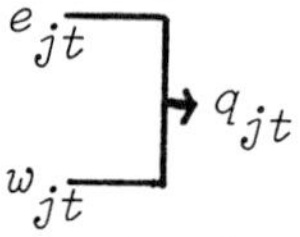

(j:2) production level selection

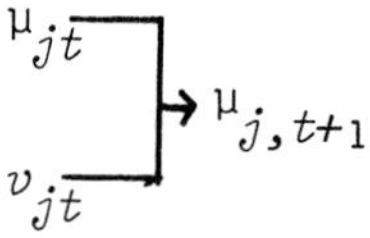

(j:3) revised expectations selection

Labor market system:

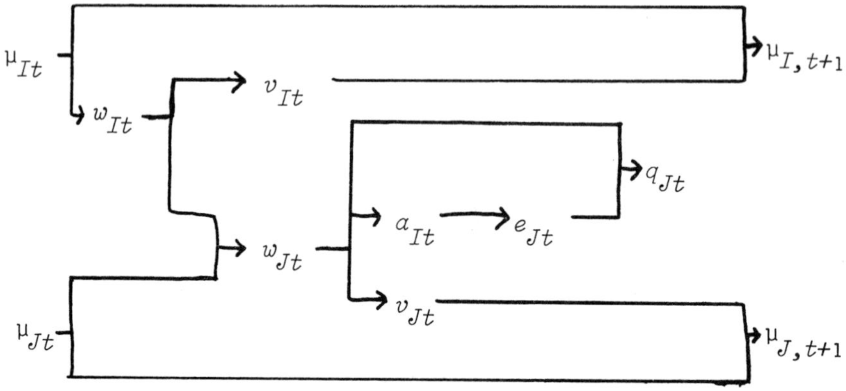

3. *The Main Results*

The model in section 2 describes a labor market with n labor applicants and m employers and incomplete information on the behavior of competing participants. The information available is summarized in subjective probability distributions over the behavior of the rival agents. As deduced in section 2, this implies the structure of a triple decision process in each trading period for each individual. The labor applicants bid under uncertainty, choose an allocation from their offers, and revise expectations. The employers make offers under uncertainty, choose a profit maximizing level of production, and revise expectations. These decision problems have optimal solutions and the process has an equilibrium, as will be proven in section 4.

Proposition 1. $\mathcal{L}_h \ \forall t \ \forall i,$

$\exists$ an optimal solution to decision problems

(i:1) wage bid selection

(I.1)-(I.14)

(i:2) job acceptance selection

(I.5)-(I.8),(I.10)-(I.11)

(i:3) revised expectations selection

(I.15)-(I.18), (I.9)

Proposition 2. $\forall \mathcal{L}_h \ \forall t \ \forall j,$

$\exists$ an optimal solution to decision problems

(j:1) wage offer selection

(E.1)-(E.17)

(j:2) production level selection

(E.6)-(E.12),(E.14)-(E.15)

(j:3) revised expectations selection

(E.18)-(E.21)

a) if $m > 1$ and $D_j(q_j) = \bar{p}_j \ \forall q_j$

(i.e. competitive firms)

b) if $m = 1$ and $D_j(q_j) = P_j(q_j)$ for $q_j \in Q_j$

(i.e. monopoly firm)

Definition 3.1. $\forall \mathcal{L}_h \in \mathcal{L}$

$\exists$ an equilibrium in $t' \in T$

if $(\mu_{It}, \mu_{Jt}) = (\mu_{I,t+1}, \mu_{J,t+1})$ for $t=t', t'+1, \ldots, t_h$.

Proposition 3. $\forall \mathcal{L}_h \in \mathcal{L}$

$\exists \; (\mu_I, \mu_J)^* \in M_I \times M_J = (\Pi_{i=1}^{n} M_I) \times (\Pi_{j=1}^{m} M_J)$
such that $(\mu_{It'}, \mu_{Jt'}) = (\mu_I, \mu_J)^*$
implies there is an equilibrium in t',
a) given competitive firms
b) given a monopoly firm.

Proposition 4. $\exists \; \mathcal{L}_h \in \mathcal{L}$
for which $\exists$ a $(\mu_I, \mu_J)^*$
such that $(\mu_{It'}, \mu_{Jt'}) = (\mu_I, \mu_J)^*$
implies there is an equilibrium in t'
and $U_h = n - \Sigma_{j=1}^{m} \Sigma_{h=1}^{k_j} (n_j)_h$
(i.e. unemployment in equilibrium).
a) given competitive firms
b) given a monopoly firm.

These results deal with issues raised in Leijonhufvud's On Keynesian Economics and the Economics of Keynes and Keynes' General Theory of Employment, Interest, and Money. These results demonstrate that under conditions of incomplete information on the actions of rival agents and wage setting based on expectations there can exist a behavioral equilibrium or steady state at wage levels which do not clear the labor market and imply unemployment and which do not change or adjust to the unemployment. Such an equilibrium will occur when each participant accepts the prior expectations on the basis of ovservations on the underlying distributions of economic behavior. Beliefs are confirmed and the behavior is repeated with each agent selecting optimal actions with respect to expected utility. This equilibrium allows wage adjustment and overcomes the conceptual problems of the fix-price analysis in Benassy and Barro-Grossman. It is also persistent and not a stage in a dynamic adjustment as in Tobin.

4. *Proofs*

Proposition 1: (i:1) wage bid selection. The proposition follows from the following lemmas:

Lemma 1: $U_i(\phi_i(\psi_i(\theta_i(b_i, w_i))))$ is a continuous function from $(B_i \times W_i, \tau(B_i) \times \tau(W_i))$ into $(R_+, \tau(R_+))$.

Proof: θ_i is a continuous function from $(B_i \times W_i, \tau(B_i) \times \tau(W_i))$ into $(O_i, \tau(O_i))$ by the assumption on (I.12). $U_i(\phi_i(\psi_i(O_i)) = max\{U_i(a_i) : a_i \in O_i\} = max\{U_i((j_i)_1, (O_j)_1), \ldots, U_i((j_i)_{k_i}, (O_j)_{k_i})\}$ by (I.10)-(I.11). Let $O_i \to O_i^*$ in $\tau(O_i)$, then $U_i((j_i)_h, (O_j)_h) \to U_i((j_i)_h, (O_j^*)_h)$ for $h = 1,2,\ldots,k_i$, so $max\{U_i(a_i) : a_i \in O_i\} \to max\{U_i(a_i) : a_i \in O_i^*\}$ since this is the maximum function defined on k_i continuous functions, and $U_i(\phi_i(\psi_i(O_i))$ is a continuous function from $(O_i, \tau(O_i))$ into $(R_+, \tau(R_+))$. Since the composition of continuous functions is continuous, the lemma follows. QED

Lemma 2: $EU_i(w_i) = \int_B U_i(\phi_i(\psi_i(\theta_i(b_i, w_i))))d\mu_i(b_i)$ is a continuous function from $(W_i, \tau(W_i))$ into $(R_+, \tau(R_+))$.

Proof: $B_i \times W_i$ is compact by the deductions on (I.3) and (I.4), and $U_i(b_i, w_i)$ is continuous by Lemma 1, so $\exists$ a finite m_i such that $U_i(b_i, w_i) \leq m_i$ $\forall$ $(b_i, w_i) \in B_i \times W_i$. U_i is continuous, so it is measurable. $\mu_i(B_i) = 1$, so $\int_{B_i} U_i(b_i, w_i)d\mu_i(b_i) \leq m_i < +\infty$, $U_i(b_i, w_i)$ is integrable for every given $w_i \in w_i$, and $EU_i(w_i)$ is defined for every $w_i \in W_i$. Let $w_i^k \to w_i^*$. $U_i(b_i, w_i)$ is a continuous function from $(W_i, \tau(W_i))$ to $(R_+, \tau(R_+))$ by Lemma 1, so $U_i(b_i, w_i^k) \to U_i(b_i, w_i^*)$ $\forall$ given $b_i \in B_i$. $U_i(b_i, w_i) \geq 0$ by the definition of (I.8), so $|U_i(b_i, w_i^k)| \leq m_i$, which is a constant and therefore integrable with respect to μ_i, $\forall_k$. Hence, by the Lebesque Dominated Convergence Theorem, $U_i(b_i, w_i^*)$ is integrable and $\int_{B_i} U_i(b_i, w_i^*)d\mu_i(b_i) = \lim_{k \to +\infty} \int U_i(b_i, w_i^k)d\mu_i(b_i)$, so $EU_i(w_i)$ is a continuous function from $(W_i, \tau(W_i))$ into $(R_+, \tau(R_+))$. QED

Lemma 3: $\forall$ $\mu_i \in M_i$,

$\chi_i(\mu_i) = \{w_i^! \in W_i : EU_i(w_i^!) \geq EU_i(w_i) \ \forall \ w_i \in W_i\} \neq \{\emptyset\}$.

Proof: $EU_i(w_i)$ is a continuous function over W_i by Lemma 2, W_i is a compact set with respect to $\tau(W_i)$, so there is a $w_i^* \in W_i$ such that w_i^* is a maximum for EU_i over W_i. (e.g. Russel Optimization Theory).

Since $\chi_i(\mu_i)$ is non-empty there is a wage bid selection from this set by (I.14), so (i:1) of Porposition 1 follows. QED

Proposition 1: (i:2) job acceptance selection

Proof: For any $0_i \in 0_i$ there is a finite set of possible job allocations or no job offers. $\psi_i(0_i)$ is non-empty since there is at most a finite number of choices. If there are no job offers the selection is no job. If there are job offers there is a choice by (I.11). QED

Proposition 1: (i:3) revised expectations selection

Lemma 4: M_i is compact with respect to $\tau(M_i)$.

Proof: Consider $M_i \subset M_i$. M_i is its own closure with respect to $\tau(M_i)$. If a subset of the space of probability measures is tight, then its closure is tight (Billingsley, App. III, Thm. 6) and compact in the topology of weak convergence. Given a family of finite measures on the Borel sets of a metric space Ω, the family M is tight iff for each $\varepsilon > 0$ there is a compact set $K \subset \Omega$ such that for all $\mu \in M$ $\mu(\Omega - K) < \varepsilon$. (Ash, Ch. 8, Def. 8.2.3) M_i is a family of probability measures on the Borel sets of B_i which is a metric space with respect to the standard metric on Euclidean space. B_i is compact, so M_i is tight, and it follows that M_i is compact with respect to $\tau(M_i)$. QED

Given any observation $v_i \in V_i$, and $\mu_{it} \in M_i$ in t, L_j is a continuous function from $(M_i, \tau(M_i))$ into $(R_+, \tau(R_+))$ and M_i is compact by Lemma 4, so $\lambda_i(0_{it}, \mu_{it}) \neq \{\emptyset\} \ \forall \ (0_{it}, \mu_{it}) \in 0_i X M_i$. Hence there is a revised expectations selection by (I.18). QED

Proposition 2: (j:1) wage offer selection (a) competitive firms

Variable relations in Π_j, given $\bar{p}_j$:

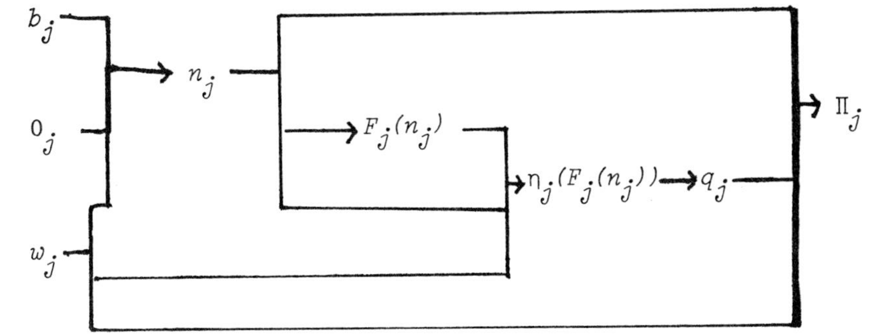

Figure 4.1

Proposition 2: (j:1) wage offer selection (a) competitive firms

Lemma 5.a. $\Pi_j(S_j(b_j,0_j,w_j), w_j, \alpha_j(\eta_j(n_j,w_j,(F_j(S_j(b_j,0_j,w_j))),\bar{p}_j)$ is a continuous function from $(B_j \text{ X } 0_j \text{ X } W_j, \tau(B_j) \text{ X } \tau(0_j) \text{X} \tau(W_j))$ into $(R, \tau(R))$.

Proof: S_j is a continuous function from $(B_j \text{ X } 0_j \text{ X } W_j, \tau(B_j) \text{ X } \tau(0_j) \text{ X } \tau(W_j))$ into $(N_j, \tau(N_j)$ by the assumption on (E.12). F_j is a continuous function from $(N_j, \tau(N_j))$ into $(2^{Q_j}, \tau(2^{Q_j}))$ by the assumption on (E.8). $F_j(n_j)$ is compact by assumption, so $\eta_j(n_j,w_j,F_j(n_j))$ is non-empty $\forall\ (n_j,w_j) \in N_j \text{X} W_j$.

$$\Pi_j(n_j,w_j,\alpha_j(\eta_j(n_j,w_j,F_j(n_j)))) = \max_{q_j \in F_j(n_j)} \{\Pi_j(n_j,w_j,q_j) \quad \forall\ (n_j,w_j)$$

so Π_j is a continuous function from $(N_j \text{X } W_j, \tau(N_j) \text{X } \tau(W_j))$ into $(R, \tau(R))$. The composition of continuous functions is continuous, so the lemma follows. QED

Lemma 6.a. $E\Pi_j(w_j) = \int_{0_j} \Pi_j(b_j,0_j,w_j)d\mu_j(0_j)$ is a continuous function from $(W_j, \tau(W_j))$ into $(R, \tau(R))$ for all given $b_j \in B_j$.

Proof: $0_j \text{ X } W_j$ is compact so there is an $m_j < +\infty$ such that $\Pi_j(b_j,0_j,w_j) \leq m_j\ \forall\ (0_j,w_j) \in 0_j \text{ X } W_j$. Π_i is continuous and μ_j

is a probability measure, so Π_i is integrable for every given $w_j \in W_j$, and $E\Pi_j(w_j)$ is defined for all $w_j \in W_j$. Let $w_j^k \to w_j^*$. Π_j is a continuous fcn. from $(W_j, \tau(W_j))$ into $(R, \tau(R))$ for every given b_j and 0_j, so $\Pi_j(b_j, 0_j, w_j^k) \to \Pi_j(b_j, 0_j, w_j^*)$ for every $0_j \in 0_j$. $|\Pi_j(b_j, 0_j, w_j^k)| \leq m_j \; \forall k$, so by the Lebesque Dominated Convergence Theorem $\Pi_j(b_j, 0_j, w_j^*)$ is integrable and

$$\int_{0_j} \Pi_j(b_j, 0_j, w_j^*) d\mu_j(0_j) = \lim \int_{0_j} \Pi_j(b_j, 0_j, w_j^k) d\mu_j(0_j),$$

so $E\Pi_j(w_j)$ is continuous. QED

Lemma 7.a. $\forall \mu_j \in M_j$,

$$\Delta_j(\mu_j) = \{w_j' \in W_j : E\Pi_j(w_j') \geq E\Pi_j(w_j) \; \forall w_j \in W_j\} \neq \{\emptyset\}.$$

Proof: W_j is compact by assumption, $E\Pi_j(w_j)$ is continuous from $(W_j, \tau(W_j))$ into $(R, \tau(R))$ by lemma 6.a, so there is a maximal element in W_j with respect to $E\Pi_j$. QED

Since $\Delta_j(\mu_{jt}) \neq \{\emptyset\} \; \forall t$, there is a wage offer selection by (E.17), so Prop. 2 (j:1) (a) follows. QED

Proposition 2 (j:1). wage offer selection (b) monopoly firm

Variable relations in Π_j, given a monopoly firm:

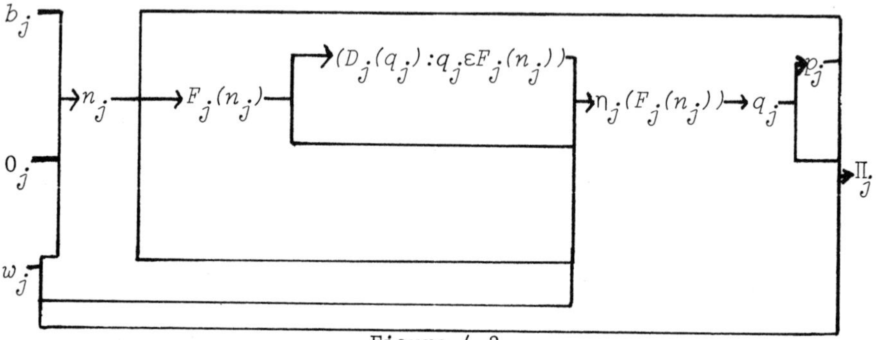

Figure 4.2

Lemma 5.b. $\Pi_j(S_j(b_j, 0_j, w_j), w_j, \alpha_j(\eta_j(n_j, w_j, F_j(S_j(b_j, 0_j, w_j)))),$
$D_j(\alpha_j(\eta_j(n_j, w_j, F_j(S_j(b_j, 0_j, w_j)))))$

is a continuous function from $(B_j \times 0_j \times W_j, \tau(B_j) \times \tau(0_j) \times \tau(W_j))$ into $(R, \tau(R))$.

Proof: D_j is a continuous function from $(Q_j, \tau(Q_j))$ into $(P_j, \tau(P_j))$ so there is a continuous function from $(2^{Q_j}, \tau(2^{Q_j}))$ into $(2Rj, \tau(2Rj))$ given by $\Pi_j(n_j, w_j, 2^{q_j}_j(n_j)) =$

$$\max_{q_j \varepsilon 2^{q_j}(n_j)} \Pi_j(n_j, w_j, q_j, D_j(q_j)) \;\forall\; (n_j, w_j) \;\varepsilon\; N_j \;X\; W_j.$$

Hence, by an argument analogous to the proof of Lemma 5.a, Lemma 5.b follows. There are no rival firms so 0_j is an empty set and there is no uncertainty about the outcome from any action. For any given b_j, Π_j is therefore a continuous function from $(W_j, \tau(W_j))$ into $(R, \tau(R))$. W_j is a compact set, so there is a $w_j' \;\varepsilon\; W_j$ such that $\Pi_j(w_j') \geq \Pi_j(w_j) \;\forall\; w_j \;\varepsilon\; W_j$ so (i:1) of Proposition 2 (b) follows. QED

Proposition 2 (j:2) production level selection (a) competitive firms.

Proof: For every $n_{jt} \;\varepsilon\; N_j$, $F_j(n_{jt}) \;\varepsilon\; 2^{Q_j}$ is compact by assumption. Since for every $(w_{jt}, n_{jt}) \;\varepsilon\; W_j \;X\; N_j$ the maximum function

$$\Pi_j(w_{jt}, n_{jt}, F_j(n_{jt})) = \max_{q_j \varepsilon F_j(n_{jt})} \Pi_j(w_{jt}, n_{jt}, q_j, \overline{p}_j)$$

is the profit function, and Π_j is a continuous function from $(Q_j, \tau(Q_j))$ into $(R, \tau(R))$ for every given $(w_{jt}, n_{jt}) \eta_j(n_{jt}) \neq \{\emptyset\}$, so there is a production level selection by (E.15) so Prop. 2 (j:2) (a) follows, and the aggregate supply curve is determined. QED

Proposition 2 (j:2) production level selection (b) monopoly firm

Proof: For every $(n_{jt}, w_{jt}) \;\varepsilon\; N_j \;X\; W_j$, D_j is a continuous function from $(Q_j, \tau(Q_j))$ into $(P_j, \tau(P_j))$, and Π_j is a continuous function from $(Q_j \;X\; P_j, \tau(Q_j) \;X\; \tau(P_j))$ into $(R, \tau(R))$, so $\Pi_j(n_j, w_j, q_j, D_j(q_j))$ is a continuous function from $(Q_j, \tau(Q_j))$ into $(R, \tau(R))$. $\forall\; n_j \;\varepsilon\; N_j$, $F_j(n_j)$ is compact by assumption, so $\eta_j(n_{jt}) \neq \{\emptyset\}$. Hence there is a production level selection by (E.15). QED

Proposition 2 (j:3) revised expectations selection (a) competitive firms

Proof: M_j is compact by an argument analogous to the proof of Lemma 4. $\forall\, (\mu_{jt}, v_{jt}) \in M_j \times V_j, L_j$ is a continuous function from $(M_j,\ \tau(M_j))$ into $(R_+,\ \tau(R_+))$, so $\lambda_j(v_{jt}, \mu_{jt}) \neq \{\emptyset\}$, and there is a revised expectations selection by (E.21). QED

Proposition 3 (j:3) revised expectations selection (b) monopoly firm

Proof: Analogous to Prop. 2 (j:3) (a). QED

Definition 3.1: Lemma 8 demonstrates that an equilibrium of expectations as given in Definition 3.1 implies that every variable in the labor market is fixed in all subsequent time intervals.

Lemma 8: If $(\mu_{It}, \mu_{Jt}) = (\mu_{I,t+1}, \mu_{j,t+1})$ for $t = t',\ t'+1, \ldots, t_h$, then $(w_{It}, 0_{It}, a_{It}, v_{It}, b_{It}, b_{Jt}, w_{Jt}, 0_{Jt}, n_{Jt}, q_{Jt}, p_{Jt}, v_{Jt}) = (w_{It+1}, 0_{Jt+1}, a_{It+1}, v_{It+1}, b_{It+1}, b_{Jt+1}, w_{Jt+1}, 0_{Jt+1}, n_{Jt+1}, q_{Jt+1}, p_{Jt+1}, v_{Jt+1})$ for $t = t',\ t'+1, \ldots, t_h$,

where I denotes the Cartesian product for the n labor applicants, and J denotes the Cartesian product for the m employers.

Proof: $\mu_{It} \to w_{It}$, $w_{It} \to b_{It}$, $b_{It} \to b_{Jt}$, $b_{It} \to v_{It}$, $\mu_{Jt}, b_{Jt} \to w_{Jt}$, $w_{Jt} \to 0_{Jt}$, $0_{Jt} \to v_{Jt}$, $0_{Jt} \to 0_{It}$, $0_{It} \to a_{It}$, $a_{It} \to n_{Jt}$, $n_{Jt}, w_{Jt}, D_J, F_J \to q_{Jt}$, $q_{Jt} \to p_{Jt}$,

where $\to$ denotes a function, so every variable is fixed if expectations are fixed. QED

Proposition 3 (a) competitive firms

Lemma 9: The systems response function,

$$R_h : (M_I \times M_J,\ \tau(M_I) \times \tau(M_J)) \to (M_I \times M_J,\ \tau(M_I) \times \tau(M_J)),$$

is a continuous function $\forall\, \mathcal{L}_h \in \mathcal{L}$.

Proof: Let $\mu_i^k \overset{w}{\to} \mu_i^*$, then $\int_{B_i} U_i(b_i, w_i)\, d\mu_i^k(b_i) \to \int_{B_i} U_i(b_i, w_i)\, d\mu_i^*(b_i)$ (Ash, Thm. 4.5.1), so χ_i is a continuous function from $(M_i,\ \tau(M_i))$ into $(2^{W_i},\ \tau(2^{W_i}))$. Let $\mu_j^k \to \mu_j^*$, then $\int_{0_j} \Pi(b_j, 0_j, w_j)\, d\mu_j^k(0_j) \to \int_{0_j} \Pi_j(b_j, 0_j, w_j)\, d\mu_j^*($

so $\chi_j(\mu_j^k) \to \chi_j(\mu_j^*)$, so χ_i is a continuous fcn. from $(M_j,\ \tau(M_j))$ into $(2^{W_j},\ \tau(2^{W_j}))$, for every given $(b_j, w_j) \in B_j \times W_j$.

It also follows that the function from $(M_j, \tau(M_j))$ into $(W_j, \tau(W_j))$ is continuous since (E.17) is continuous. This implies that the function from $(M_J, \Pi_{j=1}^{m} \tau(M_j))$ into $(0_J, \Pi_{j=1}^{m} \tau(0_j))$ is continuous. Since (E.18.a) is continuous, the function from $(M_J, \tau(M_J))$ into $(V_J, \tau(V_J))$ is continuous.

$(v_{jt}, \mu_{jt}) \in V_j \times M_j$, $\lambda_j(v_{jt}, \mu_{jt}) \neq \{\emptyset\}$ by the proof of Prop. 2 (j:3) (a). Let $(v_{jt}, \mu_{jt})^k \to (v_{jt}, \mu_{jt})^*$. L_j is continuous, so

$$\lim_{k\to+\infty} \lambda_j((v_{jt}, \mu_{jt})^k) = \lambda_j((v_{jt}, \mu_{jt})^*),$$

so the function from $(V_J \times M_J, \tau(V_J) \times \tau(M_J))$ into $(2^{M_j}, \tau(2^{M_j}))$ is continuous. δ_j is a continuous function from $(2^{M_j}, \tau(2^{M_j}))$ into $(M_j, \tau(M_j))$, so, since the composition of continuous functions is continuous, the function from $(M_J, \tau(M_J))$ into $(M_J, \tau(M_J))$ is continuous. By a similar argument, the function from $(M_I, \tau(M_I))$ into $(M_I, \tau(M_I))$ is continuous, hence R_h is a continuous function from $(M_I \times M_J, \tau(M_I) \times \tau(M_J))$ into $(M_I \times M_J, \tau(M_I) \times \tau(M_J))$ is defined by the labor market system.

QED

Lemma 10: Let S be a non-empty compact convex subset of a Hausdorff Topological vector space X. Let $G : S \to S$ be a point-to-point mapping which satisfies:

(1) $G(x)$ is continuous,

Then there exists a fixed point $x_0 = G(x_0)$ with $x_0 \in S$.

Proof: This lemma follows as a corollary of the Glicksberg fixed-point theorem (Glicksberg; Parthasarathy and Raghavan):

Let S be a nonempty compact convex subset of a Hausdorff Topological vector space x. Let $G : S \to S$ be a point-to-set mapping satisfying the following conditions:

(1) $G(x)$ is non-empty, closed, convex for each $x \in S$,

(2) $G(x)$ is upper semi-continuous,

Then there exists a fixed point $x_0 \in G(x_0)$ with $x_0 \in S$. since a point is a non-empty, closed, convex set, and a continuous function is upper semi-continuous by definition.

QED

Let C_i denote the set of all finite charges on $(B_i, \mathcal{B}(B_i))$.

Let C_j denote the set of all finite charges on $(0_j, \mathcal{B}(0_j))$.

$(C_I, \tau(C_I))$ and $(C_J, \tau(C_J))$ are Hausdorff spaces when the topology is the topology of weak convergence on the set of possible finite charges on the Borel sets of a metric space for both spaces (Billingsley, App. III, The Top. of Wk. Cnvgnce.-analogous defn.)

This implies that $(C_I \times C_J, \tau(C_I) \times \tau(C_J))$ is a Hausdorff topological vector space, since $(c_I, c_J) \in (C_I \times C_J)$ and $(c'_I, c'_J) \in (C_I, C_J)$ implies $(c_I + c'_I,\ c_J + c'_J)$, i.e., $(c_I + c'_I)(s) = (c_I)(s) + (c'_I)(s)$, is a finite charge on $(B_i \times 0_j, \mathcal{B}(B_i) \times \mathcal{B}(0_i))$, and $(d)\cdot(m_I, m_J)$, i.e., $(d)\cdot(m_I, m_J)(s) = (d\cdot m_I(s), d\cdot m_J(s))$, is a finite charge for all $d \in R$.

$(M_I \times M_J)$ is a non-empty subset of $(C_I \times C_J)$ which is convex by the definition of the set of all possible probability measures on the Borel sets of s metric space.

R_h is a continuous function from $(M_I \times M_J, \tau(M_I) \times \tau(M_J))$ into itself by Lemma 9, so the existence of a fixed $(\mu_I, \mu_J) \in M_I \times M_J$ follows by Lemma 10 and Proposition 3 (a) follows.

QED

Proposition 3 (b) monopoly firm

Proof: The monopoly firm decisions are a function of the actions of the labor applicants, so the system response function reduces to $R_h : (M_I, \tau(M_I)) \to (M_I, \tau(M_I))$. The continuity of R_h follows by an argument analogous to the proof of Lemma 9, and the existence of an equilibrium follows from the existence of a fixed point by an argument analogous to the proof of Prop. 3 (a), using Lemma 10.

QED

Proposition 4. (unemployment in equilibrium)

Proof: The existence of a subclass of $\mathcal{L}$ with unemployment in equilibrium follows from the existence of an example. An example is provided by a monopoly firm and two labor applicants bidding for two different jobs. If the monopolist maximizes profits by hiring only one labor applicant at the lower wage bid then there will exist a pair of labor applicant expectations such that there is a behavioral equilibrium by Prop. 3 (b).

The class of labor markets with unemployment in equilibrium is very large and includes monopoly firm markets and competitive firm markets.

The presence of a labor union in the labor market can be represented by labor applicants who know the behavior of the competing labor applicants since they have a unified response. The labor applicant wage bid decisions are then decisions without uncertainty. Similar results on the existence of an equilibrium in all such markets, and the existence of equilibria with unemployment follow.

Uncertainty on the behavior of participants that are not direct rivals can be introduced for labor applicants by considering a subjective probability measure space on the Borel sets implied by the set of possible wage offers by employers. This would express expectations with respect to the demand for labor services. For employers a subjective probability space on the measurable space implied by the possible combinations of acceptances would express expectations on the supply of labor services.

A stochastic model can be considered if the observations are noisy and there is a conditional distribution over the observations for every set of wage bids and wage offers. This opens the question of the existence of a stochastic equilibrium or invariant measure over the observations.

REFERENCES

[1] Arrow, K., and F. Hahn, *General Competitive Analysis*, Holden-Day, San Francisco, 1971.

[2] Ash, R., *Real Analysis and Probability*, Academic Press, New York, 1972.

[3] Barro, R., and H. Grossman, *A General Disequilibrium Model of Income and Employment*, The American Economic Review, 61(1971), 82-93.

[4] Bartles, R., *The Elements of Integration*, John Wiley and Sons, New York, 1966.

[5] Benassy, J., *Disequilibrium Exchange in Barter and Monetary Economies*, Economic Inquiry 13(1975), 131-156.

[6] ________, *Neo-Keynesian Disequilibrium Theory in a Monetary Economy*, Review of Economic Studies 42(1975), 503-523.

[7] Billingsley, P., *Convergence of Probability Measures*, John Wiley and Sons, 1968, New York.

[8] Cadzow, J., *Discrete-time Systems*, Prentice-Hall, Inc., Englewood Cliffs, N.J., 1973.

[9] Glicksberg, *A Further generalization pf the Kakutani fixed point theorem, with application to Nash Equilibrium points*, Proceedings of the American Mathematical Society 3(1952), 170-174.

[10] Leijonhufvud, A., *On Keynesian Economics and the Economics of Keynes*, University Press, 1968.

[11] Mesarovic, M., and Y. Takahara, *General Systems Theory: Mathematical Foundations*, Academic Press, New York, 1975.

[12] Mirman, L., and W. Porter, *A Microeconomic Model of the Labor Market under Uncertainty*, Economic Inquiry, 12(1974), 135-145.

[13] Parthasarathy, T., and T. Raghavan, *Existence of Saddle Points and Nash Equilibrium Points for Differential Games*, SIAM Journal of Control, 1975.

[14] Porter, W., *Market Adjustment with Adaptive Supply and Pricing Decisions*, in Day and Gross, Editors, Adaptive Economic Models, Academic Press, New York, 1975.

[15] Radner, R., *Market equilibrium and uncertainty: concepts and problems*", in Intriligator and Kendrik, editors, Frontiers of Quantitative Economics, North-Holland Publishing Co., Amsterdam, 1974.

[16] Russel, D., *Optimization Theory*, W. A. Benjamin, New York, 1970.

[17] Schweppe, F., *Uncertain Dynamic Systems*, Prentice-Hall, Inc., Englewood Cliffs, N.J., 1973.

TIME DELAYS IN PREDATOR-PREY SYSTEMS

J. M. Cushing
University of Arizona

The integrodifferential system

$$N_1' = b_1 N_1 (1 - c_{11} N_1 - c_{12} \int_0^\infty N_2(t - u) k_1(u) du)$$

$$(1) \qquad N_2' = b_2 N_2 (-1 + c_{21} \int_0^\infty N_1(t - u) k_2(u) du)$$

$$c_{ij} > 0, \quad b_i > 0, \quad k_i(u) \geq 0, \quad \int_0^\infty k_i(u) du = 1$$

describes the dynamics of a predator-prey interaction where N_1 and N_2 measure (in some appropriate units) the population sizes of the prey and predator species respectively. Here the interaction terms represented by the integrals account for delay effects (over possibly all past times) which interactions with the opposite species have on the growth rate of each species. The coefficient c_{11} accounts for density effects within the prey population and $1/c_{11}$ is called the "carrying capacity" of the prey. The coefficients b_1 and b_2 are the natural birth and death rates of N_1 and N_2 respectively in the absence of all constraints.

Volterra first considered such delay systems in his well-known book [5]. Although not a great deal is known about the solutions of (1), some recent work has been concerned with its study as well as with more general systems (see [1-4]) and it is now possible to describe rather completely certain aspects of the qualitative behavior of solutions (as $t \to +\infty$) as they are functions of the parameters in the system.

The first simple observation we make is that all nonidentically zero solutions of (1) are obviously of one sign for all t. We will only be concerned with positive solutions here. We also observe that the only equilibria $(N_1, N_2) = (e_1, e_2) =$ constant lying in the right half plane are $E_1 : e_1 = 1/c_{21} > 0$,

$e_2 = (c_{21} - c_{11})/c_{12}c_{21}$ and $E_2 : e_1 = 1/c_{11}$, $e_2 = 0$. We describe the asymptotic behavior of solutions of (1) by means of four different cases which depend on the relative values of c_{ij} and b_i.

(i) First of all, in a manner very similar to a special case studied in [1], it can be proved that *if* $c_{11} > c_{21}$ *then* $(N_1(t), N_2(t)) \to (1/c_{11}, 0)$ *as* $t \to +\infty$ *for all positive solutions of* (1). In ecological terms, if the carrying capacity of the prey is less than the critical value $1/c_{21}$ then the predator goes extinct while the prey tends to this carrying capacity. (This can in fact be proved if $k_i(u)du$ is replaced by $dh_i(u)$ and hence is also true for systems with constant time lags.)

(ii) *If* c_{11} *is less than, but close to* c_{21} *then the equilibrium* E_1 *is locally asymptotically stable.* (Technically, given $\varepsilon > 0$ there exists a $\delta > 0$ such that $|N_i(t)| < \delta$ for all $t < 0$ implies $|N_i(t)| < \varepsilon$ for all $t \geq 0$ and a constant $\gamma > 0$ exists such that $|N_i(t)| < \gamma$ for all $t < 0$ implies $N_i(t) \to e_i$ as $t \to +\infty$.) This is proved by linearizing (1) about E_1 and by showing that the "eigenvalues" (i.e. the roots of the Paley-Weiner determinant [4]) of the resulting linear integrodifferential system have negative real parts provided c_{11} is close to, but less that c_{21}. This linearization procedure is formally justified for integrodifferential systems in [2]. Actually several numerically solved examples carried out by the author have all indicated that every positive solution tends to E_1 in this case and hence that the asymptotic stability of E_1 is in fact global in the first quadrant.

(iii) *If* $c_{11} > 0$ *is small then the equilibrium* E_1 *is "usually" unstable.* If again (1) is linearized about E_1 and if c_{11} is taken to be zero, then one finds that E_1 is unstable if the Paley-Wiener determinant, which turns out to be $f(s) = s^2 + K(s)$, has roots in the right half plane; here $K(s) = b_1 b_2 k_1^*(s) k_2^*(s)$ where $k_i^*(s)$ is the Laplace transform of $k_i(u)$. In this event some roots will also lie in the right half plane for c_{11} small. Often $K(s)$ is a rational function and $f(s)$ can be investigated by means of the Routh-Hurwitz criteria. For a more general criterion it is also possible to show by means of the argument principle that if $f(s)$ has no purely imaginary roots then it is unstable if and only if $\arg f(+\infty i) \neq \pi$. Inasmuch as it is easy to show that this limit equals $(2n+1)\pi$ for some n 0, ± 1, $\pm 2, \ldots$ we see that all but one of infinitely many cases lead to instability. (The remaining case $n = 0$ turns out to yield asymptotic stability as in (ii).) In all cases worked out by the author (for specific kernels such as $k(u) = (\alpha u + \beta) \exp(-\delta u)$) instability has always been found. See [1] and [2] for examples.

(iv) The loss of the stability of E_1 as c_{11} decreases suggests the possible existence of limit cycles. The existence of periodic solutions for general two species interactions with delays was studied by means of bifurcation theory in [3]. If the techniques in [3] are applied to (1) one obtains the following results.

Define

$$S_{ij}(n) = C_{ij} \int_0^\infty k_i(u) \sin 2n\pi p^{-1} du, C_{ij}(n) = C_{ij} \int_0^\infty k_i(u) \cos 2n\pi p^{-1} du$$

$$\sum_1 (n) = S_{12}C_{21} + S_{21}C_{12}, \quad \sum_2 (n) = S_{12}S_{21} - C_{12}C_{21}.$$

Consider the hypotheses:

(H1) $\sum_1(n) > 0$, $\sum_2(n) < 0$ for some integer $n \geq 1$ and period $p > 0$

(H2) $C_{21}(n) \neq 0$ for n in (H1)

(H3) Either $n\sum_1(m) \neq m\sum_1(n)$ or $n^2\sum_1(n)\sum_2(m) \neq m^2\sum_1(m)\sum_2(n)$ for all $m \neq n$ (n as in (H1)), m an integer ≥ 1.

The condition (H1) *for some* n *and* p *is necessary and the conditions* (H1)-(H3) *are sufficient for the bifurcation of p-periodic solutions of* (1) *from the equilibrium* E_1 *as the birth and death rates* b_1 *and* b_2 *pass through the critical values*

$$b_1^0 = -\ 2\pi np^{-1}c_{21}\Sigma_1(n)/c_{11}\Sigma_2(n), \quad b_2^0 = 2\pi np^{-1}c_{11}/(c_{21} - c_{11})\Sigma_1(n)$$

for all other constants fixed and $0 < c_{11} < c_{21}$.

This result also holds for the more general case of Stieltjes integrals $k_i(u)du = dh_i(u)$ in (1) and hence for systems with constant time lags [3].

All four cases above have been fully demonstrated numerically by the author for selected kernels. Similar cases were predicted by different means for a special case of (1) in [1], where other numerical results can be found and related to the cases above.

REFERENCES

[1] J. M. Bownds, and J. M. Cushing, *On the behavior of solutions of predator-prey equations with hereditary terms*, Math. Biosci. 26(1975), 41-54.

[2] J. M. Cushing, *An operator equation and bounded solutions of integrodifferential systems*, SIAM J. Math. Anal. 6(1975), No. 3, 433-445.

[3] J. M. Cushing, *Periodic solutions of two species interaction models with lags*, Math. Biosci. 31(1976), 143-156.

[4] R. K. Miller, *Asymptotic stability and perturbations for linear Volterra integrodifferential systems*, appearing in *Delay and Functional Differential Equations and their Applications*, K. Schmitt, editor, New York, Academic Press, 1972.

[5] V. Volterra, *Lecons sur la Théorie Mathématique de la Lutte par la Vie*, Paris, Gauthier-Villars, 1931.

ON SELECTING A RESPONSE FUNCTION IN NONLINEAR REGRESSION

J. Wanzer Drane
Southern Methodist University and
University of Texas Health Sciences Center at Dallas

Generic Functions and Hierarchies

Whether employing difference or differential equations to describe the phenomenon under investigation the solution set invariably gives rise to a family of solutions. Witness the differential equation

$$(D - \alpha_1)(D - \alpha_2)Y = 0, \tag{1}$$

whose solution is one of the following:

$$Y = A_1 e^{\alpha_1 t} + A_2 e^{\alpha_2 t} \tag{2A}$$

if $\alpha_1 \neq \alpha_2 \neq 0$; or

$$Y = (A_1 + A_2 t)e^{\alpha t} \tag{2B}$$

if $\alpha_1 = \alpha_2 \neq 0$; or

$$Y = A_1 + A_2 e^{\alpha_2 t} \tag{2C}$$

if $\alpha_1 = 0$, $\alpha_2 \neq 0$; or

$$Y = A_1 + A_2 t$$

if $\alpha_1 = \alpha_2 = 0$. (2D)

I will refer, to (1) as the generic equation and (2A) as the generic solution, as (2B), (2C), and (2D) are all special cases of (2A) or special solutions of (1). Pictorially Figure 1 displays two hierarchies of the solutions of (1).

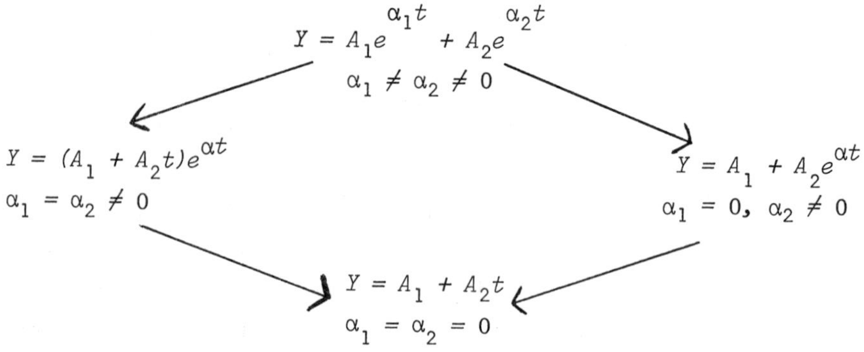

Figure 1: The hierarchy of four solutions to Equation (1) depending on the various values of α_1 and α_2.

Many papers have been written on determining the number of exponentials in a function such as (2A) (See Bailey, Eadie and Schmidt, 1974), but I prefer to view the problem in terms of determining the number of operators in the differential equation

$$\prod_{i=1}^{k} (D - \alpha_i)Y = 0. \quad (3)$$

That is, what is the value of k?

From a nonlinear regression standpoint, wherein the tool is the fitted function and the residual sum of squares or the sum of squared error, the problem is two-fold. Firstly, one must determine the value of k in (3), then he must determine the place in an hierarchy much akin to that illustrated in Figure 1.

As a second example let's consider an adaptation of the Burr distribution (See Burr 1942, 1967, and Burr and Cislak, 1968) as a dose-response function. Its form is

$$F = 1 - [1 + (\alpha + \beta x)^C]^{-K}; \quad C,k > 0. \quad (4)$$

This simple algebraic function is very flexible within the C, K space as is illustrated in Figure 2. These same functions in quite different parametric form were derived by Turner, et al. (1975) as special solutions to the generic growth differential

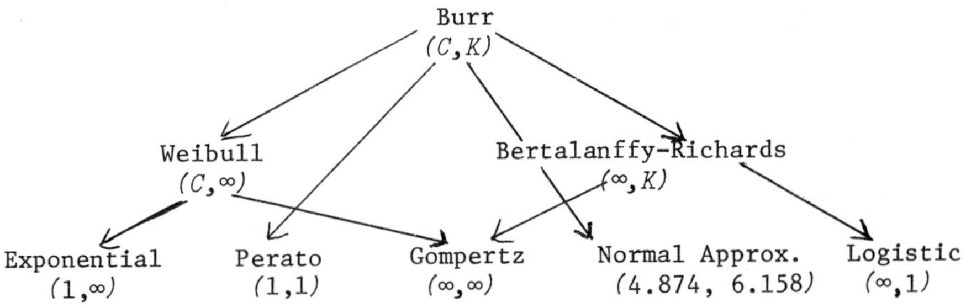

Figure 2: Hierarchies formed by varying C and K in the Burr distribution.

equation

$$\frac{dX}{dt} = \frac{\beta}{K^n} X^{1-np} (K^n - X^n)^{1+p} \tag{5}$$

where

$$X = \frac{K}{\{1 + [1 + \beta np(t - \tau)]^{-1/p}\}^{1/p}} \tag{6}$$

Statistical Methodology

Once the generic function is determined it is a matter of establishing the most tenable special case. Any one of the several hierarchies of either Figure 1 or 2 defines in statistics an hierarchy of hypotheses, with each succeeding one embedded in all previous ones. That is, from a point set view, $W_1 \supset W_2 \supset \ldots \supset W_m$ where the W's represent the spaces spanned by the parameters of the various hypothesized functions, and hence hypotheses H_1, $H_2, \ldots, H_m$. Hogg (1961) introduced independent test statistics for choosing the most tenable of the various candidates when the regression is linear. Drane and Harrist (1974) extended this to include quantal response experiments and multiple contingency tables.

Let SSE_i be the sum of squared error after fitting the ith function with p_i parameters to n points; $i = 1,2,\ldots,m$.

The likelihood ratio comparing H_1 and H_i, $i > 1$, is

$$\lambda_{1i} = \left(\frac{SSE_i}{SSE_1} \right)^{n/2}, \quad \lambda_{1i} \geq 1. \tag{7}$$

The monotone function of λ_{1i}

$$\frac{n - p_1}{p_1 - p_i} (\lambda_{1i}^{2/n} - 1) = \frac{(SSE_i - SSE_1)/(p_1 - p_i)}{SSE_1/(n - p_1)} \tag{8}$$

is asymptotically an F statistic with $p_1 - p_i$ and $n - p_1$ degrees of freedom. Reject H_i in favor of H_1 for large values of the statistic in (8).

Test H_1 successively against H_i, $i = 2,3,\ldots,m$ until, say, H_j is rejected. Accept H_{j-1} as the most tenable. Or, if H_2 is accepted, test H_2 against H_3, etc. until a rejection occurs at H_j and then accept H_{j-1}. This author prefers to use the two in combination as the following example will illustrate.

If confidence intervals are the object of your pursuit, then your most tenable hypothesis has been obtained and I refer you to a paper by Hartley (1964).

Numerical Example

On succeeding days a patient is subjected to an histamine challenge test. That is, he is given different doses of histamine in response to which his stomach produces acid. Traditionally, the Michaelis-Menten kinetic equation (rectangular hyperbola) has been used as a dose response function. It was first suggested that the basal level of acid secretion in some patients was different on different days and that it would be more reasonable to consider a generalization of the Michaelis-Menten kinetics, namely that of a multiple binding site enzyme kinetic equation. To account for differing basal levels A_{i+3} was introduced as the basal equivalent dose for the basal secretion level on the ith day. A_3 is the exponent that accounts for the multiple site

H_1: Generic Function

$$\hat{Y}_{ij} = A_1 \frac{(X_{ij}+A_{i+3})^{A_3}}{A_2 + (X_{ij}+A_{i+3})^{A_3}}$$

$i = 1, \ldots, 12$

$j = 1, \ldots, n_i$

$n = \Sigma n_i = 26$

$p_1 = 15$

$SSE_1 = 2147.2766$

H_2: $A_3 = 1$

$$\hat{Y}_{ij} = A_1 \frac{X_{ij} + A_{j+3}}{A_2 + X_{ij} + A_{j+3}}$$

$p_2 = 13$

$SSE_2 = 4652.6110$

H_3: $A_4 = \cdots = A_{15}$

$$\hat{Y}_{ij} = A_1 \frac{(X_{ij}+A_4)^{A_3}}{A_2 + (X_{ij}+A_4)^{A_3}}$$

$p_3 = 4$

$SSE_3 = 2226.2450$

H_4: $A_3 = 1$

$A_4 = \cdots = A_{15} = 0$

$$\hat{Y}_{ij} = A_1 \frac{X_{ij}}{A_2 + X_{ij}}$$

$p_4 = 2$

$SSE_4 = 5479.9744$

Figure 3: Hierarchies of some functions which have the rectangular hyperbola through the origin as one of their limit forms. In this context, this rectangular hyperbola is known as the Michaelis-Menten function.

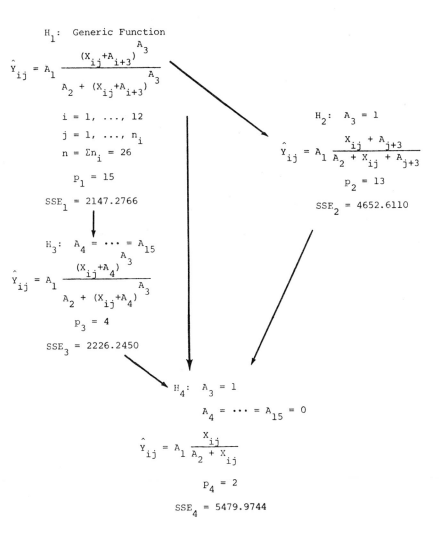

analogue. Thus

$$\hat{Y}_{ij} = A_1 \frac{(X_{ij} + A_{i+3})^{A_3}}{A_2 + (X_{ij} + A_{i+3})^{A_3}} \tag{9}$$

where

$$i = 1,2,\dots ; \quad j = 1,2,\dots,n_i; \quad n = \Sigma n_i$$

is our generic model with one A_{i+3} for each day the patient visited, all other parameters remaining constant day to day.

Figure 3 depicts the entire set of possible degenerations of interest including p, the number of parameters and SSE the sum of squared error fitting each hypothesized model by nonlinear least squares. $H_1 \supset H_2 \supset H_4$ or $H_1 \supset H_3 \supset H_4$ are the two hierarchies.

Testing H_1 against H_2, $F(2,11) = 6.417$ which tells us to reject H_2 and do not compare H_2 with H_4. Testing H_1 against H_3 and H_4 gives us $F(11,11) = 0.037$ and $F(13,11) = 1.313$ respectively, both of which are not significant. However, testing H_3 against H_4 gives $F(2,22) = 16.077$ which is highly significant. The summary of these tests is that

$$\hat{Y}_{ij} = A_1 \frac{(X_{ij} + A_4)^{A_3}}{A_2 + (X_{ij} + A_4)^{A_3}} \tag{10}$$

where A_1 is the maximal response, A_2 a location constant used in calculating ED-50, A_4 the average basal equivalent dose and A_3 a shape constant measuring the departure of the model from the Michaelis-Menten model.

REFERENCES

[1] Bailey, R. C., Eadie, G. S., and Schmidt, F. H., *Estimation procedures for consecutive first order irreversible reactions*, Biometrics 30(1974), 67-75.

[2] Burr, I. W., *Cumulative frequency functions*, Annals of Mathematical Statistics, 13(1942), 215-32.

[3] ______, *A useful approximation to the normal distribution*

function, with application to simulation, Technometrics 9(1967), 647-51.

[4] Burr, I. W., and Cislak, P. J., *On a general system of distributions: I. Its curve shape characteristics. II. The sample median*, Journal of the American Statistical Association 63(1968), 627-35.

[5] Drane, J. W., and Harrist, R., *Resolving hypotheses with successive chisquares*, Statistische Heft 15(1974), 171-80.

[6] Hartley, H. O., *Exact confidence regions for parameters in non-linear regression laws*, Biometrika 51(1964), 347-53.

[7] Hogg, R. V., *On the resolution of statistical hypotheses*, Journal of the American Statistical Association 56(1961), 978-89.

[8] Turner, M. E., Jr., Bradley, E. L., Jr., Kirk, K. A., and Pruitt, K. M., *A theory of growth. Preprint*, Department of Biostatistics, University of Alabama in Birmingham, Al., 25294, U.S.A.

ON THE MOMENTS OF EXPONENTIAL DECAY

by

J. Eisenfeld [1,2,3], D. J. Mishelevich [1],

R. L. Senger [1], and S. R. Bernfeld [1,2,3,4]

1. *Introduction*

Assume we are given discrete values for a function $f(t)$ which may be expressed as a finite exponential sum

$$f(t) = \sum_{i=1}^{N} \alpha_i \; exp(-t/\tau_i), \quad t \geq 0 \tag{1.1}$$

where $\alpha_i \geq 0$, $\tau_i \geq 0$, $i = 1,2,\ldots,N$ and $\tau_i \neq \tau_j$ for $i \neq j$. The classical problem of computing the number of components N, the amplitudes α_i and the time constants τ_i arises in virtually all areas of mathematical science. The problem, for example, occurs in exponential decay where τ_i measures the half life of substance i and α_i measure the initial quantity. In compartment analysis α_i and τ_i determine the transfer parameters between compartments. In mathematical modelling α_i and τ_i determine the law governing a process. In the theory of differential equations one needs to determine the differential equation

$$Y^{(N)} + c_1 Y^{(N-1)} + \ldots c_N Y(t) = 0 \tag{1.2}$$

of which $f(t)$ is a solution. This is called the identification problem.

1 Department of Medical Computer Science, University of Texas Health Science Center at Dallas, Dallas, Texas 75235.

2 Department of Mathematics, University of Texas at Arlington, Arlington, Texas 76019.

3 Partially supported by UTA Organized Research Grant.

4 Partially supported by N.I.H. Grant HL16678.

In spite of an abundance of available theory, the problem of computing a unique set of parameters α_i, τ_i from discrete data is difficult if not impossible. Examples ([1],[2]) show how a 3-exponential sum can be fitted very well (in the sense of root mean square) by a 2-exponential sum having completely different time constants and amptitudes. Many scientists seem unaware of this lack of uniqueness of fit and some seem content to obtain a solution which is consistent with other experimental or theoretical results.

A useful technique in solving the above-mentioned problem is to consider the moments

$$mom_n(f(t)) = \int_0^\infty t^n f(t)dt, \quad n = 0,1,2,... \tag{1.3}$$

of a function of the form (1.1). Moment theory has a mathematical history dating back to the nineteenth century but its application to exponential separation is surprisingly limited (see Isenberg's account in [3]). Isenberg and co-workers [2],[3],[4],[5] have developed the method of moments for analysis of fluorescence decay. In this application the function $f(t)$ is given implicitly as a solution of an integral equation of Volterra type

$$g(t) = \int_0^t K(t - x)f(x)dx.$$

To apply the method of moments one computes the moments of $g(t)$ and $K(t)$, which are given in discrete form, and from these one can find the moments of $f(t)$. Thus in this application the method of moments has the added advantage of solving an integral equation. Another advantage which arises in this particular application is that certain nonrandom errors occur predominently in the early moments (i.e. small n) and it turns out these errors may be "automatically eliminated" when considering higher moments (see [3] and [5] for further details).

No matter which application one considers it is clear that

the moments of a function are more reliable than the observed data points. Indeed moments are areas under a curve which act as filtering devices. Thus instead of trying to compute a function to fit prescribed values it is better to ask for a function with prescribed moments.

Instead of working with the moments it is more convenient to work with the related constants

$$G(n+1) = mom_n(f(t))/n!, \quad n = 0,1,\ldots \, .$$

It is easy to see using integration by parts that in the case of functions of the form (1.1)

$$G(K) = \sum_{i=1}^{N} \alpha_i \tau_i^K, \quad K = 1,2,\ldots \, . \tag{1.4}$$

We will also have occasion to consider the analytic continuation of (1.4)

$$G(s) = \sum_{i=1}^{N} \alpha_i \tau_i^s, \quad -\infty < s < \infty \, . \tag{1.5}$$

Returning to the question of exponential separation, we may pose the problem as follows: for given $G(K)$, $K = 1,2,\ldots,$ find N, α_i, τ_i, $i = 1,2,\ldots,N$, such that the set of eq's (1.4) is satisfied. However these equations are not independent. We will show, in fact, that for any $N+1$ consecutive constants $G(s)$, $G(s+1),\ldots, G(s+N)$ there is a linear relationship which can be written in either one of the following forms:

$$G(s) = \sum_{i=1}^{N} a_i G(s+i), \quad -\infty < s < \infty, \tag{1.6}$$

$$G(s+N+1) = \sum_{i=1}^{N} b_i G(s+i), \quad -\infty < s < \infty, \tag{1.7}$$

where a_i, $1 \leq i \leq N$, are constants which depend only on the τ_i (not on α_i or the index s), the b_i are related to the a_i as

$$b_1 = a_N^{-1}, \quad b_K = -a_{N-1}/a_N, \quad K \geq 2, \tag{1.8}$$

and $a_N \neq 0$. We will refer to (1.6) as the backward formula and (1.7) as the forward formula. Observe that in these formulas s may be any real number.

In view of the relationship (1.6) or (1.7) it suffices to satisfy (1.4) for $K = s + 1, \ldots, s + N$ where s is arbitrary in order to obtain $G(s + i)$ for any integer i. The method of moments gives us a procedure for calculating the α_i and τ_i from a given set of $2N$ values of $G(K)$, $K = s + 1, \ldots, s + 2N$. Our main concern here is not with determining the α_i and τ_i which, as we have pointed out above, are suspect, but rather with developing algebraic relationships between the moments from which one may determine the coefficients c_i in the differential equation (1.2). In fact, we show

$$c_i = (-1)^{i+1} a_i \tag{1.9}$$

so that once the a_i are determined we have the differential equation as well as the moment formulas. Moreover, by means of the backward formula one can compute $G(0), G(-1), \ldots, G(-N + 1)$ and since the initial values of the derivatives are related by

$$f^{(K)}(0) = (-1)^K G(-K) \tag{1.10}$$

one has all the necessary information for computing $f(t)$ as a solution of an initial value problem. The a_i $i = 1, \ldots, N$ are solutions of the system of equations

$$G(s + j) = \sum_{i=1}^{N} G(s + j + i) X_i, \quad j = 1, 2, \ldots, N \tag{1.11}$$

where s is arbitrary and the a_i will be shown to be independent of s and j. Recall that we are assuming knowledge of the moments.

We have been treating the integer N as if it were known when in practice it is not. In fact one of the most difficult tasks in exponential separation is to determine N. However in most applications an upper bound for N, say $M \geq N$, is known. We then consider what happens when we replace N by M in equations (1.11). First of all, in theory, we do not have a unique solution to (1.11) for $M > N$ although in practice, because of errors, the solution may be unique. In any case we will show that if $\{a_i\}_{i=1}^M$ is any solution of

$$G(s + j) = \sum_{i=1}^{M} G(s + j + i)X_i, \quad j = 1,2,\ldots,M \tag{1.12}$$

(with $M \geq N$) then the moment formulas, for arbitrary s, are still valid and so is the Cauchy problem (1.2)-(1.10) where the coefficients and initial data are computed in the manner prescribed above (with N replaced by M). Thus this technique does not require the determination of the number of decay components.

2. *Analysis*

The relationship between the backward formula and Cauchy problem can be expressed by the following theorem.

THEOREM 1: Let $f(t)$ be of the form (1.1) and let $G(s)$ be the corresponding moment function. Let $\{a_i\}_{i=1}^M$ be any set of constants where $M \geq N$. The following statements are equivalent:

(i) $$G(s + j) = \sum_{i=1}^{M} a_i G(s + j + i), j = 1,\ldots,M, \text{ for some } s,$$

(ii) $$G(s) = \sum_{i=1}^{M} a_i G(s + i), \quad -\infty < s < \infty$$

(iii) τ_i^{-1}, $1 \leq i \leq N$, are roots of the polynomial

$$P_M(x) = X^M - a_1 X^{M-1} - \dots - a_M , \tag{2.1}$$

(iv) $f(t)$ is a solution of the differential equation

$$Y^{(M)} + a_1 Y^{(M-1)} - a_2 Y^{(M-2)} + a_3 Y^{(M-3)} + \dots + (-1)^{M+1} a_M Y = 0. \tag{2.2}$$

Proof: It is obvious that (iii) and (iv) are equivalent since $P_M(-x)$ is the characteristic polynomial for (2.2). Also it is clear that (ii) implies (i). It suffices to show that (iii) implies (ii) and (i) implies (iii).

(iii) → (ii): Define the vectors

$$\underline{a} = (-1, a_1, a_2, \dots, a_M)$$
$$\underline{T}_i = (1, \tau_i^1, \tau_i^2, \dots, \tau_i^M)$$
$$\underline{G}(s) = (G(s), G(s+1), \dots, G(s+M)).$$

We use the usual scalar product

$$\underline{b} \cdot \underline{c} = b_1 c_1 + b_2 c_2 + \dots + b_{M+1} c_{M+1}$$

in $M + 1$ Euclidean space. We wish to show that

$$\underline{a} \cdot \underline{T}_i = 0, \quad i = 1, \dots, N \tag{2.3}$$

implies

$$\underline{a} \cdot \underline{G}(s) = 0, \quad -\infty < s < \infty. \tag{2.4}$$

But this follows since

$$\underline{G}(s) = B_1(s)\underline{T}_1 + \dots + B_N(s)\underline{T}_N \tag{2.5}$$

where $B_i(s) = \alpha_i \tau_i^s$. Suppose (iii) holds. Then, for each τ_i, $\tau_i^{-M} = a_1 \tau_i^{1-M} + \dots + a_M$. Multiplying by τ_i^M one obtains $-1 + a_1 \tau_i + \dots + a_M \tau_i^M = 0$ i.e., equations (2.3). On the other hand, (2.4) is equivalent to (ii) and the proof is complete.

(i) → (iii): When we interpret (i) and (iii) as orthogonality relationships as in the preceding argument our task is to show that

$$\underline{a} \cdot \underline{G}(s+j) = 0, \quad j = 1, 2, \dots, M, \quad \text{for some } s \tag{2.6}$$

implies eqs. (2.3). It suffices to show that the linear subspace U spanned by $\underline{T}_i$, $i = 1,2,\ldots,N$ is identical to the space V spanned by $\underline{G}(s+j)$, $j = 1,2,\ldots,M$. Clearly, by (2.5), $V \subset U$. Our problem is to show that the dimension of V is N. The dimension of V is the rank of the matrix, G whose rows are the vectors $\underline{G}(s+j)$, $j = 1,2,\ldots,M$. Consider the $N \times N$ principle minor of G which we denote by H and observe that $H = Q^T DQ$ where Q^T is the Vandermonde matrix whose columns are the vectors $(1,\tau_i,\ldots,\tau_i^{N-1})$, $i = 1,2,\ldots,N$ and D is the diagonal matrix whose diagonal entries are $\alpha_i \tau_i^{s+1}$. Since $\tau_i > 0$, $\alpha_i > 0$, and $\tau_i \neq \tau_j$, $i \neq j$ it follows that

$$det\ H = det\ Q^T \cdot det\ D \cdot det\ Q = \prod_{i=1}^{N} \alpha_i \tau_i^{s+1} \cdot \prod_{i \neq j} (\tau_i - \tau_j)^2 > 0.$$

Thus the rank of G is N. This completes the proof of Theorem 1.

Observe that each of the four statements has special significance. Statement (i) is the algorithm for computing the a_i; statement (ii) is the backward formula; statement (iii) is essentially the method of moments formula for computing τ_i when $M = N_j$ statement (iv) gives the decay "law".

As for the forward formula one can prove in a similar manner the following result.

THEOREM 2: Let $f(t)$ be of the form (1.1) and let $G(s)$ be the corresponding moment function. Let $\{b_i\}_{i=1}^{M}$ be any set of constants where $M \geq N$. The following statements are equivalent:

(i) $$G(s+N+1) = \sum_{i=1}^{N} b_i G(s+i), \quad j = 1,\ldots,M, \text{ for some } s$$

(ii) $$G(s+N+1) = \sum_{i=1}^{N} b_i G(s+i), \quad -\infty < s < \infty$$

(iii) τ_i, $1 \leq i \leq N$, is a root of the polynomial

$$Q_M(x) = x^M - b_M x^{M-1} - \ldots - b_1.$$

We caution however that if $M > N$ the relationships (1.8) require that $a_M \neq 0$. To insure this we observe $P_M(x)$ can be represented as

$$P_M(x) = R_{M-N}(x)\ P_N(x)$$

where $R_{M-N}(x)$ is a polynomial of order $M - N$ with arbitrary roots and $P_N(x)$ is the polynomial of order N whose roots are τ_i^{-1} $1 \leq i \leq N$. Thus as long as we select the roots of R_{M-N} to be non-zero then $a_M \neq 0$ and (1.8) holds.

3. *Problems for Future Research*

The technique might be expanded to include systems of equations nonhomogeneous (forcing) terms and complex roots. Problems also arise (eg. [7]) where one encounters an infinite spectrum of roots, i.e., (1.1) is replaced by

$$f(t) = \int_0^\infty \alpha(s)\ exp(-t/\tau(s))ds \tag{3.1}$$

The corresponding differential equation

$$x'(t) = Ax \tag{3.2}$$

must be considered in a Banach space where $\{\tau(s) \mid 0 < s < \infty\}$ is the spectrum of the operator A. The form (3.1) includes (1.1) where $\alpha(s)$ is a linear combination of Dirac delta functions

$$\alpha(s) = \sum_{i=1}^{N} \alpha_i \delta(s - s_i) \tag{3.3}$$

and $\tau(s_i) = \tau_i$. When N is unknown it might be more advantageous to use (3.1) and hope that the estimation of the distribution of $\alpha(s)$ approximates a sum of delta functions.

REFERENCES

[1] C. Lanczos, *Applied Analysis*, Prentice-Hall, Inc., Englewood Cliffs, New Jersey, 1956.

[2] I. Isenberg, R. D. Dyson, and R. Hanson, *Studies on the analysis of fluorescence decay data by the method of moments*, Biophys. J. 13(1973), 1090.

[3] I. Isenberg, *On the theory of fluorescence decay experiments. I. Nonrandom distortions*, J. Chem. Phys. 59(1973), 5696.

[4] I. Isenberg, *On the theory of fluorescence decay experiments. II. Statistics*, J. Chem. Phys. 59(1973), 5708.

[5] I. Isenberg, and R. D. Dyson, *The analysis of fluorescence decay by a method of moments*, Biophys. J. 9(1969), 1337.

[6] J. Eisenfeld, and D. J. Mishelevich, *On nonrandom errors in fluorescence decay experiments*, J. Chem. Phys., (to appear).

[7] M. M. Judy, and S. R. Bernfeld, *On the characterization of macromolecular length distributions by analysis of electrical birefringence decay, (to appear)*.

ON THE STEADY STATE OF AN AGE DEPENDENT MODEL FOR MALARIA*

R. H. Elderkin,† D. P. Berkowitz, F. A. Farris, C. F. Gunn,
F. J. Hickernell, S. N. Kass, F. I. Mansfield, & R. G. Taranto
Pomona College

0. *Introduction.*

We present here some basic analysis of the steady state equations of an age-dependent model of malaria, suggested by Klaus Dietz in cooperation with others at the World Health Organization. Although a great deal has been written on epidemiological modelling, and on malarial modelling in particular, relatively little has been written on models including a dependency on the age of the hosts (see [1], [2], [4], [5]). Our model is as follows:

$$\begin{aligned} &\frac{\partial y}{\partial t} + \frac{\partial y}{\partial a} = V \int_0^\infty \mu e^{-\mu x} g(t,x)dx - ry \\ \text{(0.1a)} \quad &\frac{\partial r}{\partial t} + \frac{\partial r}{\partial a} = \alpha y - \beta(r - r_0) \\ &\frac{\partial g}{\partial t} + \frac{\partial g}{\partial a} = \gamma y e^{-rT} - \delta g \end{aligned}$$

considered with boundary conditions of the form

$$\begin{aligned} &y(t,0) = f_1(t) \\ \text{(0.1b)} \quad &r(t,0) = f_2(t) \\ &g(t,0) = f_3(t) \qquad -\infty < t < \infty \end{aligned}$$

or of the form

$$\begin{aligned} &y(t,0) = f_1(t) \qquad y(0,a) = h_1(a) \\ \text{(o.1c)} \quad &r(t,0) = f_2(t) \qquad r(0,a) = h_2(a) \\ &g(t,0) = f_3(t) \qquad g(0,a) = h_3(a) \qquad t,\ a \geq 0 \end{aligned}$$

† Research supported in part by a grant from the National Science Foundation, #MCS72-04965A04.

* This research was funded in part by the National Science Foundation Undergraduate Research Participation Grant #SMI76-03038.

subject to various conditions on the boundary functions f_i, h_i. Of course it is the integral in the first equation which gives the problem its mathematical interest. Indeed, this interest is maintained by the time (t) independent equations to an extent such that this paper is devoted to the following equations and initial conditions which are derived from (0.1).

(0.2a)
$$\frac{dy}{da} = V\int_0^\infty \mu e^{-\mu x}\, g(x)dx - ry$$
$$\frac{dr}{da} = \alpha y - \beta(r - r_0)$$
$$\frac{dg}{da} = \gamma y e^{-rT} - \delta g$$

(0.2b)
$$\frac{dy}{da} = VG - ry$$
$$\frac{dr}{da} = \alpha y - \beta(r - r_0)$$
$$\frac{dg}{da} = \gamma y e^{-rT} - \delta g$$

(0.2c)
$$y(0) = y^0 \geq 0$$
$$r(0) = r^0 \geq 0$$
$$g(0) = g^0 \geq 0$$

In both sets of equations, (0.1) and (0.2) y is the population density of the sporozoite (asexual) malarial parasite in a host; r is the death rate of the sporozoites; and g is the population density of the gametocyte (sexual) malarial parasite in a host. The sporozoite is involved directly in the effect of the disease on the host, but not in the transfer of disease; while the gametocyte has little symptomatic effect, but is the key to the spread of the disease via the mosquito. A more detailed statement of the assumptions of this model is given at the end of the introduction.

Our fundamental result is Theorem (1.1), that solutions to

the initial value problem (0.2a,c) always exist. For a triple $(y(a),r(a),g(a))$ to be a solution, we demand that y, r, g all be non-negative and defined for all $a \geq 0$, in addition to the usual conditions. For trivial initial conditions $(y^0 = g^0 = 0)$, there is only a trivial solution $(y = g = 0)$ when V is small, but there are both trivial and nontrivial solutions when V is large, and the situation is not known for some intermediate situations. There is numerical evidence to suggest that nontrivial solutions are unique whenever they exist. In section 2 some additional properties are investigated. In particular, the map which takes (y^0,r^0,g^0,G) to $\int_0^\infty \mu e^{-\mu a} g(a)da$ where (y,r,g) solves (0.2b,c) is shown to be continuously differentiable. This is a first step towards establishing the differentiability of solutions with respect to initial conditions, but the full proof remains elusive. Finally, in section 3, we raise some further questions based on our numerical experience.

The assumptions incorporated in the model are as follows. The sporozoite density (y) changes with the density of gametocytes (g) transferred to mosquitos, or according to the "biting rate" V. A negative exponential distribution is used to approximate the number of humans surviving to an age, x, so the expected (average) density of gametocytes transferred to mosquitos is then given by $V \int_0^\infty \mu e^{-\mu x} g(x,t)dx$. If r is the death rate of sporozoites, then y is reduced at a rate r. The equation for y follows in the usual manner. The sporozoite death rate, r, can also be interpreted as human immunity, or the density of human defense cells. It changes proportionally with the threat of disease as evidenced by the value of y, increasing at a rate α. If there are no sporozoites present $(y = 0)$, r will tend to decrease to a value r_0, the immunity of a newborn, at a rate β. Hence we obtain the second equation. Increases in g, the gametocyte density, are in proportion to the sporozoites surviving long enough to develop into the gametocytes. The probability of surviving for a time T is approximated by e^{-rT}. If the rate of reproduction is γ then $\gamma y e^{-rT}$ is the rate of change of y to g. Since gametocytes die at rate δ, the third equation is established. It is reasonable to assume that a newborn human has no parasites in its blood and has a minimum level of immunity, r_0. This leads to the boundary conditions of primary biological interest:

$$y(0,t) = 0, \quad g(0,t) = 0, \quad r(0,t) = r_0, \quad t \geq 0.$$

The constants $\alpha, \beta, \gamma, \delta, r_0, \mu, T$, and V are all assumed to be positive. Some approximate values for these constants that were used are: $\alpha = .2$, $\beta = .005$, $\gamma = .5$, $\delta = .4$, $r_0 = .66$, $\mu = .0001$, $T = 3.5$. Values for V vary over several orders of magnitude. The effect of this variation is noticeable in Theorem (1.1).

1. *Existence and Non-uniqueness of Solutions*-

We emphasize that we are interested here only in solutions $(y(a), r(a), g(a))$ to (0.2a) defined for $a \geq 0$ and such that $y(a)$, $r(a)$, $g(a)$ are non-negative. "Solution" is henceforth restricted to this sense. Let $R^3_+ = \{(y,r,g) | y \geq 0, r \geq 0, g \geq 0\}$. A solution will be called called <u>trivial</u> if $y(a) = g(a) \equiv 0$ for all $a \geq 0$.

<u>THEOREM 1.1</u>. The initial value problem (0.2a,c) has at least one solution. If $y^0 = g^0 = 0$, there is a trivial solution, and furthermore

(i) if $r^0 \leq r_0$, then for $V > \gamma^{-1}(\delta + \mu)(r_0 + \mu)e^{r_0 T}$ there is a non-trivial solution, while for $V \leq \gamma^{-1}(\delta + \mu)(r_0 + \mu)e^{r^0 T}$ there is only a trivial solution; and

(ii) if $r^0 \geq r_0$, then for $V > \gamma^{-1}(\delta + \mu)(r_0 + \mu)e^{r^0 T}$ there is a non-trivial solution, while for $V \leq \gamma^{-1}(\delta + \mu)(r_0 + \mu)e^{r_0 T}$ there is only a trivial solution.

<u>PROOF:</u> We need an elementary fact about solutions to (0.2b).

<u>LEMMA 1.2</u>. Let G be non-negative and fixed in (0.2b). Then solutions to (0.2b,c) exist in the above sense and are unique. Fur-

thermore, there is a point $(\overline{y},\overline{r},\overline{g}) \in \mathbb{R}^3_+$ such that every solution $(y(a), r(a), g(a))$ of (0.2b,c) tends to $(\overline{y},\overline{r},\overline{g})$ as $a \to \infty$.

PROOF: To show that solutions exist in our restricted sense it is sufficient to note that $\dot{y} \geq 0$ when $y = 0$, $\dot{r} > 0$ when $r = 0$, and $\dot{g} \geq 0$ when $g = 0$. Hence the vector field corresponding to (0.2b) either points into $\mathbb{R}^3_+$ or is tangent to the boundary planes (similarly one accounts for the corners), so that trajectories cannot leave $\mathbb{R}^3_+$. Standard uniqueness results apply here.

Upon setting the right hand side of (0.2b) to zero, one finds a unique rest point $(\overline{y},\overline{r},\overline{g})$ in $\mathbb{R}^3_+$. We will show that it attracts every solution in $\mathbb{R}^3_+$. The first two equations of (0.2b) may be decoupled from the third, and g may be solved for explicitly in terms of y,r using the variation of parameters formula. Hence if we show that $(y(a), r(a)) \to (\overline{y},\overline{r})$ as $a \to \infty$, it will follow immediately that $g(a) \to \overline{g}$. Our analysis of the (y,r) phase plane is simple. The case when $G = 0$ is trivial, as a sketch of the phase plane shows. If $G > 0$, then the situation is as in Figure 1, and it is easily checked that the following Lyapunov function is strictly decreasing to zero along noncritical solutions in the first quadrant:

$$(1.3)\quad V(y,r) = max\{\alpha(y-\overline{y}), \beta(r-\overline{r}), \beta VG((1/y)-(1/\overline{y})), \alpha VG((1/r)-(1/\overline{r})\}.$$

Hence $(\overline{y},\overline{r})$ is the ω-limit of every trajectory which enters the first quadrant, and the lemma is established.

<u>COROLLARY 1.4</u>. If (y,r,g) is a solution of an initial value problem (0.2b,c), the integral $\int_0^\infty \mu e^{-\mu x} g(x)dx$ converges.

Returning to the proof of (1.1) let (y^0,r^0,g^0) be fixed for the remainder of this proof. Since trivial solutions are easily found explicitly when $y^0 = g^0 = 0$, we will henceforth limit our attention to nontrivial solutions. Hence we consider only positive G in (0.2b). Let

$$(1.5)\qquad F(G) = \int_0^\infty \mu e^{-\mu a}\, g(a)da, \qquad G > 0$$

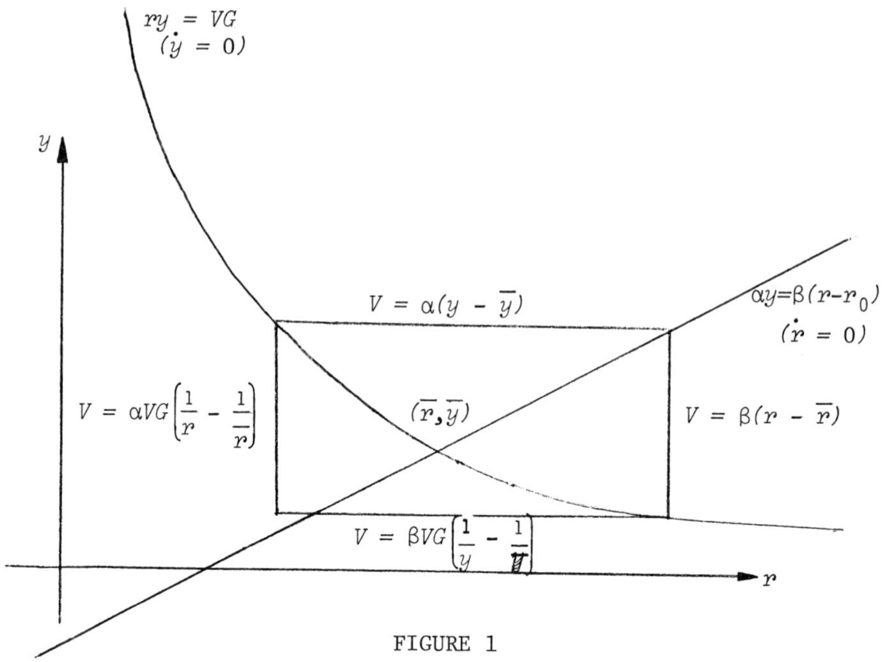

FIGURE 1

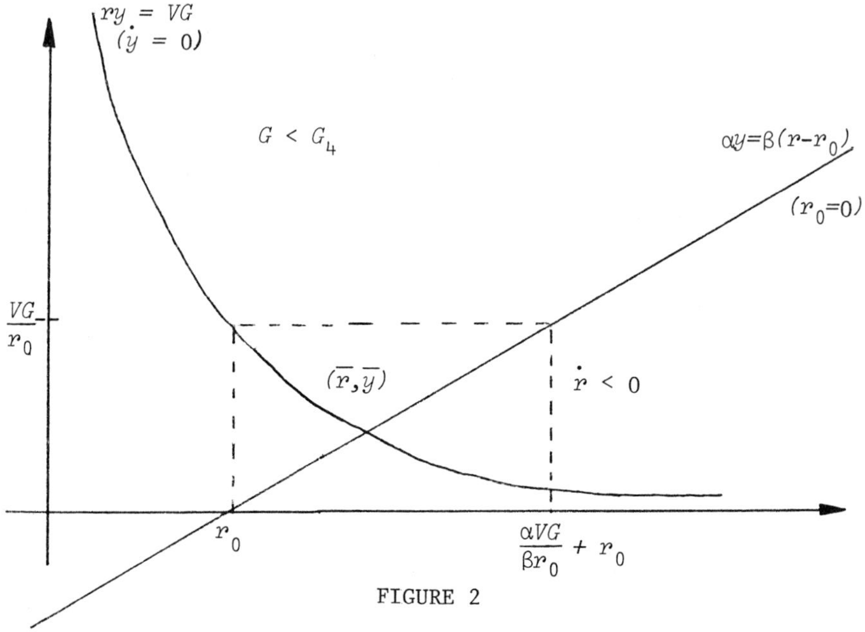

FIGURE 2

where $g(a)$ is obtained from the solution $(y(a),r(a),g(a))$ to the initial value problem (0.2b,c). Using the variation of parameters formula to solve for $g(a)$ explicitly in terms of y and r, and integrating once by parts, one obtains

(1.6) $$F(G) = \mu(g^0 + \gamma \int_0^\infty e^{-\mu a} y(a)e^{-r(a)T} da)/(\delta + \mu).$$

A necessary and sufficient condition that $(y(a),r(a),g(a))$ solve the steady state problem (0.2a,c) nontrivially is

(1.7) $$F(G) = G, \qquad G > 0.$$

We will demonstrate that F has a fixed point in three steps. First we show F to be continuous. Second, we show that $F(G) < G$ for all sufficiently large G. Finally we show that $F(0) > 0$ if either y^0 or g^0 is nonzero, and we show that $F(G) > G$ for sufficiently small positive G when V is large as specified in (i) and (ii) of the statement of this theorem. The intermediate value theorem then guarantees at least one solution to (1.7) in these cases. The establishment of these ideas concerns us through Lemma (1.16), and Lemma (1.19) completes the proof of (1.1).

Some notation will be useful. Let $C^0_\infty(\mathbb{R}_+,\mathbb{R}^n)$ denote the Banach space of bounded continuous $\mathbb{R}^n$-valued functions of a nonnegative real variable with the uniform norm. Let $\mathbb{R}^n_+$ be analogous to $\mathbb{R}^3_+$ $(\mathbb{R}_+ = \mathbb{R}^1_+)$ and let $S : \mathbb{R}^4_+ \to C^0_\infty(\mathbb{R}_+,\mathbb{R}^3)$ be given by declaring that $S(y^0,r^0,g^0,G)$ solve (0.2b,c). Hence $F(G)$ can be viewed as the composition of S with a projection and an integration functional. It is easily seen that continuity of S implies continuity of F.

<u>LEMMA 1.8</u>. Consider the parameterized family of differential equations of

$$\dot{x} = X_a(x)$$

where $X_a : \mathbb{R}^n \to \mathbb{R}^n$ is C^1 for each real a near zero and where the parameterization by a is continuous in the compact-open topology on vector fields. Denote the solution for X_a through x

at $t = 0$ by $f_a(x,t)$. Assume that each X_a has a rest point w_a varying continuously with a and that w_0 is globally uniformly asymptotically stable. Then for each compact subset K of $\mathbb{R}^n$ there is a positive number A such that the map $\Sigma : (-A,A) \times K \to C^0_\infty(\mathbb{R}^+,\mathbb{R}^n)$ given by $\Sigma(a,x)(t) = f_a(x,t)$ is continuous.

PROOF. By continuously translating w_a to the origin, we may assume $w_a \equiv 0$. Let $\varepsilon > 0$ be given and set $B = \{x \in \mathbb{R}^n \,|\, ||x|| < \varepsilon/2\}$. There is a smooth Lyapunov function for X_0 (see [3] for existence, [6] for smoothness), that is, there is a smooth function $V : \mathbb{R}^n \to \mathbb{R}^+$ such that $V^{-1}(0) = \{0\}$ and $\dot{V}(x) = \frac{d}{dt} V o f_0(x,t)|_{t=0} = grad\, V(x)\cdot X_0(x) < 0$ if $x \neq 0$. Since $V^{-1}(0) = \{0\}$ there is a $c > 0$ such that $V^{-1}([0,c]) \subset B$. Choose $d > V(x)$ for all $x \in K$. The level surfaces $V^{-1}(s)$ for $s > 0$ are all diffeomorphic (push along trajectories of X_0) and hence are all compact since $V^{-1}(c)$ is. It follows that $K' = V^{-1}([c,d])$ is compact and $K' \supset K$. Because $grad\, V(x)\cdot X_0(x)$ is bounded away from 0 on K' and because X_a is continuous in a in the compact-open topology, there is an $A_1 > 0$ such that $grad\, V(x)\cdot X_a(x)$ is bounded away from 0 for all $x \in K'$, $|a| < A_1$. It follows that there is a $T > 0$ such that $V o f_a(x,t) < c$ if $x \in K'$, $|a| < A_1$, $t > T$ and hence $d(f_a(x,t), f_a{}'(x',t)) < \varepsilon$ if $x, x' \in K'$, $a, a' \in (-A_1,A_1)$ and $t > T$. Finally, the usual theorems on continuity of solutions to $\dot{x} = X_a(x)$ in a and x on the compact interval yield the desired result, namely that for some A, $0 < A \leq A_1$ and for some $\delta > 0$, we have $d(f_a(x,t), f_{a'}(x',t)) < \varepsilon$ whenever $a, a' \in (-A,A)$ and $x, x' \in K$ with $d(x,x') < \delta$.

Consider a nonautonomous differential equation

$$\dot{x} = X(t,x), \qquad X(t,0) \equiv 0$$

with X continuously differentiable and let $f(t;t_0,x)$ denote the solution with $f(t_0;t_0,x) = x$. The origin is called globally uniformly asymptotically stable if for every $\varepsilon > 0$ there is a

$\delta > 0$ such that $||x|| < \delta$ implies $||f(t;t_0,x)|| < \varepsilon$ for $t \geq t_0$ for every t_0 and if for every compact set K in $\mathbb{R}^n$ and for every $\varepsilon > 0$ there is a T such that $x \in K$ implies $||f(t;t_0,x)|| < \varepsilon$ for every $t \geq t_0 + T$. In section 2 we will need

COROLLARY 1.9. Let $X(t,x)$ be as above and globally uniformly asymptotically stable. Then the map $\Sigma : R^n \to C^0_\infty(\mathbb{R}_+,\mathbb{R}^n)$ given by $\Sigma(x)(t) = f(t;0,x)$ is continuous.

PROOF. The proof here is a direct adaptation of the ideas in the proof of the lemma, but the use of Lyapunov functions is not necessary. Solutions beginning close together remain close together for bounded times, but there is a uniform time after which they get close to the origin, and hence they remain close to each other for all time.

COROLLARY 1.10. The map $S : \mathbb{R}^4_+ \to C^0_\infty(\mathbb{R}_+,\mathbb{R}^3)$ is continuous, and hence so is $F(G)$.

To establish the inequality results for F it will be convenient to use some geometry of the (r,y) phase plane for solutions of (0.2b), using abscissa r and ordinate y.

LEMMA 1.11. Trajectories in the (r,y) phase plane lying above the $\dot{r} = 0$ line and below the $\dot{y} = 0$ hyperbola are concave down whenever $y > \beta^2/4$.

PROOF. For $r > \beta$, $\frac{d}{dr} \cdot \frac{dy}{dr} = \frac{d}{da} \cdot \frac{dy}{dr} \cdot \frac{da}{dr} = \frac{\ddot{y}\dot{r} - \ddot{r}\dot{y}}{(\dot{r}^3)} =$

$$- \frac{(\dot{r})^2\, y + \dot{r}\dot{y}(r - \beta) + \alpha(\dot{y})^2}{(\dot{r})^3} = - \frac{y}{(\dot{r})^3}\left[(\dot{r})^2 + (\dot{r}\dot{y}(r - \beta))/y + \alpha(\dot{y})^2/y\right]$$

is negative since in the specified region $\dot{r} > 0$ and $\dot{y} > 0$. For $0 \leq r \leq \beta$, we rewrite $\frac{d}{dr} \cdot \frac{dy}{dr}$ as

$$\left[-y/(\dot{r}^3)\left[(\dot{r} - \dot{y}(\beta - r)/2y)^2 + ((\dot{y})^2/y)(\alpha - (\beta - r)^2/4y)\right)\right],$$

which is negative whenever $\alpha > (\beta - r)^2/4y$, or whenever

$y > \beta^2/4\alpha$ since $0 \leq r \leq \beta$ implies $(\beta - r)^2 \leq \beta^2$.

LEMMA 1.12. There exists $G_0 > 0$ such that $F(G) < G$ when $G_0 < G$.

PROOF. By Lemma (1.2) we may consider $y(a)$ and $r(a)$ to be a continuously defined on the compact interval $[0,\infty]$. Hence we may choose $a^* \varepsilon [0,\infty]$ such that $y^* = y(a^*)$ and $r^* = r(a^*)$ maximize the value of $y(a)e^{-r(a)T}$ over $a \ \varepsilon \ [0,\infty]$. So if $g^0 \leq G$ we obtain directly from (1.6) that

$$F(G) \leq [\mu G + \gamma y^* e^{-r^*T}]/(\delta + \mu).$$

Hence it suffices to show that

(1.13) $$\gamma y^* e^{-r^*T} < \delta G$$

for sufficiently large $G \geq g^0$.

Consider the curve in the plane described by $y = (\delta/\gamma)Ge^{rT}$. Let ℓ_1 be the line through $(0,\delta G/\gamma)$ tangent to this curve. The slope of ℓ_1 is obviously proportional to G. Now, let ℓ_2 be the line through $(0,\bar{y})$ whose slope is equal to $\dot{y}/\dot{r}$ evaluated at that point. Since $\bar{y} = [-r_0 + (r_0^2 + 4(\alpha/\beta)VG)^{\frac{1}{2}}]/(2\alpha/\beta)$ and $\dot{y}/\dot{r} = VG/(\alpha\bar{y} + \beta r_0)$, both the y-intercept and the slope of ℓ_2 are nearly proportional to $\sqrt{G}$ for large G. So we can find a $G_1 > 0$ such that for $G > G_1$, ℓ_1 lies above ℓ_2 everywhere in the first quadrant.

Next, consider the trajectory τ from $(0,\bar{y})$. τ must lie above the line $y = \bar{y}$ whenever τ is below the $\dot{y} = 0$ hyperbola. It is geometrically (and algebraically) clear that $\bar{y}$ increases monotonically without bound with G. Choose G_2 so that $\bar{y} > \beta^2/4\alpha$ when $G \geq G_2$. Then by Lemma (1.12), τ must be concave down whenever it lies below the $\dot{y} = 0$ hyperbola for all $G \geq G_2$.

Finally, note that if $y^0 \geq \beta(r^0 - r_0)/\alpha$, there is some G_3 such that $\bar{y} > y^0$ when $G \geq G_3$. Pick

(1.14) $$G_0 \geq max\{g^0, G_1, G_2, G_3, r^0y^0/V\}.$$

Case 1: $y^0 \geq (\beta/\alpha)(r^0 - r_0)$. For $G > G_0$, (r^0,y^0) lies below the $\dot{y} = 0$ hyperbola and above the $\dot{r} = 0$ line. Since $y^0 < \bar{y}$, the trajectory from (r^0,y^0) lies below τ, which lies below ℓ_2 because $G > G_2$. Since $G > G_1$, ℓ_2 lies below ℓ_1, which lies below the curve $y = (\delta/\gamma)Ge^{rT}$. Therefore $\gamma y^* e^{-r^*T} < \delta G$.

Case 2: $y^0 < (\beta/\alpha)(r^0 - r_0)$. The trajectory from (r^0,y^0) must lie below the $\dot{y} = 0$ hyperbola for $y < (\beta/\alpha)(r - r_0)$ and below the trajectory τ for $y \geq (\beta/\alpha)(r - r_0)$. Hence, it lies below ℓ_2 and therefore ℓ_1 so that $\gamma y^* e^{-r^*T} < \delta G$.

In every case (1.13) holds, and the lemma has been proved.

LEMMA 1.15. If $y^0 > 0$ or $g^0 > 0$, then $F(0) > 0$.

PROOF. For $G \geq 0$, $y \geq 0$, as can be seen from writing the first equation of (0.2b) as an integral equation:

$y = y^0 exp\left(-\int_0^a r da_1\right) + VG\int_0^a exp\left(-\int_{a_1}^a r da_2\right)da_1$. Hence, for any $G \geq 0$, $\gamma\int_0^\infty e^{-\mu x} y e^{-rT} dx \geq 0$ and $\gamma\int_0^\infty e^{-\mu x} y e^{-rT} dx > 0$ if $y^0 > 0$. So, if $y^0 > 0$ or $g^0 > 0$, $F(0) = (\mu/\mu + \delta)[g^0 + \gamma\int_0^\infty e^{-\mu x} y e^{-rT} dx\ ] > 0$.

LEMMA 1.16. If $y^0 = g^0 = 0$ then $F(G_1) > G_1$ for some $G_1 > 0$ if either

i) $r^0 \leq r_0$ and $V > (\delta + \mu)(r_0 + \mu)e^{r_0T}/\gamma$, or

ii) $r^0 > r_0$ and $V > (\delta + \mu)(r^0 + \mu)e^{r^0T}/\gamma$.

PROOF. Considering the case $r^0 > r_0$, refer to Figure 2. Find $G_4 > 0$ small enough that $(\alpha VG/\beta r_0) + r_0 \leq r^0$ when $0 < G < G_4$, so that $r(a) \leq r^0$ for every $a \geq 0$. Then

$$y(a) = y^0 exp\left(-\int_0^a r(a_1)da_1\right) + VG\int_0^a exp\left(-\int_{a_1}^a r(a_2)da_2\right)da_1$$
$$\geq VG(1 - e^{-r^0a})/r^0.$$

Using first this inequality and then the appropriate hypothesis on

V we obtain

(1.17) $$F(G) \geq VG\gamma\left(\int_0^\infty \mu e^{-\mu x}\,(1 - e^{-r^0 x})e^{-r^0 T}dx\right)/(\delta + \mu)r^0$$

$$= VG\gamma/((\delta + \mu)(r^0 + \mu)e^{r^0 T})$$

$$> G.$$

Next consider the case $r^0 \leq r_0$. Let $P(G) = (\alpha VG/\beta r_0) + r_0$. Considerations similar to those with Figure 2 show that $r(a) \leq P(G)$ and $y(a) \geq VG(1 - e^{-P(G)a})/P(G)$ $(a \geq 0)$. In analogy with (1.17) we obtain

(1.18) $$F(G) \geq VG\gamma\left(\int_0^\infty \mu e^{-\mu x}(\ \ - e^{-P(G)x})e^{-P(G)T}dx\right)/(\delta + \mu)P(G)$$

$$= VG\gamma/[(\delta + \mu)(P(G) + \mu)e^{P(G)T}].$$

Now applying the hypothesis on V we find that the derivative of the last expression in (1.18) with respect to G, evaluated at $G = 0$ is strictly greater than one. Hence $F(G) > G$ for sufficiently small $G > 0$. The proof of (1.16) is complete.

To complete the proof of Theorem (1.1), it remains to be seen that only trivial solutions exist in certain circumstances. For this, it suffices to prove

LEMMA 1.19. $F(G) < G$ for all $G > 0$ whenever $[V \leq \gamma^{-1}(\delta + \mu)(r^0 + \mu)e^{r^0 T}]$ and $r^0 \leq r_0$ [or whenever $V \leq \gamma^{-1}(\delta + \mu)(r_0 + \mu)e^{r_0 T}$ and $r^0 > r_0$.

PROOF. Since the proofs for the two cases are similar, we present only the latter case. If $r^0 \leq r_0$ it is easily seen that $r(a) > r^0$ when $a > 0$ and so

$$y(a) < VG \int_0^a exp\left(-\int_{a_1}^a r^0 dx\right)da_1 = VG(1 - e^{-r^0 a})/r^0.$$

Therefore

(1.20) $$F(G) < \gamma\left(\int_0^\infty \mu e^{-\mu x}(1 - e^{-r^0 x})e^{-r^0 T}dx\right)/(\delta + \mu)r^0$$

$$= VG\gamma/((\delta + \mu)(r^0 + \mu)e^{r^0 T})$$

For nontrivial solution of (0.2a) it is necessary that $F(G) = G$ for some $G > 0$, which in conjunction with (1.20) implies the negation of our hypothesis on V.

The proofs of Lemma (1.19) and Theorem (1.1) are complete.

2. *Properties of Solutions*

LEMMA 2.1. If $V \leq \delta r_0 e^{r_0 T}/\gamma$ there is a unique equilibrium solution of (0.2a) in the first octant $\mathbb{R}^3_+$, namely the trivial solution $(y(a),r(a),g(a)) \equiv (0,r_0,0)$. If $V > \delta r_0 e^{r_0 T}/\gamma$ there are exactly two equilibria in $\mathbb{R}^3_+$: the trivial one and one other, $(y(a),r(a),g(a)) \equiv (y_e,r_e,g_e)$.

PROOF: It is trivial to check that $(0,r_0,0)$ is an equilibrium state for every V. Substituting $(y(a),r(a),g(a)) \equiv (y_e,r_e,g_e)$ into (0.2a) and setting the right hand side to zero, we obtain the equivalent condition

$$0 = Vg_e - r_e y_e$$
$$0 = \alpha y_e - \beta(r_e - r_0)$$
$$0 = \gamma y_e e^{-r_e T} - \delta g_e .$$

Solving the first two equations for y_e and r_e in terms of g_e and substituting into the third equation we see that equilibrium states are determined by solutions to the equation

$$Vg_e = g_e \gamma^{-1}\delta \tfrac{1}{2}(r_0+(r_0^2+4\alpha Vg_e/\beta)^{\frac{1}{2}}) exp(\tfrac{1}{2}(r_0+(r_0^2+4\alpha Vg_e/\beta)^{\frac{1}{2}}T)$$

(only positive square roots are of interest).

The solution $g_e = 0$ yields the trivial equilibrium. For $g_e > 0$, one finds that V is monotone increasing in g_e and hence has its minimum value given by setting $g_e = 0$. The stated inequalities follow as a direct result.

In Lemma (1.2), we showed that the limit of every solution $(y(a),r(a),g(a))$ exists as $a \to \infty$. For an ordinary system of differential equations, such a limit would be an equilibrium solution. For the problem (0.2a,c) this is usually not the case. In fact the following lemma and corollary show that if the asymptotic limit of any solution to (0.2a,c) is an equilibrium solution then the asymptotic limit of any solution with perturbed g_0 is not an equilibrium solution.

LEMMA 2.2. Let $y_i(a)$, $r_i(a)$, $g_i(a))$, $i = 1,2$, solve (0.2a) subject to the initial conditions $y_i(0) = y^0 \geq 0$, $r_i(0) = r^0 \geq 0$ and $g_1(0) = g_1^0$, $g_2(0) = g_2^0$ with $g_1^0 \neq g_2^0$. Then $\int_0^\infty \mu e^{-\mu a} g_1(a)da \neq \int_0^\infty \mu e^{-\mu a} g_2(a)da$.

PROOF. If the integrals are equal, then it is seen from (0.2a) that $y_1(a) = y_2(a)$ and $r_1(a) = r_2(a)$ $(a \geq 0)$. But then the variation of parameters formula shows that $g_1(a) < g_2(a)$ $(a \geq 0)$ and hence the integrals are different.

COROLLARY 2.3. In the notation of (2.2) let $(\bar{y}_i, \bar{r}_i, \bar{g}_i) = lim\ (y_i(a), r_i(a), g_i(a))$ as $a \to \infty$, $i = 1,2$. If $(\bar{y}_1, \bar{r}_1, \bar{g}_1)$ is an equilibrium solution to (0.2a) then $(\bar{y}_2, \bar{r}_2, \bar{g}_2)$ is not, with the possible exception of at most one value of g_2^0.

PROOF. Depending on the magnitude of V, there are at most two equilibrium states. Suppose $(\bar{y}_1, \bar{r}_1, \bar{g}_1)$ is one of them. Then $\int_0^\infty \mu e^{-\mu a} g_1(a)da \neq \int_0^\infty \mu e^{-\mu a} g_2(a)da$ and substitution of these values into (0.2b) shows that $(\bar{y}_1, \bar{r}_1) \neq (\bar{y}_2, \bar{r}_2)$. It follows that there is at most one value of g_2^0 distinct from g_1^0 for which $(\bar{y}_2, \bar{r}_2, \bar{g}_2)$ can be an equilibrium solution.

A considerable amount of information about a solution $(y(a), r(a), g(a))$ of (0.2a) can be obtained by substituting $\int_0^\infty \mu e^{-\mu a} g(a)da$ into (0.2b) for G and viewing $(y(a), r(a))$ as a trajectory in the (r,y) phase plane for the first two equations of (0.2b). For example, one finds that for values of the parameters $\alpha,\beta,\gamma,\delta$ near those given in the introduction, the topological type of the rest point $(\bar{r},\bar{y})$ is a node, which gives us information of the asymptotic nature of $(y(a), r(a), g(a))$. A result of similar use is the following:

PROPOSITION 2.4. Assume that $r_0 \geq 4\beta$ and $G > 0$ in the system given by the first two lines of (0.2b), and let $(r(a), y(a))$ be any solution. Let B be the compact box of Figure 3, that is, $r^* \leq r \leq \bar{r}$ where $r^* = 4\alpha VG/(\bar{r} - \beta)^2$ and

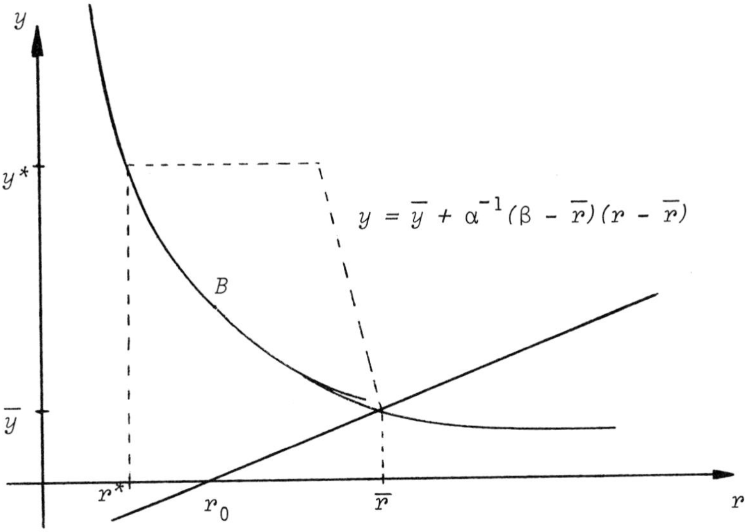

FIGURE 3

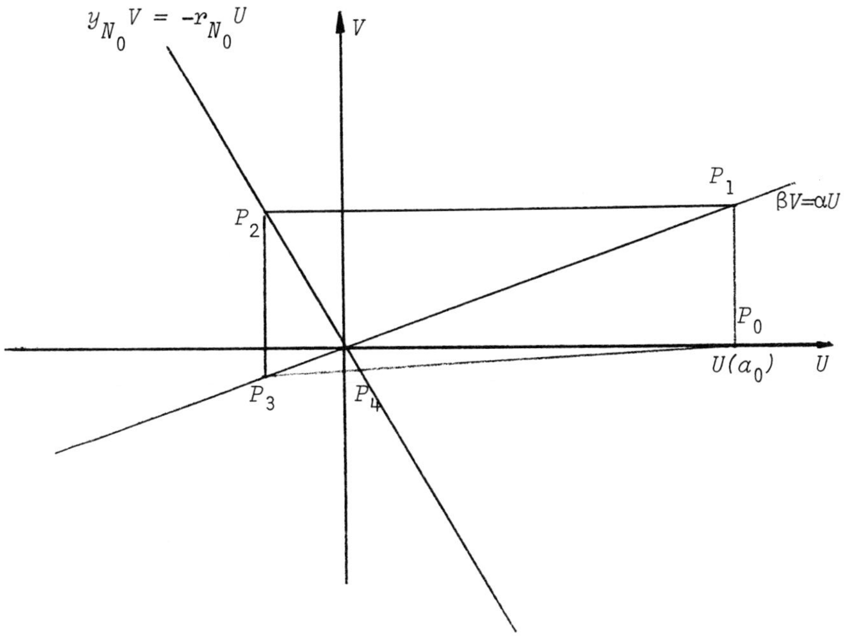

FIGURE 4

$0 \le y \le min\{y^*, \bar{y} + (\beta - \bar{r})(r - \bar{r})/2\alpha\}$ where $y^* = (\bar{r} - \beta)^2/4\alpha$. Then $(r(a), y(a))$ is in B whenever $a \ge a_0$ and $(r(a_0), y(a_0)) \in B$.

PROOF. It is clear that trajectories are trapped by the vertical and horizontal boundary segments of B if $r^* < r_0$. But $\bar{r} = r_0/2 + ((r_0^2/4) + (\alpha VG/\beta))^{\frac{1}{2}}$ and $r_0/2 > \beta$ so $(\bar{r} - \beta)^2 > (r_0^2/4) + (\alpha VG/\beta) > \alpha VG/\beta$. Hence $r^* = 4VG\alpha/(r - \beta)^2 < 4\beta \le r_0$. Now we must show that the vector field points into B along its boundary along $y = \bar{y} + \alpha^{-1}(\beta - \bar{r})(r - \bar{r})$. This amounts to verifying that $\frac{dy}{da}/\frac{dr}{da} \le (\beta - \bar{r})/2\alpha$ when $y = \bar{y} + \alpha^{-1}(\beta - \bar{r})(r - \bar{r})$ and $\bar{y} \le y \le y^*$. But on that locus

$$\frac{dy}{da}/\frac{dr}{da} = \frac{VG - ry}{\alpha y - \beta(r - r_0)} = \frac{\overline{ry} - \bar{r}(\bar{y} + [(\beta-\bar{r})(r-\bar{r})/2\alpha])}{\alpha(\bar{y} + [(\beta-\bar{r})(r-\bar{r})/2\alpha]) - \beta(r - r_0)}$$

$$= \frac{(\bar{r} - r)(\bar{y} + r(\beta - \bar{r})/2\alpha)}{\beta(\bar{r} - r_0) + [(\beta - \bar{r})(r - \bar{r})/2\alpha] - \beta(r - r_0)}$$

$$= \frac{2\bar{y} + \alpha^{-1}r(\beta - \bar{r})}{(\beta + \bar{r})}$$

From the description of the locus

$$y = \bar{y} + (\beta - \bar{r})(r - \bar{r})/2\alpha$$

$$r = 2\alpha(\beta - \bar{r})^{-1}(y - \bar{y}) + \bar{r}$$

$$\le 2\alpha(\beta - \bar{r})^{-1}(y^* - \bar{y}) + \bar{r}$$

since $y \le y^*$ and $\beta < r_0 \le \bar{r}$. So

$$\frac{dy}{da}/\frac{dr}{da} \le \frac{2\bar{y} + \alpha^{-1}(\beta - \bar{r})[2\alpha(\beta - \bar{r})^{-1}(y^* - \bar{y}) + \bar{r}]}{\beta + \bar{r}}$$

$$= \frac{2y^* + \bar{r}\alpha^{-1}(\beta - \bar{r})}{\beta + \bar{r}}$$

$$= \frac{[(\bar{r} - \beta)^2/2\alpha] + \bar{r}\alpha^{-1}(\beta - r)}{\beta + \bar{r}}$$

$$= \frac{\beta - \bar{r}}{2\alpha}$$

as desired.

Finally, we establish a proposition which should be useful in proving the uniqueness of nontrivial solutions and continuity of such solutions in their initial conditions. Expanding the notation (1.5), now define

$$F(y^0,r^0,g^0,G) = \int_0^\infty \mu e^{-\mu a} \; g(a)da \; . \tag{2.5}$$

As before, the question of existence is equivalent to solving $F(y^0,r^0,g^0,G) - G = 0$. To obtain a result for continuity in initial conditions, one would like to use an implicit function theorem. As a first step we have

THEOREM 2.6. $F(y^0,r^0,g^0,G)$ is continuously differentiable on the domain of y^0,r^0,g^0,G all positive.

PROOF. Rewrite F in the form of (1.6),

$$F(y^0,r^0,g^0,G) = \mu(g^0 + \gamma\int_0^\infty \mu e^{-\mu a} \; ye^{-rT}da)/(\delta + \mu)$$

where $y = y(y^0,r^0,g^0,G,a)$ and $r = r(y^0,r^0,g^0,G,a)$. Differentiating F involves the partial derivatives of y and r, all of which satisfy variational equations. For example

$$\begin{aligned} \frac{d}{da}\frac{\partial y}{\partial G} &= V - r\frac{\partial y}{\partial G} - y\frac{\partial r}{\partial G} \\ \frac{d}{da}\frac{\partial r}{\partial g} &= \alpha\frac{\partial y}{\partial G} - \beta\frac{\partial r}{\partial G} \; . \end{aligned} \tag{2.7}$$

Partial derivatives with respect to y^0,r^0,g^0 satisfy exactly analogous equations, but with V replaced by 0.

In every case, the following lemma applies to show that for fixed (y^0,r^0,g^0,G) the partial derivatives are bounded as functions of $a \geq 0$. A translation of coordinates, varying with a, is necessary for application to (2.7). Since (y^0,r^0,g^0,G) will be assumed fixed, we write $y = y(a)$, $r = r(a)$.

LEMMA 2.8. Solutions of the system

$$\frac{du}{da} = -ru - yv$$

(2.9)

$$\frac{dv}{da} = \alpha u - \beta v$$

exist and are bounded for all $a \geq 0$.

PROOF: It is evident from the boundedness of r and y that solutions exist for all $a \geq 0$, and that if a solution were to become unbounded, it would spiral counterclockwise around the origin, intersecting the positive u axis at arbitrarily large values. We will show that at such a crossing at sufficiently large u, a compact "box" can be constructed which the solution must enter and cannot leave. Refer to Figure 4.

For each $N > 0$ let $r_N = \inf r(a)$ and $y_N = \sup y(a)$ taken over $a \geq N$. Notice that r_N and y_N exist and $r_N > 0$. Then for large N, y_N/r_N is close to $\bar{y}/\bar{r}$ $((r(a),y(a)) \to (\bar{r},\bar{y})$ as $a \to \infty)$; and since $\alpha\bar{y} = \beta(\bar{r} - r_0)$ we have $y_N/r_N < \beta/\alpha$ for N larger than some N_0.

Now suppose a solution $(u(a),v(a))$ crosses the u-axis at some $a_0 \geq N_0$. We claim it is trapped in a "box" described in Figure 4. The abscissa of P_4 is $(\alpha y_{N_0}/\beta r_{N_0})^2 u(a_0)$ which is to the left of $u(a_0)$ because of the choice of N_0. Because of the choice of r_{N_0} and y_{N_0}, $\frac{du}{da} < 0$ for $a \geq N_0$ everywhere to the right of the line $y_{N_0} v = -r_{N_0} u$ and $\frac{du}{da} > 0$ for $a \geq N_0$ to the left of the line. Furthermore $\frac{dv}{da} < 0$ above $\beta v = \alpha u$ and $\frac{dv}{da} > 0$ below $\beta v = \alpha u$, for every a. Hence the box traps the solution after is passes through $(u(a_0),0)$, and the solution is therefore bounded.

LEMMA 2.10. The identically zero solution of the system (2.9) is globally uniformly asymptotically stable.

PROOF. We compare the a-varying system (2.4) with the constant-coefficient system obtained by setting $r = \bar{r}$, $y = \bar{y}$ in (2.9). In the latter case the eigenvalues have strictly negative real part, so there is a smooth Lyapunov function $V(u,v)$ such that $\dot{v}(u,v) \leq 0$ and equality holds only when $u = v = 0$. As in (1.8), for any $\varepsilon > 0$ there is a $c > 0$ such that all points of $V^{-1}([0,c])$ are within ε of the origin, and $\dot{V}$ is bounded away from 0 on each compact set $V^{-1}([c,d])$, say $\dot{V} < -\delta < 0$. By compactness and continuity, there is a number A such that $\nabla V(u,v)\cdot(-r(a)u - y(a)v, \alpha u - \beta v) < -\delta/2$ when $a \geq A$ and $V(u,v) \in [c,d]$. Hence there is a T such that if $V(u(a),v(a)) \leq d$ for $a \geq A$ then $||(u(a+t),v(a+t))|| < \varepsilon$ for $t \geq T$. The claim of the lemma is now immediate.

Returning to the proof of (2.6), we are now in a position to apply Corollary (1.9) to find that the map which takes initial data into the functions y, r, $\frac{\partial y}{\partial G}$, $\frac{\partial r}{\partial G}$, $\frac{\partial y}{\partial y^0}$, etc. are continuous. By a slight extension of the argument, we obtain the continuity of the map

$$\Sigma : \mathbb{R}^4 \to C^0_\infty(\mathbb{R}_+, \mathbb{R}^{10})$$

given by

$$\Sigma(y^0,r^0,g^0,G)(a) = \left(y, r, \frac{\partial y}{\partial y^0}, \frac{\partial r}{\partial y^0}, \frac{\partial y}{\partial r^0}, \frac{\partial r}{\partial r^0}, \frac{\partial y}{\partial g^0}, \frac{\partial r}{\partial g^0}, \frac{\partial y}{\partial G}, \frac{\partial r}{\partial G}\right)$$

where $y = y(y^0,r^0,g^0,G,a)$, $\frac{\partial y}{\partial y^0} = \frac{\partial}{\partial y^0} y(y^0,r^0,g^0,G,a)$, and so forth. We now show $\frac{\partial F}{\partial G}(y^0,r^0,g^0,G)$ exists and is continuous on positive values of the variables, leaving the similar derivations of the other derivatives to the reader. Setting $G = G_0 > 0$ and letting $f(y^0,r^0,g^0,G,a) = ye^{-rT}$ we have

$$\frac{\partial F}{\partial G}(y^0,r^0,g^0,G_0)$$

$$= \lim_{h\to\infty} \int_0^\infty e^{-\mu a} \frac{f(y^0,r^0,g^0,G_0+h,a) - f(y^0,r^0,g^0,G_0,a)}{h}\, da$$

Because of the continuity of Σ, the bracketed expression in the integral is bounded uniformly in $a \in [0,\infty)$ and $h \in [-\delta,\delta]$ for some $\delta > 0$. Hence the Lebesgue Dominated Convergence Theorem applies, and we find that

$$\frac{\partial F}{\partial G}(y^0,r^0,g^0,G) = \int_0^\infty e^{-\mu a-rT}\left(\frac{\partial y}{\partial G} - y\frac{\partial r}{\partial G}T\right)da\ ,$$

which is continuous in (y^0,r^0,g^0,G). This completes the proof of Theorem (2.6).

3. *Numerical Evidence and Further Questions*

The most basic unsettled question is that of uniqueness of nontrivial solutions. The numerical evidence of many computations is that for fixed values of (y^0,r^0,g^0), the graph of $F(G)$ has essentially one of the three shapes shown in Figure 5. Graph A seems to depict the case of V small and $y^0 = g^0 = 0$; graph B, the case of V large and $y^0 = g^0 = 0$; and graph C, that of y^0 or g^0 nonzero. Hence it seems reasonable to try to show $F'(g) < 1$ whenever $F(G) = G$ for $G > 0$. This would nicely give us the desired uniqueness and continuity results.

A natural method for proving existence of solutions would be the following. Pick $G_0 = g^0$ and let $G_{k+1} = F(G_k)$. One would hope that the corresponding solutions (y_k,r_k,g_k) of (0.2b,c) would converge to a solution of (0.2a,c). But the numerical evidence is to the contrary. It appears that for many values of V,y^0,r^0,g^0, there are distinct numbers G' and G'' (depending on $V,y^0,r^0,g^0)$ such that $G_{2k+1} \to G'$ and $G_{2k} \to G''$ as $k \to \infty$.

Finally, we consider the differential equation (0.2b) with z replacing VG. A solution (y_z,r_z,g_z) to the altered problem is a solution to the problem (0.2a,c) if and only if $z = V\mu\int_0^\infty e^{-\mu a}g_z(a)da$ or equivalently if and only if $V = z/\mu\int_0^\infty e^{-\mu z}g_z(a)da$. Since the solutions (y_z,r_z,g_z) are unique for each z, there also exists a unique V for each z such that (y_z,r_z,g_z) solves (0.2a,c). Using numerical methods the altered problem was solved using values of z ranging from .0001 to .09 and $V = z/\mu\int_0^\infty e^{-\mu a}g_k(a)da$ was plotted as a function of $\mu\int_0^\infty e^{-\mu a}g_z(a)da$. The graph exhibited interesting properties which have as far eluded proof as shown in Figure 6. Finally

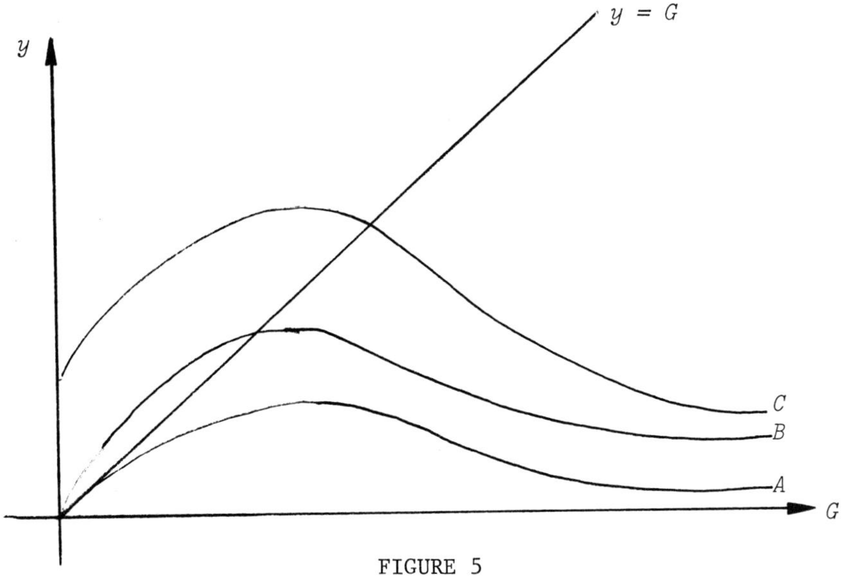

FIGURE 5

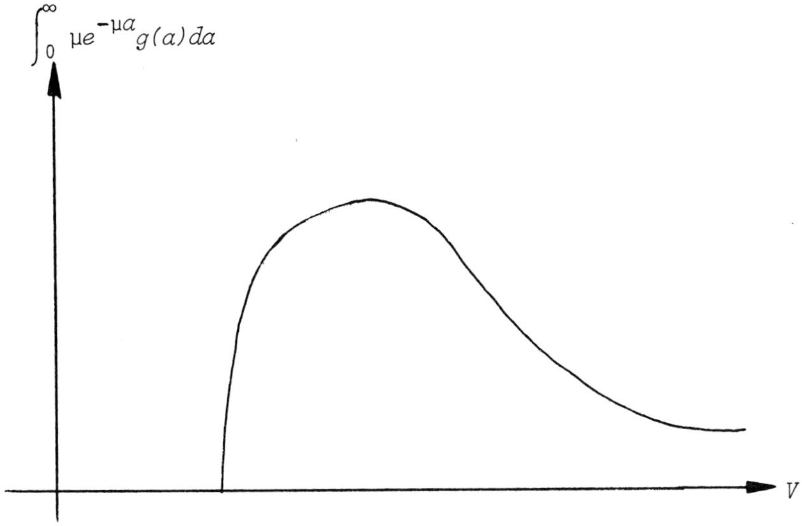

FIGURE 6

$V = z/\mu\int_0^\infty e^{-\mu a} g_z(a)\,da$ seems to be monotonic in z.

The authors have established an existence and uniqueness theorem for solutions to the partial differential equations (0.1a) with bounded, continuously differentiable boundary conditions of the form (0.1c). These results will appear elsewhere.

REFERENCES

[1] N. T. J. Bailey, *The mathematical theory of infectious diseases and its applications*, 2nd ed., Haffner Press, 1975, New York.

[2] K. Dietz, *Transmission and control of arbovirus diseases*, Epidemiology, "Proceedings of a SIMS Conference, D. Ludwig and K. L. Cooke, eds., SIAM, Philadelphia, 1975.

[3] A. Halanay, *Differential Equations*, Academic Press, New York, 1966.

[4] F. C. Hoppensteadt, *Mathematical theories of populations*, SIAM Regional Conference Series in Applied Mathematics, SIAM, Philadelphia, 1974.

[5] G. Oster, *The role of age structure in the dynamics of interacting population, "Mathematical Problems in biology"*, Lecture Notes in Biomathematics, Springer-Verlag, Berlin, 1974.

[6] F. W. Wilson, Jr., *The structure of the level surfaces of a Lyapunov function*, Jour. of Diff. Eq. 3(1967), 323-329.

ABSTRACT VOLTERRA EQUATIONS WITH INFINITE DELAY

W. E. Fitzgibbon
University of Houston

Introduction

In this paper we shall be concerned with the stability, representation, and approximation of a class of nonlinear Volterra equations which involve infinite delay and have a linear evolution operator as convolution kernel. More specifically we consider equations of the form:

$$x(\phi)(t) = W(t,\tau)\phi(0) + \int_{\tau}^{t} W(t,s)F(s,x_s(\phi))ds; \ \phi\varepsilon C \tag{1.1}$$

Our notation follows that of Hale [11] and Travis and Webb [13]. X denotes a Banach space and $W(t,s)$ is a linear evolution operator mapping X to X; C is a space of functions from the interval $(-\infty,0]$ to X which will be specified later; if u is continuous from an interval $(-\infty,b]$ to X then for $t \varepsilon [a,b]$, u_t is that element of C having pointwise definition $u_t(\theta) = u(t + \theta)$. If the linear evolution operator $W(t,s)$ is generated by a family of linear operators $\{A(t) | t \varepsilon [0,T]\}$ then equation (1.1) may be seen to provide a variation of parameters representation of solutions to the abstract functional differential equation

$$\begin{aligned} \dot{x}(\phi)(t) &= A(t)x(\phi)(t) + F(t,x_t(\phi)) \\ x_\tau(\phi) &= \phi \ \varepsilon\ C \end{aligned} \tag{1.2}$$

In the terminology of F. Browder [3] (1.2) may be considered a mild solution to (1.1). As a model of the type of problem to our theory may be applied we consider the following functional partial differential equation:

$$(1.3) \qquad \partial u(x,t)/\partial t = P(t,D)u(x,t) + \int_{-\infty}^{t} g(t-s)f(u(x,s))ds$$

$$x \in \Omega \subseteq R^n, \ t \in [0,T]$$

Here, $P(t,D)$ is generally a linear elliptic operator, g is L' and f is Lipschitz continuous. Equation (1.3) may also be viewed as a Banach space differential equation of the form (1.2) by using $P(t,D)$ to define a family of operators $\{A(t) | t \in [0,T]\}$ on a function space and using the integral to define a nonlinear delay term F (c.f. [3], [8], [13]).

Much recent work deals with abstract delay equations. Travis and Webb [13] apply the theory of nonlinear evolution equations to study finite delays in the case that the convolution kernel is a linear semigroup; a treatment of equations of the form (1.1) is contained in [9]. Dyson and Villella Bressan [6] apply the work of Crandall and Pazy [5] to associate product integrals with non-linear delay equations and thereby obtain stability criteria. The case of infinite delays is considered by Brewer [3] and the author [8].

2. The Evolution Operator

In this section we guarantee the existence of solutions to equation (1.1) and thereby define a nonlinear evolution operator $U(t,s)$. Before proceeding however we specify our space of initial functions. We consider two classes of initial functions. $C_u = C_u((-\infty,0],X)$ is the Banach space of uniformly continuous X valued functions endowed with the supremum norm $||\cdot||_{C_u}$.

We also consider the spaces of so called "fading memory" [1]. Let $r > 0$ and ρ be a positive nondecreasing function on $(-\infty,-r]$. We let C_ρ be the set of strongly measurable functions ψ from $(-\infty,0]$ to X such that ψ is continuous on $[-r,0]$ and ρ is integrable on $(-\infty,-r]$ C_ρ is a Banach space under the norm

$$(2.1)\qquad ||\psi||_{C_\rho} = \sup_{-r\leq\theta\leq 0} ||\psi(\theta)|| + \int_{-\infty}^{-r} \rho(\theta)||\psi(\theta)||d\theta$$

Our stability results shall be restricted to the space C_u. However, our existence and approximation results are valid in both C_u and C_ρ. When both spaces of initial functions are treated simultaneously we shall simply let C denote space of initial functions and $||\cdot||_C$ denote its norm.

It is now convenient to make precise our notion of a linear evolution operator.

<u>Definition</u>. A family of linear operators $\{W(t,\tau)|0 \leq \tau \leq t < \infty\}$ defined on a Banach space X is said to be a <u>linear evolution system of type E on X</u> provided the following are true:

i. $W(t,\tau)$ is jointly continuous in t and τ, $W(\tau,\tau) = I$ and $W(t,s)W(s,\tau) = I$ for $0 \leq \tau \leq s \leq t$.

ii. There exists a continuous real-valued function $\alpha(\cdot)$ defined on the nonnegative reals such that

$$(2.2)\qquad \cdot||W(t,\tau)|| \leq \exp(\int_\tau^t \alpha(s)ds)$$

If the nonlinear term is Lipschitz then the foregoing growth condition on $W(t,\tau)$ allows utilization of classical Picard techniques to obtain the following existence-uniqueness result.

<u>Proposition (2.1)</u>. Let $\{W(t,s)|0 \leq s \leq t < \infty\}$ be a linear evolution system of type E and suppose that $\beta(\cdot) : R \to R^+$ and $F : R^+ \times C \to X$ are continuous and satisfy,

$$||F(t,\phi) - F(t,\psi)|| \leq \beta(t)||\phi - \psi||_C \quad t > 0,\ \phi,\ \psi \in C.$$

If $\phi \in C$ and $\tau \geq 0$ there exists a unique $x(\phi) : (-\infty,\infty) \to X$ so that

$$(2.3)\qquad x(\phi)(t) = W(t,\tau)\phi(0) + \int_\tau^t W(t,s)F(s,x_s(\phi))dx; \quad t \geq \tau$$

$$x_\tau(\phi) = \phi.$$

We now associate a nonlinear evolution operator with solution to (2.3). If $x(\phi)(t)$ denotes the solution with initial condition $x_\tau(\phi) = \phi$. We define $U(t,\tau) : C \to C$ by the equation

$$U(t,\tau)\phi = x_t(\phi)$$

Thus for $0 \leq \tau \leq t$ we see that $U(t,\tau) : C \to C$. Examination of the integral equation and the uniqueness of solutions guarantee that $U(t,x)U(s,\tau)\phi = U(t,\tau)\phi$ for $0 \leq \tau \leq s \leq t$ and that $U(t,t) = I$ on C. Continuity properties of $U(t,\tau)$ may be deduced directly from the integral equation.

3. *Stability*

In this section we obtain a uniform stability result for the operator $U(t,\tau)$. Our stability results are limited to equations involving initial functions in C_u and are obtained by specification of the relationship between the growth rate of the linear evolution system and the Lipschitz bound on the nonlinear term.

If $\alpha(t) < 0$ and $sup\ -\beta(t)/\alpha(t) < \gamma$ and $x(\phi)(t)$ and $x(\psi)(t)$ denote solutions to (2.3) then straightforward computations yield the inequality:

$$\text{(3.1)} \qquad exp(\int_\tau^t -\alpha(s)dx)||x(\phi)(t) - x(\psi)(t)|| \leq ||\phi - \psi||_{C_u}$$
$$+ \int_\tau^t -\gamma\alpha(s)exp(\int_\tau^s -\alpha(u)du)||x_s(\phi) - x_s(\psi)||ds$$

$$\text{for } t > \tau.$$

We have the following stability result [8].

<u>Theorem 1</u>. Let $\{W(t,s)|0 \leq s \leq t\}$ and $F : R \times C_u \to X$ satisfy the conditions of Porposition 2.1 and suppose that for all $t \geq 0$, $\alpha(t) < 0$ and $\sup_{t>\tau} (-\beta(t)/\alpha(t)) \leq 1$. If $U(t,\tau)$ denotes the nonlinear evolution operator associated with solutions to (2.1) then for all $\phi, \psi \in C_u$ we have $||U(t,\tau)\phi - U(t,\tau)\psi||_{C_u} =$ $||\phi - \psi||_{C_u}$.

The proof of Theorem 1 consists of using (3.1) to estimate $||x(\phi)(t) - x(\psi)(t)||$; assuming that $||x_t(\phi) - x_t(\psi)||_{C_u} > ||\phi - \psi||_{C_u}$ and obtaining a contradiction by virtue of the following lemma:

Lemma 3.1. Suppose that $f(\cdot)$ is a continuous nonnegative function on the interval $[a,b]$. If t_0 denotes the maximum point of $f(\cdot)$ on $[a,b]$ and there exists a continuous nonnegative function $g(\cdot)$ on $[a,b]$ such that,

$$(exp \int_a^{t_0} g(s)ds)f(t\ ) \leq f(a) + \int_a^{t_0} g(s)exp(\int_a^s g(u)du)f(s)ds,$$

then $f(a) = \sup_{t\varepsilon[a,b]} f(t)$.

We remark that equality follows because ϕ and ψ are initial segments of $x_t(\phi)$ and $x_t(\psi)$. We can also prove an analogous result for well-posed totally nonlinear functional differential equations.

Theorem 2. Let $\{B(t)|t \varepsilon [0,\infty)\}$ be a family of nonlinear operators mapping Banach space X to itself and suppose that $\sigma : R \to R^+$ is a continuous function such that $B(t) + \sigma(t)I$ is dissipative for each $t \geq 0$. Let $F : R \times C_u \to X$ and $\beta(\cdot) : R \to R^+$ satisfy the conditions of Proposition (2.1) and assume $\sup_{t\geq 0} \beta(t)/\sigma(t) \leq 1$. If $x(\phi)(t)$ and $x(\psi)(t)$ satisfy the differential equation:

$$x'(t) = B(t)x(t) + F(t,x_t) \quad \text{for a.e } t \varepsilon [0,\infty)$$

with initial conditions $x_\tau(\phi) = \phi \varepsilon C_u$ and $x_\tau(\psi) = \psi \varepsilon C_u$ then

$$||x_t(\phi) - x_t(\psi)||_{C_u} = ||\phi - \psi||_{C_u} .$$

Proof. We begin by estimating the quantity $||x(\phi)(t) - x(\psi)(t)||$. Let j denote the duality map from X to 2^{X^*}. Using well known techniques [4] of differentiating the norm of vector-valued functions we see that for a.e t

(3.2) $$||x(\phi)(t) - x(\psi)(t)||\, d/dt\, ||x(\phi)(t) - x(\psi)(t)||$$

$$= Re\langle B(t)x(\phi)(t) + F(t,x_t(\phi)) - B(t)x(\psi)(t) - F(t,x_t(\psi)), f(t)\rangle$$

where $f(t) \in j(x(\phi)(t) - x(\psi)(t))$

Making use of the dissipativeness of $B(t) + \sigma(t)I$ and Lipschitz property of F we obtain from (3.2):

(3.3) $$d/dt||x(\phi)(t) - x(\psi)(t)|| \leq -\sigma(t)||x(\phi)(t) - x(\psi)(t)|| + \beta(t)||x_t(\phi) - x_t(\psi)||_{C_u}$$

Multiplying through by $exp \int_\tau^t \sigma(s)ds$ and integrating we obtain the inequality:

$$exp(\int_\tau^t \sigma(s)ds)||x(\phi)(t) - x(\psi)(t)|| \leq$$

$$||\phi - \psi||_{C_u} + \int_\tau^t \sigma(s)exp(\int_\tau^s \sigma(u)du)||x_s(\phi)-x_s(\psi)||_{C_u} ds.$$

Since we have obtained the same inequality as (3.1) $(\alpha(t) = -\sigma(t))$ we obtain our stability via Lemma 3.1.

4. An Infinitesimal Generator

In this section we define and describe a nonlinear operator $\hat{A}$ which maps C to C. For the remainder of the paper, C may be taken to be the space C_u or a space C_ρ. For convenience all computations will be made in C_u. If $T(t)$ is a strongly continuous linear semigroup acting on a Banach space X and satisfies a growth condition of the form:

(4.1) $$||T(t)|| \leq e^{\omega t} \text{ for some real } \omega$$

It is well known that $T(t)$ has a closed, linear, densely defined infinitesimal generator A. Moreover, $R(\lambda,A) = (A - \lambda I)^{-1}$ is a bounded linear operator with bound $(\lambda - \omega)^{-1}$ whenever $\lambda > \omega$.

If $F : C \to X$ is Lipschitz with constant β we define

$\hat{A} : C \to C$ by the equations:

$$\hat{A}\,\phi(\theta) = \dot{\phi}(\theta) \qquad \theta \leq 0 \tag{4.2}$$

$$D(\hat{A}) = \{\phi \in C | \dot{\phi} \in C,\ \phi(0) \in D(A) \text{ and } \dot{\phi}^{-}(0) = A\,\phi(0) + F(\phi)\}$$

In case we are in C_ρ the derivative must be taken a.e. on $(-\infty,-r]$. $\hat{A}$ is nonlinear by virtue of its nonlinear domain. In [8] $\hat{A}$ is shown to be densely defined and the infinitesimal generator of a nonlinear semigroup $U(t) : C \to C$ where $U(t)$ is the semigroup associated with autonomous equation:

$$x(\phi)(t) = T(t)\phi(0) + \int_0^t T(t - s)F(x_s(\phi))ds, \quad x_0(\phi) = \phi \in C.$$

It is further shown that if λ is chosen sufficiently small then $(I - \lambda\hat{A})^{-1}$ is everywhere defined and Lipschitz continuous with Lipschitz constant at most $1/(1 - \lambda(\omega + \beta))$ and that $U(t)$ has exponential representation.

$$U(t) = \lim_{n\to\infty} (I - t/n\hat{A})^{-1}\phi$$

We define an operator $B_\lambda : C \to C$ pointwise by the equation

$$B_\lambda\psi(\theta) = e^{\theta/\lambda}\int_\theta^0 e^{-s/\lambda}\psi(s)ds \quad \phi \in C, \quad \theta \in (-\infty,0].$$

B_λ bears the following relationship to $(I - \lambda\hat{A})^{-1}$.

$$(I-\lambda\hat{A})^{-1}\psi(\theta) = e^{\theta/\lambda}(I-\lambda A)^{-}\ (\psi(0) + \lambda F((I-\lambda\hat{A})^{-1}\phi)) + B_\lambda\psi(\theta) \tag{4.3}$$

We can use this representation to approximate $(I - \lambda\hat{A})^{-1}\phi$. Since A is an infinitesimal generator we can form the everywhere defined Lipschitz continuous Yosida approximates $A_n = A(I - 1/nA)^{-1}x$. Each A_n is the infinitesimal generator of a linear semigroup of type (4.1) and we can define $\hat{A}_n : C \to C$ by the equations

$$\hat{A}_n\phi(\theta) = \dot{\phi}(\theta) \qquad \theta \leq 0$$

$$D(\hat{A}_n) = \{\phi \in C \mid \dot{\phi} \in C,\ \dot{\phi}^{-}(0) \in A_n\phi(0) + F(\phi)\} \tag{4.4}$$

We now approximate $(I - \lambda\hat{A})\phi$.

Lemma 4.1. If λ is sufficiently small and $\phi \in C$ then $\lim_{n\to\infty} (I - \lambda\hat{A}_n)\phi = (I - \lambda\hat{A})^{-1}\phi$.

Indication of proof. The theory of linear semigroup guarantees the convergence of $(I - \lambda A_n)^{-1}x$ to $(I - \lambda A)^{-1}x$. This fact together with the representation of $(I - \lambda\hat{A})^{-1}\phi$ and $(I - \lambda\hat{A}_n)^{-1}\phi$ yield the desired convergence.

5. *Representation and Approximation*

This section provides a product integral representation for the nonlinear evolution operator associated with solutions to (2.1). This representation is then applied to insure the convergence of evolution operators associated with approximating equations. It will be necessary to place additional conditions on our nonlinear delay term and additional restrictions on the generators of our linear evolution system.

We place the following additional conditions on a time dependent family of closed densely defined linear operators.

(A.1) $D(A(t))$ is independent of t.

(A.2) There exists an real ω so that for all $t \in [0,T]$, $A(t)$ is the infinitesimal generator of a strongly continuous linear semigroup satisfying (4.1).

(A.3) There exists an $L > 0$ so that for all $x \in D(A(t))$ and $t, \tau \in [0,T]$, $||A(t)x - A(\tau)x|| \leq |t - \tau|L(||x||)$.

The foregoing conditions guarantee that $\{A(t)\ t \in [0,T]\}$ is the generator of a linear evolution system $\{W(t,s) \mid 0 \leq s \leq t \leq T\}$ on X having the properties:

$$||W(t,s)|| \leq e^{\omega t}, \quad W(t,s)x = \lim_{n\to\infty} \prod_{i=1}^{n} (I + \frac{t-s}{n}A(s + i(\frac{t-s}{n}))x^{-1}.$$

We place the following continuity requirements on $\boldsymbol{F}$; there exists a continuous $\beta : [0,T] \to R$ and an $M > 0$ so that:

(F.1) $||F(t,\phi)-F(t,\psi)|| \leq \beta(t)||\phi-\psi||_C$, $t \in [0,T]$, $\phi, \psi \in C$.

(F.2) $||F(t,\phi)-F(\tau,\psi)|| \leq |t-\tau|M||\phi||_C, \quad \phi \in C, \ t, \quad \tau \in [0,T].$

For $t \in [0,T]$ we define $\hat{A}(t) : C \to C$

$$\hat{A}(t)\phi(\theta) = \dot{\phi}(\theta) \qquad \theta \leq 0 \tag{5.1}$$

$$D(\hat{A}(t)) = \{\phi \mid \dot{\phi} \in C, \quad \phi(0) \in D(A(t)) \text{ and }$$
$$\dot{\phi}(0) = \bar{A}(t)\phi(0) + F(t,\phi)\}.$$

We obtain the following product integral.

Proposition (5.1). Let $\{A(t) \mid t \in [0,T]\}$ be family of closed densely defined linear operators on a Banach X which satisfy conditions (A.1) through (A.3) and suppose that $F : [0,T] \times C \to X$ satisfies (F.1) and (F.2). Then, there exists a nonlinear evolution system $\{V(t,\tau) \mid 0 \leq \tau \leq t \leq T\}$ mapping $C \to C$ such that for all ϕ and $t \in [0,T]$.

$$V(t,\tau)\phi = \lim_{n\to\infty} \Pi(I + \frac{t-\tau}{n}\hat{A}(t + i\frac{(t-\tau)}{n})^{-1}\phi. \tag{5.2}$$

The foregoing proposition is obtained by a minor modification of a result due to Dyson and Villella Bresson [6]. Its proof consists of showing that the operators $\hat{A}(t)$ satisfy the conditions of the Crandall-Pazy [5] result for the representation theorem for nonlinear evolution equations.

We now observe that $\{A_n(t) \mid t \in [0,T]\}$ $n \in Z^+$, the Yosida approximations of $\{A(t) \mid t \in [0,T]\}$ also satisfy condition (A.1) through (A.4). Thus we can define a sequence of operators $U_n(t,\tau) : C \to C$ by the equation:

$$U_n(t,\tau)\phi = \lim_{n\to\infty} \prod_{i=1}^{n} (I + \frac{t-\tau}{n}\hat{A}_n(\tau + i\frac{(t-\tau)}{n})^{-1}\phi. \tag{5.3}$$

Since $A_n(t)$ is continuous we have by virtue of Theorem 2. [3], $U_n(t,\tau)\phi(0)$ is the unique solution of the functional differential equation:

$$d/dt\, x^n(\phi)(t) = A_n(t)x^n(\phi)(t) + F(t,x^n_t(\phi)) \quad t \in \quad [0,T] \tag{5.4}$$

$$x_\tau^n(\phi) = \phi \in C.$$

and hence it is the unique solution of the integral equation

(5.5) $$x^n(\phi)(t) = W_n(t,\tau)\phi(0) + \int_\tau^t W_n(t,\tau)F(s,x_s^n(\phi))ds$$

where $W_n(t,\tau)$ is the linear evolution operator generated by $\{A_n(t) \mid t \in [0,T]\}$. We are now in a position to obtain our representation result.

Theorem 3. Let $\{A(t) \mid t \in [0,T]\}$ be a family of closed, densely defined linear operators which satisfy (A.1) through (A.3) and suppose that $F : R^+ \times C \to X$ satisfies (F.1) and (F.2). Then

(5.6) $$U(t,\tau)\phi = \lim_{n\to\infty} \Pi(I + \frac{t-\tau}{n} \hat{A}(\tau + i(t - \tau)/n))^{-1}\phi$$

exists for $t \in [0,T]$ and $\phi \in C$ and $U(t,\tau)\phi(0)$ provides the unique solution to the integral equation:

(5.7) $$x(\phi)(t) = W(t,\tau)\phi(0) + \int_\tau^t W(t,s)F(s,x_s(\phi))ds, \quad t \in [\tau,T]$$
$$x_\tau(\phi) = \phi.$$

Proof. Since $W(t,s)$ and $F : R \times C \to X$ satisfy the conditions of Proposition (2.1) we know that a unique solution to (5.7) exists and we can thereby define a nonlinear evolution operator $U(t,\tau)\phi = x_t(\phi)$. We further know that we can obtain a nonlinear evolution operator as the product integral (5.3). We need to argue that $V(t,\tau) = U(t,\tau)$. From Lemma (4.1) we see that for all $t \in [0,T]$ $\lim_{n\to\infty} (I - \lambda\hat{A}_n(t)^- \phi = (I - \lambda\hat{A}(t))^- \phi$ and it is readly verified that the operators $\{A_n(t) \mid t \in [0,T]\}$ and $\{\hat{A}(t) \mid t \in [0,T]\}$ satisfy the conditions of the Crandall-Pazy approximation theorem [5] and consequently $\lim_{n\to\infty} U_n(t,\tau)\phi = V(t,\tau)\phi$ in C uniformly for $t \in [0,T]$. Since $\lim_{n\to\infty} W_n(t,\tau)x = W(t,\tau)x$ we can take limits on each side of (5.5) to see that $V(t,\tau)\phi(0) =$

$U(t,\tau)\phi(0)$ and the equality of $U(t,\tau)$ and $V(t,\tau)$ immediately follows.

We conclude by applying this representation to obtain the following approximation result.

Theorem 4. For each $k \in Z^+$ let $\{A^k(t) \mid t \in [0,T]$ and $F^k : R^+ \times C \to X$ satisfy the conditions of Theorem 3 with ω, L, $\beta(t)$ and M independent of K. Also, let $\{A(t) \mid t \in [0,T]\}$ and $F : R^+ \times C \to X$ satisfy these conditions. If $U^k(t,\tau)$ and $U(t,\tau)$ denote the nonlinear evolution operators associated with

$$x^k(\phi)(t) = W_k(t,\tau)\phi(0) + \int_\tau^t W_k(t,s)F^k(s,x_s^k(\phi))ds$$

and

$$x(\phi)(t) = W(t,\tau)\phi(0) + \int_\tau^t W(t,s)F(s,x_s(\phi))ds$$

then $\lim_{k\to\infty} U^k(t,\tau)\phi = U(t,\tau)\phi$ uniformly for $t \in [0,T]$ provided that for sufficiently small λ, $x \in X$ and $t \in [0,T]$ $\lim_{k\to\infty} (I - \lambda A^k)x^{-1} = (I - \lambda A)^{-1}x$ and for $\phi \in C$, $\lim_{k\to\infty} F^k(t,\phi) = F(t,\phi)$.

Theorem 4 is proved by showing that the operators $\hat{A}^k(t) : C \to C$ satisfy the conditions of the Crandall-Pazy approximation theory and by establishing a resolvent approximation lemma similar to Lemma 4.1.

REFERENCES

[1] D. W. Brewer, *A nonlinear semigroup for a functional differential equation*, Dissertation, University of Wisconsin, 1975.

[2] H. Brezis, and A. Pazy, *Convergence and approximation of semigroups of nonlinear operators in Banach spaces*, J. Functional Analysis 9(1972), 63-64.

[3] F. Browder, *Nonlinear equations of evolution*, Ann. Math. 80 (1964), 485-523.

[4] M. Crandall, and T. Liggett, *Generation of semigroups of nonlinear transformations on general Banach spaces*, Amer. J. Math. 93(1971), 265-298.

[5] ______, and A. Pazy, *Nonlinear evolution equations in Banach spaces*, Israel J. Math 11(1972), 57-94.

[6] J. Dyson, and R. Villella Bressan, *Functional differential equations and nonlinear evolution operators*, Edinburgh J. Math. (to appear).

[7] W. Fitzgibbon, *Approximations of nonlinear evolution equations*, J. Math. Soc. Japan 25(1973), 211-221.

[8] ______, *Nonlinear Volterra equations with infinite delay*, (to appear).

[9] ______, *Stability for abstract nonlinear Volterra equations involving finite delay*, J. Math. Anal. Appl. (to appear).

[10] H. Flaschka, and M. Leitman, *On semigroups of nonlinear operators and the solution of the functional differential equation* $x(t) = F(x_t)$, J. Math. Anal. Appl. 49(1975), 649-658.

[11] J. Hale, *Functional Differential Equations*, Applied Math. Series, Vol. 3, Springer-Verlag, New York, 1971.

[12] V. Lakshmikantham, and S. Leela, *Differential and Integral Inequalities, Vol. II*, Academic Press, New York, 1972.

[13] C. Travis, and G. Webb, *Existence and stability for partial functional differential equations*, Trans. Amer. Math. Soc. 200(1974), 395-418.

[14] G. Webb, *Autonomous nonlinear functional differential equations*, J. Math. Anal. Appl. 46(1974), 1-12.

[15] ______, *Asymptotic stability for abstract nonlinear functional differential equations*, Proc. Amer. Math. Soc. (to appear).

[16] ______, *Functional differential equations and nonlinear semigroups in* L^p *spaces*, J. Diff. Equations (to appear).

THE INCIDENCE OF INFECTIOUS DISEASES UNDER THE INFLUENCE OF SEASONAL FLUCTUATIONS - ANALYTICAL APPROACH

Z. Grossman[1], I. Gumowski[1], and K. Dietz[2]

1. Introduction

A model, treated earlier numerically [1], describing in a simplified manner the dynamical behaviour of a system of susceptibles, infectives and immunes ($\underline{X}$, $\underline{Y}$ and $\underline{Z}$, respectively), is considered. The model assumes equal and constant (age independent) birth and death rates (μ), disregarding deaths caused by the infectious agent under consideration, so that the population size ($\underline{n} = \underline{X} + \underline{Y} + \underline{Z}$) remains constant. It is assumed that the number of "effective" contacts per unit time per individual susceptible with infectives is proportional to the number of infectives, the proportionality constant being β. After an effective contact, a susceptible becomes immediately infective himself and recovers at a rate γ. After recovery, an individual stays lifelong immune.

These assumptions lead to the following set of real-valued differential equations:

$$\frac{dX}{dt} = n\mu - (\beta Y + \mu)X,$$

$$\frac{dY}{dt} = \beta YX - (\gamma + \mu)Y, \tag{1}$$

$$\frac{dZ}{dt} = \gamma Y - \mu Z.$$

Two epidemiologically relevant quantities are the reproduction rate of the infection, $\underline{R} = \beta \underline{n}/(\gamma + \mu)$ and the infectious period $\underline{I} = 1/(\gamma + \mu)$.

It is now assumed that the contact rate β undergoes a simple harmonic oscillation with a period of one year.

[1] CERN, European Organization for Nuclear Research, 1211-Geneva 23, Switzerland.

[2] WHO, World Health Organization, 1211-Geneva 27, Switzerland.

Thus the constant β in (1) becomes:

$$\beta(t) = \beta_0 + \beta_1 \cos\omega t, \quad 0 \leq \beta_1/\beta_0 < 1. \tag{2}$$

The variations in β are introduced in order to represent seasonal variations of the contact rate β. The purpose of the present paper is to study this effect. Such variations are supposed to be responsible for the empirically found oscillations of the number of cases around the average endemic level in some infectious diseases, and in particular for the two-year period of measles in some large communities (see for example [2]).

For the mathematical discussion it is convenient to rewrite (1) in the form:

$$\frac{dX}{dt} = a - bX - \beta XY,$$

$$\frac{dY}{dt} = -cY + \beta XY, \tag{3}$$

$$\frac{dZ}{dt} = (c - b)Y - bZ,$$

where β is defined in (2) and all the variables and parameters are positive. The two sets of parameters are related by

$$a = n\mu, \quad b = \mu, \quad c = \gamma + \mu; \quad R = \beta_0 a/bc, \quad I = 1/c. \tag{4}$$

When $\beta_1 = 0$, the system (3) admits a constant equilibrium solution:

$$X(t) \equiv X_0 = A = c/\beta_0 > 0, \; Y(t) \equiv Y_0 = B = (\beta_0 a - bc)/\beta_0 c > 0,$$

$$Z(t) \equiv Z_0 = C = a/b - A - B, \quad \text{if} \quad R > 1.$$

The characteristic equation of $(\underline{X}_0, \underline{Y}_0)$ is

$$\lambda^2 + \delta\lambda + \nu^2 = 0, \; \nu^2 = \beta_0 a - bc = \beta_0^2 AB, \quad \delta = \beta_0 a/c. \tag{6}$$

The composite parameters ν and δ represent the internal (free) frequency and damping respectively. By inspection of (5) and (6) it is obvious that the equilibrium solution is asymptotically stable for all admissible values of the "primary" parameters a, b,

c and β_0. Since one never has $\underline{Re}(\lambda) = 0$, there exists no bifurcation solution originating from $(\underline{X}_0, \underline{Y}_0)$[3].

In this paper, periodic solutions of eq. (3) for $\nu/\omega \neq$ rational number as well as for $\nu/\omega \approx 1/2$ are established. In the latter case there appears a critical value β_{1c} of β_1 so that the periodic solutions are qualitatively different when $\beta_1 < \beta_{1c}$ and $\beta_1 > \beta_{1c}$, respectively. The value $\beta_1 = \beta_{1c}$ corresponds to a bifurcation. From a dynamic system point of view the periodic solutions for $\nu/\omega \neq$ rational, and those for $\nu/\omega \approx$ rational and $\beta_1 < \beta_{1c}$, appear as forced (or passive) oscillations around the static equilibrium (5). When $\nu/\omega \approx$ rational and $\beta_1 > \beta_{1c}$ they constitute a harmonic or subharmonic (parametric) resonance, developing from the passive oscillations at the critical excitation $\beta_1 = \beta_{1c}$.

The passive oscillations are described by means of converging series in powers of β_1 with periodic coefficients. The 1/2-subharmonic case is treated in three distinct stages: 1) identification of the nature of the bifurcation, ii) determination of β_{1c} and construction of a generating dynamic system and of its approximate fundamental solution, iii) determination of a regular perturbation scheme in powers of $\beta_1 - \beta_{1c} > 0$. The nature of the bifurcation was found to be related to the degree of degeneracy of a linear differential system with periodic coefficients. For the case $\nu/\omega \approx 1/2$ the system turns out to be singly-degenerate. This result was obtained by a combination of numerical and analytical methods. Knowing the presence of single-degeneracy and the critical value β_{1c}, the generating system and its approximate solution could be established by successive iterations of the Poincaré type. The perturbation scheme in iii) leads to a term proportional to $(\beta_1 - \beta_{1c})^{1/2}$ and a second one proportional to $\beta_1 - \beta_{1c}$.

The analytic approximations obtained are quite close to the numerical results in the case of passive oscillations and when $\beta_1 = \beta_{1c}$. The amplitude and phase of the 1/2-subharmonic oscillation agrees qualitatively with the numerically determined solutions.

The analysis of the passive oscillations is given in Section 2, the bifurcation of the 1/2-subharmonic is treated in Section 3, and its amplitude is defined in Section 4. The possible epidemiological relevance of the results is discussed in the conclusions, section 5.

2. *Passive Oscillations*

Consider the first two equations of (3). The nature of $\underline{X}(\underline{t})$, $\underline{Y}(\underline{t})$ implies that they can be sought in the form of the convergent series:

$$X(t) = A + \sum_{i=1}^{\infty} X_i(t)\beta_1^i, \quad Y(t) = B + \sum_{i=1}^{\infty} Y_i(t)\beta_1^i, \tag{7}$$

where $X_i(t)$, $Y_i(t)$ are undetermined periodic functions of period $2\pi/\omega$. Inserting (7) into (3) and identifying the coefficients of β_1^i, $i = 1,2,\ldots$, yields a linear recursive system. The first two terms are:

$$X_1(t) = A_1 \cos\omega t + B_1 \sin\omega t, \quad Y_1(t) = C_1 \cos\omega t + D_1 \sin\omega t,$$

$$A_1 = \beta_0 A^2 B\left(1 - \frac{\omega^2}{\nu^2}\right) - \frac{\delta AB}{\Delta_1}, \quad B_1 = \frac{\beta_0 A^2 B}{\omega\Delta}(\beta_0 B - \delta) - \frac{\omega AB}{\Delta_1}, \tag{8a}$$

$$C_1 = \frac{-\beta_0 B}{\omega} B_1, \quad D_1 = \frac{\beta_0 B}{\omega} A_1 + \frac{Ab}{\omega}, \quad \Delta_1 = \omega^2 \left(1 - \frac{\nu^2}{\omega^2}\right)^2 + \delta^2 > 0,$$

and

$$X_2(t) = \overline{X} + A_2 \cos 2\omega t + B_2 \sin 2\omega t, \quad Y_2(t) = \overline{Y} + C_2 \cos 2\omega t + D_2 \sin 2\omega t,$$

$$\overline{X} = \frac{-K_0}{\beta_0 B}, \quad \overline{Y} = \frac{(\delta - \beta_0 B)K_0}{\beta_0^2 AB}, \quad K_0 = \frac{\beta_0}{2}(A_1 C_1 + B_1 D_1) + \frac{1}{2}(BA_1 + AC_1), \tag{8b}$$

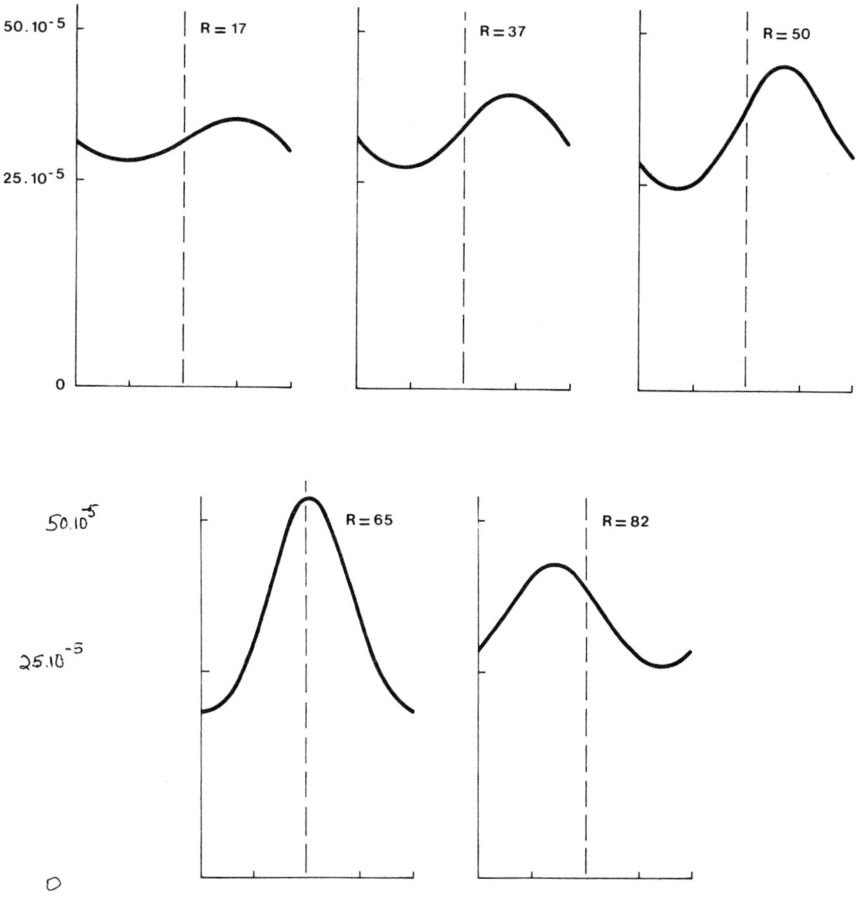

FIGURE 1

β_1/β_0 = 0.01 at different levels of *R* (*I* = 8.9 days) (reproduced from [1]).

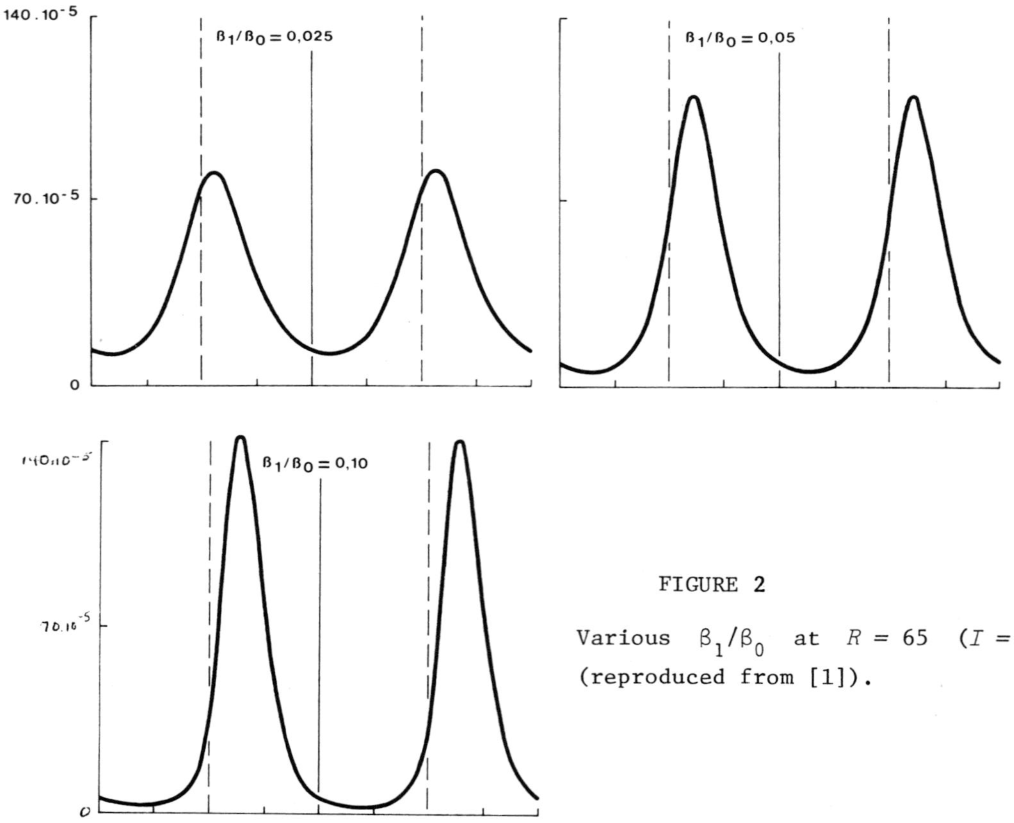

FIGURE 2

Various β_1/β_0 at $R = 65$ ($I = 8.9$ days) (reproduced from [1]).

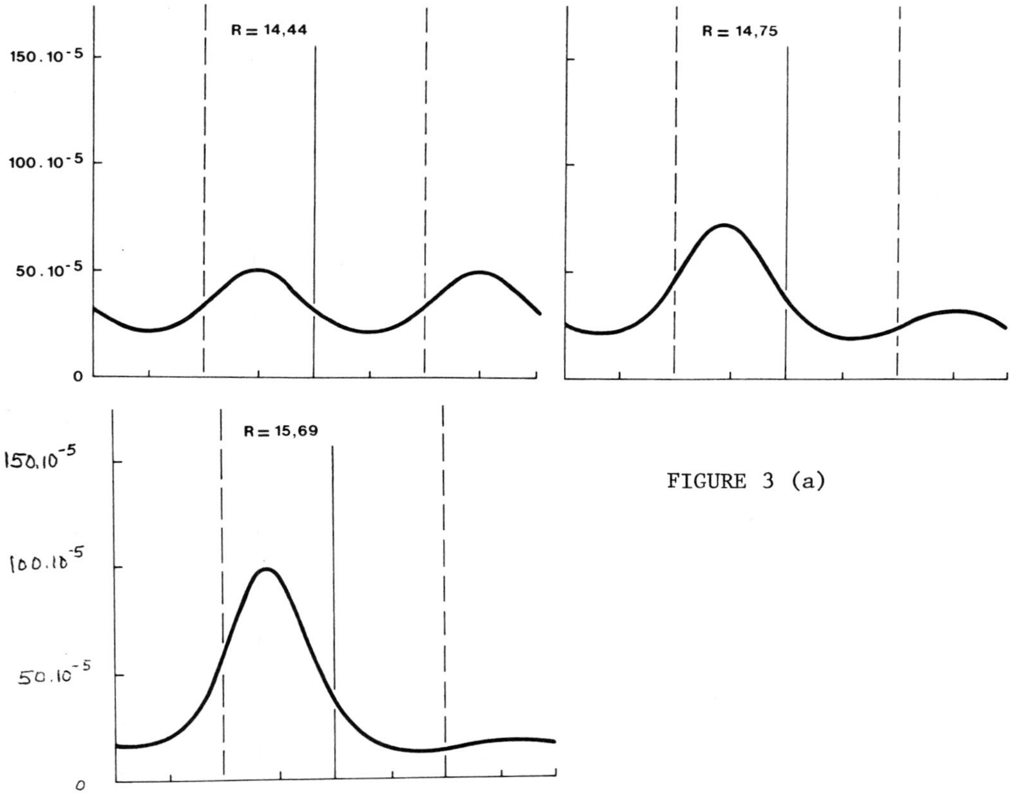
R = 14,44
R = 14,75
R = 15,69
150.10-5
100.10-5
50.10-5
0
150.10-5
100.10-5
50.10-5
0

FIGURE 3 (a)

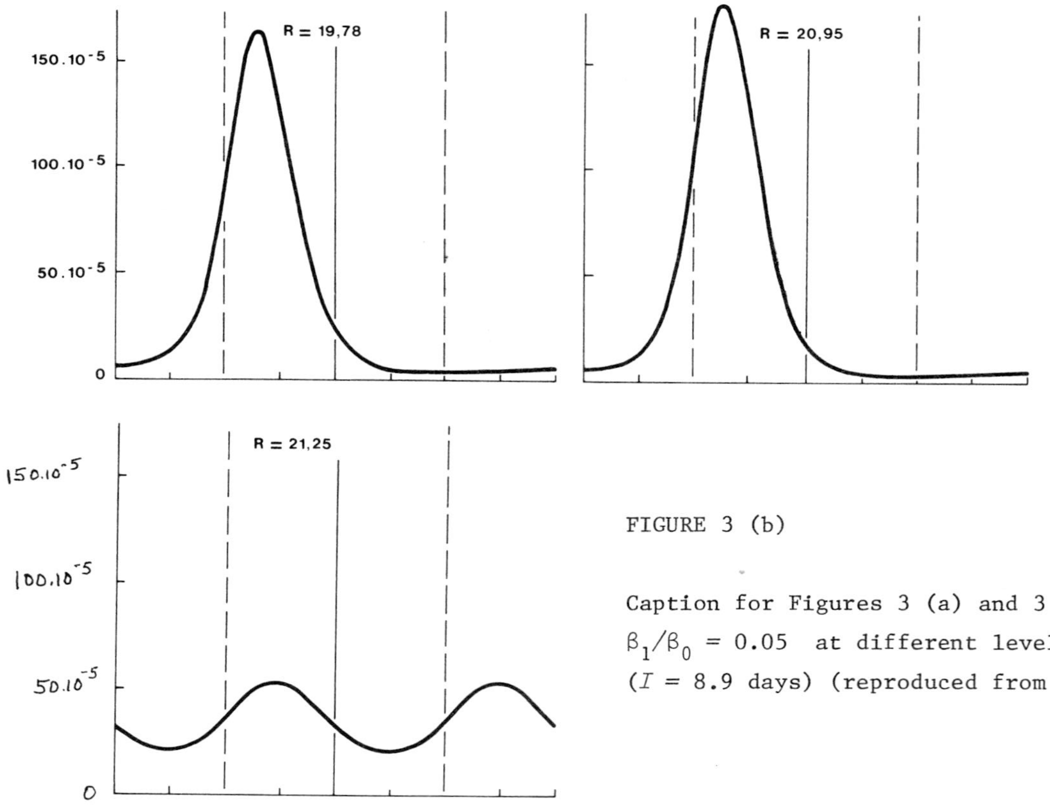

FIGURE 3 (b)

Caption for Figures 3 (a) and 3 (b)
$\beta_1/\beta_0 = 0.05$ at different levels of R
(I = 8.9 days) (reproduced from [1])

and $\underline{A}_2$, $\underline{B}_2$, $\underline{C}_2$, $\underline{D}_2$ are defined by the linear system (with non-vanishing determinant):

$$2\omega(1 + \frac{\nu^2}{4\omega^2})A_2 - \delta B_2 = \frac{\beta_0 A}{2\omega} K_1 + K_2,$$

$$\delta A_2 + 2\omega(1 + \frac{\nu^2}{4\omega^2}) B_2 = \frac{\beta_0 A}{2\omega} K_2 - K_1,$$

$$K_1 = \frac{1}{2}(BA_1 + AC_1) + \frac{\beta_0}{2} (A_1C_1 - B_1D_1),$$

$$K_1 = \frac{\beta_0}{2} (A_1D_1 + B_1C_1) + \frac{1}{2}(BC_1 + AD_1),$$

and the relations

$$C_2 = \frac{\beta_0 B}{2\omega} B_2 - \frac{K_2}{2\omega}, \qquad D_2 = \frac{-\beta_0 B}{2\omega} A_2 + \frac{K_1}{2\omega} .$$

The radius of convergence of the series in (7) is defined by the critical value β_{1c} (whenever a bifurcation arises).

A family of non-resonant phase-space solutions is illustrated in Figure 4. For the primary parameters chosen ($\underline{a} = 0.2$, $\underline{b} = 0.1$, $\underline{c} = 1$, $\beta_0 = 1$) one finds $\underline{A} = 1$, $\underline{B} = 0.1$, $\Delta_1 = 0.85$, $\underline{A}_1 = 0.0824$, $\underline{B}_1 = -0.129$, $\underline{C}_1 = -0.0.129$, $\underline{D}_1 = 0.108$, $\underline{\overline{X}} = -0.0064$, $\underline{\overline{Y}} = -0.01\underline{\overline{X}}$, and negligibly small $\underline{A}_2$, $\underline{B}_2$, $\underline{C}_2$, $\underline{D}_2$.

For comparison with the numerically computed solution a harmonic analysis of this solution was performed, and the error in the first three terms of (7) was found to be less than 1%. The scaling of Figure 4 shows the dominance of the first harmonic in the series.

Neglecting the contribution of harmonics higher than the first, the amplitude of the oscillations is:

$$A_x^2 = \beta_1^2(A_1^2 + B_1^2) = \left(\frac{\beta_1}{\beta_0}\right)^2 \left(\frac{\nu^2}{\beta_0}\right)^2 (1 + A^2\beta_0^2)[\omega^2\left(1 - \frac{\nu^2}{\omega^2}\right)^2 + \delta^2]^{-1},$$

$$A_y^2 = \beta_1^2(C_1^2 + D_1^2) = \left(\frac{\beta_1}{\beta_0}\right)^2 \left(\frac{\nu^2}{\beta_0}\right)^2 \left[1 + \frac{(\delta - \nu^2/A\beta_0)^2}{\omega^2}\right]\left[\omega^2\left(1 - \frac{\nu^2}{\omega^2}\right)^2 + \delta^2\right]^{-1}, \qquad (9)$$

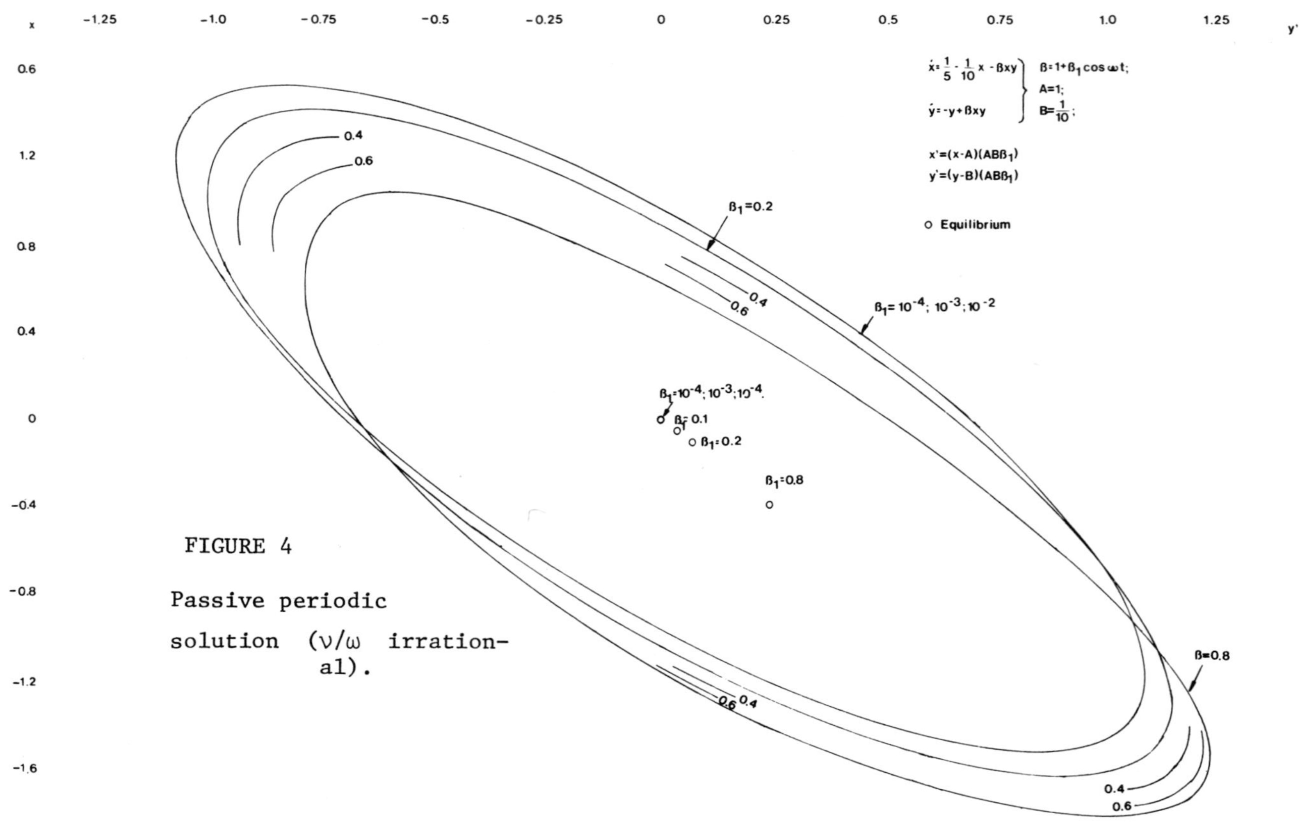

FIGURE 4

Passive periodic solution (ν/ω irrational).

$$A_x/A_y = (1 + A^2\beta_0{}^2/\omega^2)^{1/2}.$$

3. *Subharmonic Resonance*

The ratio ν/ω is assumed now to lie in the neighborhood of 1/2.

Without loss of generality it is possible to choose $\underline{A} = 1$ and $\omega = 1$. $\underline{X}$, $\underline{Y}$, $\underline{Z}$ then describe population densities (where $n = 2$) rather than the corresponding numbers, and the periods of 2π and 4π correspond to one and two years, respectively. Sample numerical solutions for the set of primary parameters $\underline{a} = 0.1$, $\underline{b} = 0.05$, $\beta_0 = \underline{c} = 5$, $\omega = 1$, implying $\underline{A} = 1$, $\underline{B} = 0.01$, $\underline{C} = 0.99$, $\nu/\omega = 1/2$ and the eigenvalues $\lambda_{1,2}(\underline{A}_1\underline{B}) = -0.05 \pm i0.5$, are shown in Figure 5. It can be seen that the passive oscillations are valid for a finite range of β_1 with a first harmonic dominance (dotted elipse). A bifurcation arises at $\beta_{1c} \approx 0.31$ with a transition from period $2\pi/\omega$ to $4\pi/\omega$ shown schematically in Figure 6. The "passive" elipse opens into two parts which close after one additional period. As $\beta_1 - \beta_{1c}$ increases the self-intersecting phase-plane curve develops a cusp and then becomes singly-connected. The presence of self-intersection represents the existence of two unequal maxima in the time representation of $\underline{X}$ and $\underline{Y}$ (the first examples in Figure 3).

The periodic solution at bifurcation is assumed to be the known passive one (eq. 8):

$$X(\beta_1 = \beta_{1c}) = \phi(t) = A + \beta_{1c}^2\bar{X} + \beta_{1c}(A_1 \cos\omega t + B_1 \sin\omega t) + \beta_{1c}^2(A_2 \cos 2\omega t + B_2 \sin 2\omega t),$$

$$Y(\beta_1 = \beta_{1c}) = \psi(t) = B + \beta_{1c}^2\bar{Y} + \beta_{1c}(C_1 \cos\omega t + D_1 \sin\omega t) + \beta_{1c}^2(C_2 \cos 2\omega t + D_2 \sin 2\omega t). \tag{10}$$

The omission of higher terms causes an error of less than 2%, as confirmed by the harmonic analysis of the directly computed solu-

FIGURE 5

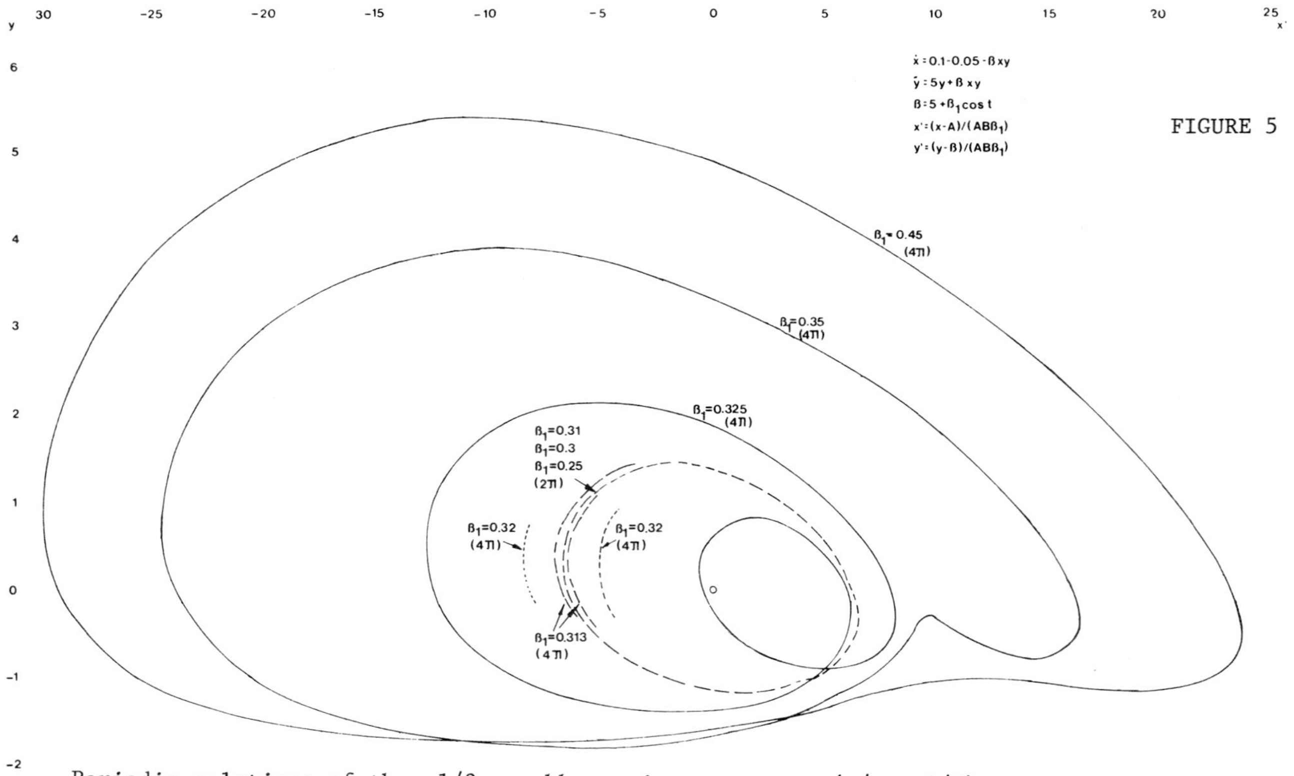

Periodic solutions of the 1/2 - subharmonic resonance ($\nu/\omega = 1/2$).

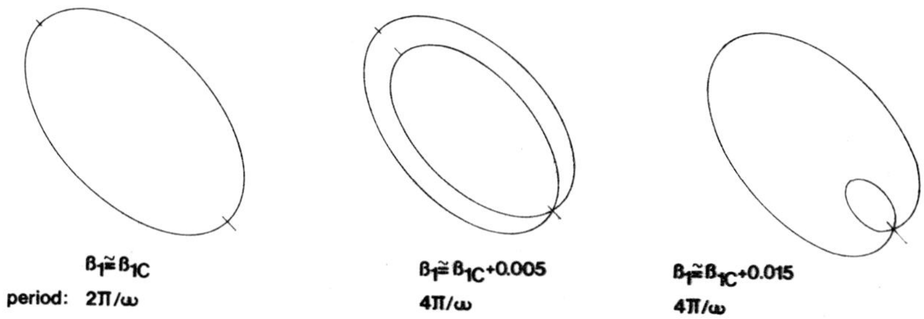

FIGURE 6

Schematic description of the development of the bifurcation.

tion.

Eliminating X from (3), the equations for Y and Z can be written:

$$\frac{dY(t)}{dt} = cY + \frac{a}{b}\beta Y - \beta(ZY + Y^2), \quad \frac{dZ}{dt} = (c - b)Y - bZ. \tag{11}$$

New variables x, y, z are defined by:

$$X = \phi(t) + x, \quad Y = \psi(t) + y, \quad Z = a/b - \phi(t) - \psi(t) + z. \tag{12}$$

Inserting (12) into (11) and omitting the nonlinear term $-\beta(t)(zy + y^2)$ yields:

$$\frac{dy}{dt} = [-c + \beta(t)(\phi(t) - \psi(t))]y - \beta(t)\psi(t)z,$$
$$\frac{dz}{dt} = (c - b)y - bz, \tag{13}$$

$$\text{with } \beta(t) = \beta_0 + \beta_1 \cos\omega t.$$

These are the variational equations of the dynamic system with respect to the passive solution at $\beta_1 = \beta_{1c}$.

The main question to be answered is whether the two characteristic numbers vanish simultaneously at bifurcation (double degeneracy) or not (single degeneracy) [4]. For a doubly-degenerate case the trace of (13), integrated over the period $2\pi/\omega$, should vanish:

$$\int_0^{2\pi/\omega} [(b + c) - \beta(t)(\phi(t) - \psi(t))]dt = 0, \tag{14a}$$

or equivalently,

$$\alpha(\beta_1) = \frac{\beta_0 a}{c} - \beta_1^2[\beta_0(\bar{X} - \bar{Y}) + \frac{1}{2}(A_1 - C_1)] = 0, \tag{14b}$$

where A_1, C_1, $\bar{X}$, $\bar{Y}$ are defined in (8). α_1 can be considered as an effective damping.

There is however no admissible β_1 which fulfills this condition. The small truncation error is unlikely to be responsible for this failure. Hence the possibility of single degeneracy is considered.

The two first-order equations in (13) can be transformed into a single second-order one. Eliminating y yields:

$$Lz = \frac{d^2z}{dt^2} + [b + c - \beta(t)(\phi(t) - \psi(t))]\frac{dz}{dt} + [bc - \beta(t)(b\phi(t) - c\psi(t))]z = 0. \tag{15}$$

Eq. (15) can be rewritten in the form

$$L_0 z = \frac{d^2z}{dt^2} + \frac{\omega^2}{4}z = [-(b + c) + \beta(t)(\phi(t) - \psi(t))]\frac{dz}{dt} + [\frac{\omega^2}{4} - bc + \beta(t)(b\phi(t) - c\psi(t))]z = f(t)\frac{dz}{dt} + g(t)z. \tag{16}$$

Since the left hand side admits a periodic solution of period $4\pi/\omega$, it is possible to regard $L_0 z = 0$ as the generating equation for Poincaré-type iterations.

The functions $f(t)$ and $g(t)$ in (16) can be separated into constant and zero-average parts:

$$f(t) = -\alpha(\beta_1) + \bar{f}(t), \quad g(t) = -\Delta(\beta_1) + \bar{g}(t), \tag{17a}$$

with α defined in (14b) and

$$\Delta(\beta_1) = \frac{\omega^2}{4} - \nu^2 + \beta_1^2\,[\beta_0(b\bar{X} - c\bar{Y}) + \frac{1}{2}(bA_1 - cC_1)] \tag{17b}$$

is the effective detuning.

The corresponding values at the bifurcation are

$$\alpha_c = \alpha(\beta_{1c}), \quad \Delta_c = \Delta(\beta_{1c}). \tag{17c}$$

Making the order of magnitude assignments

$$Z(t) = \sum_{i=0}^{\infty} Z_i(t)\varepsilon^i, \quad \alpha_c = \sum_{i=1}^{\infty} \alpha_i\varepsilon^i, \quad \Delta_c = \sum_{i=1}^{\infty} \Delta_i\varepsilon^i, \tag{18}$$

$$\bar{f}(t) = O(\varepsilon), \quad \bar{g}(t) = O(\varepsilon),$$

where ε is an auxiliary parameter, the generating solution ($\underline{i} = 0$) becomes:

$$z_0 = \gamma_0 \cos \frac{\omega}{2} t + \delta_0 \sin \frac{\omega}{2} t , \tag{19}$$

the coefficients γ_0 and δ_0 being unknown.

The absence of secular terms for $\underline{i} = 1$ leads, after some algebra, to two homogeneous equations for γ_0 and δ_0 (neglecting terms $O(\beta_1^3/\beta_0^3)$ and higher):

$$(\Delta_{c1} + R_1)\gamma_0 + (-\frac{\omega}{2}\alpha_{c1} + R_2)\delta_0 = 0,$$

$$(\frac{\omega}{2}\alpha_{c1} + R_2)\gamma_0 + (\Delta_{c1} - R_1)\delta_0 = 0,$$

with (20)

$$R_1 = bA - cB + \beta_0(bA_1 - cC_1) - \beta_0\frac{\omega}{2}(B_1 - D_1),$$

$$R_2 = \frac{\omega}{2}(A - B) + \beta_0(bB_1 - cD_1) + \beta_0\frac{\omega}{2}(A_1 - C_1).$$

For a non-trivial solution the determinant should vanish. If the contributions from $\underline{i} > 1$ to (18) are neglected, $\alpha_{1c} \approx \alpha_c$, $\Delta_{c1} \approx \Delta_c$, and the condition obtained is:

$$\omega^2\alpha_c^2 + 4\Delta_c^2 = \beta_{1c}^2(R_1^2 + R_2^2), \tag{21}$$

from which β_{1c} is determined.

For the parameter set used in Figure 5 $\beta_{1c} = 0.307$, in good agreement with the direct numerical result.

If $\delta_0 = k\gamma_0$, then from (20)

$$k = \frac{2\Delta_c + \beta_{1c}R_1}{\omega\alpha_c - \beta_{1c}R_2} . \tag{22}$$

Because of the strong cancellation in the denominator $\underline{k}$ is large, and it is difficult to determine its accurate numerical value which turns out to be sensitive to the exact value of β_{1c}.

Having eliminated the secular terms for $\underline{i} = 1$, one obtains:

$$z_1 = \gamma_1 \cos\frac{\omega}{2}t + \delta_1 \sin\frac{\omega}{2}t + k_1 \cos\frac{3\omega}{2}t + k_2 \sin\frac{3\omega}{2}t + k_3 \cos\frac{5\omega}{2}t + k_4 \sin\frac{5\omega}{2}t. \tag{23}$$

$\underline{k}_1, \underline{k}_2, \underline{k}_3, \underline{k}_4$ are determined directly as functions of $\underline{k}$ (eq. 22), and γ_1, δ_1 are fixed by the continuity of the series solution, requiring the initial condition $\underline{z}_1(\underline{t} = 0) = \dot{\underline{z}}_1(t = 0) = 0$.

The variational equations (13) at $\beta_1 = \beta_{1c}$ and the original ones (11) for $\beta_1 > \beta_{1c}$ were integrated numerically and analyzed harmonically. It was found that the phase relationships (the form of the periodic solutions) are approximately the same in both cases and change slowly with increasing β_1. The approximation for $\underline{z}_0 + \underline{z}_1$ is consistent with the numerically found results, and the dominance of the $\sin\frac{\omega}{2}\underline{t}$ - term is confirmed.

For the integration of the non-homogeneous recursive system resulting from eq. (15) (see the next section) the second linearly independent (damped) solution is also needed. This solution is of the form:

$$z(t) = e^{-\alpha_c t}\,\xi(t), \tag{24}$$

where ξ is a periodic function of period $4\pi/\omega$. Inserting (24) into (15) yields an equation for $\xi(t)$, analogous to (15). Repeating the procedure described above leads to

$$\omega^2\alpha_c^2 + 4\Delta_c^2 = \beta_{1c}^2[(R_1 - P_1)^2 + (R_2 - P_2)^2], \tag{25}$$

$$P_1 = \alpha_c[\beta_0(A_1 - C_1) + A - B], \quad P_2 = \alpha_c\beta_0(B_1 - D_1).$$

Since the additional term $\underline{P}_1^2 + \underline{P}_2^2 - 2\underline{R}_1\underline{P}_1 - 2\underline{R}_2\underline{P}_2$ vanishes the same value of β_{1c} is obtained, as required for consistency. The relations analogous to (22) are:

$$\overline{\delta}_0 = \overline{k}\gamma_0, \quad \overline{k} = \frac{2\Delta_c + \beta_{1c}R_1 - \beta_{1c}P_1}{\omega\alpha_c - \beta_{1c}R_2 + \beta_{1c}P_2}. \tag{26}$$

For the set of parameters used, $\underline{\overline{k}}$ is small, hence $\xi(t)$ is essentially proportional to $cos \frac{\omega}{2} \underline{t}$.

4. Subharmonic Resonance - the Amplitude

The full equation for $\underline{z}$ is obtained by adding $-\beta(t)(\underline{zy} +$ $\underline{y}^2)$ to the RHS of (15), with $\underline{y} = \frac{1}{(c - b)}(\frac{dz}{dt} + \underline{bz})$. The following magnitude assignments lead to a consistent iteration scheme:

$$z(t) = \sum_{i=1}^{\infty} z_i(t)\epsilon^i, \quad \rho = \beta_1 - \beta_{1c} = 0(\epsilon^2). \tag{27}$$

Since a few iterations are required, simplifications are made to reduce the amount of algebra involved.

$\underline{z}_0(\underline{t})$ and $\overline{z}_0(\underline{t})$ in (27) are approximated by the leading terms,

$$z(t) \approx \delta_0 \ sin \frac{\omega}{2} \ t, \quad \overline{z}_0(t) \approx e^{-\alpha_c t} \ \overline{\gamma}_0 \ cos \frac{\omega}{2} \ t. \tag{28}$$

This approximation depends on the validity of the relation $[\omega\alpha_c -$ $\beta_{1c}\underline{R}_2]/\omega\alpha_c << 1$ (eq. (22)). Varying the values of δ and ν, it was confirmed numerically that there is a certain region in parameter space around the point chosen in Section 4 for which this relation holds. The approximation (28) will thus be appropriate for this region.

The first iteration, $\underline{i} = 1$, leads to:

$$L z = \delta_0^2 \kappa \left[bc + \frac{\omega^2}{2} - (bc - \frac{\omega^2}{2}) \ cos\omega t + \frac{\omega}{2}(b + c) \ sin\omega t \right], \tag{29}$$

$$\kappa = \beta_0/2(c - b)^2.$$

A second simplification is now introduced by replacing the operator $\underline{L}$ by $\underline{L}_0 = \frac{d^2}{dt^2} + \frac{\omega^2}{4}$ in performing the integration of (29).

It is assumed that the small damping, detuning and time-dependent terms of $\underline{L}$ have a small effect on the response to the external strongly detuned (non-resonant) periodic excitations. The solution is:

$$z_1 \approx a_1 + b_1 \cos\omega t + c_1 \sin\omega t + d_1 \cos\frac{\omega}{2}t + \delta_1 \sin\frac{\omega}{2}t, \quad (30)$$

$$a_1 = \frac{-4}{\omega^2}\kappa(bc + \frac{\omega^2}{2}),\ b_1 = \frac{-4}{3\omega^2}\kappa(bc - \frac{\omega^2}{2}),\ c_1 = \frac{2}{3\omega}(b+c),$$
$$d_1 = \frac{16}{3\omega^2}\kappa(b + \frac{\omega^2}{4}),$$

where δ_1 is to be determined.

For $i = 3$,

$$Lz_2 \approx \rho\cos\omega t\left[(\phi(t) - \psi(t))\frac{dz_0}{dt} + (b\phi(t) - c\psi(t))z_0\right] - \beta_0(z_0 y_1 + y_0 z_1 + y_0 y_1),$$

$$y_1 = \frac{1}{(c - b)}\left(\frac{dz_1}{dt} + bz_1\right). \quad (31)$$

Requiring the vanishing of secular terms yields, after some algebra:

$$[(-bA + cB) + (bA_1 - cc_1)]\rho \approx E\delta_0^2, \quad (32)$$

$$E = \frac{-8\kappa^2}{\omega^2}\left[4bc(bc + \frac{\omega^2}{4}) + \frac{2}{3}(bc + \frac{\omega^2}{2})(bc - \frac{\omega^2}{4}) + \frac{\omega^2}{12}(b + c)^2\right].$$

Hence $\delta_0 \sim \sqrt{\rho}$. This rate of increase is too small (see Figure 5) even for quite small $\rho = \beta_1 - \beta_{1c}$. A third iteration is thus required.

Proceedings as before, the calculations give:

$$\frac{\rho\delta_1}{2}(-bA + cB + bA_1 - cC_1) - 2\kappa\delta_0\left[bc(2a_2 - b_2)\frac{\omega}{4}(b + c)c_2 - \frac{\omega^2}{2}b_2\right]$$

$$+ \frac{4\kappa^2}{3\omega^2}\delta_1\delta_0^2\left[6bc(bc + \frac{\omega^2}{4}) + (b^2 - 2bc - \frac{\omega^2}{2})(bc - \frac{\omega^2}{4})\right. \quad (34)$$

$$\left. + \frac{\omega^2}{4}c(b + c)\right] \cong \frac{64\kappa^3}{9\omega^4}\delta_0^4\left[-3\omega b(bc + \frac{\omega^2}{4}) - \frac{\omega}{2}c(bc - \frac{\omega^2}{4})\right.$$

$$\left. - (b^2 + 2bc + \frac{\omega^2}{2})\frac{\omega}{2}(b + c)\right](bc + \frac{\omega^2}{4})$$

with

$$a_2 = -\frac{8\kappa}{\omega^2}\,(bc + \frac{\omega^2}{4})\delta_0\delta_1,$$

$$b_2 = -\frac{8\kappa}{3\omega^2}\,(b_{1c} - \frac{\omega^2}{4})\delta_0\delta_1 + \frac{4\kappa}{3\omega}\,(b + c)\delta_0 d_1,$$

$$c_2 = \frac{4\kappa}{3\omega}\,(b + c)\delta_0\delta_1 + \frac{8\kappa}{3\omega^2}\,(bc - \frac{\omega^2}{4})\delta_0 d_1,$$

and d_1 given in (30).

Relation (33) takes finally the form:

$$F\delta_1 \approx G\rho, \tag{34}$$

where F and G are functions of b, c, β_0 and ω.

Numerically, there are strong cancellations in F, leading to an enhancement of δ_1. This, however, also renders the result sensitive to the details of the approximations. Specifically,

$$\delta \approx \delta_0 + \delta_1 \approx 0.1\sqrt{\rho} + s\rho, \tag{35}$$

where s is a constant considerably larger than 0.1.

This estimate is only in a qualitative and order-of-magnitude agreement with the numerical integration, for $\rho < 0.015$ (Figure 5).

5. *Conclusions*

Some explicit relationships between primary and secondary parameters are established in the preceeding sections. Epidemiologically, one is interested mainly in the dependence on two of the primary parameters which are more variable (and controllable) than the rest. These are the average contact rate, β_0, or equivalently the reproduction ratio R, and the strength of the seasonal variations, β_1.

Eq. (7) - (9) of Section 3 for the passive oscillations constitute a generalization of the approximation given by Bartlett as quoted in Bailey [5]. From eq. (9) we have, in the original notation (eq. (1)):

$$A_y^2 \cong \frac{\beta_1^2\mu^2\left(n - \frac{\gamma}{\beta_0}\right)^2}{\beta_0^2\left[\omega^2\left(1 - \frac{\beta_0\mu n}{\omega^2} + \frac{\gamma\mu}{\omega^2}\right)^2 + \frac{\beta_0\mu^2}{\gamma}\right]} \tag{36}$$

If $n \to \infty$, $\mu \to 0$, such that $\mu n \to \bar{\mu}$, where $\bar{\mu}$ is finite, then eq. (36) can be reduced to the formula (7.13) given in [5], if one takes into account that A_y has to be divided by Y_0 in order to get a comparable result. From eq. (7) and (8) one can deduce the contributions of higher harmonics and, specifically, the shift of the endemic average.

From eq. (21) one obtains, neglecting some small terms:

$$\beta_{1c}^2 \approx \frac{\omega^2\delta^2 + 4\left(\frac{\omega^2}{4} - \nu^2\right)^2}{R_1^2 + R_2^2} . \tag{37}$$

From the definitions $\delta = \beta_0 a/c$, $\nu^2 = \beta_0 a - bc$ and from (20), the dependence of β_{1c} on β_0 is established, for any choice of a, b and c. Generally, it is non-monotonous as it has a (local) minimum when $\nu \approx \omega/2$.

The approximations for the amplitude and the phase of the subharmonic resonance are valid only in the neighborhood of $\beta_1 = \beta_{1c}$. In this region the oscillations still possess a marked secondary maximum every second year (see the comments before eq. (10)). As it can be seen from the empirical data on measles [2], which clearly show the existence of such secondary maxima, this region is epidemiologically relevant.

When the model represented by eq. (3) is considered from a more general point of view, then a positive feature appears to be the presence of forced oscillations with and without resonance. This property is also found in other contexts (for example in ecology [6]). On the other hand, it has a negative feature: the sensitivity of results with respect to a small change of the form of the non-linearity in eq. (3). This sensitivity is a normal property of slightly modified Lotka-Volterra models, which constitute almost Hamiltonian dynamic systems. The "fragility" of results obtained on such a basis has also been discussed in other contexts (for example in population dynamics [7]). The sensitivity with respect to the choice of the precise form of the non-linearity may however constitute a characteristic feature of the evolution of an epidemic known to be rich in qualitative variety of possibilities.

REFERENCES

[1] Dietz, K., *The incidence of infectious diseases under the influence of seasonal variations*, to appear in Proceedings of a Workshop on Mathematical models in Medicine, Mainz, 8-9 March 1976, Lecture Notes in Biomathematics.

[2] London, W. P., and Yorke, J. A., *American J. of Epidemiology*, 98(1973), 453-482.

[3] Andronov, A. A., Leontovich, E. A., Gordon, I. I., and Maier, A. G., *Qualitative theory of second order dynamic systems*, Nauka, Moscow (1966).

[4] Malkin, I. G., *Theory of Stability of Motion*, US Atomic Energy Commission Translation Series (1952).

[5] Bailey, N. T. J., *The Mathematical Theory of Infectious Diseases and its applications*, (2nd edition) Griffin, London and High Wycombe (1975).

[6] Saunders, P. T., *Oscillations in tropical ecosystems*, *"Matters Arising"*, Nature 261(1976), 525.

[7] May, R. M., *Limit cycles in predator-prey communities*, Science 177(1972), 900-902.

SUM OF RANGES OF OPERATORS AND APPLICATIONS

Chaitan P. Gupta
Northern Illinois University

Let H denote a Hilbert space and let A,B be two operators on H. The purpose of this paper is to present an abstract analogue of Landesman-Lazer condition ([12]) for the solvability of a nonlinear functional equation $Au + Bu \ni f$. This leads to a study of the sum $R(A) + R(B)$ of the ranges $R(A)$, $R(B)$ of A and B respectively. It is well-known that, in general, $R(A) + R(B)$ is much larger than $R(A + B)$ (the range of $A + B$). It has been recently observed that $R(A) + R(B)$ is almost equal to $R(A + B)$ (in the sense that $Int\, [R(A) + R(B)] = Int\, R(A + B)$ and $cl[R(A) + R(B)] = cl\, R(A + B)$ when B is monotone and A and B satisfied certain additional conditions ([2], [3], [4]). The abstract analogue of Landesman-Lazer condition then turns out to be that $f \in Int[R(A) + R(B)]$. Our purpose is to extend this condition to the case when B is non-monotone. We apply these results to some nonlinear elliptic boundary value problems in section 2. Similar boundary value problems have been treated earlier by various methods in ([3] - [15]).

Section 1

Let H be a real Hilbert space. A subset G of $H \times H$ is said to be monotone if $(w_1 - w_2, u_1 - u_2) \geq 0$ for $[u_i,w_i] \in G$, $i = 1,2$. Let $A : H \to 2^H$ be a given mapping. We define the effective domain of A by $D(A) = \{u \in H \mid Au \neq \phi\}$, the range of A by $R(A) = \cup\{Au \mid u \in D(A)\}$ and the graph of A by $G(A) = \{[u,w] \mid u \in D(A),\, w \in Au\}$ $A : H \to 2^H$ is said to be monotone if $G(A)$ is a monotone subset of $H \times H$ and A is maximal monotone if $G(A)$ is maximal among all monotone subsets of $H \times H$, in the sense of inclusion. For a linear mapping $A : D(A) \subset H \to H$, the null-space of A is the subspace $N(A) = \{u \in H \mid Au = 0\}$ and the graph of A is the subspace $G(A) = \{[u,Au] \mid u \in D(A)\}$. A mapping $T : H \to H$ is said to be bounded if it maps bounded sets into bounded sets, it is compact if it maps bounded sets into relatively compact sets and it is demi-continuous if it is continuous from H into H endowed with weak-topology.

For a subspace F of H, the orthogonal complement $F^{\perp} = \{u \in H \mid (u,f) = 0$ for $f \in F\}$ and for a subset G of H we denote its interior by $Int\ G$, its closure by $cl\ G$.

DEFINITION 1: A mapping $T : H \to 2^H$ is boundedly-inversely-compact if for any pair of bounded subsets G, G' of H, the set $G \cap T^{-1}(G')$ is relatively compact in H. We may note that a maximal monotone mapping $T : H \to 2^H$ is boundedly-inversely-compact if and only if the resolvent mapping $(T + \lambda I)^{-1} : H \to H$ is compact for $\lambda > 0$.

THEOREM 1. *Let $A : H \to 2^H$ be a maximal monotone, boundedly-inversely-compact mapping and $B : H \to H$ a demi-continuous nonlinear mapping satisfying the following:*

"for every $k > 0$ there is a constant $C(k)$ such that

(1.1) $$(Bu,u) \geq k||Bu|| - C(k) \qquad \text{for } u \in H."$$

Then (i) $R(A) \subset cl\ R(A + B)$

(ii) $Int\ R(A) \subset R(A + B)$.

PROOF. It is easy to see using Leray-Schauder Principle that $R(A + B + \varepsilon I) = H$, for $\varepsilon > 0$. (See e.g. Proposition 1 of [8]).

(i) Let, now, $f \in R(A)$ so that $f \in Av$, for some $v \in H$. For $\varepsilon > 0$, let $u_\varepsilon \in H$ be such that

$$f \in Au_\varepsilon + Bu_\varepsilon + \varepsilon u_\varepsilon.$$

Subtracting we get $0 \in Au_\varepsilon - Av + Bu_\varepsilon + u_\varepsilon$ which gives, on multiplication by $u_\varepsilon - v$ that

(1.2) $$0 \geq (Bu_\varepsilon, u_\varepsilon - v) + \varepsilon(u_\varepsilon, u_\varepsilon - v).$$

Applying (1.1) with $k = ||v||$ we get a constant C such that

(1.3) $$(Bu_\varepsilon, u_\varepsilon - v) \geq (Bu_\varepsilon, u_\varepsilon) - ||v||\,||Bu_\varepsilon|| \geq -C.$$

(1.2) and (1.3) imply that

(1.4) $\varepsilon||u_\varepsilon||^2 \leq \varepsilon||u_\varepsilon|| \, ||v|| + C$ for $\varepsilon > 0$

(1.4) gives that $\sqrt{\varepsilon}||u_\varepsilon|| \leq const$ for $\varepsilon > 0$ so that $\varepsilon u_\varepsilon \to 0$ as $\varepsilon \to 0$. Hence $f \in cl\ R(A + B)$. This proves (i)

(ii) Let, now, $f \in Int\ R(A)$ so that there exists a $\rho > 0$ such that $h \in H$, $||h|| \leq \rho$ implies $f + h \in R(A)$. Now, for $h \in H$, $||h|| \leq \rho$, let $v_h \in H$ be such that

(1.5) $f + h \in Av_h$

and as in (i) $u_\varepsilon \in H$ be such that

(1.6) $f \in Au_\varepsilon + Bu_\varepsilon + \varepsilon u_\varepsilon$ for $\varepsilon > 0$.

Subtracting (1.5) from (1.6) we get that

$$-h \in Au_\varepsilon - Av_h + Bu_\varepsilon + \varepsilon u_\varepsilon.$$

This gives, on multiplication by $u_\varepsilon - v_h$ that

$$(-h, u_\varepsilon - v_h) \geq (Bu_\varepsilon, u_\varepsilon - v_h) + \varepsilon(u_\varepsilon, u_\varepsilon - v_h) \geq -Const(h).$$

using $\sqrt{\varepsilon}||u_\varepsilon|| \leq const$ for $\varepsilon > 0$ and condition (1.1) with $k = ||v_h||$. Hence

$$(h, u_\varepsilon) \leq Const.\ (h) \quad \text{for } \varepsilon > 0.$$

It then follows from the principle of uniform boundedness that

$$||u_\varepsilon|| \leq const. \quad \text{for } \varepsilon > 0.$$

Now, observing that condition (1.1) implies that B is a bounded mapping and A is boundedly inversely compact we get a sequence $\varepsilon_n \to 0$ and a $u \in H$ such that $u_{\varepsilon_n} \to u, B_{u_{\varepsilon_n}} \rightharpoonup Bu$ (weakly) and

$Au_{\varepsilon_n} \ni f - Bu_{\varepsilon_n} - \varepsilon u_{\varepsilon_n} \rightharpoonup f - Bu$ (weakly) as $\varepsilon_n \to 0$. It then follows from the maximal monotonicity of A that $f - Bu \in Au$, that is, $f \in Au + Bu$. Hence $f \in R(A + B)$. This proves (ii) and the proof of the theorem is complete.//

In applications, the maximal monotone operator A of Theorem 1 is usually of the form $A = A_1 + B_1$ where A_1, B_1 are maximal operators on H with $R(A_1) + R(B_1) \simeq R(A_1 + B_1)$ (almost equal in the sense that $Int[R(A_1) + R(B_1)] = Int\, R(A_1 + B_1)$ and $cl[R(A_1) + R(B_1)] = cl\, R(A_1 + B_1))$. Then the abstract analogue of Landesman-Lazer condition in our context becomes $Int[R(A_1) + R(B_1)] \subset R(A + B_1 + B)$. We now state a proposition summarizing conditions on A_1, B_1 which guarantee $R(A_1) + R(B_1) \simeq R(A_1 + B_1)$.

PROPOSITION 1: Let $A_1 : H \to 2^H$, $B_1 : H \to 2^H$ be maximal monotone operators on H with $A_1 + B_1$ maximal monotone. Suppose that one of the following conditions hold:

(i) $A_1 = \partial\phi$, $B_1 = \partial\psi$, ϕ, ψ proper, convex lower semi-continuous functions in H.

(ii) $D(A_1) \subset D(B_1)$, $B_1 = \partial\psi$, ψ a proper, convex lower-semi-continuous function on H.

(iii) $A_1 = \partial\phi$, ϕ convex lower semi-continuous function on H with $\phi((I + \lambda B_1)^{-1}u) \leq \phi(u)$ for every $u \in H$ and for every $\lambda > 0$

(iv) for $u \in D(A_1) \cap D(B_1)$ and $w_1 \in A_1 u$, $w_2 \in B_1 u$ $(w_1, w_2) \geq 0$

(v) for $u \in D(A_1)$ and $w \in A_1 u$ $(w, B_{1\lambda} u) \geq 0$ for every $\lambda > 0$ (where $B_{1\lambda}$ denotes the Yosida-approximation of B_1)

Then $R(A_1) + R(B_1) \simeq R(A_1 + B_1)$.

We omit the proof of the above Proposition which is due to Brezis [2] and [3] and Brezis-Haraux [4].

In case A is not monotone we have the following theorem.

THEOREM 2: Let $A : D(A) \subset H \to H$ be a closed, densely defined linear operator with $N(A) = N(A^*)$ and $\{u \in H \mid ||u|| \leq 1, ||Au|| \leq 1\}$ relatively compact in H.

Let $B_1 : H \to H$ be a monotone, demi-continuous mapping such that

(i) " $\delta > 0$ there exists a constant C_δ such that

$$(1.7) \quad (B_1 u_1 - B_1 v,\ u - v) \geq \frac{1}{\delta}||Bu||^2 - \delta||v||^2 - C_\delta \quad \text{for } u, v \in H."$$

and (ii) $\lim_{||u||\to\infty} \frac{||B_1u||}{||u||} = 0.$

Further, let $B_2 : H \to H$ be a demi-continuous mapping with $R(B_2)$ bounded in H and satisfying "for every $k > 0$ there is a constant $C(k)$ such that

(1.8) $(B_2u,u) \geq k||B_2u|| - C(k)$ for $u \in H.$"

Then $R(A) + R(B_1) \subset cl\, R(A + B_1 + B_2)$ and $Int[R(A) + R(B_1)] \subset R(A + B_1 + B_2).$

REMARK 1: The condition $\lim_{||u||\to\infty} \frac{||B_1u||}{||u||} = 0$ is necessary as can easily be seen in the case $H = \mathbb{R}$ by taking $Au = -ku (k > 0)$, $B_1u = ku$, $u \in \mathbb{R}$ and $B_2 \equiv 0$.

REMARK 2: In case B_1 is a subdifferential the assumption $\lim_{||u||\to\infty} \frac{||B_1u||}{||u||} = 0$ implies that B_1 satisfies condition (1.7) of Theorem 2. (see Lemma 15 of [3]).

REMARK 3: In case $B_2 = 0$, Theorem 2 is due to Brezis-Nirenberg ([3],[5]). For $B_2 \neq 0$ and $R(B_1)$ bounded, Theorem 2 is due to Gupta [7].

PROOF OF THEOREM 2: It follows easily from our assumptions that $R(A)$ is closed in H and $H = R(A) + N(A)$ and $R(A + B_1 + B_2 + \epsilon P_2) = H$ for $\varepsilon > 0$. (Here P_2 denotes the projection of H onto $N(A)$.) (see e.g. [3], [7]).

(i) Let, now, $f \in R(A) + R(B_1)$ so that

(1.9) $f = Av + B_1w$ for some $v, w \in H.$

For, $\varepsilon > 0$, let $u_\varepsilon \in H$ be such that

(1.10) $f = Au_\varepsilon + B_1u_\varepsilon + B_2u_\varepsilon + \varepsilon u_{2\varepsilon}, \; u_\varepsilon = u_{1\varepsilon} + u_{2\varepsilon}$

where $u_{1\varepsilon} \in R(A)$ and $u_{2\varepsilon} \in N(A)$. (1.10) implies that

(1.11) $$||Au_\varepsilon|| \leq const + ||B_1u_\varepsilon|| + ||B_2u_\varepsilon||$$

which in turn implies, in view of the fact that $A/R(A) \cap D(A)$ has a bounded inverse, that

(1.12) $$||u_{1\varepsilon}|| \leq C(const + ||B_1u_\varepsilon|| + ||B_2u_\varepsilon||)$$

Subtracting (1.9) and (1.10) and multiplying by $u_\varepsilon - w$ we get that

$$0 = \varepsilon||u_{2\varepsilon}||^2 + (Au_\varepsilon - Av, u_\varepsilon - w) + (B_1u_\varepsilon - B_1w, u_\varepsilon - w) + (B_2u_\varepsilon, u_\varepsilon - w) - \varepsilon(u_{2\varepsilon}, w)$$

$$\geq \varepsilon||u_{2\varepsilon}||^2 - \varepsilon||u_{2\varepsilon}||\,||w|| - C_1||B_1u_\varepsilon||^2 - C_2||B_2u_\varepsilon||^2 - C_3$$

$$+ \frac{1}{\delta}||B_1u_\varepsilon||^2 - \delta||w||^2 - C_\delta + (B_2u_\varepsilon, u_\varepsilon) - ||w||\,||B_2u_\varepsilon||$$

$$\geq \varepsilon||u_{2\varepsilon}||^2 - \varepsilon||u_{2\varepsilon}||\,||w|| + (\tfrac{1}{\delta} - C_1)||B_1u_\varepsilon||^2 - Const.$$

since $R(B_2)$ is bounded and B_2 satisfies (1.8) with $k = ||w||$, where C_i's are constants and $\delta > 0$.

We, now, take δ small enough to get

$$\varepsilon||u_{2\varepsilon}||^2 + ||B_1u_\varepsilon||^2 \leq \varepsilon||u_{2\varepsilon}||\,||w|| + Const.$$

This gives that $\varepsilon||u_{2\varepsilon}||^2 \leq const$

and $||B_1u_\varepsilon|| \leq const$ for $\varepsilon > 0$

We then have from (1.12) that $||u_{1\varepsilon}|| \leq const$ for $\varepsilon > 0$ since $R(B_2)$ is bounded.

Hence $\varepsilon u_{2\varepsilon} \to 0$ as $\varepsilon \to 0$ and so $f \in cl\, R(A + B_1 + B_2)$.

(ii) Let, now, $f \in Int\,[R(A) + R(B_1)]$ so that there exists $\rho > 0$ such that $f + h \in R(A) + R(B_1)$ for $h \in H$, $||h|| \leq \rho$. For $h \in H$, $||h|| \leq \rho$, let

(1.13) $$f + h = Av_h + B_1w_h$$ and as in (i)

for $\varepsilon > 0$, $u_\varepsilon \in H$ be such that

(1.14) $$f = Au_\varepsilon + B_1u_\varepsilon + B_2u_\varepsilon + \varepsilon u_{2\varepsilon}, \quad u_\varepsilon = u_{1\varepsilon} + u_{2\varepsilon}.$$

This gives

$$||u_{1\varepsilon}|| \leq const., ||B_1u_\varepsilon|| \leq const. \quad ||Au_\varepsilon|| \leq const.$$

and $\sqrt{\varepsilon}||u_{2\varepsilon}|| \leq const.$ as in (i). Subtracting (1.13) from (1.14) and multiplying by $u_\varepsilon - w_h$ we get that

$$(-h, u_\varepsilon - w_h) \geq \varepsilon(u_{2\varepsilon}, u_\varepsilon - w_h) + (Au_\varepsilon - Av_h, u_\varepsilon - w_h)$$
$$+ (B_2 u_\varepsilon, u_\varepsilon - w_h) \geq -const.(h)$$

in view of our assumptions and the above observations. It then follows from the Principle of Uniform Boundedness that $||u_\varepsilon|| \leq const.$ for $\varepsilon > 0$. Since, now, $\{u \in H \mid ||u|| \leq 1, ||Au|| \leq 1\}$ is relatively compact in H there exists a sequence $\varepsilon_n \to 0$ and a $u \in H$ such that

$$u_{\varepsilon_n} \to u \quad \text{as} \quad \varepsilon_n \to 0.$$

Then, $B_1 u_{\varepsilon_n} + B_2 u_{\varepsilon_n} \rightharpoonup B_1 u + B_2 u$ (weakly) in H and so, $Au_{\varepsilon_n} \rightharpoonup f - B_1 u - B_2 u$. Since A is closed we get that $f - B_1 u - B_2 u = Au$ and $f \in R(A + B_1 + B_2)$.

This completes the proof of the Theorem.//

Section 2

Let Ω be a bounded domain in a Euclidean space $\mathbb{R}^N (N \geq 1)$ with smooth boundary Γ. Let $g : \Omega \times \mathbb{R} \to \mathbb{R}$ be a function satisfying Caratheodory's conditions. Suppose that (i) there exist $a \geq 0$, $b \geq 0$ such that $|g(x,t)| \leq a + b|t|$ for $x \in \Omega$, $t \in \mathbb{R}$ and (ii) there is a $T > 0$ such that $g(x,t)\, t \geq 0$ for $|t| \geq T$, $x \in \Omega$. Let $g_+(x) = \liminf_{t \to +\infty} g(x,t)$ and $g_-(x) = \limsup_{t \to -\infty} g(x,t)$.

Let β be a maximal monotone graph in $\mathbb{R}^2$ with $0 \in \beta(0)$. For $t \in \mathbb{R}$, let $\beta^o(t) \in \beta(t)$ be the element with least absolute value in $\beta(t)$, where we set $\beta^o(t) = +\infty$ (resp. $-\infty$) in case $\beta(t) = \phi$ and $t > 0$ (resp. $t < 0$). Finally, let $\beta_{\pm} = \lim_{t \to \pm\infty} \beta^o(t)$ in the extended sense.

<u>THEOREM 3</u>: <u>Let $f \in L^2(\Omega)$ be such that</u>

(2.1) $$\text{meas}\,(\Gamma)\beta_- + \int_\Omega g_- dx < \int_\Omega f dx < (\text{meas}\,\Gamma)\beta_+ + \int_\Omega g_+ dx.$$

<u>Then the equation</u>

$$(2.2)\qquad \begin{cases} -\Delta u + g(x,u) = f & \text{a.e. on } \Omega \\ -\dfrac{\partial u}{\partial n} \in \beta(u) & \text{a.e. on } \Gamma \end{cases}$$

<u>has a solution</u> $u \in H^2(\Omega)$. (Here $\frac{\partial}{\partial n}$ denotes the outward normal derivative to Γ and Δ denotes the Laplacian).

We need the following proposition in the proof of Theorem 3.

<u>PROPOSITION 2</u>: Let $A : L^2(\Omega) \to 2^{L^2(\Omega)}$ <u>be defined by</u> $Au = -\Delta u$ <u>with effective domain</u> $D(A) = \{u \in H^2(\Omega) \mid -\frac{\partial u}{\partial u} \in \beta(u)$ a.e. on $\Gamma\}$. <u>Then</u> (i) A <u>is maximal monotone</u> (ii) <u>for</u> $f \in L^2(\Omega)$ <u>with</u>

$$(2.3)\qquad meas\ (\Gamma)\beta_- < \int_\Omega f dx < (meas\ \Gamma)\beta_+$$

<u>we have</u> $f \in Int\ R(A)$.

For a proof of Porposition 2 we refer to Brezis [1], Schatzman [13] and Gupta - Hess [8].

<u>PROOF OF THEOREM 3</u>: We decompose g into a sum $g = g_1 + g_2$ of functions on $\Omega \times \mathbb{R}$ satisfying Caratheodory's conditions such that

$$(2.4)\qquad \begin{cases} g_1(x,\cdot) \text{ is monotonically increasing on } \mathbb{R}, \\ g_1(x,0) \equiv 0, \quad \lim_{t\to\pm\infty} g_1(x,t) = g_\pm(x), \quad x \in \Omega \\ g_2(x,t)\, t \geq 0 \text{ for } |t| \geq T, \quad x \in \Omega \text{ and} \\ |g_i(x,t)| \leq c + d|t| \text{ for } x \in \Omega, \quad t \in \mathbb{R}, \quad i = 1,2. \end{cases}$$

Let $B_i : L^2(\Omega) \to L^2(\Omega)$ be given by $B_i u(x) = g_i(x,u(x))$ $(x \in \Omega,$ $i = 1,2)$. Then B_1 is a continuous, monotone, potential mapping. It then follows from Proposition 1 that $R(A) + R(B_1) \simeq R(A + B_1)$ (almost equal). Further $B_2 : L^2(\Omega) \to L^2(\Omega)$ is a continuous mapping satisfying condition (1.1) of Theorem 1. Indeed, for $u \in L^2(\Omega)$ we have

$$(B_2u,u) = \int_\Omega g_2(x,u(x))u(x)dx$$

$$= \int_{\{|u(x)|\leq T+1\}} g_2(x,u(x))u(x)dx + \int_{\{|u(x)|\geq T+1\}} g_2(x,u(x))u(x)dx$$

$$\geq \frac{1}{c+d} \int_{\{|u(x)|\geq T+1\}} g_2(x,u(x))|^2dx - const$$

$$\geq \frac{1}{c+d} ||B_2u||^2 - const$$

which immediately implies (1.1) since for every $\varepsilon > 0$, there is a $C(\varepsilon) > 0$ such that $p \leq \varepsilon p^2 + C(\varepsilon)$. That $A + B_1$ is boundedly inversely compact follows easily from the Sobolev embedding theorem and $||gradu||^2 \leq (Au,u) + (B_1u,u)$ for $u \in D(A)$. It, now, remains to verify that $f \in Int[R(A) + R(B_1)]$ to apply Theorem 1. Two cases arise:

Case 1: $\beta_- < \beta_+$. In this case we can write

$$\int_\Omega fdx = \lambda \; meas\; (\Gamma) + \mu$$

where λ and μ are such that $\beta_- < \lambda < \beta_+$ and $\int_\Omega g_-dx \leq \mu \leq \int_\Omega g_+dx$. Then there exists a $t_0 \in \mathbb{R}$ such that $\mu = \int_\Omega g_1(x,t_0)dx$. Writing $f = (f - \frac{1}{|\Omega|}\mu) + (\frac{1}{|\Omega|}\mu - g_1(x,t_0)) + g_1(x,t_0)$ we see that $f \in Int[R(A) + R(B_1)]$ since $(f - \frac{1}{|\Omega|}\mu) + (\frac{1}{|\Omega|}\mu - g_1(x,t_0)) \in Int\; R(A)$ by Proposition 2.

Case 2: $\beta_1 = \beta_+ = 0$. In this case condition (2.1) reduces to

$$\int_\Omega g_-dx < \int_\Omega fdx < \int_\Omega g_+dx .$$

Writing $k_f = \frac{1}{meas\ (\Omega)} \int_\Omega f dx$ we see that $f - k_f \in R(A)$ in view of the well-known existence theory for linear Neumann boundary value problem. Moreover, $k_f \in R(A) + R(B_1)$ as in case 1. Applying, the same decomposition to small perturbations of f, we conclude that $f \in Int[R(A) + R(B_1)]$.

This completes the proof of Theorem 3.//

In the next theorem, for reasons of simplicity we shall assume that $g(x,t)$ is independent of $x \in \Omega$, that is, $g : \mathbb{R} \to \mathbb{R}$ is a continuous function such that

(i) there exist a, $b \geq 0$ such that

$$|g(t)| \leq a + b|t|, \quad t \in \mathbb{R}.$$

(ii) there is a $T > 0$ such that $g(t)\, t \geq 0$ for $|t| \geq T$.

(iii) g can be decomposed as a sum of continuous functions $g = g_1 + g_2$ such that $g_1 : \mathbb{R} \to \mathbb{R}$ is monotonically increasing $g_1(0) = 0$, $|g_1(t)| \leq a + b|t|$ for $t \in \mathbb{R}$, $|g_2(t)| \leq const.$ for $t \in \mathbb{R}$. and $\lim_{t \to \pm\infty} g_1(t) = g_{\pm}$.

(iv) $\lim_{|t| \to \infty} \left|\frac{g(t)}{t}\right| = 0$.

Let $A : D(A) \subset L^2(\Omega) \to L^2(\Omega)$ be defined by $Au = -\Delta u - \lambda u$, $D(A) = H^2(\Omega) \cap H^1_0(\Omega)$, λ an eigen-value of $-\Delta$ (not necessarily the first eigen-value).

<u>THEOREM 4</u>: <u>Let</u> $f \in L^2(\Omega)$ <u>be such that</u>

$$\int_\Omega fv dx < \int_\Omega (g_+ v^+ - g_- v^-) dx \tag{2.5}$$

<u>for</u> $v \in N(A)$, $v \neq 0$ $(v^+ = max\ (v,0),\ v^- = -min(v,0))$. <u>Then the equation</u>

$$\begin{cases} -\Delta u - \lambda u + g(u) = f & \text{a.e. on } \Omega \\ u = 0 & \text{a.e. on } \Gamma \end{cases} \tag{2.6}$$

has a solution $u \in H^2(\Omega) \cap H_0^1(\Omega)$.

We need the following proposition in the proof of Theorem 4.

PROPOSITION 3: Let $g : \mathbb{R} \to \mathbb{R}$ be a continuous monotonically increasing function such that $\lim_{t\to\pm\infty} g(t) = g_\pm$. Let F be a finite-dimensional subspace of $L^2(\Omega)$ and let P be the orthogonal projection of $L^2(\Omega)$ onto F. Then for $f \in F$ such that for any $v \in F$, $v \neq 0$

$$\int_\Omega fv\,dx < \int_\Omega (g_+v^+ - g_-v^-)\,dx \tag{2.7}$$

there exists a $u \in F$ such that $Pg(u) = f$.

PROOF: We observe that for $v \in F$, $v \neq 0$

$$\int_\Omega (g_+v^+ - g_-v^-)\,dx = \lim_{t\to\infty} \int_\Omega g(tv)v\,dx.$$

It, then, follows from (2.7) that there exists an $R > 0$ such that

$$(Pg(v) - f, v) > 0 \quad \text{for} \quad ||v|| = R, \quad v \in F.$$

Applying Brouwer-degree theorem for continuous mappings on finite-dimensional spaces we then get a $u \in F$ such that $Pg(u) = f$. This completes the proof of the Proposition.//

PROOF OF THEOREM 4: Let $A : D(A) \subset L^2(\Omega) \to L^2(\Omega)$ be defined by $Au = -\Delta u - \lambda u$, $D(A) = H^2(\Omega) \cap H_0^1(\Omega)$, an eigen-value of $-\Delta$ (not necessarily the first eigenvalue). Then, it is well-known that A satisfies the conditions of Theorem 2 and $L^2(\Omega) = R(A) + N(A)$, with $N(A)$ finite-dimensional.

The mappings $B_i : L^2(\Omega) \to L^2(\Omega)$ defined by $B_iu(x) = g_i(u(x))$, $x \in \Omega$, $g = g_1 + g_2$ also satisfy conditions of Theorem 2. This is easily done similarly as in the proof of Theorem 3.

Let, now, P_1, P_2 denote the orthogonal projections of $L^2(\Omega)$ onto $R(A)$ and $N(A)$ respectively. Writing

$$f + h = f_{1h} + f_{2h} \quad \text{with} \quad f_{1h} \in R(A), \quad \text{and} \quad f_{2h} \in N(A)$$

and noting that $\int_\Omega (f + h)\, v dx = \int_\Omega f_{2h} v dx$, $v \in N(A)$ we see from Proposition 3 that there is a $u_h \in N(A)$ such that $f_{2h} = P_2 g_1(u_h)$. This gives that

$$f + h = f_{1h} + f_{2h} = f_{1h} + P_2 g_1(u_h)$$
$$= [f_{1h} - P_1 g_1(u_h)] + g_1(u_h) \in R(A) + R(B_1) \quad \text{for} \quad ||h|| \quad \text{small.}$$

Hence $f \in Int[R(A) + R(B_1) \subset R(A + B_1 + B_2)$ by Theorem 2 and the proof of the theorem is complete.

REMARK 4: If λ is the first eigen-value of $-\Delta$, we do not need to assume that $\lim_{|t|\to\infty} \left|\frac{g(t)}{t}\right| = 0$ and that $|g_2(t)| \leq const.$ for $t \in \mathbb{R}$, since in that case A is maximal monotone and Theorem 3 gives the existence of a solution for (2.6).

REFERENCES

[1] H. Brezis, *Monotonicity methods in Hilbert spaces and some applications to nonlinear partial differential equations*, Contributions to Nonlinear Functional Analysis, Editor, E. H. Zarantonello, Academic Press, 1971, pp. 101-156.

[2] H. Brezis, *Monotone operators, nonlinear semi-groups and applications*, Proc. Int. Cong. Math., Vancouver, 1974.

[3] H. Brezis, *Quelques proprietes des operateurs monotones et des semi-groupes nonlineaires*, Proc. Symp. NATO Conference Brussels, 1975, Springer Lecture Notes.

[4] H. Brezis, and A. Haraux, *Sur l'image d'une somme d'operateurs monotones et applications*, Israel J. Math. 23(1976), p.165-186.

[5] H. Brezis, and L. Nirenberg, *On some nonlinear operators and their ranges*, Ann. Sc. Norm. Sup. Pisa (to appear).

[6] L. Cesari, *Alternative Methods in Nonlinear Analysis*, Int. Conf. Diff. Equns., Academic Press, 1975, p. 95-148.

[7] C. P. Gupta, *Sum of ranges of operators and nonlinear elliptic boundary value problems*, (to appear).

[8] C. P. Gupta, and P. Hess, *Existence theorems for nonlinear noncoercive operators equations and nonlinear elliptic boundary value problems*, Jour. Diff. Equns., (to appear).

[9] P. Hess, *On semi-coercive nonlinear problems*, Indiana U. Math. J. 23(1974), p. 645-654.

[10] P. Hess, *On a theorem by Landesman-Lazer*, Indiana U. Math. J., 23(1974), p. 827-830.

[11] R. Kannan, *Existence of solutions of a nonlinear problem in potential theory*, Michigan Math. J. 21(1974), p. 207-213.

[12] E. Landesman, and A. Lazer, *Nonlinear perturbations of linear elliptic boundary value problems at resonance*, J. Math. Mech. 19(1970), p. 609-623.

[13] M. Schatzman, *Problemes anx limites nonlineaires, non-coercif*, Ann. Scoula. Norm. Sup. Pisa 27(1973), p. 641-686.

[14] M. Schechter, *A nonlinear elliptic boundary value problem*, Ann. Scoula. Norm. Sup. Pisa 27(1973), p. 707-716.

[15] S. Williams, *A sharp sufficient condition for solution of a nonlinear elliptic boundary value problem*, Jour. Diff. Equns. 9(1970), p. 580-586.

A VARIATION OF RAZUMIKHIN'S METHOD FOR RETARDED FUNCTIONAL DIFFERENTIAL EQUATIONS

Stephen R. Bernfeld, University of Texas at Arlington

and

John R. Haddock, Memphis State University

1. *Introduction*

Although Liapunov *functionals* are "theoretically" perhaps the natural way to study various stability properties of functional differential equations, Liapunov *functions* have played a prominent role in such investigations during the past two decades. The idea of employing Liapunov functions to functional differential equations was apparently first conceived by Razumikhin [6] in 1960. In 1962, Driver [4] helped to clarify some of the ideas of the Razumikhin method and introduced them to the non-Russian reader.

One of the basic techniques involved in this now classical method has been to examine inequalities in connection with the derivative of a Liapunov function. This derivative is often examined with respect to a certain subset C_0 of the general solution space C. This has been done in contrast to examining the entire set C itself.

The purpose of this paper is to employ a variation of the Razumikhin method in order to gain information concerning the behavior of solutions of functional differential equations. In particular, by exploiting the property that $\lim_{t\to\infty} ||x_t||$ exists for a solution $x(\cdot)$, we display conditions that guarantee the existence of the limits

$$\lim_{t\to\infty} |x(t)| \quad \text{and} \quad \lim_{t\to\infty} x(t).$$

Our method essentially requires knowledge of the derivative of a Liapunov function on a set "between" the standard sets considered.

For simplicity in exposition, the results are presented here in a finite delay setting. Proofs will be given in a subsequent paper that will allow for equations with infinite delay. However, it should be emphasized that extensions to the infinite delay setting must be handled with care.

2. *Notation*

Let $r \geq 0$ be given and let $C = C([-r,0],R^n)$ denote the space of continuous functions that map the interval $[-r,0]$ into R^n. For each $\phi \in C$, define $||\phi|| = \max_{-r\leq s\leq 0} |\phi(s)|$, where $|\cdot|$ denotes any convenient norm in R^n. Finally, if $x(\cdot)$ is a continuous function on $[-r,A)$, $A > 0$, x_t will denote an element in C defined by $x_t(s) = x(t + s)$, $-r \leq s \leq 0$, $0 \leq t < A$.

We consider the nonautonomous system of functional differential equations

$$(2.1) \qquad x' = f(t,x_t),$$

where (') denotes the right - derivative and $f : [0,\infty) \times C \to R^n$ is continuous.

By a *Liapunov* function, we mean a continuous, locally Lipschitzian function $V : [0,\infty) \times R^n \to [0,\infty)$. A Liapunov function V is said to be *monotonic* if $t_1 < t_2$ and $|x_1| < |x_2|$ together imply $V(t_1,x_1) < V(t_2,x_2)$. For each $t \geq 0$ and $\phi \in C$, the upper-right derivative of a Liapunov function V with respect to (2.1) is given by

$$(2.2) \qquad D^+_{(2.1)} V(t,\phi(0),\phi) =$$

$$\limsup_{h\to 0+} \frac{1}{h} [V(t+h,\phi(0) + hf(t,\phi)) - V(t,\phi(0))].$$

As was previously mentioned, we will be considering sets "between" the traditional sets considered. Define

$$C_0 = \{\phi \in C : ||\phi|| = |\phi(0)|\}.$$

Further, for each α, ε, $0 < \alpha < \varepsilon$, define

$$C(\alpha,\varepsilon) = \{\phi \in C : ||\phi|| - |\phi(0)| \leq \alpha, ||\phi|| \leq 2\varepsilon \ |\phi(0)| \geq \varepsilon\}.$$

Let Λ denote the set of all functions $\eta(\cdot)$ defined on $(0,\infty)$

such that $0 < \eta(\varepsilon) < \varepsilon$ for each $\varepsilon > 0$. It is easily seen that $\Lambda \neq \emptyset$. Now, for $\eta(\cdot) \in \Lambda$, define

$$C(\eta(\cdot)) = \bigcup_{\varepsilon>0} (C(\eta(\varepsilon),\varepsilon)) \cup \{0\}.$$

Finally, for a Liapunov function V and $\eta(\cdot) \in \Lambda$, define

$$C_0(V) = \{\phi \in C_0 : D^+_{(2.1)}V(t,\phi(0),\phi) \leq 0 \text{ for all } t \geq 0\};$$

$$C_0(\eta(\cdot),V) = \{\phi \in C(\eta(\cdot)) : D^+_{(2.1)}V(t,\phi(0),\phi)) \leq 0 \text{ for all } t \geq 0\}; \text{ and } C_f(\eta(\cdot),V) =$$

$$\bigcup_{\varepsilon>0} \{\phi \in C(\eta(\varepsilon),\varepsilon) : D^+_{(2.1)}V(t,\phi(0),\phi) \leq -\eta(\varepsilon)|f(t,\phi)|\} \cup \{0\}$$

Clearly, $C_0(V) \subseteq C_0(\eta(\cdot),V) \subseteq C_f(\eta(\cdot),V)$.

3. *The Main Result*

Using the previous notation, we can now give a concise statement of our main result.

THEOREM 3.1. Let V be a monotonic Liapunov function and $\eta(\cdot) \in \Lambda$. Then the following hold for any solution $x(\cdot)$ of (2.1):

(i) if $C_0(V) = C_0$, then $\lim_{t\to\infty} ||x_t||$ exists;

(ii) if $C_0(\eta(\cdot),V) = C(\eta(\cdot))$, then $\lim_{t\to\infty} |x(t)|$ exists;

(iii) if $C_f(\eta(.),V) = C(\eta(\cdot))$, then $\lim_{t\to\infty} x(t)$ exists.

PROOF. Let $x(\cdot)$ be any solution of (2.1). The proof of (i) follows from standard Razumikhin arguments. Likewise, the proof of (ii) is similar to that of (iii) and will be omitted here. So, assume $C_f(\eta(\cdot),V) = C(\eta(\cdot)$ and that $\lim_{t\to\infty} x(t)$ does not exist. From (i) it follows that $||x_t|| \to \alpha$ as $t \to \infty$ for some $\alpha > 0$. Let $\varepsilon > 0$ be chosen such that $\varepsilon < \alpha < 2\varepsilon$ and let $\beta_0 = \min\{\eta(\varepsilon)/3, (\alpha - \varepsilon)/3\}$. Since $\lim_{t\to\infty} x(t)$ does not exist, there exist $\beta > 0$ and sequences $\{t_n\}$, $\{t'_n\}$ tending monotonically to ∞ as $n \to \infty$ such that

$$|x(t_n) - x(t'_n)| = \beta \quad \text{and} \tag{3.2}$$

$$|x(t_n) - x(t)| \le \beta < \beta_0 \tag{3.3}$$

for $t'_n \le t \le t_n < t'_{n+1}$. Let T be chosen sufficiently large that $|||x_t|| - \alpha| \le \eta(\varepsilon)/3$, $||x_t|| \le 2\varepsilon$, $|x(t)| \ge \varepsilon$ and $|\alpha - |x(t)|| \le \eta(\varepsilon)/3$ for $t \ge T$. This can be accomplished since $|x(t)|$ tends to α from (ii). Now, choose N large such that t'_n, $t_n \ge T$ for $n \ge N$. Then, for $n \ge N$ and $t \in [t'_n, t_n]$, we have

$$|||x_t|| - |x_t(0)|| \le |||x_t|| - \alpha| + |\alpha - |x(t_n)|| + ||x(t_n)| - |x(t)||$$
$$\le |||x_t|| - \alpha| + |\alpha - |x(t_n)|| + |x(t_n) - x(t)| \le \eta(\varepsilon).$$

Thus, $x_t \in C(\eta(\cdot))$ and, therefore,

$$V'(t, x_t(0), x_t) \le -\eta(\varepsilon)|f(t, x_t)|$$

for $n \ge N$ and $t \in [t'_n, t_n]$. It follows that

$$V(t_n, x(t_n)) - V(t'_n, x(t'_n)) \le \int_{t'_n}^{t_n} V'(t, x_t(0), x_t)dt$$
$$\le -\eta(\varepsilon) \int_{t'_n}^{t_n} |f(t, x_t)|dt \le -\eta(\varepsilon)\left|\int_{t'_n}^{t_n} f(t, x_t)dt\right|$$
$$= -\eta(\varepsilon)|x(t_n) - x(t'_n)| = -\eta(\varepsilon)\beta.$$

Therefore, $\limsup_{n \to \infty} [V(t_n, x(t_n)) - V(t'_n, x(t'_n))] \le -\eta(\varepsilon)\beta < 0$, a contradiction to the monotonicity of V and the fact, from (ii), that $\lim_{n\to\infty} |x(t_n)| = \lim_{n\to\infty} |x(t'_n)| = \alpha$. This concludes the proof.

As was mentioned, this result can also be given in an "infinite delay" setting. However, the notation is slightly more cumbersome and the proof will be given elsewhere. Likewise, there are interesting examples for the finite delay case that also warrant this restriction here.

In [2], the authors studied the scalar equation

$$(3.1) \qquad x' = -a(t)x^{\partial}(t) + b(t)x^{\partial}(t - r),$$

where $a(\cdot)$ and $b(\cdot)$ are continuous, $a(t) \geq 0$ and $|b(t)| \leq a(t)\theta$ for all $t \geq 0$ and some $0 \leq \theta < 1$ and $\partial > 0$ is the quotient of odd integers. In [2], the authors were able to improve some asymptotic stability results of Winston [7] related to (3.1). One of the motivations of the theorem presented here was to arrange the ideas of [2] into a more general setting. The results here are also related to a recent paper of the second author [5]. Upon examining the various applications of the above theorem (cf. [1]), it would appear at a glance that it is essentially an asymptotic stability result. However, this turns out not to be the case as can be seen by the simple, scalar example

$$(3.2) \qquad x' = -2x(t)G(x_t) + x(t - r)G(x_t),$$

where $G(\phi) \overset{def}{=} \left[\int_{-r}^{0} \phi(s)ds - \phi(0)\right]^2$. It can be shown that (iii) (hence (ii) and (i)) are satisfied in Theorem 3.1. However, any constant function is a solution so asymptotic stability cannot be concluded.

A further examination of the applications of Theorem 3.1 indicates that there is usually an "ordinary" part of the differential equation that dominates a "functional" part. Unfortunately, we cannot as yet dispell this rumor. Thus, this leads to an interesting question: What happens if a functional differential equation has an ordinary and functional part, but the ordinary part does not dominate the functional part? As a starting point, what can be said about the equation

$$(3.3) \qquad x' = -x^{\partial}(t) + x^{\partial}(t - r)$$

where $\partial > 0$ is the quotient of odd integers? It is easy to see that for any such ∂, any constant function is a solution. For $\partial \geq 1$, one can often construct a Liapunov functional and employ an invariance principle to show that each solution tends to a constant as $t \to \infty$. Also, a clever result of Cooke and Yorke [3] can be employed to obtain this result. However, for $\partial < 1$ neither technique seems to work. After having made several attempts at examining the case $\partial < 1$, we are lead to the following conjecture.

CONJECTURE. Each solution of the scalar equation

$$x' = -x^{1/3}(t) + x^{1/3}(t - r) \tag{3.4}$$

tends to a constant as $t \to \infty$.

The authors would certainly be interested in seeing the above conjecture proved or disproved.

ACKNOWLEDGEMENT: The second author would like to express appreciation to G. Hinson and K. Perkins for several valuable discussions.

REFERENCES

[1] S. R. Bernfeld, and J. R. Haddock, *A variation of the Liapunov-Razumikhin method applied to nonlinear functional differential equations with finite and infinite delay*, (in preparation).

[2] ______, *Liapunov-Razumikhin-functions and convergence of solutions of scalar functional differential equations.*

[3] K. L. Cooke, and J. A. Yorke, *Some equations modelling growth processes and gonorrhea epidimics*, Math. Biosci. 16(1973), 75-101.

[4] R. D. Driver, *Existence and stability of solutions of a delay-differential system*, Arch. Rat. Mech. Anal., 10(1962), 401-426.

[5] J. R. Haddock, *Asymptotic behavior of solutions of nonlinear functional differential equations in Banach space*, Trans. Amer. Math. Soc., to appear.

[6] B. S. Razumikhin, *Application of Liapunov's method to problems in the stability of systems with a delay*, Avtomat. i. Telemch., 21(1960), 740-748.

[7] E. Winston, *Asymptotic stability for ordinary differential equations with delayed perturbations*, SIAM J. Math. Anal., 5(1974), 303-308.

A DEGENERATE PROBLEM OF SINGULAR PERTURBATIONS

V. Dragan and A. Halanay
Bucharest University

Motivated by a linear-quadratic optimization problem in the case of small time constants, the following singular perturbations problem is considered:

$$\varepsilon \frac{dP}{dt} = -\varepsilon\,[A_{11}^*(t)P + PA_{11}(t) - PM(t)P + F_{11}(t)] - $$

$$-A_{21}^*(t)Q^* - QA_{21}(t)$$

$$\varepsilon \frac{dQ}{dt} = -\varepsilon\,[A_{11}^*(t)Q + PA_{12}(t) - PM(t)Q + F_{12}(t)]$$

$$-A_{21}^*(t)R - QA_{22}(t)$$

$$\varepsilon \frac{dR}{dt} = -\varepsilon\,[A_{12}^*(t)Q + Q^*A_{12}(t) - Q^*M(t)Q + F_{22}(t)] - $$

$$- A_{22}^*(t)R - RA_{22}(t)$$

$$P(T) = G_{11},\quad Q(T) = G_{12},\quad R(T) = G_{22}$$

$$A_{ij} : [t_0,T] \to L(R^{n_i},R^{n_j}),\quad F_{ij} : [t_0,T] \to L(R^{n_i},R^{n_j})$$

F_{11}, F_{22} symmetric, $M : [t_0,T] \to L(R^{n_1},R^{n_1})$, $M(t)$ symmetric, G_{ij} $n_i \times n_j$ matrices, G_{11}, G_{22} symmetric. The degenerate character of the problem is expressed in the fact that for $\varepsilon = 0$ we get no information on P. We shall denote solutions of the problem by $P(t,\varepsilon)$, $Q(t,\varepsilon)$, $R(t,\varepsilon)$ and study the behavior of these functions for $\varepsilon \to 0$.

A fundamental assumption in what follows is that for all $t \in [t_0,T]$ the eigenvalues of $A_{22}(t)$ lie in $Re\, z \leq -\alpha < 0$.

We assume that A_{ij}, F_{ij}, M are Lipschitz and the main result is that

$$P(t,\varepsilon) = P_0(t) + \tilde{P}_0\left(\frac{T-t}{\varepsilon},\varepsilon\right) + \varepsilon\,\tilde{P}_1\left(\frac{T-t}{\varepsilon},\varepsilon\right) + O(\varepsilon)$$

$$Q(t,\varepsilon) = \tilde{Q}_0\left(\frac{T-t}{\varepsilon},\varepsilon\right) + \varepsilon\,\tilde{Q}_1\left(\frac{T-t}{\varepsilon},\varepsilon\right) + O(\varepsilon^2)$$

$$R(t,\varepsilon) = \tilde{R}_0\left(\frac{T-t}{\varepsilon},\varepsilon\right) + \varepsilon\,\tilde{R}_1\left(\frac{T-t}{\varepsilon},\varepsilon\right) + O(\varepsilon^2)$$

where $\lim_{\varepsilon\to 0} \tilde{P}_0\left(\frac{T-t}{\varepsilon},\varepsilon\right) = 0$

$$\lim_{\varepsilon\to 0} \tilde{Q}_0\left(\frac{T-t}{\varepsilon},\varepsilon\right) = 0, \quad \lim_{\varepsilon\to 0} \tilde{R}_0\left(\frac{T-t}{\varepsilon},\varepsilon\right) = 0,$$

$$\lim_{\varepsilon\to 0} \tilde{R}_1\left(\frac{T-t}{\varepsilon},\varepsilon\right) = L(t),$$

$$\lim_{\varepsilon\to 0} \tilde{Q}_1\left(\frac{T-t}{\varepsilon},\varepsilon\right) = -P_0(t)A_{12}(t)A_{22}^{-1}(t) - F_{12}(t)A_{22}^{-1}(t) - A_{21}^*(t)L(t)A_{22}^{-1}(t)$$

$$\left|\tilde{P}_1\left(\frac{T-t}{\varepsilon},\varepsilon\right)\right| \le k_0 \quad \text{for all} \quad 0 < \varepsilon < \varepsilon_0, \quad t_0 \le t \le T;$$

P_0 is the solution of the matrix Riccati equation

$$\frac{dP_0}{dt} = -\,[A_{11}^*(t) - A_{21}^*(t)A_{22}^{*-1}(t)A_{12}^*(t)]P_0 - P_0[A_{11}(t) - A_{12}(t)A_{22}^{-1}(t)A_{21}(t)] + P_0M(t)P_0 - F_{11}(t) + A_{21}^*(t)A_{22}^{*-1}(t)F_{12}^*(t) + F_{12}(t)A_{22}^{-1}(t)A_{21}(t) - A_{21}^*(t)A_{22}^{*-1}(t)F_{22}(t)A_{22}^{-1}(t)A_{21}(t)$$

with the condition

$$P_0(T) = G_{11} - A_{21}^*(T)A_{22}^{*-1}(T)G_{12}^* - G_{12}A_{22}^{-1}(T)A_{21}(T) +$$

$$+ A_{21}^*(T)A_{22}^{*-1}(T)G_{22}A_{22}^{-1}(T)A_{21}(T)$$

which we assume defined on all of $[t_0,T]$. $L(t)$ is defined by the Liapunov equation

$$A_{21}^*(t)L(t) + L(t)A_{22}(t) = - F_{22}(t).$$

The functions $\tilde{P}_0$, $\tilde{Q}_0$, $\tilde{R}_0$, $\tilde{P}_1$, $\tilde{Q}_1$, $\tilde{R}_1$ are defined by the equations

$$\tilde{P}_0'(\tau,\varepsilon) = A_{21}^*(T - \varepsilon\tau)\tilde{Q}_0^*(\tau,\varepsilon) + \tilde{Q}_0(\tau,\varepsilon)A_{21}(T - \varepsilon\tau)$$

$$\tilde{Q}_0'(\tau,\varepsilon) = \tilde{Q}_0(\tau,\varepsilon)A_{22}(T - \varepsilon\tau) + A_{21}^*(T - \varepsilon\tau)\tilde{R}_0(\tau,\varepsilon)$$

$$\tilde{R}_0'(\tau,\varepsilon) = A_{22}^*(T - \varepsilon\tau)\tilde{R}_0(\tau,\varepsilon) + \tilde{R}_0(\tau,\varepsilon)A_{22}(T - \varepsilon\tau)$$

$$\tilde{P}_1'(\tau,\varepsilon) = A_{11}^*(T - \varepsilon\tau)\tilde{P}_0(\tau,\varepsilon) + \tilde{P}_0(\tau,\varepsilon)A_{11}(T - \varepsilon\tau) +$$

$$+ P_0(T - \varepsilon\tau)M(T - \varepsilon\tau)P_0(T - \varepsilon\tau) -$$

$$- [P_0(T - \varepsilon\tau) + \tilde{P}_0(\tau,\varepsilon)]M(T - \varepsilon\tau)[P_0(T - \varepsilon\tau)+\tilde{P}_0(\tau,\varepsilon)] +$$

$$+ P_0(T - \varepsilon\tau)A_{12}(T - \varepsilon\tau)A_{22}^{-1}(T - \varepsilon\tau)A_{21}(T - \varepsilon\tau) +$$

$$+ A_{21}^*(T - \varepsilon\tau)A_{22}^{*-1}(T - \varepsilon\tau)A_{12}^*(T - \varepsilon\tau)P_0(T - \varepsilon\tau) +$$

$$+ A_{21}^*(T - \varepsilon\tau)A_{22}^{*-1}(T - \varepsilon\tau)F_{12}^*(T - \varepsilon\tau) +$$

$$+ F_{12}(T - \varepsilon\tau)A_{22}^{-1}(T - \varepsilon\tau)A_{21}(T - \varepsilon\tau) -$$

$$- A_{21}^*(T - \varepsilon\tau)A_{22}^{*-1}(T - \varepsilon\tau)F_{22}(T - \varepsilon\tau)A_{22}^{-1}(T - \varepsilon\tau)A_{21}(T - \varepsilon\tau) +$$

$$+ A_{21}^*(T - \varepsilon\tau)\tilde{Q}_1^*(\tau,\varepsilon) + \tilde{Q}_1(\tau,\varepsilon)A_{21}(T - \varepsilon\tau)$$

$$Q_1'(\tau,\varepsilon) = A_{11}^*(T - \varepsilon\tau)Q_0(\tau,\varepsilon) + P_0(T - \varepsilon\tau)A_{12}(T - \varepsilon\tau) -$$

$$- [P_0(T - \varepsilon\tau) + \tilde{P}_0(\tau,\varepsilon)]M(T - \varepsilon\tau)\tilde{Q}_0(\tau,\varepsilon) + \tilde{P}_0(\tau,\varepsilon)A_{12}(T - \varepsilon\tau) +$$

$$+ A_{21}^*(T - \varepsilon\tau)\tilde{R}_1(\tau,\varepsilon) + \tilde{Q}_1(\tau,\varepsilon)A_{22}(T - \varepsilon\tau) + F_{12}(T - \varepsilon\tau)$$

$$R_1'(\tau,\varepsilon) = A_{12}^*(T - \varepsilon\tau)\tilde{Q}_0(\tau,\varepsilon) + \tilde{Q}_0^*(\tau,\varepsilon)A_{12}(T - \varepsilon\tau) -$$

$$- \tilde{Q}_0^*(\tau,\varepsilon)M(T - \varepsilon\tau)\tilde{Q}_0(\tau,\varepsilon) + A_{22}^*(T - \varepsilon\tau)\tilde{R}_1(\tau,\varepsilon) +$$

$$+ \tilde{R}_1(\tau,\varepsilon)A_{22}(T - \varepsilon\tau) + F_{22}(T - \varepsilon\tau)$$

and the initial condition

$$\tilde{P}_0(0,\varepsilon) = A_{21}^*(T)A_{22}^{*-1}(T)G_{12}^* + G_{12}A_{22}^{-1}(T)A_{21}(T) -$$

$$- A_{21}^*(T)A_{22}^{*-1}(T)G_{22}A_{22}^{-1}(T)A_{21}(T)$$

$$\tilde{Q}_0(0,\varepsilon) = G_{12}, \quad \tilde{R}_0(0,\varepsilon) = G_{22}$$

$$\tilde{P}_1(0,\varepsilon) = 0, \quad \tilde{Q}_1(0,\varepsilon) = 0, \quad \tilde{R}_1(0,\varepsilon) = 0.$$

Since P_0 is defined in all of $[t_0,T]$ all these functions are defined in $[0, \frac{T - t_0}{\varepsilon}]$.

To prove the result we had to use the following lemma: Let $C(t,s,\varepsilon)$ be the fundamental matrix solution of the linear system

$$\varepsilon \frac{dw}{dt} = A_{22}(t)w$$

Let $F : [t_0, T] \quad L(R^{n_2}, R^{n_2})$ be continuous. Then

$$\lim_{\varepsilon\to 0} \frac{1}{\varepsilon} \int_t^T C^*(s,t;\varepsilon)F(s)C(s,t;\varepsilon)ds = L(t)$$

$$\lim_{\varepsilon\to 0} \frac{1}{\varepsilon} \int_t^T C^*(T,s;\varepsilon)F(s)C(T,s;\varepsilon)ds = L(T)$$

$$\lim_{\varepsilon\to 0} \frac{1}{\varepsilon} \int_t^T C(s,t;\varepsilon)F(s)ds = -A_{22}^{-1}(t)F(t)$$

$$\lim_{\varepsilon\to 0} \frac{1}{\varepsilon} \int_t^T C(T,s;\varepsilon)F(s)ds = -A_{22}^{-1}(T)F(T);$$

The convergence is uniform on $[t_0, t - \delta]$ for all $\delta \in (0, T - t_0)$, $L(\tau)$ is defined by the Liapunov equation. The only place where Lipschitz property of the functions has been used is to prove boundedness of $\tilde{P}_1$, for all other properties continuity being sufficient. The main idea underlying the asymptotic expression for the solutions has been suggested by the lecture of A. B. Vasilieva at the International Conference on Nonlinear Oscillations in Berlin in September, 1975.

BÔCHER-OSGOOD TYPE THEOREMS FOR THIRD ORDER DIFFERENTIAL EQUATIONS

Gary D. Jones
Murray State University

Oscillation properties of third order differential equations have been studied extensively recently. Many results analogous to results for second order equations have been obtained for the third order case. See [1] or [6] for good sources of references. It is the purpose of this note to obtain some results analogous to the Bôcher-Osgood theorem for second order equations. One such result is due to A. C. Lazer.

THEOREM 1. [4]. If $q(x) > 0$ (< 0) and

$$2p(x)/q(x) + \frac{d^2}{dx^2}(q^{-1}(x)) \leq 0 \quad (\geq 0)$$

the absolute values of a solution of

$$y''' + p(x)y' + q(x)y = 0 \tag{1}$$

at its successive maxima and minima form a nondecreasing (nonincreasing) sequence.

We will assume p, p' and q, q' continuous on some ray $[\alpha,\infty)$. We will also assume that (1) is oscillatory.

THEOREM 2. Suppose $p(x) \leq 0$, $q(x) > 0$ and $q'(x)/q(x) \leq 1/x$. Let y be a solution of (1) with $y(a) = y'(b) = y'(c) = 0$ where $a < b < c$. Then $|y(c)| \geq |y(b)|$.

PROOF. Let z be a solution of (1) independent of y with a zero at a and chosen such that $N''(a) > 0$, where

$$N(x) \equiv y(x)z'(x) - z(x)y'(x).$$

Then $N(x) > 0$, $N'(x) > 0$, $N''(x) > 0$, $(N'' + pN)' = qN > 0$

and y is a solution of

$$(y'/N)' + [(N'' + pN)/N^2]y = 0 \tag{2}$$

for $x > a$. [3]. Let $t = \int_a^x N(s)ds$. Then (2) becomes

$$\ddot{y} + [(N'' + pN)/N^3]y = 0 \tag{3}$$

where $(\cdot)$ represents differentiation with respect to t. To prove the theorem, it is enough to show that the derivative of $(N'' + pN)/N^3$ with respect to x is negative [5]. Now

$$[(N'' + pN)/N^3]' = [N^2q - (N'' + pN)3N']/N^4.$$

Let $f(x) = N^2q - (N'' + pN)3N'$. Then

$$f'(x) = N^2q' - NN'q - 3N''(N'' + pN)$$

$$\leq Nq[N/x - N'] - 3N''(N'' + pN).$$

But $N(x) \leq N(a) + xN'(x)$ since $N''(x) > 0$. Thus $f'(x) \leq 0$, and since $f(a) = 0$ the result follows.

THEOREM 3. Suppose $p(x) \leq 0$, $q(x) \leq 0$ and $q'(x) \leq 0$. Let y be an oscillatory solution of (1) with $y'(a) = y'(b) = 0$ where $a < b$. Then $|y(a)| \geq |y(b)|$.

PROOF. There is a solution N of

$$(y'' + py)' = qy \tag{4}$$

such that $N(x) > 0$, $N'(x) < 0$, $N'' + pN > 0$ such that every oscillatory solution of (1) is a solution of (2) [2]. Proceeding as in proof of Theorem 2, it is enough to show that the derivative of $(N'' + pN)/N^3$ with respect to x is positive [5]. Now

$$[(N'' + pN)/N^3]' = [N^2q - (N'' + pN)3N']/N^4.$$

Letting $f(x) = N^2q - (N'' + pN)3N'$ then

$$f'(x) = N^2q' - NN'q - 3N''(N'' + pN).$$

Thus f is decreasing. Thus either f is eventually negative or always positive. Suppose f is eventually negative. Then since f and f' are negative $(N'' + pN)/N^3$ is eventually negative, which is not possible. Thus f is positive and the result follows.

REFERENCES

[1] J. Barrett, *Oscillation theory of ordinary linear differential equations*, Advances in Math. 3(1969), 415-509.

[2] G. Jones, *Properties of solutions of a class of third-order differential equations*, J. Math. Anal. Appl. 48(1974), 165-169.

[3] G. Jones, *An Asymptotic property of solutions of* $y''' + py' + qy = 0$, Pacific J. Math. 48(1973), 135-138.

[4] A. Lazer, *The behavior of solutions of the differential equation* $y''' + py' + qy = 0$, Pacific J. Math. 17(1966), 435-466.

[5] W. Leighton, *Ordinary Differential Equations*, Wadsworth Publishing Company, Belmont, California, 1967.

[6] C. Swanson, *Comparison and Oscillation Theory of Linear Differential Equations*, Academic Press, New York, 1968.

ON THE CHARACTERIZATION OF MACROMOLECULAR LENGTH DISTRIBUTIONS BY ANALYSIS OF ELECTRICAL BIREFRINGENCE DECAY

by

M. M. Judy [1,4], and S. R. Bernfeld [2,3,5,6]

Introduction

A number of large polymeric molecules of biological importance consist of a regular array of identical repeating subunits or monomers. Electron and x-ray diffraction studies have given insight into some of the ways in which subunits are arranged to form a macromolecule. For example, the long fibrous proteins F-actin and tropomyosin of the basic contractile unit of muscle were found to be comprised, respectively, of two intertwining helical strands [1] and an end-to-end linear array [2] of monomers. The tubular polymers of sickle-cell hemoglobin which form upon loss of bound oxygen were shown [3,4] to be formed of parallel side-by-side helical strands of monomers. Early work showed [5] that a helical coil is one of the stable arrangements of a chain of repeating amino acids.

Knowledge of the size and shape of the molecule and the distribution function describing a molecular dimension can give useful information about the mechanism which governs the assembly of the monomers to produce the biological macromolecule. Theoretical

1 Department of Biophysics, University of Texas Health Science Center at Dallas, Dallas, Texas.

2 Department of Mathematics, University of Texas at Arlington, Arlington, Texas.

3 Department of Medical Computer Science, University of Texas Health Science Center at Dallas, Dallas, Texas.

4 Partially supported by grant from the Texas Affiliate of the American Heart Association.

5 Partially supported by a University of Texas at Arlington Organized Research Grant.

6 Partially supported by N.I.H. Grant HL16678.

[6,7] and experimental [8,9] work has shown that, at thermodynamic equilibrium in solution, the length of linear polymers formed by end-to-end addition of monomers and of helically twisted polymers formed by addition of monomers to a nucleus of critical size is distributed exponentially. Metastable Poisson length distributions [6,7,8] of helically-twisted polymers can arise by aggregation of monomers under transient non-equilibrium conditions, such as by condensation of monomers on nuclei present at supersaturated concentration. The metastable distribution decays with time [7,8] at the expense of the development of the equilibrium exponential length distribution. Thermodynamically stable Poisson length distributions of helical amino and chains or polypeptides can be formed under some reaction conditions [10].

Direct visualization using electron micrographic techniques [8] or determination of the molecular hydrodynamic properties [8, 9] have been used to estimate the parameters which characterize the length distribution of long macromolecules. Hydrodynamic measurements which depend on molecular size and shape offer some advantage in that possible artifacts arising from sample preparation for microscopy are avoided and the measurements may be made sufficiently rapidly so that the kinetics of the formation of the length distribution can be followed. The rate of the evolution of the length distribution has been shown to depend on the growth mechanism [e.g. References 6,7].

The rotational diffusion constant about a diameter of a macromolecule behaving hydrodynamically as a long rigid cylinder varies as the inverse cube of the length [11,12]; thus, this molecular constant is a sensitive measure of molecular length. One way of measuring this hydrodynamic property is by electrical birefringence decay measurements [9,13,14,15]. Pulsed electrical fields are used to induce molecular orientation through electrical dipole-field interaction. If the molecules are optically anisotropic the optical polarizability tensor of the solution becomes anisotropic. Determination of the anisotropy of the real part of the solution polarizability tensor or in the refractive index arising from molecular orientation in the external electric field constitutes the electrical birefringence measurement. The decay of the measured optical anisotropy follows randomization of molecular orientation by Brownian collisions when the electrical field is impulsively switched to zero. For solutions having identical cylindrical macromolecules of uniform length the electrical birefringence decay $\Delta n(t)$ is characterized by the exponential [13]

$$\Delta n(t) = \Delta n_0 \ exp(-6Dt) \tag{1}$$

in which D is the molecular rotational diffusion constant and Δn_0 is the value of the birefringence at steady-state molecular orientation at the instant the electrical field is switched off.

If the cylindrical macromolecules are identical except for having lengths described by the continuous distribution function $h(l)$, $a \leq l \leq d$, the decay of the electrical birefringence can not be described by a single exponentially varying process and equation (1) then becomes

$$\Delta n(t) = \int_a^d \Delta n_0(l) \ exp[-6D(l)t]dl \tag{2}$$

where $\Delta n_0(l)dl$ is the contribution to the decay signal of molecules in $h(l)$ having lengths between l and $l + dl$. In using a continuous function $h(l)$ to describe the length distribution of polymers comprised of discrete subunits, we make the assumption that the length distribution is sufficiently broad and mean polymer length sufficiently large that a point distribution function can be replaced by a continuous one.

Expressing $\Delta n_0(l)$ as a function of $h(l)$ and molecular electrical and optical parameters, we obtain from equation (2)

$$\Delta n(t) = (g_1 - g_2) \int_a^d \frac{h(l)}{L} l\Phi(l) \ exp[-6D(l)t] \ dl \tag{3}$$

where $g_1 - g_2$ is the molecular optical anisotropy [16] which sensibly is independent of molecular length; $h(l)l/L$ is the volume fraction of molecules in the length range l and $l + dl$; and $\Phi(l)$ describes the dependence of the degree of molecular orientation on the molecular electrical permanent and induced dipole moments and the magnitude of the electrical field [17]. If the molecular dipole moment is a continuous function of molecular length, then so is $\Phi(l)$.

The quantity $\Phi(l)$ varies between 0 and 1, approaching unity as the electrical field $E \to \infty$, whereas at sufficiently small electrical fields

$$\Phi(l) = kl^cE^2 \tag{4}$$

where for a molecule having a longitudinal permanent electrical dipole moment per unit length μ_0

$$\begin{aligned} k &= \mu_0^{\ 2} \\ c &= 2, \end{aligned} \tag{5}$$

and for a molecule having an electronic polarizability anisotropy per unit length α_0 giving rise to an induced dipole $\alpha_0 lE$

$$\begin{aligned} k &= \alpha_0 \\ c &= 1. \end{aligned} \tag{6}$$

If molecular orientation is due to either a permanent or induced dipole we may use equation (4) in order to write equation (3) in normalized form

$$\frac{\Delta n(t)}{\Delta n_0} = \frac{\int_a^d l^{c+1} h(l) \ exp[-6D(l)t] dl}{\int_a^d l^{c+1} h(l) dl} \tag{7}$$

where c is given by either of equations (5) and (6).

Equation (7) has been derived independently by a number of workers [18,19,20,21] seeking to use electrical birefringence decay as a tool for characterizing molecular length distributions. O'Konski [18] and Matsumoto and his coworkers [19] recasted equation (7) as the Laplace transform of the relaxation time $\tau(l) = 6D(l)$ to obtain

$$\frac{\Delta n(t)}{\Delta n_0} = \int_0^\infty F(\alpha)\, exp[-\alpha t] d\alpha \tag{8}$$

where $\alpha = 1/\tau(l)$. Matsumoto and coworkers [19,22] used

$$F(\alpha) = \sum_{i=1}^{N} A_i\, \alpha^i\, exp(-B\alpha) \tag{9}$$

to obtain from equation (8) the expression

$$\frac{\Delta n(t)}{\Delta n_0} = \sum_{i=1}^{N} \frac{A_i\, \alpha^i}{[B + t]^{i+1}} \tag{10}$$

and obtain A_i, and B by least square fit to data in $0 \leq t < \infty$. From equation (8) evaluated at $t = 0$ they obtain for a molecule having a longitudinal permanent dipole moment

$$\int_0^\infty F(\alpha) d\alpha = \int_0^\infty l^3 h(l) dl$$

from which they obtain

$$h(l) = \frac{F(\alpha)}{l^3} \frac{d\alpha}{dl} \tag{11}$$

where $d\alpha/dl$ is obtained from the dependence of $D(l)$ on molecular length [11,12]. This approach was applied by the Japanese workers to obtain the characteristics of the Poisson length distribution of the polypeptide poly(L-Glutamic acid) in various solvents [22]. The decreasing signal to noise ratio of the decay signal may adversely affect the precision with which the parameters which characterize $h(l)$ may be estimated by solution of the decay data for the kernal $F(\alpha)$.

Seeking to use the region of the birefringence decay signal with the maximum available signal to noise ratio, Schweitzer and Jennings [20] considered the behavior of the first time derivative of the birefringence decay (equation (7)) evaluated at the onset of decay, $t = 0$, i.e.,

$$[d[\Delta n(t)/\Delta n_0]/dt]_0 = \frac{-6\int_a^d l^{c+1}\ h(l)D(l)dl}{\int_a^d l^{c+1}\ h(l)dl} \tag{12}$$

for macromolecules with Gaussian length distribution. By numerical integration of equation (12) they obtained values of the first derivative which vary uniquely with both the nature of the orienting dipole moment through dependence on c (equations (4) through (6)) and on the magnitude of the standard deviation σ of the length distribution. Their results show monotonic decrease of the magnitude of the derivative for both $c = 1$ and $c = 2$ with increasing values of σ and a monotonic increase of this quantity with σ at sufficiently large electrical fields such that $\Phi \rightarrow 1$.

In this paper we report the development of equations for the n-th derivative $n = 1,2,3,...$ of the birefringence decay of macromolecules having a Poisson or exponential length distribution. The expressions were derived by direct integration of equations of the form of equation (12). This work was motivated by two considerations.

The Poisson and exponential length distributions characterize some macromolecules of biological importance. As we reported previously [21], knowledge of the first three derivatives of the birefringence decay suffice in principle to determine the three parameters which characterize these distributions (also see later in this paper). The equations we derive herein serve as a fundamental basis for subsequent development of data management and error analysis procedures for the treatment of birefringence decay data of rigid macromolecules with lengths distributed according to an exponential or Poisson distribution.

If the macromolecules behave as rigid cylinders having rotational diffusion constant $D(l)$, application of equations (9) through (11) to analysis of birefringence data will give, within a precision imposed by the signal to noise ratio, an $h(l)$ uniquely related to molecular length. Because this approach is fundamentally based on treatment of the decay as a Laplace integral,any dispersion in the signal arising from internal molecular motion such as lateral flexing or changes in the diffusion constants of the molecule due to distortion of its shape will be treated as arising from dispersion in molecular length. Thus, the length distribution derived from use of equations (9) through (11) in treating the birefringence decay, in general, will not be uniquely related to the molecular contour length of flexible macromolecules. Equations and analysis procedures with which to treat the birefringence decay of flexible macromolecules with distributed lengths in solution are currently not available. Because our research focuses on the structure and assembly of the long flexible rod-like proteins [23,24,25] of muscle F-actin, myosin and its subunits and on the muscle protein tropomyosin which may be flexible, we are working on development of the required equation and analysis. Our approach to the problem is to directly integrate the diffusion constants for rotational and flexural motion (taking due considerations for coupling) over the length distribution as in equation (12) for the rigid cylindrical molecule.

Derivation of Equations

We use the normalized distribution function

$$h(l) = \frac{l^s \, exp[-(l - a)/b]}{b^{s+1}[\Gamma(s + 1) - \gamma(s + 1, a/b)]exp(a/b)} \qquad (13)$$

where

$$a \leq l < \infty \; .$$

For

$$s = 0$$

$h(l)$ is exponential, and for

$$s \neq 0$$

$h(l)$ is a continuous approximation to the Poisson distribution. Physically the quantity a is the minimum length of the macromolecule in solution.

Substituting equation (13), Burger's approximation for rigid cylinders [26]

$$D(l) = \frac{3kT}{\pi\eta l^3} \; ln \, \frac{2l}{\delta d} \tag{14}$$

where

η = solvent viscosity
d = molecular diameter
δ = $exp(0.8)$
kT = (Boltzmann's constant)(absolute temperature)

and equation (4) for $\Phi(l)$ into equation (7) for the birefringence decay, we obtain

$$\frac{\Delta n(t)}{\Delta n_0} = \frac{\int_a^\infty l^{s+c+1} exp[-(l-a)/b] exp - \frac{18kT}{\pi\eta l^3}\left[ln\left(\frac{2l}{d}\right) - 0.80\right] t \, dl}{\int_a^\infty l^{s+c+1} exp[-(l-a)/b] dl} \tag{15}$$

Defining

$$q - 1 = s + c + 1$$

$$r - 1 = s + c - 3m + 1$$

and

$$[d^m[\Delta n(t)/\Delta n_0]/dt^m]_{t=0} = D^{(m)}$$

we use the transformations

$$x = \frac{l}{b}$$

$$K = \frac{18kT}{\pi\eta}$$

and the binomial expansion of

$$\left[ln \frac{2b}{\delta d} x \right]^m$$

to obtain for the m-th derivative evaluated at $t = 0$ of the birefringence decay (equation (15))

$$D^{(m)} = \frac{(-K)^m}{b^{3m}} \frac{\sum_{p=0}^{m} \left[ln \frac{2b}{\delta d} \right]^{m-p} \binom{m}{p} \int_{a/b}^{\infty} l^{r-1} (ln\ x)^p exp(-x) dx}{\int_{a/b}^{\infty} l^{q-1} exp(-x) dx} .$$

Now we use

$$\int_{a/b}^{\infty} l^{r-1} (ln\ x)^p exp(-x) dx = \frac{d^p}{dr^p} \Gamma(r, a/b)$$

with the definition

$$\frac{d^p}{dr^p} = D_r^{(p)}$$

to write the equation for the m-th derivative of the birefringence decay in terms of the incomplete gamma function $\Gamma(\alpha, x)$ [26]

$$D_{(c)}^{(m)} = \frac{(-K)^m}{b^{3m}} \frac{\sum_{p=0}^{m} \binom{m}{p} \left[ln \frac{2b}{\delta d} \right]^{m-p} D_r^{(p)} \Gamma(r, a/b)}{\Gamma(q, a/b)} \qquad (16)$$

$$m = 1, 2, 3, \ldots$$

$$r - 1 = s + c - 3m + 1 \neq -1, -2, -3, \ldots$$

$$q - 1 = s + c + 1.$$

For $r - 1$ equal to zero or a negative integer the integrals in the numerator of equation (15) can be evaluated in a straight forward manner using integration by parts.

Discussion of the Equations

Using [26]

$$\Gamma(a,x) = \Gamma(a) - \gamma(a,x)$$
$$= \Gamma(a)\left[1 - x^{a}\exp(-x)\sum_{n=0}^{\infty}\frac{1}{\Gamma(a+n)}x^{n}\right] \qquad (17)$$

and the following relationships between $\Gamma(a)$ and its derivatives [27]

$$\frac{d\Gamma(a)}{da} = \Gamma(a)\psi(a) \qquad (18)$$

$$\frac{d^{m}}{da^{m}}\psi(a) = \overset{(m)}{\psi}(a), \quad m = 1,2,3 \qquad (19)$$

where

$$\psi(a) = \frac{d[\ln \Gamma(a)]}{da}$$

we can expand (16), for $D^{(m)}_{(c)}$.

Also, using equations (17) through (19) and the following properties of $\Gamma(a)$ and its derivatives [27]

$$\Gamma(n + a) = (n - 1 + a)(n - 2 + a)\ldots(a + 1)\,\Gamma(a)$$

$$\Gamma(a)\Gamma(1 - a) = \pi\csc(\pi a)$$

$$\psi(n + a) = \frac{1}{n - 1 + a} + \frac{1}{n - 2 + a}\cdots\frac{1}{a} + \psi(a) \qquad (20)$$

$$\psi(1 - a) = \psi(a) + \pi\operatorname{ctn}(\pi a)$$

we can show that the $D^{(m)}_{(c)}$, $r - 1 \neq -1,-2,\ldots$ are linearly independent; direct comparison of the equations obtained for $r = 0,-1,-2,$ through integration by parts shows them also to be linearly independent. Thus, using the first three derivatives, $D^{(m)}_{(c)}$ $(m = 1,2,3)$ we obtain a set of three independent equations which, aside from effects on precision arising from the experimental signal to noise ratio, specify uniquely the parameters

a, b, s of the Poisson or exponential length distribution functions provided that the molecular diameter is known through independent measurement. Frequently this information is available; if it is not, then additional knowledge of the fourth derivative allows solution for the four parameters of interest.

We have briefly examined the behavior of the first two derivatives of the birefringence decay as the quantity s in equation (13) is varied over the range $0 \leq s < \infty$ for distributions having the minimum macromolecular length $a << b$. Hence, for all s the average length is $\langle l \rangle \approx (s + 1)b$. This can be ascertained by evaluating the first moment of equation (13). With the substitution of equations (17) through (19) into equation (16) and neglecting the contribution of terms arising from operating on $\gamma(a,x)$ which can be shown to be of $o(\frac{a}{b})$ as $\frac{a}{b} \to 0$ we obtain

$$D_{(c)}^{(1)} = \frac{-K}{\langle l \rangle^3} \frac{(s + 1)^3 \Gamma(s + c - 1)}{\Gamma(s + c + 2)} \left[\psi(s + c - 1) - ln(s + 1) + ln \frac{2\langle l \rangle}{\delta d}\right] \tag{21}$$

$$s + c - 1 > 0$$

and

$$D_{(c)}^{(2)} = \frac{K^2}{\langle l \rangle^6} \frac{(s + 1)^6 \Gamma(s + c - 4)}{\Gamma(s + c + 2)} \left[\left[\psi(s + c - 4) - ln(s + 1) + ln \frac{2\langle l \rangle}{\delta d}\right]^2 + \psi^{(1)}(s + c - 4)\right] \tag{22}$$

$$s + c - 4 > 0.$$

Using [27]

$$\psi(x) = ln\, x + \int_0^\infty [t^{-1} - (1 - e^{-t})^{-1}]e^{-xt}\, dt$$

$$x > 0,$$

and [27]

$$\psi^{(n)}_{(x)} = (-1)^{n+1}\, n! \sum_{r=0}^{\infty} (x + r)^{-n-1}$$

it can be shown that the bracketed terms in equations (21) and (22) containing $\psi(x)$ and its first derivative approach zero as s increases without bound. Since, the variable coefficients of the bracketed terms approach unity as $s \to \infty$

$$\lim_{s\to\infty} D^{(1)}_{(c)} = D_0 \,,$$

and

$$\lim_{s\to\infty} D^{(2)}_{(c)} = (D_0)^2,$$

$$c = 1,2,$$

where D_0 is the rotational diffusion constant (equation 14) for a molecule having length $<l>$. This result is expected because the width of the distribution narrows with increasing s, approaching zero as $s \to \infty$. Also, as s becomes large the difference between $D^{(1)}_{(1)}$ and $D^{(1)}_{(2)}$ and that between $D^{(2)}_{(1)}$ and $D^{(2)}_{(2)}$ approaches zero. This behavior for the first derivative was also found for a Gaussian length distribution by Schweitzer and Jennings [20]. Furthermore, in parallel with their result, $D^{(1)}_{(2)} > D^{(1)}_{(1)}$.

With $s \to 0$, it can be shown that the magnitude of $D^{(1)}_{(c)}$, $c = 1,2,$ takes on its minimum value, whereas $D^{(2)}_{(c)}$, $c = 1,2,$ assumes its maximum value. Schweitzer and Jennings [20] also noted that the magnitude of the first derivative increased for both $c = 1$ and 2 as the standard deviation of the Gaussian distribution increases.

Plans for Future Work

Our current research on the topic of this paper is devoted to:

1. Derivation of the $D^{(m)}_{(c)}$ using the higher order approxi-

mation to $D(l)$ due to Breorsma [28]. This equation is more accurate as $l/d \to 1$ than the Burgers equation [14] used in the work reported here.

2. Error analysis of the system $D^{(m)}_{(c)}$ to ascertain the precision which we may expect in the application of these equations to determination of the characteristic parameters s, a, and b.

REFERENCES

[1] Hanson, J., and Lowey, J., *The structure of f-Actin and of actin filaments isolated from muscle*, J. Mol. Biol. 6(1963), 46.

[2] Bailey, K., *A new asymmetric protein component of muscle*, Nature 157(1946), 368.

[3] Finch, J. T., Perutz, M. F., Bertles, J. F., and Dobler, J., *Structure of sickled erythrocytes and of sickle-cell hemoglobin fibers*, Proc. Nat. Acad. Sci. USA 70(1973), 718.

[4] Hofrichter, J., Ross, P. D., Eaton, W. A., *Kinetics and mechanism of deoxy hemoglobin S geleton: A new approach to understanding sickle cell disease*, Proc. Nat. Acad. Sci. USA 71(1974), 4864.

[5] Pauling, L., and Corey, R. B., *The structure of proteins: two hydrogen bonded helical configurations of the polypeptide chain*, Proc. Nat. Acad. Sci. USA 37(1951), 205.

[6] Oosawa, F., and Kasai, M., *A theory of linear and helical aggregations of macromolecules*, J. Mol. Biol. 4(1962), 10.

[7] Oosawa, F., *Size distribution of protein polymers*, J. Theor. Biol. 27(1970), 69.

[8] Kawamure, M., and Maruyama, K., *Electron microscopic particle length of f-actin polymerized in vitro*, J. Biochem. 67(1970), 437.

[9] Asai, H., *Electric birefringence of rabbit tropomyosin*, J. Biochem. 50(1961), 182.

[10] Flory, P. J., *Molecular size distribution in ethylene oxide polymers*, J. Am. Chem. Soc. 62(1940), 1561.

[11] Burgers, J. M., "*Verhandel. Koninkl. Ned. Akad. Wetenschap. Afdeel. Natuurk.*", Sec. 1, No. 4, 16(1938), 113.

[12] Broersma, S., J. Chem. Phys. 32(1960), 1626.

[13] Benoit, H., Ann. Phys. 6(1951), 561.

[14] O'Konski, C. T., and Haltner, A. J., *Characterization of the monomer and diamer of tobacco mosaic virus by transient electrical birefringence*, J. Amer. Chem. Soc. 78(1956), 3604.

[15] Matsumoto, M., Watanabe, H., and Yochioka, K., Biopolymers 6(1968), 929.

[16] Peterlin, A., and Stuart, H. A., Z. Physik 112(1939), 129.

[17] Holcomb, D. N., and Tinoco, I., J. Phys. Chem. 67(1963), 2691.

[18] O'Konski, C. T., "*Encyclopedia of Polymer Science and Technology*, Vol. 9, p. 551, Interscience, 1966.

[19] Matsumoto, M., Watanabe, H., Yoshioka, K., Kolloid-Z.u. Z., *A method for determining the relaxation spectrum from the decay curve of electric birefringence of macromolecular solution*, Polymere 250(1972), 298.

[20] Schweitzer, J., and Jennings, B. R., Biopolymers 11(1972), 1077.

[21] Judy, M., Campbell, G., and Dowben, R. M., Biophysical J. Abstr. 16(1976), 166a.

[22] Matsumoto, M., Watanabe, H., and Yoshioka, K., Biopolymers 11(1972), 1711.

[23] Fujime, S., J. Phys. Soc. Japan 29(1970), 751.

[24] Burke, M., Himmelfarb, S., Harrington, W. F., Biochemistry 12(1973), 701.

[25] Tanaka, H., Biochem. Biophys. Acta 278(1972), 556.

[26] Erdelyi, A., Ed., "*Higher Transcendental Functions*", Vol. II, McGraw-Hill Book Company, Inc., New York (1953).

[27] Erdelyi, A., Ed., "*Higher Transcendental Functions*", Vol. I, McGraw-Hill Book Company, Inc., New York (1953).

PERIODIC SOLUTIONS AND PERTURBED SEMIGROUPS OF LINEAR OPERATORS

James H. Lightbourne, III
Pan American University

Let X be a Banach space with norm $|\cdot|$. Suppose $\{T(t) : t > 0\}$ is a strongly continuous linear semigroup on the Banach space X and D is a closed subset of X such that if $x \in D$ then $T(t)x \in D$ for all $t \geq 0$. Consider the integral equation:

$$(IE)\ u(t) = T(t)z + \int_0^t T(t - r)F(r,u(r))dr, \quad z \in D \text{ where}$$

$$F : [0,\infty) \times D \to X \text{ is continuous.}$$

We say that u is a solution to (IE) on $[0,b)$ provided $u : [0,b) \to D$ is continuous and satisfies (IE) for each $t \in [0,b)$. Let A be the generator of $T(t)$. If u is a solution to (IE) and $u(t)$ is in the domain of A for all $t \in [0,b)$ with $t \to Au(t)$ continuous, then u is differentiable on $[0,b)$ and satisfies

(0.1) $$u'(t) = Au(t) + F(t,u(t)), \quad u(0) = z.$$

Consequently, a solution to (IE) is referred to as a mild solution to (0.1). In this paper we consider the existence of solutions and periodic solutions to (IE). One may establish existence of solutions to (IE) with additional criteria on F (see e.g., Martin [4] and Webb [8]). Pazy [5], using a different approach than that given here, obtains existence to (IE) when $T(t)$ is a compact semigroup. We shall restrict our discussion to a compactness condition on the composition of the semigroup with the function F which is satisfied when $T(t)$ is compact. Section 1 is devoted to the existence of solutions to (IE). In section 2, periodic solutions to (IE) are discussed and in section 3, examples involving systems of non-linear perturbations of parabolic equations are considered.

Section 1. Preliminaries and Existence.

A family $T(t) = \{T(t) : t \geq 0\}$ of bounded linear maps from X into X is said to be a c_0 - semigroup of type ω provided (i) $T(0) = I$, the identity on X; (ii) $T(t + s) = T(t)T(s)$ for all $t, s \geq 0$; (iii) for each $x \in X$ $\lim_{h \to 0^+} T(h)x = x$; and (iv) $|T(t)x| \leq e^{\omega t}|x|$ for all $(t,x) \in [0,\infty) \times X$ with ω the least such real number. Since for $T(t)$ satisfying (i) - (iii) there exists a norm equivalent to $|\cdot|$ and a real number ω for which $T(t)$ is of type ω, we may assume for our purposes that $T(t)$ is of type ω.

Throughout this paper we shall refer to the following:

(E_1) $\{T(t) : t \geq 0\}$ is a c_0 - semigroup of type ω;

(E_2) $D \subset X$ is closed and has the property that if $x \in D$, then $T(t)x \in D$ for all $t \geq 0$;

(E_3) $F : [0,\infty) \times D \to X$ is continuous with $\liminf_{h \to 0^+} |x + hF(t,x);D|/h = 0$ for all $(t,x) \in [0,\infty) \times D$, where for $y \in X$ $|y;D| = \inf\{|y - z| : z \in D\}$.

We shall also employ the following notion of the measure of non-compactness. For each bounded set $E \subset X$ define the measure of non-compactness of $E, \alpha[E]$, to be the infimum of $\varepsilon > 0$ such that E has a finite cover of sets with diameter ε. A few useful properties of the measure of non-compactness are given in the following theorem. For E_1, $E_2 \subset X$ let

$$E_1 + E_2 = \{x + y : x \in E_1; y \in E_2\}.$$

Proposition 1. Let E_1, $E_2 \subset X$ be bounded. Then:

(1) $\alpha[E_1 + E_2] \leq \alpha[E_1] + \alpha[E_2]$;

(2) $\alpha[E_1] = 0$ if and only if E_1 is totally bounded;

(3) if $E_1 \subset E_2$ then $\alpha[E_1] \leq \alpha[E_2]$;

(4) $\alpha[co(E_1)] = \alpha[E_1]$, where $co(E_1)$ is the convex hull of E_1; and

(5) $\alpha[cl(E_1)] = \alpha[E_1]$, where $cl(E_1)$ is the closure of E_1.

Essential to our discussion are approximate solutions to (IE). In Proposition 2 we define approximate solutions and give sufficient conditions for their existence. For a proof see Martin [4].

Proposition 2. Suppose (E_1) - (E_3) hold. Let $\{\varepsilon_m\}_1^\infty \subset (0,1)$ be a sequence with $\lim_{n\to\infty} \varepsilon_n = 0$, let $z \in D$, and let $R > 0$. Then there exists $b = b(z,R) > 0$ and for each n there exists a positive integer $N = N(n)$, a partition

$\{t_i\}_1^N$ of $[0,b]$ with $t_{i+1}^n - t_i^n \leq \varepsilon_n$, and $u_n : [0,b] \to X$

with the properties:

(1) $u_n(0) = z$, $u_n(t_i^n) \in D$ with $|u_n(t_i^n) - z| \leq R$ and if $t \in [t_i^n, t_{i+1}^n)$ then

$$u(t) = T(t - t_i^n)u(t_i^n) + \int_{t_i^n}^{t} T(t - r)F(t_i^n, u_n(t_i^n))dr;$$

(2) $|u_n(t_{i+1}^n -) - u_n(t_i^n)| \leq \varepsilon_n(t_{i+1}^n - t_i^n)$ and if $t \in [t_i^n, t_{i+1}^n)$ then $|u_n(t);D| \leq \varepsilon_n(t - t_i^n)$ and $|(T(t - t_i^n) - I)u_n(t_i^n)| \leq \varepsilon_n$.

Furthermore, if $\lim_{n\to\infty} u_n(t) = u(t)$ uniformly on $[0,b]$ then u is continuous from $[0,b]$ into D and satisfies (IE) on $[0,b]$.

By imposing a condition on the composition of the semigroup $T(t)$ and F, we insure local existence of solutions to (IE).

THEOREM 1. In addition to (E_1) - (E_3) suppose for each $z \in D$ there exists $M_z > 0$ and $c_z > 0$ such that

(E_4) for each $t > 0$ there is a compact set $K_t \subset X$ such that $T(t)F(s,x) \in K_t$ for all $s \in [0,c_z]$ and $|x - z| < M_z$.

Then (IE) has a local solution for each $z \in D$.

Remark 1. Note that (E_4) holds when $T(t)$ is a compact semigroup (i.e., $T(t) : X \to X$ is completely continuous for each $t > 0$) or when $F : [0,\infty) \times D \to X$ is completely continuous.

Proof. Let $z \in D$. Let $c_z > 0$ and $M_z > 0$ be such that (E_4) is satisfied. As in Proposition 2, construct $\{u_n\}_1^\infty$ on $[0,b]$ where $b < c_z$ and $|u_n(t_i^n) - z| \leq M_z$. Also since F is continuous we may assume there exists $M > 0$ such that $|F(t_i^n, u_n(t_i^n))| \leq M$ for all $1 \leq i \leq N$. It is convenient to consider the continuous approximate solutions $w_n : [0,b] \to X$ defined by

$$w_n(t) = T(t)z + \int_0^t T(t - r)F(\gamma_n(r), u_n((\gamma_n(r)))dr \quad \text{where}$$

$$\gamma_n : [0,b] \to \{t_i^n\}_{i=1}^N \quad \text{with} \quad \gamma_n(t) = t_i^n \text{ for } t_i^n \leq t < t_{i+1}^n$$
$$\text{and} \quad \gamma_n(b) = b.$$

Note that $|w_n(t) - u_n(t)| \leq b\varepsilon_n e^{\omega b}$ for all $t \in [0,b]$. Thus if we show that $\{w_n\}^\infty$ has a uniformly convergent subsequence, then $\{u_n\}^\infty$ has a uniformly convergent subsequence and the theorem follows from Proposition 2.

If $t = 0$ then $\alpha[\{w_n(t)\}_1^\infty] = \alpha[\{z\}] = 0$. Let $t \in (0,b]$ and $0 < \varepsilon < t$. It follows from (E_4) that there exists a compact set K such that

$$T(t)F(s,x) \in K \text{ for all } \varepsilon \leq t \leq b,\ s \in [0,b], \text{ and } |x - z| \leq M_z.$$

Thus $\alpha[\{w_n(t)\}^\infty] = \alpha[\{T(t)z + \int_0^t T(t - r)F(\gamma_n(r), u_n(\gamma_n(r)))dr\}_1^\infty]$

$$\leq \alpha[\{\int_0^{t-\varepsilon} T(t - r)F(\gamma_n(r), u_n(\gamma_n(r)))dr\}_1^\infty]$$
$$+ \alpha[\{\int_{t-\varepsilon}^t T(t - r)F(\gamma_n(r), u_n(\gamma_n(r)))dr\}_1^\infty]$$
$$\leq \varepsilon Me^{\omega b}.$$

Since $\varepsilon > 0$ is arbitrary we conclude that $\{w_n(t)\}_1^\infty$ is totally bounded for each $t \in [0,b]$.

Clearly, the family $\{w_n(t)\}_1^\infty$ is equicontinuous at $t = 0$. Let $t \in (0,b]$ and suppose $s \in (0,b)$ with $t > s$. Let $0 < \varepsilon < s$. Then

$$|w_n(t) - w_n(s)| = |T(t)z + \int_0^t T(t - r)F(\gamma_n(r), u_n(\gamma_n(r)))dr$$

$$- T(s)z - \int_0^s T(s - r)F(\gamma_n(r),u_n(\gamma_n(r)))dr|$$

$$\leq |T(t)z - T(s)z| + \int_s^t |T(t - r)F(\gamma_n(r),u_n(\gamma_n(r)))|dr$$

$$+ \int_0^{s-\varepsilon} |[T(t - s) - I]T(s - r)F(\gamma_n(r),u_n(\gamma_n(r))|dr$$

$$+ \int_{s-\varepsilon}^s |[T(t - r) - T(s - r)]F(\gamma_n(r),u_n(\gamma_n(r)))|dr$$

$$\leq |[T(t - s) - I]T(\varepsilon)z| + |t - s|Me^{\omega b} +$$

$$+ \int_0^{s-\varepsilon} |[T(t - s) - I]T(s - r)F(\gamma_n(r),u_n(\gamma_n(r)))|dr$$

$$+ \varepsilon Me^{\omega b}.$$

Since $\{T(s - r)F(\gamma_n(r), u_n(\gamma_n(r))) : 0 \leq r \leq s - \varepsilon\}$ is totally bounded and $\lim_{h\to 0^+} T(h) = I$ uniformly on compact subsets of X we have that $\{w_n\}_1^\infty$ is equicontinuous at t. By Ascoli's theorem $\{w_n\}_1^\infty$ has a subsequence uniformly convergent on $[0,b]$ and the theorem follows.

Section 2. Periodic Solutions

Let $\sigma > 0$ and suppose F is σ-periodic in t. Define U from $[0,\infty) \times D$ into the subsets of D by

$U(t,x) = \{u(t) : u$ is a solution to (IE) with $u(0) = x\}$.

Then the existence of σ-periodic solution to (IVP) is equivalent to the existence of $x \in D$ for which $x \in U(\sigma,x)$.

Our results rest on the following fixed point theorem which may be found in Sadovskii [6] (one should also see Darbo [1]). Let $E \subset X$. A continuous function $f : E \to X$ is said to be condensing if for each bounded $E_0 \subset E$ with $\alpha[E_0] > 0$ we have $\alpha[f(E_0)] < \alpha[E_0]$.

Proposition 3. Suppose $E \subset X$ is closed, bounded, and convex and $f : E \to E$ is condensing. Then the set $\{z \in E : f(z) = z\}$ is non-empty and compact.

For the existence of fixed points for $U(\sigma,\cdot)$ we apparently must assume the non-local version of (E_4):

$(E_4)^1$ For each bounded subset $J \subset [0,\infty)$ and $t \in (0,\sigma]$ there exists a compact subset K_t such that $T(t)F(s,x) \in K_t$ for all $(s,x) \in J \times D$.

Let $BC([0,\infty);X)$ denote the space of bounded continuous functions $u : [0,\infty) \to X$ with $||u|| = sup\{|u(t)| : t \in [0,\infty)\}$.

THEOREM 2. Suppose D is bounded, (E_1) - (E_3) and $(E_4)^1$ hold, and F is bounded on bounded subsets of $[0,\infty) \times D$. Then for each $t \in [0,\infty)$ and $E \subseteq D$

$$\alpha[\{U(t,x) : x \in E\}] \leq \alpha[\{T(t)x : x \in E\}].$$

In addition, suppose D is convex, $T(\sigma)$ is condensing, and (IE) has non-continuable solutions uniquely determined by initial values. Then if F is σ-periodic in t, the family of σ-periodic solutions to (IE) is non-empty and compact in $BC([0,\infty);X)$.

Lemma 1. Let the suppositions of Theorem 2 hold. Then non-continuable solutions to (IE) exist on $[0,\infty)$ and solutions depend continuously on initial values.

Proof of Lemma 1. With local existence assured, global existence follows by standard techniques when the set D is bounded and F is bounded on bounded subsets of $[0,\infty) \times D$. Let $\{z_n\}_1^\infty \subset D$ with $\lim_{n\to\infty} z_n = z$ and let u_n denote the unique solution to (IE) with $u_n(0) = z_n$ and u the solution to (IE) with $u(0) = z$. Using the techniques employed in the proof of Theorem 1 one can show that there exists $b > 0$ and a subsequence $\{u_{n_k}\}_{k=1}^\infty$ of $\{u_n\}_1^\infty$ uniformly convergent on $[0,b]$. For $t \in [0,b]$ let $v(t) = \lim_{k\to\infty} u_{n_k}(t)$. Then v is continuous on $[0,b]$ and since $\{(r,v(r)) : 0 \leq t \leq b\}$ is compact we have for each $t \in [0,b]$ that

$$v(t) = \lim_{k\to\infty} T(t)z_{n_k} + \int_0^t T(t - r)F(r,u_{n_k}(r))dr$$

$$= T(t)z + \int_0^t T(t - r)F(r,v(r))dr$$

Since solutions to (IE) are assumed to be unique, $v(t) = u(t)$ on $[0,b]$ and it readily follows that $\lim_{n\to\infty} u_n(t) = u(t)$ uniformly on compact subsets of $[0,\infty]$.

Proof of Theorem 2. Let $E \subset D$ and define

$\Delta = \{u : u$ is a solution to (IE) with $u(0) \in E\}$.

Then for each $t \in (0,\infty)$ and $0 < \varepsilon < t$ we have

$$\alpha[\{U(t,x) : x \in E\}]$$

$$= \alpha[\{T(t)u(0) + \int_0^t T(t - r)F(r,u(r))dr : u \in \Delta\}]$$

$$\leq \alpha[\{T(t)x : x \in E\}] + \alpha[\{\int_{t-\varepsilon}^t T(t - r)F(r,u(r))dr : u \in \Delta\}]$$

$$\leq \alpha[\{T(t)x : x \in E\}] + 2\varepsilon Me^{\omega\varepsilon}$$

Where $M = \sup\{|F(s,x)| : (s,x) \in [0,t] \times D\}$. Thus we have

$$\alpha[\{U(t,x) : x \in E\}] \leq \alpha[\{T(t)x : x \in E\}].$$

If the solutions to (IE) are uniquely determined by initial values it is immediate from Lemma 1 that $U(\sigma,\cdot) : D \to D$ is a continuous function. Thus the remaining assertions of the theorem are consequences of Proposition 3.

When F is autonomous, Theorem 2 may be applied to critical values of $A + F$. (Recall that the generator of $T(t)$ is defined by $\lim_{h\to 0^+} (T(h)x - x)/h = Ax$ where the domain of A, $D(A)$, is the set of all x for which the limit exists. Also, A is a densely defined closed linear operator on X.)

THEOREM 3. In addition to the suppositions of Theorem 2, suppose $F(t,x) = F(x)$ for all $(t,x) \in [0,\infty) \times D$ and $T(t)$ is condensing for each $t > 0$. Then the set

$$Z = \{z \in D(A) \subset D : Az + Fz = \theta\}$$

is non-void and compact, where θ denotes the zero of X and $D(A)$ the domain of the generator A of the semigroup $T(t)$.

<u>Proof</u>. By Theorem 2 for each positive integer n there exists $x_n \in D$ and a solution u_n to (IE) such that $x_n = u_n(0) = u_n(n^{-1})$. By uniqueness of solutions we have $u_n(\ell n^{-1}) = u_n(0)$ for each $\ell \in \{0,1,2,\ldots\}$ and, in particular, $u_n(1) = u_n(0)$. Thus for each n we have x_n is a fixed point for $U(1,\cdot) : D \to D$ and since $U(1,\cdot)$ is condensing we have from Proposition 3 that $\alpha[\{x_n\}_1^\infty] = 0$. Therefore, there exists $x \in D$ and a subsequence

$\{x_{n_k}\}_{k=1}^\infty$ of $\{x_n\}_1^\infty$ for which $\lim_{k\to\infty} x_{n_k} = x$. Thus

$$x_{n_k} = u_{n_k}(n_k^{-1}) = T(n_k^{-1})x_{n_k} + \int_0^{n_k^{-1}} T(n_k^{-1} - r)F(u_{n_k}(r))dr$$

or

$$0 = n_k[T(n_k^{-1})x_{n_k} - x_{n_k}] + n_k \int_0^{n_k^{-1}} T(n_k^{-1} - r)F(u_{n_k}(r))dr$$

Consequently, since $\lim_{k\to\infty} n_k \int_0^{n_k^{-1}} T(n_k^{-1} - r)F(u_{n_k}(r))dr = F(x)$ we have $\lim_{k\to\infty} n_k[T(n_k^{-1})x_{n_k}] = -F(x)$. It is an elementary property that $\int_0^t T(r)w\,dr \in D(A)$ and $T(t)w - w = A\int_0^t T(r)w\,dr$ for all $t > 0$ and $w \in X$. Thus

$$n_k[T(n_k^{-1})x_{n_k} - x_{n_k}] = n_k A \int_0^{n_k^{-1}} T(r)x_{n_k}\,dr$$

$$\lim_{k\to\infty} n_k\, A \int_0^{n_k^{-1}} T(r)x_{n_k}\,dr = -F(x).$$

However, since $\lim_{k\to\infty} n_k \int_0^{n_k^{-1}} T(r)x_{n_k}\,dr = x$ and A is a closed operator, we conclude that $x \in D(A)$ with $Ax = -F(x)$. Thus Z is non-empty. Each $z \in Z$ is a fixed point of $U(1,\cdot)$ and so by Proposition 3, $\alpha[Z] = 0$. That Z is closed and hence compact follows from A being a closed operator.

Remark 2. Versions of the preceding results may be obtained under the assumptions that the semigroup $T(t)$ be analytic and F be continuous "relative to a fractional power of A;" i.e., the map defined by $x \to F(t,A^{-\alpha}x)$ is continuous for some $\alpha \in (0,1)$. The reader is referred to Lightbourne and Martin [3].

Section 3. Examples

The following two situations were chosen as examples primarily because the semigroups considered are not compact.

Example 1. Suppose $X = BUC(\mathbb{R};\mathbb{R}^n)$ is the Banach space of bounded uniformly continuous functions x from $\mathbb{R}$ into $\mathbb{R}^n$ with $|x| = sup\{|x(r)| : r \in \mathbb{R}\}$. Let $D(A) = \{x \in BUC(\mathbb{R};\mathbb{R}^n) : x', x'' \in BUC(\mathbb{R};\mathbb{R}^n)\}$. Then $A : D(A) \to X$ defined by $Ax = x''$ is the generator of the c_0 -semigroup of type $\omega = 0$ defined by

$$(3.1)\quad [T(t)x](s) = \begin{cases} (2\pi t)^{-1/2}\displaystyle\int_{-\infty}^{+\infty} exp(-(s-r)^2/2t)x(r)dr, & \text{if } t > 0 \\ x(s), & \text{if } t = 0 \end{cases}$$

(see e.g., Yosida [7]). Also for $\lambda \in \mathbb{R}$, $A + \lambda I$ is the generator of the semigroup $T_\lambda(t) = e^\lambda T(t)$ and thus if $\lambda < 0$ we have that $T_\lambda(t)$ is condensing for $t > 0$. Let $\Lambda \subset \mathbb{R}^n$ be closed and convex and define

$$K(\Lambda) = \{x \in BUC(\mathbb{R};\mathbb{R}^n) : x(r) \in \Lambda \text{ for all } r \in \mathbb{R}\}.$$

Then $K(\Lambda)$ is closed and convex and since

$$(\ \pi t)^{-1/2}\int_{-\infty}^{+\infty} exp(s - r)^2/\ t)dr = 1 \text{ we have for}$$

$x \in K(\Lambda)$ that $T(t)x \in K(\Lambda)$ for all $t \geq 0$.

Suppose $f = (f_i)_1^n : [0,\infty) \times \mathbb{R} \times \Lambda \to \mathbb{R}^n$ is continuous and for each $t \in [0,\infty)$ and compact set $\Omega \subset \Lambda$, $f(t,\cdot,\cdot)$ is bounded and uniformly continuous on $\mathbb{R} \times \Omega$. Then F defined on $[0,\infty) \times K(\Lambda)$ by $[F(t,x)](s) = f(t,s,x(s))$ for all $s \in \mathbb{R}$ is a continuous function from $[0,\infty) \times K(\Lambda)$ into $BUC(\mathbb{R};\mathbb{R}^n)$ and is bounded on bounded subsets of $[0,\infty) \times K(\Lambda)$. Note also that for compact subsets $\Omega \subset \Lambda$, $g_\Omega : [0,\infty) \times \mathbb{R} \to [0,\infty)$ defined by $g_\Omega(t,s) = max\{|f(t,s,\xi)| : \xi \in \Omega\}$ is continuous and for each t, $g_\Omega(t,\cdot)$ is uniformly continuous.

Proposition 4. Let $\lambda \in \mathbb{R}$. In addition to the above, suppose for each compact set $\Omega \subset \Lambda$ and $t \in [0,\infty)$, that $\int_{-\infty}^{+\infty} g_\Omega(t,r)dr < \infty$. Then for each $z = (z_i)_1^n \in K(\Lambda)$ there exists a mild solution $u = (u_i)_1^n$ to

$$(3.2) \quad \frac{\partial u_i}{\partial t}(t,s) = \frac{\partial^2 u_i}{\partial s^2}(t,s)+\lambda u_i(t,s)+f_i(t,s,u_1(t,s),\dots,u_n(t,s))$$

satisfying $u_i(0,\cdot) = z_i(\cdot)$ and $u(t,s) = (u_i(t,s))_1^n \in \Lambda$ for all $(t,s) \in [0,\infty) \times \mathbb{R}$. Furthermore, if Λ is bounded, $\lambda < 0$, $f = (f_i)_1^n$ is σ-periodic in t, and mild solutions to (3.2) are unique, then the family of mild solutions to (3.2) which are σ-periodic in t is non-empty.

Proof. Let $x \in K(\Lambda)$ and Ω_1 denote the closure of $\{x(s) : s \in \mathbb{R}\}$. Since $g_{\Omega_1}(t,\cdot)$ is uniformly continuous on $\mathbb{R}$, $\int_{-\infty}^{+\infty} g_\Omega(t,r)dr < \infty$ implies that $\lim_{|r|\to\infty} g_{\Omega_1}(t,r) = 0$. Thus $\lim_{|r|\to\infty} |[F(t,x)](r)| = 0$ and we have that $\lim_{h\to 0^+} |x + hF(t,x); K(\Lambda)|/h = 0$ follows from Martin [4]. Let $E \subset K(\Lambda)$ be bounded and Ω_1 denote the closure of $\{x(s) : s \in \mathbb{R}, x \in E\}$. Let $\varepsilon > 0$ and $t > 0$. Since $\int_{-\infty}^{+\infty} g_\Omega(t,r)dr < \infty$ there exists $N > 0$ such that $(2\pi t)^{-1/2} \int_{|r|\geq N} exp(-(s - r)^2/2t)|f(t,r,x(r))|dr \leq \varepsilon$

for all $x \in E$ and $s \in \mathbb{R}$. Also, there exists $M > 0$ such that for

$$|s| \geq M, \ (2\pi t)^{-1/2} \int_{-N}^{+N} exp(-(s - r)^2/2t)|f(t,r,x(r))|dr \leq \varepsilon.$$

Thus, since $\{(2\pi t)^{-1/2}\int_{-N}^{+N} exp(-(s - r)^2/2t)f(t,r,x(r))dr : x \in E\}$

is a uniformly equicontinuous family in s on $[-M,M]$ we have that $(E_4)^1$ is satisfied (see e.g., Dunford and Schwartz [2, Theorem 5, p. 266]). The assertions now follow from Theorems 1 and 2.

Example 2. This example is included only as a remark. For the details of the example the reader is referred to Lightbourne and Martin [3]. Let Y be a Banach space and $X = L^p([0,1];Y)$, $1 < p < \infty$, be the Banach space of all measurable functions $x : [0,1] \to Y$ with $|x| = [\int_0^1 |x(r)|^p dr]^{1/p} < \infty$. Let $\Lambda \subset Y$ be closed and convex with $\theta \in \Lambda$ and define

$$K(\Lambda) = \{x \in L^p([0,1];Y) : x(r) \in \Lambda \text{ for almost all } r \in [0,1]\}.$$

Let $D(A) = \{x \in L^p([0,1];Y) : x, x'$ are abs. cont., $x'' \in L^p([0,1];Y)$ and $x'(0) = x(1) = \theta\}$

and define $Ax = x''$ for $x \in D(A)$. Then A is the generator of a c_0-semigroup $T(t)$ (in fact, $T(t)$ is analytic) of type $\omega < 0$ and hence for $t > 0$ $T(t)$ is condensing. Furthermore, using the maximal principle one can show that

$$T(t) : K(\Lambda) \to K(\Lambda) \quad \text{for all} \quad t \geq 0.$$

Let $f : [0,\infty) \times [0,1] \times \Lambda \to Y$ be completely continuous and satisfy $\lim_{h \to 0^+} |\xi + hf(t,s,\xi);\Lambda|/h = 0$ for all $(t,s,\xi) \in [0,\infty) \times [0,1] \times \Lambda$. Then defining $F : [0,\infty) \times K(\Lambda) \to L^p([0,1];Y)$ by $[F(t,x)](s) = f(t,s,x(s))$ for all $s \in [0,1]$ we have that (E_3) and (E_4) are satisfied with $D = K(\Lambda)$ and if in addition Λ is bounded $(E_4)^1$ is satisfied. Thus our results apply to the "infinite system:"

$$(3.3) \qquad \frac{\partial u}{\partial t}(t,s) = \frac{\partial^2 u}{\partial s^2}(t,s) + f(t,s,u(t,s))$$

with conditions $u(0,s) = z(s)$, $\frac{\partial u}{\partial s}(t,0) = u(t,1) = \theta$ and

$u(t,s) \in \Lambda$ for all $(t,s) \in [0,\infty) \times [0,1]$.

Acknowledgements

The results in this paper are essentially found in the author's thesis which was written under the direction of R. H. Martin, Jr.

REFERENCES

[1] G. Darbo, *Punti uniti in transformazioni a condomino non compatto*, Rend. del Sem. Mat. Univ. Padova, 24(1955), 84-92.

[2] N. Dunford and J. T. Schwartz, *Linear Operators, Part I*, Wiley, New Yor, 1958.

[3] J. Lightbourne and R. H. Martin, Jr., *Relatively continuous non-linear perturbations of analytic semigroups*, submitted.

[4] R. H. Martin, Jr., *Invariant sets for perturbed semigroups of linear operators*, Annali Mat. Pura. Appl. 105(1975), 221-39.

[5] A. Pazy, *A class of semi-linear equations of evolution*, Israel J. Math. 20(1975), 23-36.

[6] B. N. Sadovskii, *On a fixed point principle*, Funct. Anal. Appl 1(1967), 74-76.

[7] K. Yosida, *Lectures on semigroup theory and its application to Cauchy's problem in partial differential equations*, Lecture Notes, Tata Institute of Fundamental Research, Bombay, India, 1957.

[8] G. Webb, *Continuous non-linear perturbations of linear accretive operators in Banach spaces*, J. Funct. Anal. 10(1972), 191-203.

SPECULATIVE DEMAND WITH SUPPLY RESPONSE LAG

A. A. Francis[1], I. H. Herron[2,3], and Clement McCalla[2,3]

1. *Introduction*

In the well known "cobweb" model (Ezekiel [7]), the demand is assumed to be dependent on the current price, while the supply depends upon the price in the previous period. A number of authors, Goodwin [9], Nerlove [13] and Carlson [3] have extended the basic cobweb model by considering more elaborate relations between the current supply and the past history of the price.

The objective of this paper is to extend the cobweb model to allow demand to depend upon the current price and its time derivative so as to take speculation into account. As a result of these assumptions, the model is described by a differential - difference equation which exhibits a richer variety of solutions than the first order difference equation governing the basic cobweb model. Consequently the model is much more versatile in its range of behavior than the basic cobweb model. For example, in the case of rising price restraining demand and the typical upward sloping supply curve and downward sloping demand curve, the model yields oscillatory and nonoscillatory price movements about the equilibrium price and it is possible to obtain price stability for the entire range of relative slopes of supply and demand curves.

2. *The Model*

The basic equations of the model are:

$$D(t) = -ap(t) + bp'(t) + c \tag{2.1}$$

[1] Department of Economics, University of the West Indies, Kingston, Jamaica.

[2] Department of Mathematics, Howard University, Washington, D.C.

[3] Supported in part by the Faculty Research Program in the Social Sciences, Humanities and Education at Howard University.

$$S(t) = \alpha p(t - \theta) + \gamma, \tag{2.2}$$

$$D(t) = S(t), \tag{2.3}$$

where $D(t)$, $S(t)$ and $p(t)$ are respectively the demand, supply and price at time t. We assume that the demand curve slopes downwards so that $a > 0$. In section 3, we assume that the slope curve slopes upwards and in section 4 that it slopes downwards or is parallel to the price axis. We take $b \neq 0$ and the time lag $\theta > 0$.

The inclusion of the price derivative $p'(t)$ in (2.1) reflects an element of speculation in that the buyers' demand reacts to the rate at which the price is changing. Rising price stimulates demand if the coefficient of speculation $b > 0$ and suppresses demand if $b < 0$. The time lag in (2.2) arises out of the fact that there is a lapse of time between the decision to produce and the completion of the finished product.

The static equilibrium price p^* is given by

$$p^* = (c - \gamma)/(\alpha + a) \tag{2.4}$$

and the moving equilibrium price $p(t)$ is given by

$$-ap(t) + bp'(t) + c = \alpha p(t - \theta) + \gamma. \tag{2.5}$$

To solve (2.5) for $t > 0$, we have to specify the history of the price $p(t)$ over the interval $[-\theta, 0]$,

$$\text{i.e. } p(t) = P(t) \quad \text{for} \quad -\theta \leq t \leq 0. \tag{2.6}$$

The price deviation $x(t) = p(t) - p^*$ is described by the differential - difference equation

$$x'(t) = A_0 x(t) + A_1 x(t - \theta), \quad t > 0, \tag{2.7}$$

$$x(t) = h(t), \quad -\theta \leq t \leq 0,$$

where

$$A_0 = ab^{-1}, \; A_1 = \alpha b^{-1} \quad \text{and} \quad h(t) = P(t) - p^*. \tag{2.8}$$

Other economic models which reduce to differential - difference equations can be found in Allen [1].

3. *Mathematical Results*

The solution of (2.7) is given by

$$x(t) = \Phi(t)h(0) + A_1 \int_{-\theta}^{0} (t - \theta - \alpha)h(\alpha)d\alpha, \tag{3.1}$$

where the fundamental solution is given by

$$\Phi(t) = \sum_{n=0}^{\infty} \frac{A_1^n}{n!} (t - n\theta)_+^n e^{A_0(t - n\theta)}, \tag{3.2}$$

and where

$$(t)_+ = max(t,0).$$

The stability and oscillatory behavior of (2.7) have been considered in a number of papers, for example [2], [5], [6], [8], [11] and [12], and can be described in terms of the two parameters u and v, where

$$u = a\theta b^{-1}, \quad \text{and} \quad v = \alpha\theta b^{-1}. \tag{3.3}$$

In the (u,v) plane (see diagram 1), we define the curves C_1, C_2, C_3 and C_4 as follows:

$$C_1 = \{(u,v);\ u + v = 0,\ \ u < 1\},$$

$$C_2 = \{(u,v);\ v + e^{u-1} = 0,\ \ u < 1\},$$

$$C_3 = \{(u,v);\ u = p \cot p,\ v = -p \operatorname{cosec} p,\ \ 0 < p < \pi\},$$

$$C_4 = \{(u,v);\ v + e^{u-1} = 0,\ \ u \geq 1\}.$$

The regions R_1, R_2, R_3 and R_4 are defined as follows: R_i is the region bounded by C_i and C_{i+1} for $i = 1,2,3$ and R_4 is the region bounded by C_4 and C_1.

It is well known (see for example Bellman and Cooke [2], El'sgol'ts and Norkin [6]) that the system described by (2.7) is asymptotically stable for (u,v) lying in the regions R_1, R_2

and on the curve C_2 and is stable without being asymptotically stable on the curves C_1 and C_3. The system (2.7) is unstable for (u,v) lying in the regions R_3, R_4 and on the curve C_4.

For (u,v) lying in the regions R_2 and R_3 and on the curve C_3 it follows from the work of Myshkis [12] that every solution of (2.7) is oscillatory and from the work of McCalla [11] that the distance between successive zeros of a solution with positive initial data $h \geq 0$ is greater than the time lag θ. An asymptotic characterization of the solutions of (2.7) due to Driver [5] and extended using a result of DeBruijn [4] shows that for (u,v) lying in the regions R_1, R_4 and on the curves C_2 and C_3, practically no solution of (2.7) is oscillatory. For a more precise statement, the reader should consult Driver [5]. In any case, the distance between successive zeros of an oscillatory solution will have length less than the time lag θ, see McCalla [11].

It is convenient to define a new parameter $\xi = -\theta b^{-1}$. If $b > 0$, so that rising price stimulates demand and falling price suppress demand, the point $(a\theta b^{-1}, \alpha\theta b^{-1})$ lies in the first quadrant and the price movement is explosive without oscillation.

The case where $b < 0$, so that using price suppresses demand and falling price stimulates demand, is of greater interest. If the slope of the demand curve is as steep or steeper than that of the supply curve, i.e. $\frac{\alpha}{a} \leq 1$, then the price movement is stable but is oscillatory or non-oscillatory depending paramter ξ is greater than or less than ξ_1, where

$$\xi_1 = f^{-1}(\frac{\alpha}{a}) \tag{3.4}$$

and

$$f(x) = x^{-1}e^{-(x+1)}, \quad x > 0. \tag{3.5}$$

If the slope of the supply curve is steeper than that of the demand curve, i.e. $\frac{\alpha}{a} > 1$, the price movement is stable or unstable according as to whether the parameter ξ is less than or greater than ξ_2, where

$$\xi_2 = x^2(2\ cosec^2 x - 1)(a^2 + \alpha^2)^{-1}, \tag{3.6}$$

and

$$x = cos^{-1}(-\frac{a}{\alpha}), \quad 0 < x \leq \pi, \tag{3.7}$$

and is oscillatory or non-oscillatory depending upon whether ξ is greater than or less than ξ_1.

4. *Atypical Supply Curves*

If the supply curve slopes downwards, i.e. $\alpha < 0$, the point $(a\theta b^{-1}, \alpha\theta b^{-1})$ lies in the second quadrant if $b < 0$ and in the fourth quadrant if $b > 0$. If the supply curve is parallel to the price axis, the point $(a\theta b^{-1}, \alpha\theta b^{-1})$ lies on the negative u-axis if $b < 0$. In both cases, the nature of the price movement is readily obtained from diagram 1.

When the slope of the demand curve is greater than that of the supply curve i.e. $|\frac{\alpha}{a}| < 1$, a positive coefficient of speculation leads to an explosive non-oscillatory price movement. In this case, price stability is obtained for a negative coefficient of speculation and all values of the time lag. When $|\frac{\alpha}{a}| > 1$, the opposite is true and price stability is attained only for a positive coefficient of speculation and the size of the parameter smaller in size than ξ_2.

5. *Concluding Remarks*

1. If the market structure satisfies the conditions of stability for the cobweb model, i.e. the slope of the demand curve is steeper than that of the supply curve, the market will be unstable if rising price stimulates demand and stable if rising price suppress demand. In the latter case the market will be stable for all values of the time lag and the coefficient of speculation. The price movement will be nonoscillatory if the supply curve slopes downwards and nonoscillatory or oscillatory depend-

DIAGRAM 1

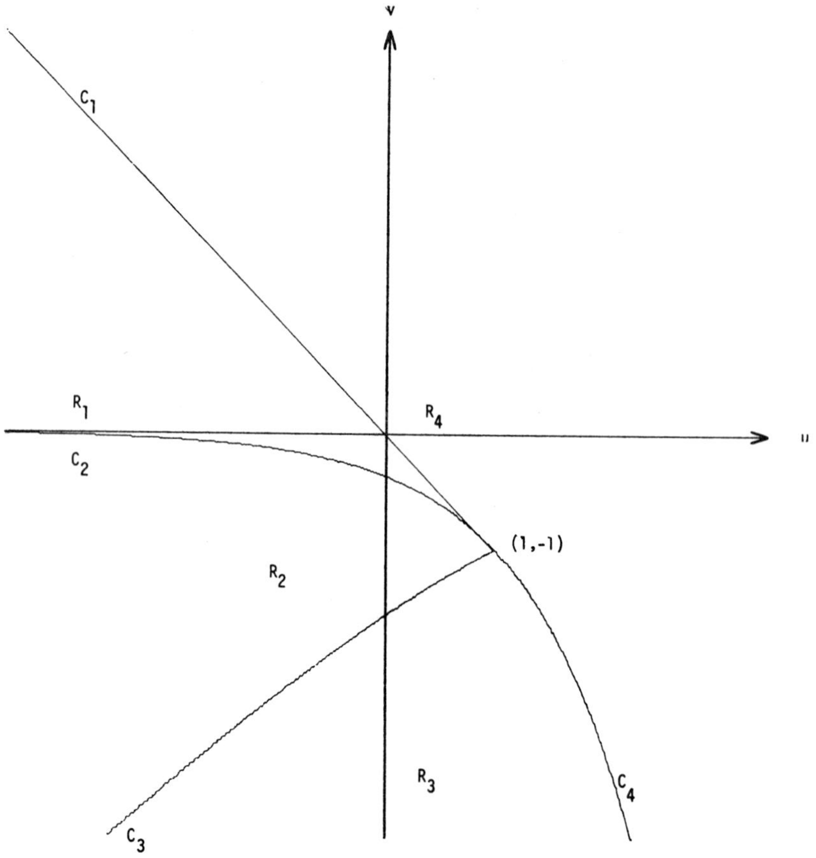

ing upon the size of the ratio of the time lag to the coefficient of speculation if supply curve slopes upwards.

2. If the market structure does not satisfy the conditions of stability for the cobweb model, and the supply curve slopes upwards, the market will be unstable if rising price stimulates demand and stable only if rising price suppress demand and the size of the ratio of the time lag to the coefficient of speculation is sufficiently small. When the supply curve slopes downwards, the market is unstable if rising price suppresses demand and stable only if rising price stimulates demand and the size of the ratio of the time lag of the coefficient of speculation is sufficiently small.

3. In the case where the parameters of the model are such that $(a\theta b^{-1}, \alpha\theta b^{-1})$ lies in the intersection of the half spaces $u + v < 0$ and $u - v > 0$, increasing the time lag (decreasing the extent of the speculation) causes the price movement to change from stable to unstable and nonoscillatory to oscillatory. When $(a\theta b^{-1}, \alpha\theta b^{-1})$ lies in the intersection of the half space $v < 0$ and $u - v > 0$, the change in the price movement is from nonoscillatory to oscillatory. This is in stark contrast to the cobweb model where the time lag does not effect the qualitative nature of the price movement which is oscillatory (nonoscillatory) when the supply curve slopes upwards (downwards).

4. We have considered only the case where the demand curve slopes downwards. There are no new special features of the model in the case where the demand curves slopes upwards except for a change in the direction of the speculation if market stability is to be achieved. The results for the special case of the demand curve being parallel to the price axis is readily determined from diagram 1.

REFERENCES

[1] R. G. D. Allen, *Mathematical Economics*, MacMillan, London, 1956.

[2] R. Bellman, and K. L. Cooke, *Differential - Difference Equations*, Academic Press, New York, 1963.

[3] J. A. Carlson, *An invariably stable cobweb model*, Review of Economic Studies 35(1968), 360-362.

[4] N. G. De Bruijn, *On some linear functional equations*, Publicationes Mathematicae Debrecen 1(1950), 129-134.

[5] R. D. Driver, *Linear differential equations with small delays*, Journal of Differential Equations 21(1976), 148-166.

[6] L. E. El'sgol'ts, and S. B. Norkin, *Introduction to the Theory and Application of Differential Equations with Deviating Arguments*, Academic Press, New York, 1973.

[7] M. Ezekiel, *The cobweb theorem*, Quart. J. of Economics 52 (1938), 255-280.

[8] R. Frish, and H. Holme, *The characteristic solutions of a mixed difference and differential equation occurring in economic dynamics*, Econometrica 3(1935), 225-239.

[9] R. M. Goodwin, *Dynamical coupling with especial reference to markets having production lags*, Econometrica, 15(1947), 181-204.

[10] C. McCalla, *The exact solutions of some functional differential equations*, *Dynamical Systems: An International Symposium, Vol. 2*, Edited by L. Cesari, J. K. Hale and J. P. La Salle, Academic Press, New York, 1976, 163-168.

[11] C. McCalla, *Zeros of the solutions of first order functional differential equations*, Howard University Mathematical Report No. 10.

[12] A. D. Myshkis, *Linear Differential Equations with Retarded Argument*, (Russian) GITTL, Moscow, 1972.

[13] M. Nerlove, *Adaptive expectations and cobweb phenomenon*, Quarterly J. of Economics 73(1958), 227-240.

A GEOMETRICAL STUDY OF B-CELL STIMULATION AND HUMORAL IMMUNE RESPONSE

Stephen J. Merrill*
The University of Iowa

Introduction

The immune response to the presence of a foreign substance, called an antigen, consists of many complex biochemical interactions involving several cell types. The response itself is usually divided into two parts: the humoral (antibody) response and the cell-mediated response.

The goal of this paper is to construct a model of the humoral response to a challenge of non-replicating antigen. The technique used will be similar to that used in the model of the heartbeat and nerve response proposed by Zeeman [12] who used catastrophe theory as developed by René Thom [10] to justify the form of the model.

The humoral response varies between species and even individuals within a species. The response also depends on the particular antigen studied. For this reason we will present a very flexible model that can easily be adopted to the species and antigen under study and modified as new information concerning the response is discovered. The form of the model also provides a possible tool for identifying some of the complex interactions, suggesting how they interact by studying the geometry of the entire interaction. This may be of special use in studying T-cell influences. The model also provides a framework for integrating the seemingly diverse parts of the response into one unit.

Models of antibody production in response to non-replicating antigen have been proposed by Bell [1], Bruni, Giovenco, Koch and Strom [2], Waltman and Butz [11], and others. Models studying the network properties of the cell types and specificities have been

* This research was supported by grant No. 1RO1CA 18639, awarded by the National Cancer Institute, DHEW.

proposed by Hoffman [5] and Richter [9]. A survey and summary of these models may be found in Merrill [7].

Acknowledgement. The author would like to thank Drs. Alan Perelson, George Bell, and Byron Goldstein at the University of California Los Alamos Scientific Laboratory and Dr. Louis Hoffmann and Dr. Thomas Feldbush of the University of Iowa Department of Microbiology for valuable discussions and suggestions.

Section 1. Description of Humoral Immune Response

The antibody producing cell is derived from B-lymphocytes (B-cells) after they are stimulated by antigen. The mechanism of this stimulation is not yet fully understood but it involves antibody on the surface of the B-cell which acts as a receptor for a certain pattern found on the antigen. If an antigen with that pattern is nearby, the receptor captures it. If only a small percentage of the thousands of receptors on the cell capture antigen no stimulation will result (we call this low dose unresponsiveness). On the other hand, if sufficient antigen is captured the B-cell will be stimulated. This stimulation results in blast transformation which involves a great increase in size and a corresponding increase in the size and numbers of the internal structures needed for protein synthesis and distribution of the product (antibody) to the surface,as well as increases in rates of macromolecular synthesis (Eisen [3], p. 468). The transformation is also marked by an increased rate of mitosis, the unstimulated B-cell dividing approximately once in 5 years to the large cell's 6-48 hour doubling time. The resulting mass of cells is called a clone. The cells of the clone then differentiate into two types of cells, the plasma cell and the memory cell. The plasma cell produces large amounts of antibody (not used as receptors) and this antibody binds to antigen hastening its removal from the body. The plasma cell lives for 10-14 days and then dies without division. The memory cell eventually shrinks to the size of the small B-cell after the response and is able to respond to antigen after essentially all the free antibody has disappeared from the system (Eisen [3], p. 486). The memory cell is responsible for the secondary response. (The first response to the antigen when only "virgin" B-cells respond is called the primary response.) The secondary response is characterized by a lower threshold dose of antigen, a quicker response, higher total antibody production, and longer persistence of antibody synthesis as well as higher peak concentrations of antibody then in a primary response (Eisen [3], p. 486). The secondary response is mainly responsible for what is usually called "immunity."

Another property of the immune response is tolerance or the selective unresponsiveness of the system to presentation of antigen in doses usually sufficient for initiation of the response. Tolerance may be induced in two ways (Eisen [3], p. 493):

1) Injection of small quantities of antigen at short intervals for a period of time (low zone tolerance); and

2) Injection of large quantities of antigen at short intervals for a period of time (high zone tolerance).

The mechanisms for tolerance, as for stimulation, are not yet fully understood.

Section 2. A Model of B-Cell Stimulation

This model describes the dynamics of the stimulation of B-cells due to presence of antigen.

Let y = number of unstimulated B-cells which can respond to a particular antigen

and z = number of stimulated B-cells which have been stimulated by the presence of that antigen.

Then $x = \frac{y - z}{y + z}$ is a dimensionless number, $-1 \leq x \leq 1$, which measures the amount of stimulation of the system (+1 when unstimulated, -1 when stimulated).

Let a = concentration of free antibody

b = concentration of unbound antigen (sites).

We assume that the rate of change in free antibody is directly proportional to the proportion of stimulated B-cells. Also free antibody becomes bound to antigen at a rate given by mass action and that antibody decays at a rate proportional to the amount present. Thus

$$(1) \qquad \frac{da}{dt} = \frac{\delta}{2}(1 - x) - a - \gamma_1 ab$$

where time unit is scaled to eliminate the constant before the "a" term and δ, $\gamma_1 > 0$. In this paper we restrict $\delta < \delta_0$ where δ_0 is the positive root of $\frac{\delta^2}{4} - \delta - 1 = 0$ $(4 < \delta_0 < 5)$. Similarly, the elimination of antigen is governed by

$$(2) \qquad \frac{db}{dt} = -\gamma_1 ab - \gamma_2 b$$

where γ_1, $\gamma_2 > 0$, γ_2 is the "nonspecific removal rate" or the rate of breakdown of antigen not due to free antibody.

The equation governing x is

$$(3) \qquad \varepsilon \frac{dx}{dt} = -(x^3 + (a - 1/2)x + b - 1/2)$$

where $1 >> \varepsilon > 0$ and indicative of the fact that the stimulation itself is taking place on a much faster time scale than the time scale of the response as indicated in (1) by the rate of exponential decay of antibody (minutes as opposed to days).

The model becomes $(\cdot = \frac{d}{dt})$

$$(4)_\varepsilon \qquad \begin{aligned} \varepsilon\dot{x} &= -(x^3 + (a - 1/2)x + b - 1/2) \\ \dot{a} &= \delta/2(1 - x) - a - \gamma_1 ab \\ \dot{b} &= -\gamma_1 ab - \gamma_2 b \,. \end{aligned}$$

The only equilibrium in the region $a \geq 0$, $b \geq 0$, $-1 \leq x \leq 1$ for $0 \leq \delta \leq \delta_0$ is $(x_0, a_0, b_0) = (1,0,0)$. This corresponds to all B-cells unstimulated, no free antibody, and no free antigen. We call this the "virgin state". Injection of antigen will correspond to initial conditions $x(0) = 1$, $a(0) = 0$, $b(0) = \tilde{b} > 0$ where $\tilde{b}$ is the concentration of antigen injected.

The surface $f = x^3 + (a - 1/2)x + b - 1/2 = 0$ is called the "cusp singularity" and is pictured in figure 1. The cusp singularity takes its name from the projection of the surface to the $a - b$ space (see figure 2). The fold curves of the cusp singularity are the places on the surface $f = 0$ where the normal is parallel to the $a - b$ plane, that is, where $\frac{\partial f}{\partial x} = 3x^2 + a - 1/2 = 0$. These fold curves when projected to $a - b$ space become the "cusp", the curve $\frac{(a - 1/2)^3}{27} + \frac{(b - 1/2)^2}{4} = 0$. The cusp divides the $a - b$ plane into two regions; inside the cusp f associates three x values with every pair (a,b) and outside f associates only one.

Since ε is small, solutions of $(4)_\varepsilon$ would be expected to stay near the surface $f = 0$.

The biological assumptions which correspond to choosing equation (3) are the following:

1) A B-cell has an internal regulatory mechanism which is responsible for reading information from the surface and responding with stimulation and antibody production if needed.

2) This mechanism defines a clustering of allowable states in the x-a-b space.

3) This clustering occurs about the surface $f = 0$.

The point $a = 1/2$, $b = 1/2$ was chosen arbitrarily and we consider $0 \leq b(0) \leq 1$.

Concerning the behavior of the solutions of the initial value problem $(4)_\varepsilon$ with initial conditions $x(0) = 1$, $a(0) = 0$, $0 < b(0) = \tilde{b} \leq 1$ we prove the following:

<u>THEOREM 1</u>. For all $\varepsilon > 0$, $0 < \delta < \delta_0$, unique solutions of the initial value problem $(4)_\varepsilon$ exist on $[0,\infty)$, all solutions are bounded and $\lim_{t\to\infty} (x(t),a(t),b(t)) = (1,0,0)$. Moreover, $-1 \leq x(t) \leq 1$, $0 \leq a(t) \leq \delta$, $0 \leq b(t) \leq \tilde{b} \leq 1$ for all $t \in [0,\infty)$.

The proof of the theorem depends on the following lemmas.

<u>Lemma 1</u>. For all $\varepsilon > 0$ and $0 < \delta < \delta_0$, $x = 1$, $a = 0$, $b = 0$

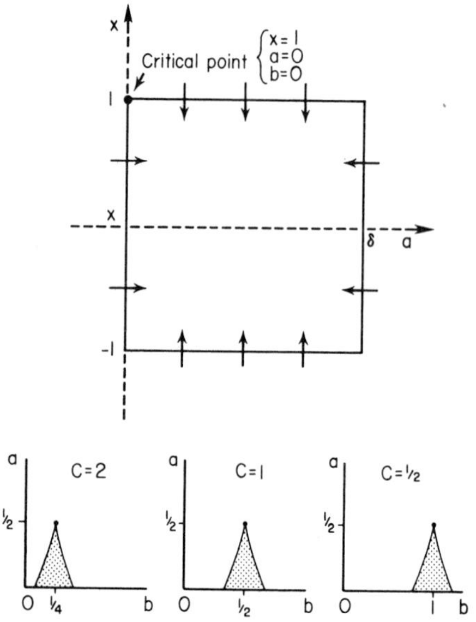

FIGURE 1

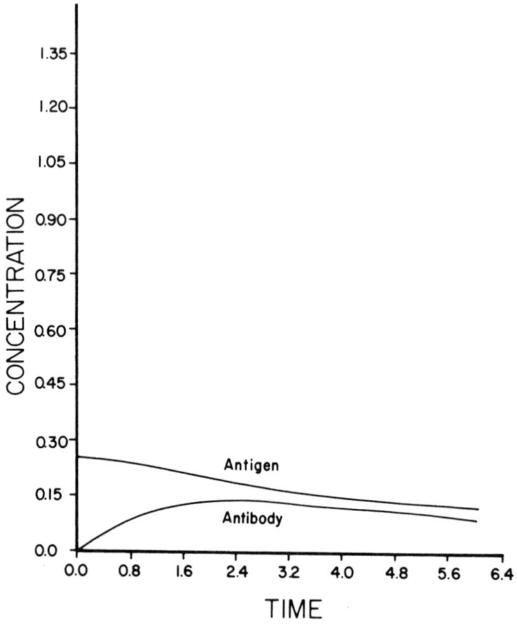

FIGURE 2

is an asymptotically stable critical point of $(4)_\varepsilon$.

Proof: Linearizing $(4)_\varepsilon$ about (1,0,0) we have

$$\begin{pmatrix} x \\ a \\ b \end{pmatrix}' = A \begin{pmatrix} x \\ a \\ b \end{pmatrix} \quad \text{where} \quad A = \begin{pmatrix} -\frac{5}{2\varepsilon} & -\frac{1}{\varepsilon} & -\frac{1}{\varepsilon} \\ -\frac{\delta}{2} & -1 & 0 \\ 0 & 0 & -\gamma_2 \end{pmatrix}$$

The eigenvalues of A are the roots, λ, of the equation

$$0 = -(\frac{5}{2\varepsilon} + \lambda)(1 + \lambda)(\gamma_2 + \lambda) + (\gamma_2 + \lambda)\frac{\delta}{2\varepsilon}$$

$$= -(\gamma_2 + \lambda)[(\frac{5}{2\varepsilon} + \lambda)(1 + \lambda) - \frac{\delta}{2\varepsilon}]$$

$$= -(\gamma_2 + \lambda)[\lambda^2 + (1 + \frac{5}{2\varepsilon})\lambda + \frac{5 - \delta}{2\varepsilon}] .$$

For $\delta > 5$, all three eigenvalues have negative real parts.

Lemma 2. The region $\mathcal{D}$ given by:

$$\mathcal{D} = \{(x,a,b) \mid -1 \le x \le 1, \quad 0 \le b \le \tilde{b}, \quad 0 \le a \le \delta\}$$

is positively invariant under $(4)_\varepsilon$. That is, if $(x(0), a(0), b(0)) \in \mathcal{D}$, then $(x(t),a(t),b(t)) \in \mathcal{D}$ for all $t \ge 0$ where $(x(t),a(t),b(t))$ is the solution of $(4)_\varepsilon$ with initial condition $(x(0),a(0),b(0))$.

Proof: We will show no solution of $(4)_\varepsilon$ in D or on $\partial\mathcal{D}$ can leave $\mathcal{D}$ by showing all faces of $\mathcal{D}$ are composed of strict entrance points (the vector field points in) or the faces themselves are invariant, and from uniqueness of solutions deduce that no solution which intersects that face can leave.

face (i): When $b = \tilde{b}, \quad -1 \le x \le 1, \quad 0 \le a \le \delta$.

On (i), $\dot{b} = (-\gamma_1 a - \gamma_2)\tilde{b} < -\gamma_2\tilde{b} < 0$. Thus b is decreasing at each point of the face.

face (ii): $x = 1, \quad 0 \le b \le \tilde{b}, \quad 0 \le a \le \delta$.

On (ii), $\dot{x} = -\frac{1}{\varepsilon}(1 + a - 1/2 + b - 1/2) = -\frac{1}{\varepsilon}(a + b) \le 0$ and

$\dot{x} = 0$ if and only if $a = b = 0$ which is the critical point.

face (iii): $x = -1, \quad 0 \le b \le \tilde{b}, \quad 0 \le a \le \delta.$

On (iii), $\dot{x} = -\frac{1}{\epsilon}(-1 + (a - 1/2)(-1) + b - 1/2) = -\frac{1}{\epsilon}(-1 - a + b) \ge 0$ since $\tilde{b} \le 1$. Further, $\dot{x} = 0$ if and only if $b = 1$ and $a = 0$. When $b = 1$, $a = 0$, and $x = -1$ we have $\dot{a} = \delta > 0$, $\dot{b} = -\gamma_2 < 0$, so any solution through $(-1,0,1)$ must be tangent to face (iii). Also,

$$\epsilon\ddot{x} = -[(3x^2 + a - 1/2)\dot{x} + x\dot{a} + \dot{b}].$$

Evaluated at $(-1,0,1)$, we have

$$\epsilon\ddot{x}\Big|_{(-1,0,1)} = \delta + \gamma_2 > 0.$$

face (iv): $a = \delta, \quad -1 \le x \le 1, \quad 0 \le b \le \tilde{b}.$

On (iv), $\dot{a} = \frac{\delta}{2}(1 - x) - \delta(1 + \gamma_1 b) \le \frac{\delta}{2}(2) - \delta = 0$. Now $\dot{a} = 0$ if and only if $b = 0$, $x = -1$, $a = \delta$, and $\ddot{a}\Big|_{(-1,\delta,0)} = -\frac{\delta}{2\epsilon}(1 + \delta) < 0$, $\epsilon\dot{x}\Big|_{(-1,\delta,0)} = 1 + \delta > 0$, $\dot{b}\Big|_{(-1,\delta,0)} = 0$.

face (v): $a = 0, \quad -1 \le x \le 1, \quad 0 \le b \le \tilde{b}.$

On (v), $\dot{a} = \frac{\delta}{2}(1 - x) \ge 0$ and $\dot{a} = 0$ if and only if $x = 1$, $0 \le b \le \tilde{b}$. $(1,0,0)$ is the critical point, so assume $b > 0$. Along the edge $x = 1$, $a = 0$, $0 < b \le \tilde{b}$, $\dot{x} < 0$, and $\dot{b} < 0$, thus any solution through $(1,0,b)$ must be tangent to face (v) at $t = 0$. Also

$$\ddot{a}\Big|_{(1,0,b)} = \frac{\delta}{2\epsilon}\, b > 0.$$

The final face is invariant.

face (vi): $b = 0, \quad -1 \le x \le 1, \quad 0 \le a \le \delta.$

With any initial condition having $b(0) = 0$, the (unique) solution of $(4)_\varepsilon$ has the form $(x(t),a(t),0)$. Moreover, from faces (ii), (iii), (iv) and (v) we have the vector field always pointing in except at the critical point $(1,0,0)$ (see figure 3). We have face (vi) invariant. This completes the proof.

Proof of Theorem 1: Since $\mathcal{L}$ is invariant by Lemma 2 and solutions of $(4)_\varepsilon$ with initial condition $x(0) = 1$, $a(0) = 0$, $1 \ge b(0) = \tilde{b} > 0$ start on $\mathcal{L}$, and since unique local solutions exist

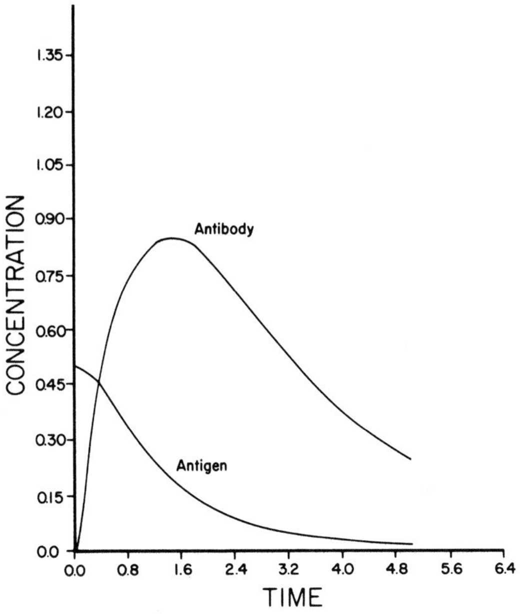

FIGURE 3

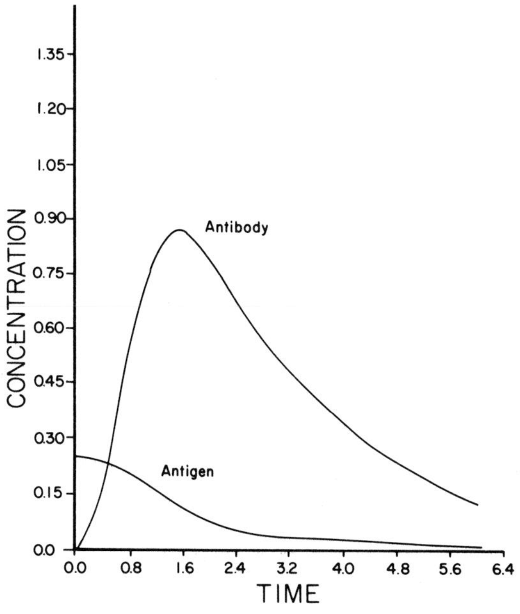

FIGURE 4

to $(4)_\varepsilon$ for all initial values in $\mathcal{D}$; all solutions are bounded and exist on $[0,\infty)$. Moreover, if $(x(t),a(t),b(t))$ is a solution of $(4)_\varepsilon$ in $\mathcal{D}$ then the ω-limit set of $(x(t),a(t),b(t))$ is contained in face (vi) as $0 \leq b(t) \leq e^{-\gamma_0 t}$. Also since face (vi) is invariant, any solution there satisfies

$$(5)_\varepsilon \qquad \begin{aligned} \dot{x} &= -(x^3 + (a - 1/2)x - 1/2) \\ \dot{a} &= \frac{\delta}{2}(1 - x) - a \end{aligned}$$

$$-1 \leq x(0) \leq 1, \quad 0 \leq a(0) \leq \delta.$$

Let $(\tilde{x}(t),\tilde{a}(t))$ be the solution of $(5)_\varepsilon$ through $(\tilde{x}(0), \tilde{a}(0)) \in \mathcal{D}$ then $\lim_{t\to\infty} (\tilde{x}(t),\tilde{a}(t)) = (1,0)$ by the Poincaré-Bendixson theorem, since there can be no periodic orbits (no critical points in the interior of face (vi)) and since $(1,0)$ is an asymptotically stable critical point of $(5)_\varepsilon$ (see proof of Lemma 1).

Now let $(x(t),a(t),b(t))$ be a solution of $(4)_\varepsilon$ and let M be the ω-limit set of $(x(t),a(t),b(t))$ and $(\tilde{x},\tilde{a},0) \in M$, then since M is invariant and closed by the above we have $(1,0,0) \in M$. Since $(1,0,0)$ is an attractor of $(4)_\varepsilon$ by Lemma 1, $M = \{(1,0,0)\}$ and $\lim_{t\to\infty} (x(t),a(t),b(t)) = (1,0,0)$.

<u>Corollary</u>. If $(x(t_0),a(t_0),b(t_0)) \in \mathcal{D}$, then the solution of $(4)_\varepsilon$ through $(x(t_0),a(t_0),b(t_0)) \to (1,0,0)$ as $t \to \infty$. Moreover, $(x(t),a(t),b(t)) \in \mathcal{D}$ for all time $t \geq 0$.

Section 3. Construction of a Model of Humoral Immune Response

A model of the primary and secondary response will be constructed by modifying the model of B-cell stimulation by adding a differential equation governing the numbers of small B and memory cells. The following assumptions will be used:

The presence of antigen in excess of antibody will cause differentiation of cells in the clone to (terminal) plasma cells proportional to the numbers of B-cells. We also assume there is a limit, c_{max}, in the number of small B-cells and memory cells. As the number of cells approaches that limit, we assume logistic growth. If

$$c = \frac{\text{number of small B-cells + memory cells}}{\text{original number of small B-cells}},$$

we have

$$(6) \qquad \frac{dc}{dt} = \gamma_3 c(a - b)(c_{max} - c), \quad \gamma_3 > 0.$$

Since the secondary response $(c(0) >> 1)$ is characterized by a lower threshold dose of antigen (see section 1), we will modify equation $(3)_\varepsilon$ to become

$$(7)_\varepsilon \qquad \varepsilon \frac{dx}{dt} = -(x^3 + (a - 1/2)x + b \cdot c - 1/2).$$

The virgin state becomes $(x_0, a_0, b_0, c_0) = (1,0,0,1)$ and the model of humoral immune response is

$$(8)_\varepsilon \qquad \begin{aligned} \varepsilon\dot{x} &= -(x^3 + (a - 1/2)x + b \cdot c - 1/2) \\ \dot{a} &= \frac{\delta}{2}(1 - x) - a - \gamma_1 ab \\ \dot{b} &= -\gamma_1 ab - \gamma_2 b \\ \dot{c} &= \gamma_3 c(a - b)(c_{max} - c) \end{aligned}$$

$$x(0) = 1, \quad a(0) = 0, \quad b(0) = \tilde{b} \leq 1, \quad c(0) = \tilde{c}, \quad 0 \leq \tilde{c} \leq c_{max}$$

We can also replace $a - b$ in (6) by $g(a,b)$ where $g \in C^1(\mathbb{R}^2, \mathbb{R})$, $g(0,0) = 0$ is arbitrary with almost no change in most of the arguments. Note that every point of the form $x = 1$, $a = 0$, $b = 0$, $c = \tilde{c}$ is a critical point, so we have an infinite number of possible rest states between $c = 0$ (no defense) and $c = c_{max}$.

The effect of the product $b \cdot c$ in $(7)_\varepsilon$ is to move the fold surves (the threshold points) nearer the starting state $(1,0,0,\tilde{c})$ for $\tilde{c} > 1$ and farther for $\tilde{c} < 1$ than in the primary response (see figure 4).

In this model, x is still considered a stimulation parameter but is no longer restricted to $[-1,1]$. x will now be restricted to $[h(c_{max}),1]$ where $h(c_{max})$ is the unique root of the equation $x^3 - 1/2\, x + c_{max} - 1/2 = 0$ (we assume $c_{max} > 1$). Since values of x will now depend on c, the differential equation governing $a(t)$ need not contain c explicitly.

As in the first model we prove

THEOREM 2: For all $\varepsilon > 0$, and $0 < \delta < \delta_0$, unique solutions of the initial value problem $(8)_\varepsilon$ exist on $[0,\infty)$, all solutions are bounded and the ω-limit set of any solution is contained in the segment of critical points $x = 1$, $a = 0$, $b = 0$, $c = \tilde{c}$ where $0 \le \tilde{c} \le c_{max}$. Moreover, $h(c_{max}) \le x(t) \le 1$, $0 \le a(t) \le \frac{\delta}{2}(1 - h(c_{max}))$, $0 \le b(t) \le \tilde{b} \le 1$ and $0 \le c(t) \le c_{max}$ for all $t \in [0,\infty)$.

The proof is very similar to Theorem 1 and depends on the following lemmas:

Lemma 3. The region $\mathcal{R}$ given by

$$\mathcal{R} = \{(x,a,b,c) \in \mathbb{R}^4 \,|\, h(c_{max}) \le x \le 1,\ 0 \le a \le \frac{\delta}{2}(1 - h(c_{max})),$$

$0 \le b \le \tilde{b} \le 1$, $0 \le c \le c_{max}\}$ is invariant.

Proof: As before, the flow is analyzed on the eight faces. The faces $c = 0$, $c = c_{max}$, and $b = 0$ are themselves invariant; the others follow as before.

Definition. Let $A : \dot{x}_i = f_i(x,t)$, $A_\infty : \dot{x}_i = g_i(x)$ for $i = 1, 2, \ldots, n$ be first order systems of differential equations with f_i continuous for $(x,t) \in G \times [t_0,\infty)$ and g_i continuous for $x \in G$ where G is an open subset of $\mathbb{R}^n$. A is asymptotic to A_∞ $(A \to A_\infty)$ in G if for $t > t_0$, f_i and g_i are locally Lipschitz in x and for each compact set $K \subset G$ and $\varepsilon' > 0$ there is a $T = T(K,\varepsilon') > t_0$ such that $|f_i(x,t) - g_i(x)| < \varepsilon'$ for all $i = 1,2,\ldots,n$, all $x \in K$ and all $t > T$.

THEOREM (Markus [6]). Let $A \to A_\infty$ in G and P be an asymptotically stable critical point of A_∞, then there is a neighborhood N of P and a time T such that the ω-limit set of every solution $x(t)$ of A which intersects N at a time later than T is P.

We use this to prove the following:

Lemma 4. Let $(x(t), a(t), b(t), c(t))$ be a solution of $(8)_\varepsilon$ with $x(0) = 1$, $a(0) = 0$, $b(0) = \tilde{b} \leq 1$, and $0 \leq c(0) \leq c_{max}$ on $t \in [0,\infty)$, then there exists a neighborhood N of $x = 1$, $a = 0$ and a time T such that every solution $(\tilde{x}(t), \tilde{a}(t))$ of

$$(9) \quad \begin{aligned} \varepsilon\dot{\tilde{x}} &= -(\tilde{x}^3 + (\tilde{a} - 1/2)x + c(t)\cdot b(t) - 1/2) \\ \dot{\tilde{a}} &= \frac{\delta}{2}(1 - \tilde{x}) - \tilde{a} - \gamma_1 \tilde{a} b(t) \end{aligned}$$

which intersects N at time $t \geq T$ has the property $\lim_{t\to\infty} (x(t), a(t)) = (1,0)$.

Proof: We will show (9) is asymptotic to

$$(10) \quad \begin{aligned} \varepsilon\dot{\tilde{x}} &= -(x^3 + (a - 1/2)x - 1/2) \\ \dot{\tilde{a}} &= \frac{\delta}{2}(1 - \tilde{x}) - \tilde{a} \end{aligned}$$

(which is $(5)_\varepsilon$ in proof of Theorem 1) in a neighborhood G of $\tilde{x} = 1$, $\tilde{a} = 0$.

As the linearization of (10) about $\tilde{x} = 1$, $\tilde{a} = 0$ has two eigenvalues with negative real part, there is a neighborhood G of (1,0) so that (1,0) is an attractor of (10). In G, assume $|\tilde{a}| \leq \frac{\gamma_2}{\gamma_1}$. Now

$$|b(t)| \leq |b(0)|e^{-\gamma_2 t} \quad \text{for } t \geq 0.$$

Let K be a compact subset of G and $\varepsilon' > 0$. Choose $t_0 > 0$ such that

$$|b(0)|e^{-\gamma_2 t_0} < \min\left\{\frac{\varepsilon\cdot\varepsilon'}{c_{max}}, \frac{\varepsilon}{\gamma_2}\right\},$$

then

$$\begin{aligned} &\left| -\frac{1}{\varepsilon}(\tilde{x}^3 + (\tilde{a} - 1/2)\tilde{x} + c(t)\cdot b(t) - 1/2) \right. \\ &\qquad \left. + \frac{1}{\varepsilon}(\tilde{x}^3 + (\tilde{a} - 1/2)\tilde{x} - 1/2) \right| \\ &= \left| \frac{1}{\varepsilon} c(t) b(t) \right| \leq \left| \frac{1}{\varepsilon} c_{max}\, b(t) \right| < \varepsilon' \quad \text{for all } t \geq t_0 \end{aligned}$$

and $|[\frac{\delta}{2}(1-\tilde{x}) - \tilde{a} - \gamma_1\tilde{a}b(t)] - [\frac{\delta}{2}(1-\tilde{x}) - \tilde{a}]| = |\gamma_1\tilde{a}b(t)| < \varepsilon'$ for all $t \geq t_0$. Thus (9)→(10). An application of the Markus theorem completes the proof.

Proof of Theorem 2: As in Theorem 1, solutions of $(8)_\varepsilon$ which start in $\mathcal{R}$ stay in $\mathcal{R}$ and exist on $[0,\infty)$. They are therefore bounded and satisfy the estimates in the statement of the theorem by Lemma 3.

Let $(x(t),a(t),b(t),c(t))$ be a solution of $(8)_\varepsilon$ with initial value in $\mathcal{R}$. Let M be the ω-limit set of $(x(t),a(t),b(t),c(t))$. Then $M \subset \{(x,a,b,c) \in \mathcal{R} | b = 0\}$. Let $(x_0,a_0,0,c_0) \in M$. Since M is invariant under $(8)_\varepsilon$, the first two components of the solution, $(x(t),a(t),b(t),c(t))$ through $(x_0,a_0,0,c_0)$ must satisfy

$$\text{(11)} \quad \begin{aligned} \varepsilon\dot{\tilde{\tilde{x}}} &= -(\tilde{\tilde{x}}^3 + (\tilde{\tilde{a}} - 1/2)\tilde{\tilde{x}} - 1/2) \\ \dot{\tilde{\tilde{a}}} &= \frac{\delta}{2}(1-\tilde{\tilde{x}}) - \tilde{\tilde{a}}. \end{aligned}$$

The two equations are system $(5)_\varepsilon$ in the proof of Theorem 1; thus solutions of $(11)_\varepsilon$ through points of M satisfy $\tilde{\tilde{x}} \to 1$, $\tilde{\tilde{a}} \to 0$ as $t \to \infty$, so that M contains at least one point of the form $(1,0,0,c^*)$ as M is closed. Since $(1,0,0,c^*) \in M$, there exists a sequence $\{t_n\} \to \infty$ such that $||(x(t_n) - 1, a(t_n), b(t_n), c(t_n) - c^*|| < 1/n$ for all $n > 0$. This implies that $(x(t),a(t))$ must enter any neighborhood N of $(1,0)$ for a whole sequence $\{t_n\} \to \infty$ and from Lemma 4, $(x(t),a(t)) \to (1,0)$ as $t \to \infty$. Thus we have that if $(x,a,b,c) \in M$, $x = 1$, $a = 0$, $b = 0$, or $M \subset \{x = 1,\ a = 0,\ b = 0,\ 0 \leq c \leq c_{max}\}$.

The following result uses that in this model, $g(a,b)$ is of the form $g(a,b) = a - b$, although the representation for $c(t)$ holds for all $g(a,b) \in C^1[\mathbb{R}^2,\mathbb{R}^1]$.

THEOREM 3. Assume the same hypotheses as in Theorem 2. Then $\lim_{t\to\infty} (x(t),a(t),b(t),c(t)) = (1,0,0,c^*)$ for some $c^* \in [0,c_{max}]$.

Proof: Since $\dot{c} = \gamma_3(c_{max} - c)c(a(t) - b(t)) = c_{max}\gamma_3(a(t) - b(t))c - c^2\gamma_3(a(t) - b(t))$, we have

$$c(t) = \frac{e^{\int_0^t c_{max}\gamma_3(a(s)-b(s))ds}}{\frac{1}{c(0)} + \int_0^t \left[\gamma_3(a(s)-b(s))e^{\int^s c_{max}\gamma_3(a(u)-b(u))du}\right] ds}$$

$$= \frac{e^{\int_0^t c_{max}\gamma_3(a(s)-b(s))ds}}{\left[\frac{1}{c_{max}} e^{+\int_0^t c_{max}\gamma_3(a(s)-b(s))ds} - \frac{1}{c_{max}}\right] + \frac{1}{C(0)}}$$

$$= \frac{1}{e^{-\int_0^t c_{max}\gamma_3(a(s)-b(s))ds}\left[\frac{1}{c(0)} - \frac{1}{c_{max}}\right] + \frac{1}{c_{max}}}$$

or

$$c(t) = \frac{c_{max}}{e^{-\int_0^t c_{max}\gamma_3(a(s)-b(s))ds}\left[\frac{c_{max}}{c(0)} - 1\right] + 1}. \tag{12}$$

If $\int_0^\infty(a(s)-b(s))ds<\infty$, $\lim_{t\to\infty} c(t)$ exists. If $c(0) = c_{max}$, $c(t) \equiv c_{max}$ for all $t \in [0,\infty)$. Note that if $\int_0^\infty(a(s)-b(s))ds = +\infty$, $\lim_{t\to\infty} c(t) = c_{max}$. Further, $\int_0^t(a(s)-b(s))ds > \int_0^\infty -b(s)ds > -\infty$ as $b(t) \in L_1$, $b(t) \geq 0$. Let

$$N_t = \{0 \leq s \leq t \mid a(s) - b(s) \leq 0\}$$

and

$$P_t = \{0 \leq s \leq t \mid a(s) - b(s) \geq 0\}.$$

Then

$$\int_0^t (a(s)-b(s))ds = \int_{P_t} (a(s)-b(s))ds + \int_{N_t} (a(s)-b(s))ds$$

and for all t, $0 \geq \int_{N_t} (a(s)-b(s))ds \geq \int_0^\infty -b(s)ds > -\infty$.

Since $F(t) = \int_{P_t} (a(s)-b(s))ds$ is a nondecreasing function of t, $\lim_{t\to\infty} \int_0^t [a(s)-b(s)]ds$ exists.

COROLLARY. For the initial value problem $(8)_\varepsilon$ with initial condition $x(0) = 1$, $a(0) = 0$, $b(0) = \tilde{b} > 0$, $0 < c(0) \leq c_{max}$,

$$\lim_{t\to\infty} c(t) \geq \frac{c_{max}}{e^{\frac{1}{\gamma_2} c_{max}\gamma_3\, b(0)}\left[\frac{c_{max} - c(0)}{c(0)}\right] + 1} > 0 .$$

Moreover, $c_{max} \geq c(t) > 0$ for all $t \in [0,\infty)$.

Proof: Since

$$\int_0^t c_{max}\gamma_3(a(s) - b(s))ds \geq \int_0^t c_{max}\gamma_3(-b(s))ds$$

then

$$e^{-\int_0^t c_{max}\gamma_3(a(s) - b(s))ds} \leq e^{\int_0^t c_{max}\gamma_3(b(s)ds}$$

$$\leq e^{\int_0^t c_{max}\gamma_3 b(0)e^{-\gamma_2 s} ds}$$

$$= e^{-\frac{1}{\gamma_2} c_{max}\gamma_3 b(0)\left[e^{-\gamma_2 t} - 1\right]} \leq e^{\frac{1}{\gamma_2} c_{max}\gamma_3 b(0)} .$$

Using this estimate in (12) yields

$$c(t) = \frac{c_{max}}{e^{-\int_0^t c_{max}\gamma_3(a(s)-b(s))ds}\left[\frac{c_{max} - c(0)}{c(0)}\right] + 1}$$

$$\geq \frac{c_{max}}{e^{\frac{1}{\gamma_2} c_{max}\gamma_3 b(0)}\left[\frac{c_{max} - c(0)}{c(0)}\right] + 1} .$$

Section 4. Numerical Experiments.

Simulations of model $(8)_\varepsilon$, $\varepsilon = .01$, were made using a standard fourth order Runge-Kutta with stepsize and error control. Some properties of the model not covered by Theorems 2 and 3 were demonstrated for

$$\delta = 2, \quad \gamma_1 = 1, \quad \gamma_2 = .01, \gamma_3 = 1, \quad \text{and} \quad c_{max} = 3.$$

A primary response is shown in figure 6 $(c(0) = 1.5)$. A secondary response is shown in figure 7 $(c(0) = c_{max} = 3)$. Low dose unresponsiveness is shown in figure 5. Figure 8 shows low zone tolerance induced by stopping the program after 1 time unit and restarting with initial conditions $x_0 = x(1)$, $a_0 = a(1)$, $b_0 = b(1) + \tilde{b}$, $c_0 = c(1)$ and repeating for 6 time units. Further results can be seen in Table 1 as well as the effect of dosage on induction of tolerance.

The nature of the threshold in $\tilde{b}$ is controlled by the value of ε. Even though the stimulation parameter x is probably hard to measure experimentally, ε can be measured indirectly by comparing the numerical experiments to data on the primary response.

The model suggests a general way to induce tolerance, both low and high zone. For low zone tolerance, the primary response should never be triggered, this says the doses must be small. The mathematical idea is to keep $a - b$ negative. Every time a approaches b, inject more antigen. Since $c(t)$ will then be

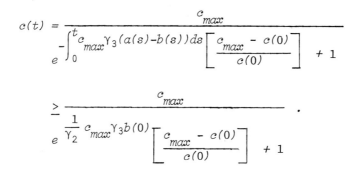

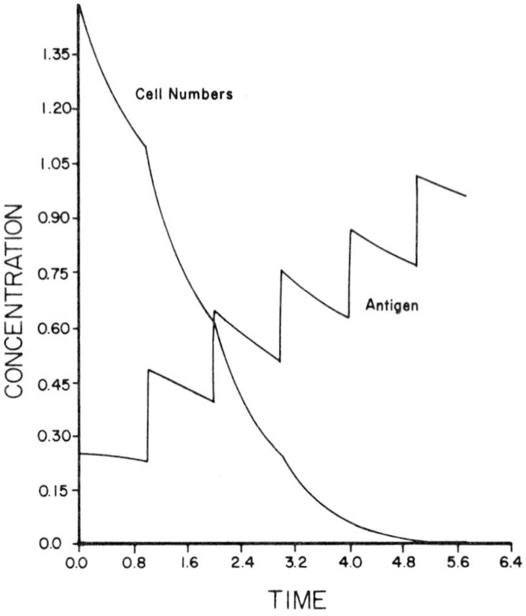

FIGURE 5

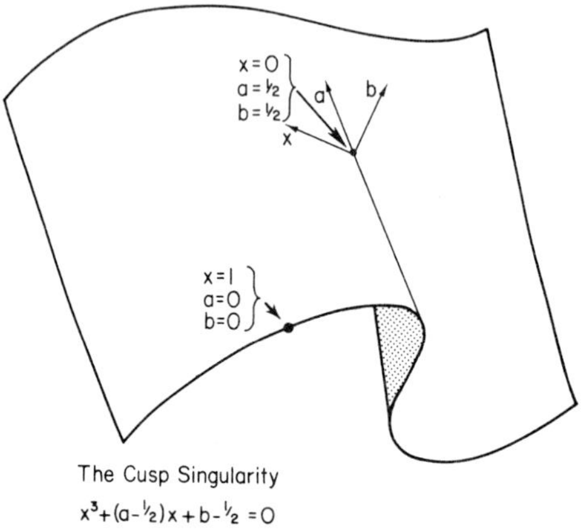

FIGURE 6

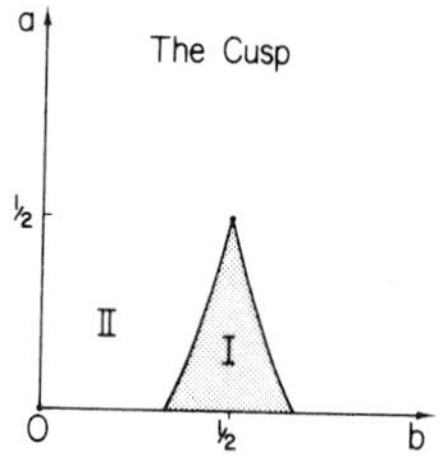

FIGURE 7

monotone decreasing, the threshold will move to larger b values. Thus the process may then be made more rapid by gradually increasing the dosage with each injection, driving $c(t)$ to very small values and thus tolerance.

High zone tolerance may also be induced by injecting an initial large antigen dose and re-injecting when a approaches b. Because the threshold has already been exceeded, dosage is not as important as timing in inducing high zone tolerance. Notice that no tolerance of any type can be induced if $c(0) = c_{max}$.

REFERENCES

[1] Bell, G.I., *Mathematical model of clonal selection and antibody production*, in three parts, Journal of Theoretical Biology 29(1970), 191-232; 33(1971), 339-378; 33(1971), 379-398.

[2] Bruni, C.,Giovenco, M. A., Koch, G., Strom, R., *A dynamical model of humoral immune response*, Math. Biosciences 27(1975), 191-211.

[3] Eisen, H. N., *Immunology* (reprinted from Davis, Dulbecco, Eisen, Ginsberg and Wood, Microbiology, 2nd ed.), Harper & Row (1974).

[4] Hale, Jack, *Ordinary Differential Equations*, Wiley-Interscience (1969).

[5] Hoffman, G. W., *A theory of regulation and non-self discrimination in an immune network*, Eur. J. Immunol. 5(1975), 638-647.

[6] Markus, L., *Asymptotically autonomous differential systems*, in Contributions to the Theory of Nonlinear Oscillations, Vol. 3(1956), 17-29.

[7] Merrill, S. J., *Mathematical models of humoral immune response*, Technical Report IM76-1, Department of Mathematics, University of Iowa, Jan. 1976.

[8] Neminskii, V. V. and Stepanov, V. V., *Qualitative Theory of Differential Equations*, Princeton University Press (1960).

[9] Richter, P. H., *A network theory of the immune system*, Eur. J. Immunol. 5(1975), 350-354.

[10] Thom, René, *Structural Stability and Morphogenesis*, W. A. Benjamin (1975).

[11] Waltman, P. and Butz, E., *A threshold model of antigen-antibody dynamics*, to appear, Journal of Theoretical Biology.

[12] Zeeman, E. C., *Differential equations for the hearbeat and nerve impulse*, in Dynamical Systems, M. M. Peixoto, Ed., Academic Press (1973).

TABLE 1

Virgin state $c_0 = 1.5$

antigen dose $\tilde{b}$	time to $a = b$	a_{max}	$c(6)$
.1	>6	.0713	1.17
.25	>6	.142	.68
.50	.38	.856	2.99
.75	.48	1.01	2.99
1.00	.60	1.03	2.99

Secondary response $c = 3.0$

$\tilde{b}$	$a = b$	a_{max}	$c(6)$
.1	.84	.172	3.0
.25	.50	.870	3.0
.50	.27	1.11	3.0
.75	.36	1.25	3.0

Induction of Tolerance $c_0 = 1.5$

$\tilde{b}$ (injected every 1 time unit)	$c(1)$	$c(2)$	$c(3)$	$c(4)$
.1	1.33	1.21	1.16	1.11
.25	1.09	.621	.251	.059

GENERALIZED STABILITY OF MOTION AND VECTOR LYAPUNOV FUNCTIONS

Roger W. Mitchell and Deborah A. Pace
University of Texas at Arlington

1. *Introduction*

The direct theory for stability of motion in terms of vector Lyapunov functions and the general comparison method is well-developed [4,5,6,7] and effectively applied for large scale dynamical systems [1,3]. However, the problem of constructing vector Lyapunov functions and developing appropriate perturbation theory has seen very little progress. Perhaps one of the reasons for this situation is that the stability definitions by means of the standard norm are not flexible enough for such a venture.

It is known [2,5,8] that a generalized space where the norm is vector valued offers a more flexible mechanism than the one which is normally used and is advantageous in certain situations. In this paper we work with generalized spaces, thus allowing us to introduce the concepts of generalized stability of motion and also boundedness. To avoid monotony we will only give a sampling of definitions and theorems which can be obtained using generalized norms.

In Section 3 we will state and prove three theorems concerning different properties of certain subspaces of R^n. One should note that these properties can be exhibited using generalized norms but could easily be unobservable if an ordinary norm were employed in the investigation. Finally, we will state a theorem combining the three results, thus emphasizing the advantages of using the generalized norm.

2. *Preliminaries*

In the following definitions and theorems we will use the vectorial inequality with the understanding that the same inequalities hold between corresponding components.

Definition 2.1. A generalized norm from R^n to R^k is a mapping $||\cdot||_G : R^n \to R^k$ denoted by $||x||_G = (\alpha_1(x),\dots,\alpha_k(x))$ such that

(a) $||x||_G \geq 0$, (that is, $\alpha_i(x) \geq 0$ for $i = 1,\dots,k$);

(b) $||x||_G = 0$ if and only if $x = 0$, (that is $\alpha_i(x) = 0$ for $i = 1,\dots,k$ if and only if $x = 0$);

(c) $||\lambda x||_G = |\lambda|\,||x||_G$, (that is $\alpha_i(\lambda x) = |\lambda|\alpha_i(x)$ for $i = 1,\dots,k$);

(d) $||x + y||_G \leq ||x||_G + ||y||_G$, (that is $\alpha_i(x + y) \leq \alpha_i(x) + \alpha_i(y)$).

It is important to note that all generalized norms (which includes all norms) generate the same topology on R^n. In particular, it is easy to show that the α_i's are ordinary norms when restricted to appropriate subspaces K_i of R^n. Furthermore, these K_i's span R^n.

Consider the generalized norm space $E = (R^n, ||\cdot||_G)$ and the following differential systems:

$$x' = f(t,x), \quad x(t_0) = x_0 \tag{2.1}$$

$$u' = g(t,u), \quad u(t_0) = u_0 \geq 0 \tag{2.2}$$

where $f \in C[R_+ \times S(p), R^n]$ and $g \in C[R_+ \times R_+^m, R^m]$. Here $R_+ = [0,\infty)$, $C[R_+ \times S(p), R^n]$ is the class of all continuous functions from $R_+ \times S(p)$ to R^n and $S(p) = \{x \in E \mid ||x||_G < p\}$ with $p \in R_+^k$ and $0 < p_i \leq \infty$ for $i = 1,\dots,k$.

Definition 2.2. The trivial solution $x = 0$ of (2.1) is equi-stable if, for each $\varepsilon \in R_+^k$, $\varepsilon > 0$, and $t_0 \in R_+$, there exists a function $0 < \delta = \delta(t_0,\varepsilon)$ in R_+^k that is continuous in t_0 for each ε such that if $||x_0||_G \leq \delta$, then $||x(t,t_0,x_0)||_G < \varepsilon$ for $t \geq t_0$.

Definition 2.3. The trivial solution $x = 0$ of (2.1) is equi-asymptotically stable if $x = 0$ is equistable and for each $\varepsilon \in R_+^k$, $\varepsilon > 0$ and $t_0 \in R_+$, there exists $\delta_0 = \delta_0(t_0)$ in R_+^k, $\delta_0 > 0$ and $T = T(t_0,\varepsilon)$ such that if $t \geq t_0 + T$ and $||x_0||_G < \delta_0$ then $||x(t,t_0,x_0)||_G < \varepsilon$.

Definition 2.4. If all solutions of (2.1) exist globally, then the system (2.1) is equibounded if, for each $\gamma \in R_+^k$ and $t_0 \in R_+$, there exists a function $0 < \beta = \beta(t_0,\gamma)$ in R_+^k that is continuous in t_0 for each γ such that if $||x_0||_G \leq \gamma$, then $||x(t,t_0,x_0)||_G < \beta$ for $t \geq t_0$.

Definition 2.5. Let $(\alpha_1(x),\ldots,\alpha_r(x))$ be a generalized norm when restricted to the subspace K of R^n. (That is, a projection of $||x||_G$ yields a generalized norm on the subspace K). Then the trivial solution of (2.1) is partially equistable with respect to the subspace K if, for each $\varepsilon \in R_+^k$, (or $\varepsilon \in R_+^r$), $\varepsilon > 0$ and $t_0 \in R_+$, there exists a function $0 < \delta = \delta(t_0,\varepsilon)$ in R_+^k that is continuous in t_0 for each ε such that if $||x_0||_G \leq \delta$, then $(||x(t,t_0,x_0)||_G)_i < \varepsilon_i$ for $t \geq t_0$, $i = 1,\ldots,r$. (That is $\alpha_i(x(t,t_0,x_0)) < \varepsilon_i$ for $t \geq t_0$, $i = 1,\ldots,r$).

We can formulate similar definitions for partial equiboundedness of (2.1) and partial equi-asymptotic stability of the solution $x = 0$ of (2.1). Also we can make similar definitions for the system (2.2) by using the generalized norm $||u||_{\hat{G}} = (|u_1|,|u_2|,\ldots,|u_m|)$. Since $u \in R_+^m$, then $||u||_{\hat{G}} = (u_1,u_2,\ldots,u_m)$. Finally, we can make similar definitions for equistability of $u = 0$ in the first s coordinates $(s \leq m)$, equi-asymptotic stability of $u = 0$ in the first s coordinates $(s \leq m)$, and for equiboundedness of (2.2) in the first s coordinates $(s \leq m)$ by restricting ε, δ, γ, and β to being in R_+^s and only comparing in the first s coordinates (or only comparing in the first s coordinates and chosing ε, δ, γ, and β

as before).

We will also need a generalization of a function of class K (see [5]).

Definition 2.6. A function $\phi \in C[R_+^r, R_+^s]$ is said to belong to class $\hat{K}$ if $\phi(v) = 0$ if and only if $v = 0$ and $v \leq w$ implies $\phi(v) \leq \phi(w)$.

Finally, when considering the system (2.1), if $V \in C[R_+ \times S(p), R_+^m]$, then throughout this paper we define $D^+V(t,x)$ for all $(t,x) \in R_+ \times S(p)$ by

$$D^+V(t,x) = \limsup_{h \to 0^+} \frac{1}{h} [V(t + h, x + h f(t,x)) - V(t,x)].$$

3. *Main Results*

The statements of the first three theorems considered below appear to be quite complicated. However, this is primarily due to the fact that we are really working with subspaces. For clarity a corollary is given after each theorem.

Theorem 3.1. Assume that

(1) $f \in C[R_+ \times S(p), R^n]$, $f(t,0) \equiv 0$;

(2) $g \in C[R_+ \times R_+^m, R^m]$, $g(t,0) \equiv 0$, g is quasimonotone nondecreasing;

(3) there exists $V \in C[R_+ \times S(p), R_+^m]$, $V(t,0) \equiv 0$, $V(t,x)$ is locally Lipschitzian in x;

(4) if $||x||_G = (\alpha_1(x), \ldots, \alpha_k(x))$, then $(\alpha_1(x), \ldots, \alpha_r(x))$ is a generalized norm when restricted to the subspace K of R^n;

(5) there exists $b \in \hat{K}$, $b : R_+^r \to R_+^s$ such that for $(t,x) \in R_+ \times S(p), b_i((\alpha_1(x), \ldots, \alpha_r(x))) \leq V_i(t,x)$, $i = 1, \ldots,$ $s \leq m$;

(6) $D^+V(t,x) \leq g(t, V(t,x))$, $(t,x) \in R_+ \times S(p)$.

Then the equistability (or partial equistability) in the first s coordinates of the trivial solution of (2.2) implies the trivial solution of (2.1) is partially equistable with respect to the subspace K.

Proof. We will do the case for equistability noting that an almost identical proof will work for the partial equistability case. Let $0 < \varepsilon < p$, $t_0 \in R_+$ be given. By Hypothesis (5) there exists a function $b(v)$ of class $\hat{K}$ such that

$$b_i((\alpha_1(x),\ldots,\alpha_r(x))) \leq V_i(t,x), \quad (t,x) \in R_+ \times S(p),$$
$$i = 1,\ldots,s.$$

Let $\psi = min\{b_i((0,\ldots,0,\varepsilon_j,0,\ldots,0)) \mid 1 \leq i \leq s,\ 1 \leq j \leq r$, and $b_i((0,\ldots,0,\varepsilon_j,0,\ldots,0)) \neq 0\}$.

If η is the s-dimensional vector with all entries equal to ψ, then $\eta > 0$. Also, for each i, there exists $k(i)$ such that $\eta_{k(i)} \leq b_{k(i)}((0,\ldots,0,\varepsilon_i,0,\ldots,0))$.

Now the equistability in the first s coordinates of $u = 0$ of (2.2) implies that given $\eta > 0$ and $t_0 \in R_+$, there exists a function $0 < \hat{\delta} = \hat{\delta}(t_0,\varepsilon)$ in R^m_+ that is continuous in t_0 for each ε, such that

$$(u(t,t_0,u_0))_i < \eta_i, \quad t \geq t_0 \quad \text{if} \quad (u_0)_i \leq (\hat{\delta})_i,$$
$$i = 1,\ldots,s.$$

Choose $u_0 = V(t_0,x_0)$. Since $V(t,x)$ is continuous and $V(t,0) \equiv 0$, there exists a function $0 < \delta = \delta(t_0,\varepsilon)$ in R^k_+ that is continuous in t_0 for each ε such that if $||x_0||_G \leq \delta$, then $V_i(t_0,x_0) \leq \hat{\delta}_i$, $i = 1,\ldots,s$.

We claim that if $||x_0||_G \leq \delta$, then $\alpha_i(x(t,t_0,x_0)) < \varepsilon_i$ for $t \geq t_0$ and $i = 1,\ldots,r$. Suppose this is not true. Then there exists a solution $x(t) = x(t,t_0,x_0)$ with $||x_0||_G \leq \delta$ and a $t_1 > t_0$ such that $\alpha_i(x(t)) < \varepsilon_i$ for $t \in [t_0,t_1)$ and $i = 1,\ldots,r$, and there exists an h such that $\alpha_h(x(t_1)) = \varepsilon_h$

where $1 \leq h \leq r$. Now by hypothesis (5)

$$b_i((\alpha_1(x(t_1)),\ldots,\alpha_r(x(t_1))) \leq V_i(t_1,x(t_1)), \quad i = 1,\ldots,s.$$

Since $||x(t)||_G < p$ for $t \in [t_0,t_1]$ and $u_0 = V(t_0,x_0)$, then using hypothesis (6) and applying Theorem 4.1.1 in [5], we obtain

$$V(t,x(t)) \leq r(t,t_0,u_0), \quad t \in [t_0,t_1],$$

where $r(t,t_0,u_0)$ is the maximal solution of (2.2). Thus, we have

$$V_i(t_1,x(t_1)) < \eta_i, \quad i = 1,\ldots,s.$$

However, for the h such that $\alpha_h(x(t_1)) = \varepsilon_h$ recall that

$$\eta_{k(h)} \leq b_{k(h)}((0,\ldots,0,\varepsilon_h,0,\ldots,0)).$$

This leads to the contradiction

$$\eta_{k(h)} \leq b_{k(h)}((0,\ldots,0,\varepsilon_h,0,\ldots,0)) \leq V_{k(h)}(t_1,x(t_1)) < \eta_{k(h)}.$$

proving that the trivial solution of (2.1) is partially equistable with respect to the subspace K.

Corollary 3.1. Assume that

(1) $f \in C[R_+ \times S(p),R^n]$, $f(t,0) \equiv 0$;

(2) $g \in C[R_+ \times R^m_+,R^m_+]$, $g(t,0) \equiv 0$, g quasimonotone nondecreasing in u;

(3) there exists $V \in C[R_+ \times S(p),R^m_+]$, $V(t,0) \equiv 0$, $V(t,x)$ is locally Lipschitzian in x;

(4) there exists $b \in \hat{K}$, $b : R^k_+ \to R^m_+$ such that for $(t,x) \in R_+ \times S(p)$, $b(||x||_G) \leq V(t,x)$;

(5) $D^+V(t,x) \le g(t,V(t,x))$ for $(t,x) \in R_+ \times S(p)$.

Then the equistability of the trivial solution of (2.2) implies the trivial solution of (2.1) is equistable.

We shall next prove a typical result covering boundedness of solutions.

Theorem 3.2. Assume that

(1) $f \in C[R_+ \times R^n, R^n]$ and all solution of (2.1) exist globally;

(2) $g \in C[R_+ \times R_+^m, R^m]$, g is quasimonotone nondecreasing in u;

(3) there exists $V \in C[R_+ \times R^n, R_+^m]$, $V(t,0) \equiv 0$, $V(t,x)$ is locally Lipschitzian in x;

(4) if $||x||_G = (\alpha_1(x),\ldots,\alpha_k(x))$, then $(\alpha_1(x),\ldots,\alpha_r(x))$ is a generalized norm when restricted to the subspace K of R^n;

(5) there exists $b \in \hat{K}$, $b : R_+^r \times R_+^s$ such that for $(t,x) \in R_+ \times R^n$, $b_i((\alpha_1(x),\ldots,\alpha_r(x))) \le V_i(t,x)$, $i = 1,\ldots,$ $s \le m$;

(6) for each i such that $1 \le i \le r$ there exists $j(i)$ such that $\lim_{v_i \to \infty} b_{j(i)}((0,\ldots,0,v_i,0,\ldots,0)) = \infty$;

(7) $D^+V(t,x) \le g(t,V(t,x))$, $(t,x) \in R_+ \times R^n$.

Then the equiboundedness (or partial equiboundedness) in the first s coordinates of the system (2.2) implies the system (2.1) is partially equibounded with respect to the subspace K.

Proof. We will do the case for equiboundedness again noting that an almost identical proof will work for the partial equiboundedness case. Let $\gamma \in R_+^k$ and $t_0 \in R_+$ be given, and let $||x_0||_G \le \gamma$. By hypothesis (3) there exists a number $\hat{\gamma} = \hat{\gamma}(t_0,\gamma)$ such that $V(t_0,x_0) \le \hat{\gamma}$ whenever $||x_0||_G \le \gamma$. Assume that the system (2.2) is equibounded in the first s coordinates. Then given $\hat{\gamma} \ge 0$ and $t_0 \in R_+$, there exists a positive vector $\beta = \beta(t_0,\gamma) = (\beta_1,\ldots,\beta_s)$ that is continuous in t for each γ

such that $u(t,t_0,u_0) < \beta$ for $t \geq t_0$ whenever $(u_0)_i \leq \hat{\gamma}_i$, $i = 1,\ldots,s$ and $u(t,t_0,u_0)$ is a solution of (2.2).

Let $\eta_i = \sup\,\{\alpha_i \mid b_j((\alpha_1,\ldots,\alpha_i,\ldots,\alpha_r)) < \beta_j$ for all $j = 1,\ldots,s\}$ for each $i = 1,\ldots,r$. Note that since $b \in \hat{K}$, then $\eta_i = \sup\,\{\alpha_i \mid b_j((0,\ldots,0\,,\alpha_i,0,\ldots,0)) < \beta_j$ for all $j = 1,\ldots,s\} > 0$. Furthermore, hypothesis (6) yields $\eta_i < \infty$. Also for each i, there exists some $k(i)$ such that

$$b_{k(i)}((0,\ldots,0,\eta_i,0,\ldots,0)) = \beta_{k(i)}$$

Let $\eta = (\eta_1,\ldots,\eta_r)$ and $u_0 = V(t_0,x_0)$. Then hypothesis (7) and Theorem 4.1.1 in [5] give us that

$$V(t,x(t,t_0,x_0)) \leq r(t,t_0,u_0), \quad t \geq t_0$$

where $r(t,t_0,u_0)$ is the maximal solution of (2.2).

We claim that if $||x_0||_G \leq \gamma$, then $(\alpha_1(x),\ldots,\alpha_r(x)) \leq \eta$. Suppose this is not true. Then there exists a solution $x(t) = x(t,t_0,x_0)$ with $||x_0||_G \leq \gamma$ and a $t_1 > t_0$ such that $(\alpha_1(x(t)),\ldots,\alpha_r(x(t))) < \eta$ for $t \in [t_0,t_1)$ and there exists an h such that $\alpha_h(x(t_1)) = \eta_h$. However, for this h, there exists some $k(h)$ such that

$$b_{k(h)}((0,\ldots,0,\eta_h,0,\ldots,0)) = \beta_{k(h)}.$$

This leads to the contradiction

$$\begin{aligned}\beta_{k(h)} &= b_{k(h)}((0,\ldots,0,\eta_h,0,\ldots,0))\\ &\leq V_{k(h)}(t_1,x(t_1)) \leq r_{k(h)}(t_1,t_0,u_0) < \beta_{k(h)}.\end{aligned}$$

Hence, we have the system (2.1) is partially equibounded with respect to the subspace K.

Corollary 3.2. Assume that

(1) $f \in C[R_+ \times R^n, R^n]$ and all solutions of (2.1) exist globally;

(2) $g \in C[R_+ \times R_+^m, R^m]$, $g(t,0) \equiv 0$, g is quasimonotone nondecreasing in u;

(3) there exists $V \in C[R_+ \times R^n, R_+^m]$, $V(t,0) \equiv 0$, $V(t,x)$ is locally Lipschitzian in x;

(4) there exists $b \in \hat{K}$, $b : R_+^k \to R_+^m$ such that for $(t,x) \in R_+ \times S(p)$, $b(||x||_G) \leq V(t,x)$;

(5) for each i such that $1 \leq i \leq k$ there exists $j(i)$ such that $\lim_{v_i \to \infty} b_{j(i)}((0,\ldots,0,v_i,0,\ldots,0)) = \infty$;

(6) $D^+V(t,x) \leq g(t,V(t,x))$, $(t,x) \in R_+ \times R^n$.

Then the equiboundedness of the system (2.2) implies the system (2.1) is equibounded.

The next result deals with asymptotic stability.

Theorem 3.3. Assume that

(1) $f \in C[R_+ \times S(p), R^n]$, $f(t,0) \equiv 0$;

(2) $g \in C[R_+ \times R_+^m, R^m]$, $g(t,0) \equiv 0$, g is quasimonotone nondecreasing in u;

(3) there exists $V \in C[R_+ \times S(p), R_+^m]$, $V(t,0) \equiv 0$, $V(t,x)$ is locally Lipschitzian in x;

(4) if $||x||_G = (\alpha_1(x),\ldots,\alpha_k(x))$, then $(\alpha_1(x),\ldots,\alpha_r(x))$ is a generalized norm when restricted to the subspace K of R^n;

(5) there exists $b \in \hat{K}$, $b : R_+^r \to R_+^s$ such that for $(t,x) \in R_+ \times S(p)$, $b_i((\alpha_1(x),\ldots,\alpha_k(x))) \leq V_i(t,x)$, $i = 1,\ldots,$ $s \leq m$;

(6) $D^+V(t,x) \leq g(t,V(t,x))$, $(t,x) \in R_+ \times S(p)$.

Then the equi-asymptotic stability (or partial equi-asymptotic stability) in the first s coordinates of the trivial solution of (2.2) implies the trivial solution of (2.1) is partially equi-asymptotically stable with respect to the subspace K.

Proof. Again we will do the case for equi-asymptotic stability noting the partial equi-asymptotic stability case has an almost identical proof. Suppose the trivial solution of (2.2) is equi-asymptotically stable in the first s coordinates. Then by Theorem 3.1 we know the trivial solution of (2.1) is partially equistable with respect to K. Let $0 < \varepsilon < p$, $t_0 \in R_+$ be given. By hypothesis (5), there exists a function $b(v)$ of class $\hat{K}$ such that

$$b_i((\alpha_1(x),\ldots,\alpha_r(x))) \le V_i(t,x), \quad (t,x) \in R_+ \times S(p),$$
$$i = 1,\ldots,s.$$

Let $\psi = \min \{b_i((0,\ldots,0,\varepsilon_j,0,\ldots,0)) \mid 1 \le i \le s, \ 1 \le j \le r,$ and $b_i((0,\ldots,0,\varepsilon_j,0,\ldots,0)) \ne 0\}$.

If η is the s-dimensional vector with all entries equal to ψ, then $\eta > 0$. Also, for each i, there exists $k(i)$ such that $\eta_{k(i)} \le b_{k(i)}((0,\ldots,0,\varepsilon_i,0,\ldots,0))$.

Now the trivial solution of (2.2) is equi-asymptotically stable in the first s coordinates, so it follows that given $\eta > 0$, $t_0 \in R_+$, there exists a positive number $T = T(t_0,\varepsilon)$ and a function $0 < \hat{\delta} = \hat{\delta}(t_0)$ in R_+^m such that

$$(u(t,t_0,u_0))_i < \eta_i, \quad t \ge t_0 + T, \quad i = 1,\ldots,s$$

if

$$(u_0)_i \le \hat{\delta}_i, \quad i = 1,\ldots,s.$$

Choose $u_0 = V(t_0,x_0)$. Since $V(t,x)$ is continuous and $V(t,0) \equiv 0$, there exists a function $0 < \delta = \delta(t_0)$ such that if $||x_0||_G \le \delta$, then $V_i(t_0,x_0) \le \hat{\delta}_i$, $i = 1,\ldots,s$.

We claim that if $||x_0||_G \le \delta$, then $\alpha_i(x(t,t_0,x_0)) < \varepsilon_i$ for $t \ge t_0 + T$ and $i = 1,\ldots,r$. Suppose this is not true. Then there exists a solution $x(t,t_0,x_0)$ with $||x_0||_G \le \delta$ and there exists a sequence $\{t_n\}_{n=1}^{\infty}$ where $t_n \ge t_0 + T$ and $\lim_{n\to\infty} t_n = \infty$ such that $\alpha_h(x(t_n,t_0,x_0)) \ge \varepsilon_h$ for some h where $1 \le h \le r$. Now proceeding as in the proof of Theorem 3.1 we are lead to the contradiction

$$\eta_{k(h)} \le b_{k(h)}((0,\ldots,0,\varepsilon_h,0,\ldots,0)) \le V_{k(h)}(t_n,x(t_n,t_0,x_0)) < \eta_{k(h)} \quad \text{for all } t_n.$$

Thus, the trivial solution is partially equi-asymptotically stable with respect to the subspace K.

Corollary 3.3. Assume that

(1) $f \in C[R_+ \times S(p), R^n]$, $f(t,0) \equiv 0$;

(2) $g \in C[R_+ \times R_+^m, R^m]$, $g(t,0) \equiv 0$, g is quasimonotone nondecreasing in u;

(3) there exists $V \in C[R_+ \times S(p), R_+^m]$, $V(t,0) \equiv 0$, $V(t,x)$ is locally Lipschitzian in x;

(4) there exists $b \in \hat{K}$, $b : R_+^k \to R_+^m$ such that for $(t,x) \in R_+ \times S(p)$, $b(||x||_G) \le V(t,x)$;

(5) $D^+V(t,x) \le g(t,V(t,x))$, $(t,x) \in R_+ \times S(p)$.

Then the equi-asymptotic stability of the trivial solution of (2.2) implies the trivial solution of (2.1) is equi-asymptotically stable.

Combining the ideas of the foregoing three theorems, we can now state a result which demonstrates the greater sensitivity of the generalized norm. We merely state the result since the proof is a direct consequence of Theorems 3.1, 3.2, and 3.3.

Theorem 3.4. Assume that

(1) $f \in C[R_+ \times R^n, R^n]$, $f(t,0) \equiv 0$, and all solutions of (2.1) exist globally;

(2) $g \in C[R_+ \times R_+^m, R^m]$, $g(t,0) \equiv 0$, g is quasimonotone nondecreasing in u;

(3) there exists $V \in C[R_+ \times R^n, R_+^m]$, $V(t,0) \equiv 0$, $V(t,x)$ is locally Lipschitzian in x;

(4) if $||x||_G = (\alpha_1(x),\ldots,\alpha_{r_1}(x),\ldots,\alpha_{r_1+r_2}(x),\ldots,$

$\alpha_{r_1+r_2+r_3}(x),\ldots,\alpha_k(x))$, then $(\alpha_1(x),\ldots,\alpha_{r_1}(x))$,

$(\alpha_{r_1+1}(x),\ldots,\alpha_{r_1+r_2}(x))$, and $(\alpha_{r_1+r_2+1}(x),\ldots,$ $\alpha_{r_1+r_2+r_3}(x))$ are generalized norms when restricted respectively to subspaces K_1, K_2, and K_3 of R^n;

(5) there exists b^j, $j = 1,2,3$, such that $b^j : R_+^{r_j} \to R_+^{s_j}$ and $b^j \in \hat{K}$ for $j = 1,2,3$; and for $(t,x) \in R_+ \times R^n$

$$b_i^1((\alpha_1(x),\ldots,\alpha_{r_1}(x))) \le V_i(t,x), \quad i = 1,\ldots,s_1$$

$$b_i^2((\alpha_{r_1+1}(x),\ldots,\alpha_{r_1+r_2}(x))) \le V_{s_1+i}(t,x), \quad i = 1,\ldots,s_2$$

$$b_i^3((\alpha_{r_1+r_2+1}(x),\ldots,\alpha_{r_1+r_2+r_3}(x))) \le V_{s_1+s_2+i}(t,x), \quad i = 1,\ldots,s_3$$

(6) for each i such that $1 \le i \le r_2$ there exists $j(i)$ such that $\lim_{v_i \to \infty} b_{j(i)}^2((0,\ldots,0,v_i,0,\ldots,0)) = \infty$;

(7) $D^+V(t,x) \le g(t,V(t,x))$, $(t,x) \in R_+ \times R^n$.

Then, if the trivial solution of (2.2) is equistable in the first s_1 coordinates, the system (2.2) is equibounded in the coordinates $s_1 + 1,\ldots,s_1 + s_2$, and the trivial solution of (2.2) is equi-asymptotically stable in the coordinates $s_1 + s_2 + 1,\ldots,$ $s_1 + s_2 + s_3$, then the trivial solution of (2.1) is partially equistable with respect to the subspace K_1, the system (2.1) is partially equibounded with respect to the subspace K_2, and partially equi-asymptotically stable with respect to the subspace K_3.

4. *Example*

Consider the system

$$(4.1)\qquad \begin{aligned} x_1' &= x_1 e^{-t} + x_2 \sin t - x_1 x_4^2 \\ x_2' &= x_1 \sin t + x_2 e^{-t} - x_2 x_4^2 \\ x_3' &= e^{-t} - x_2^2 x_3 \\ x_4' &= (e^{-t} - 2) x_4 - x_1^2 x_4 \\ x_5' &= -3x_5 - x_2^2 x_5 , \end{aligned}$$

for $t \geq 0$.

Suppose we choose the vector Lyapunov function

$$V(t,x) = \begin{bmatrix} V_1(t,x) \\ V_2(t,x) \\ V_3(t,x) \\ V_4(t,x) \end{bmatrix} = ||x||_G^2$$

where $||x||_G^2$ means square component-wise,

$$x = \begin{bmatrix} x_1 \\ x_2 \\ x_3 \\ x_4 \\ x_5 \end{bmatrix}, \quad \text{and} \quad ||x||_G = \begin{bmatrix} |x_1 + x_2| \\ |x_1 - x_2| \\ |x_3| \\ \sqrt{x_4^2 + x_5^2} \end{bmatrix}$$

Let $b^1\left(\begin{bmatrix} |x_1 + x_2| \\ |x_1 - x_2| \end{bmatrix}\right) = \begin{bmatrix} (x_1 + x_2)^2 \\ (x_1 - x_2)^2 \end{bmatrix}$, $b^2(|x_3|) = x_3^2$,

and $b^3(\sqrt{x_4^2 + x_5^2}) = x_4^2 + x_5^2$. Then

$$V_1 \geq b_1^1, \; V_2 \geq b_2^1, \; V_3 \geq b_1^2, \; \text{and} \; V_4 \geq b_1^3.$$

Now,

$$D_+V_1(t,x) = 2(x_1 + x_2)(x_1' + x_2')$$

$$= 2(x_1 + x_2)^2(e^{-t} + \sin t) - 2x_4^2(x_1 + x_2)^2$$

$$\leq 2\, V_1(t,x)(e^{-t} + \sin t).$$

Similarly, $D_+V_2(t,x) \leq 2V_2(t,x)(e^{-t} - \sin t)$,

$$D_+V_3(t,x) \leq 2\sqrt{V_3}\, e^{-t},$$

$$D_+V_4(t,x) \leq -2V_4\,.$$

Consider the comparison equation

$$u' = g(t,u) \quad \text{where} \quad u = \begin{bmatrix} u_1 \\ u_2 \\ u_3 \\ u_4 \end{bmatrix} \quad \text{and} \quad t \geq 0,$$

given by

$$(4.2)\qquad \begin{aligned} u_1' &= 2u_1(e^{-t} + \sin t) \\ u_2' &= 2u_2(e^{-t} - \sin t) \\ u_3' &= 2\sqrt{u_3}\, e^{-t} \\ u_4' &= -2u_4\,. \end{aligned}$$

Now the solution $u \equiv 0$ of (4.2) is equistable in u_1 and u_2, equibounded in u_3, and equi-asymptotically stable in u_4.

Applying Theorem 3.4, the solution $x \equiv 0$ of (4.1) is equistably in x_1 and x_2, equibounded in x_3, and equi-asymptotically stable in x_4 and x_5.

REFERENCES

[1] Bailey, F. N., *The application of Lyapunov's second method to interconnected systems*, J. SIAM Control, 3(1966), 443-462.

[2] Bernfeld, S., and Lakshmikantham, V., *An introduction to nonlinear boundary value problems*, Academic Press: New York, 1974.

[3] Grujic, L. T., and Siljak, D. D., *Asymptotic stability and instability of large scale systems*, IEEE Trans. Automatic Control (1973).

[4] Lakshmikantham, V., *Several Lyapunov functions*, Proc. Int. Conf. on Nonlinear Oscillations, Kiev, U.S.S.R., (1970).

[5] Lakshmikantham, V., and Leela, S., *Differential and integral inequalities, Vol. 1*, Academic Press: New YOrk, 1969.

[6] Lakshmikantham, V., and Leela, S., *Global Results and stability of motion*, Proc. Camb. Phi. Soc. 70(1971), 95-102.

[7] Matrasov, V. M., *Vector Lyapunov functions in the analysis of nonlinear interconnected systems*, Sympos. Math. Roma, 6, Mecaniea Nonlineare Stabilita 1970, (1971), 209-262.

[8] Mitchell, Roger W., and Moore, Marion E., *Vector Lyapunov functions and perturbations of nonlinear systems*, University of Texas at Arlington, Arlington, Texas, Technical Report #23.

SIMPLE ANALOGS FOR NERVE MEMBRANE EQUATIONS

Richard E. Plant
University of California at Davis

The Hodgkin-Huxley (1952) equations for the voltage dependent behavior of the squid giant axon in the absence of spatial variation may be written

$$
\begin{aligned}
\dot{V} &= \frac{1}{C}[\bar{g}_{Na}m^3h[V_{Na}-V] + g_K n^4[V_K-V] + \bar{g}_L[V_L-V] + I] \\
\dot{m} &= \frac{1}{\tau_m(V)}[m_\infty(V)-m] \\
\dot{h} &= \frac{1}{\tau_h(V)}[h_\infty(V)-h] \\
\dot{n} &= \frac{1}{\tau_n(V)}[n_\infty(V)-n].
\end{aligned}
\tag{1}
$$

Values for the various functions and parameters are given by FitzHugh (1969). Equations(1) represent the best currently available model for the behavior of excitable tissue and have been successfully modified to describe the vertebrate purkinje fiber (Noble, 1962), the node of Ranvier (Frankenhauser and Huxley, 1964), the *Aplysia* R15 cell (Plant and Kim, 1976) and other preparations.

In an earlier paper (Plant, 1976a) it was shown that for suitably chosen values of the constant $\hat{h}$, equations(1) describe a system whose behavior is very similar to that of the second-order system

$$
\begin{aligned}
\dot{V} &= \frac{1}{C}[\bar{g}_{Na}m_\infty{}^3(V)\hat{h}[V_{Na}-V] + \bar{g}_K n^4[V_K-V] + \bar{g}_L[V_L-V] + I \\
\dot{n} &= \frac{1}{\tau_n(V)}[n_\infty(V)-n].
\end{aligned}
\tag{2}
$$

Specifically, the solutions of equations (2) possess the properties normally associated with nerve membrane equations (e.g. threshold, oscillatory solutions, etc.). Another second-order system which exhibits the behavior associated with the nerve membrane is equations (4) of Plant and Kim (1975), used in an early report to model the *Aplysia* R15 cell when in a bathing medium containing the poison tetrodotosin (TTX). These equations may be written

$$\dot{V} = \frac{1}{C}[g_T[V_{Na}-V] + [g_A s_A(V) z_A(V) + g_P x][V_K-V] + I]$$

$$\dot{x} = \frac{1}{\tau_x}[x_\infty(V) - x]. \qquad (3)$$

Once again, the solutions of equations (3) display the qualitative properties associated with excitable membranes.

The object of the present paper is to compare and classify second-order systems of the type (2) and (3). In order to facillitate this project, we will consider only the simplest possible versions of such systems, discarding all terms in the equations which are not necessary for the retention of nerve-like properties in the solutions. Accordingly, as an approximation to the equations of the Hodgkin-Huxley (HH) type we shall consider the system

$$\varepsilon\dot{v} = c_1 s_1(v)[1 - v] - c_2 wv + I$$

$$\dot{w} = b[s_2(v) - w] \qquad (4)$$

where v represents the normalized voltage, ε is a small positive parameter, b and the c_i are positive constants, and the $s_i(v)$ are sigmoidal s-functions of Plant and Kim (1976). As an approximation to the Plant-Kim (PK) equations we consider the system

$$\varepsilon\dot{v} = c_1[1 - v] - [c_2 z(v) + c_3 w]v + I$$

$$\dot{w} = b[s(v) - w] \qquad (5)$$

where $z(v)$ represents a z-shaped function in the sense of Plant and Kim (1976). We will make use of the theory of constrained differential equations (Takens, 1974) so that we substitute zero for the term $\varepsilon\dot{v}$ in equations (4) and (5) and allow certain discontinuities in the solution. For a complete description of this method see Plant (1976b).

In the HH case we have

$$0 = c_1 s_1(v)[1 - v] - c_2 wv + I \tag{6a}$$

$$\dot{w} = b[s_2(v) - w]. \tag{6b}$$

Solving equations (6a) for w yields

$$w = G_{HH}(v) = \frac{1}{c_2 v}[c_1 s_1(v)[1 - v] + I]. \tag{7}$$

It then follows from an earlier result (Plant, 1976a) that the system (6) will have a unique stable discontinuous limit cycle if the following three conditions are satisfied:

(i) $I < c_1 s_1(0)$

(ii) $c_1 s_1'(v)[1 - v]v - c_1 s_1(v) + I > 0$ if and only if $0 < v_1 < v < v_2 < 1$ for some v_1 and v_2

(iii) $G_{HH}(\hat{v}) = s_2(\hat{v})$ for a unique $v \in (v_1, v_2)$.

Condition (i) implies that $\lim_{v \to 0} G_{HH}(v) = +\infty$, condition (ii) implies that G_{HH} has a unique region of positive slope, and condition (iii) places the equilibrium point of the system in this region. The "phase plane" of system (6) then resembles that of Figure 1.

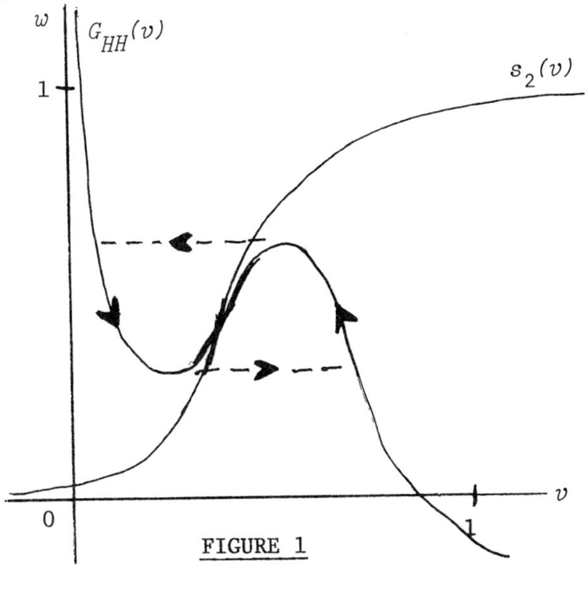

FIGURE 1

A discontinuous limit cycle of the type shown in Figure 1 will be called a simple limit cycle.

Before proceeding to the PK equations, it is interesting to compare equation (6) with the Bonhoeffer-Van der Pol (BVP) equations of FitzHugh (1969). In constrained differential equation form, these may be written:

$$0 = -\frac{v^3}{3} + v - w + I \tag{7a}$$

$$\dot{w} = b\left[\frac{V + a}{b} - w\right]. \tag{7b}$$

Figure 2 shows a sketch of the "phase plane" of this system. It is evident that although equations (6) and (7) are very different in form, their geometry is very similar. This explains why the BVP equations provide such an excellent analog to the HH equations.

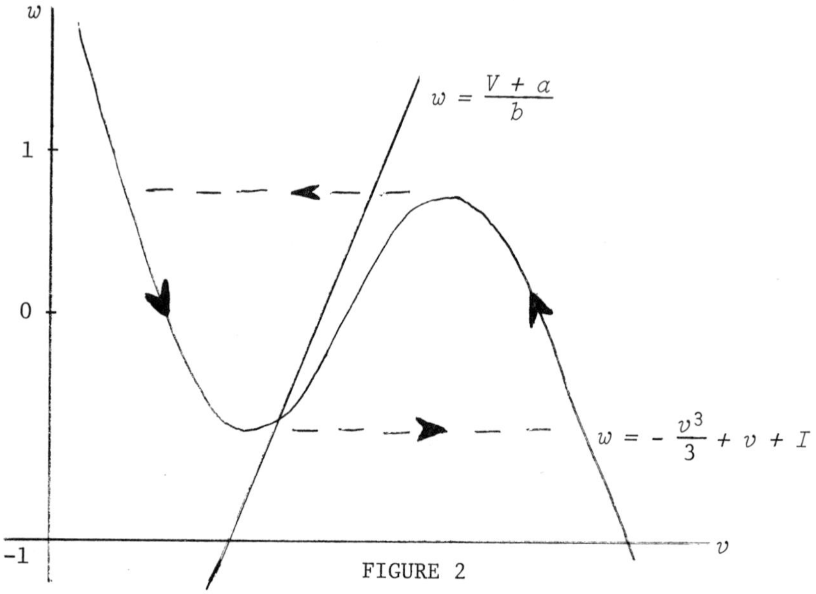

FIGURE 2

Redrawn from Fitzhugh (1969)

The PK equations (5) in constrained form are given by

$$0 = c_1[1 - v] - [c_2 z(v) + c_3 w]v + I \tag{8a}$$

$$\dot{w} = b[s(v) - w]. \tag{8b}$$

Analogously to the HH case, we may solve (8a) for $w = G_{Pk}(v)$ where

$$G_{PK}(v) = \frac{1}{c_3 v}[c_1[1 - v] + I] - \frac{c_2}{c_3} z(v) \tag{9}$$

and obtain the following three conditions for the existence of a simple limit cycle:

(i) $|I| < c_1$.

(ii) $\dfrac{I - c_1}{c_3 v^2} - \dfrac{c_2}{c_3} z'(v) > 0$ if and only if $0 < v_1 < v < v_2 < 1$.

(iii) $G_{PK}(\hat{v}) = s(\hat{v})$ for a unique $\hat{v} \in (v_1, v_2)$.

Comparison of equations (6) and (8) suggest the consideration of the following system:

$$0 = f_1(v,w)[1 - v] - f_2(v,w)v + I \tag{10a}$$

$$\dot{w} = b[s(v) - w]. \tag{10b}$$

Before proceeding we note that any results obtained for this system may also be applied to a system whose steady state function is of the form $z(v)$. Indeed, letting $u = 1 - v$ we obtain

$$s(v) = s(1 - u) = z(u),$$

$$z(v) = z(1 - u) = s(u)$$

and since $\dot{u} = -\dot{v}$, equation (10a) becomes

$$0 = f_2(1 - u,w)[1 - u] - f_1(1 - u,w)u - I \tag{11}$$

yielding

$$0 = \hat{f}_1(u,w)[1 - u] - \hat{f}_2(u,w)u + \hat{I} \tag{12a}$$

$$\dot{w} = b[z(u) - w], \tag{12b}$$

where $\hat{f}_1(u,w) = f_2(1 - u,w)$, $\hat{f}_2(u,w) = f_1(1 - u,w)$ and $\hat{I} = -I$.

Returning to equations (10), we have already considered the following functions f_1 and f_2:

$$HH:\quad f_1(v,w) = c_1 s(v),\quad f_2(v,w) = c_2 w$$

$$PK:\quad f_1(v,w) = c_1,\quad f_2(v,w) = c_2 z(v) + c_3 w.$$

This suggests that we also consider the following systems

$$\overline{HH}:\quad f_1(v,w) = c_1 w,\qquad f_2(v,w) = c_2 q(v)$$

$$\overline{PK}:\quad f_1(v,w) = c_1 r(v) + c_2 w,\quad f_2(v,w) = c_3$$

where q and r may be either s or z functions. It is evident that these four are the total set of simple functions of this form. It is easy to show that the systems $\overline{\text{HH}}$ and $\overline{\text{PK}}$ cannot have simple limit cycle solutions.

We will demonstrate this only for $\overline{\text{HH}}$ since the arguments for $\overline{\text{PK}}$ are identical. The system we must consider is

$$\begin{aligned} 0 &= c_1 w[1 - v] - c_2 q(v)v + I \\ \dot{w} &= b[s(v) - w]. \end{aligned} \tag{13}$$

Letting $u = 1 - v$ yields

$$0 = c_2\hat{q}(u)[1 - u] - c_1 wu - I \tag{14a}$$

$$\dot{w} = b[z(u) - w]. \tag{14b}$$

Solving (14a) for $w = G_{\overline{HH}}(u)$ yields

$$G_{\overline{HH}}(u) = \frac{1}{c_1 u}[c_2\hat{q}(u)[1 - u] - I]$$

and in order to have a simple limit cycle we must have the situation shown in Figure 3.

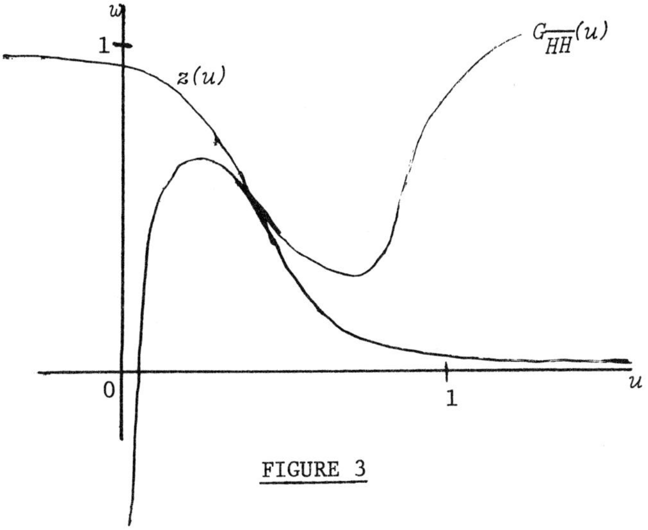

FIGURE 3

From the figure we need

$$\lim_{u \to 0+} G_{\overline{HH}}(u) = -\infty$$

which from (15) implies

$$c_2 \hat{q}(0) - I < 0. \tag{16}$$

In addition, $G_{\overline{HH}}(1)$ must be positive, implying that $-I/c_1$ is positive. Since the c_i are positive, I must be negative. Therefore, from (16), we have

$$c_2 \hat{q}(0) < I < 0. \tag{17}$$

But (17) cannot occur since both c_2 and $\hat{q}(u)$ are always positive. Thus the system $\overline{\text{HH}}$ cannot have a simple limit cycle.

In conclusion, we note that oscillations are possible in systems in which w is governed by a z-type function. Namely, letting $u = 1 - v$ in the original HH and PK equations yields

$$HH: \quad 0 = c_2 w[1 - u] - c_1 z(u) u - I \tag{18}$$

$$\dot{w} = b[z_2(u) - w]$$

$$PK: \quad 0 = [c_2 s(u) + c_3 w][1 - u] - c_1 u - I \qquad (19)$$
$$\dot{w} = b[z(u) - w].$$

Oscillations which are apparently governed by processes such as this have been observed experimentally in the mealworm and eel electroplaque (Grundfest, 1972). Such oscillations are known in the biological literature as potassium spikes and it is evident from (18) and (19) that their existence could have been predicted mathematically.

REFERENCES

[1] FitzHugh, R., *Mathematical models of excitation and propagation in nerve*, Biological Engineering, edited by H. P. Schwan, New York: McGraw-Hill, 1969, 1-85.

[2] Frankenhauser, B., and A. F. Huxley, *The action potential in the myelinated nerve of Zenopus laevis as computed on the basis of voltage clamp data*, J. Physiol 171(1964), 302-315.

[3] Grundfest, H., *The varieties of excitable membrane, Chapter 22*, Biophysics and physiology of Excitable Membranes, W. J. Adelman, ed. New Y'rk, Van Nostrand Reinhold, 1971.

[4] Hodgkin, A. L., and A. F. Huxley, *A quantitative description of membrane current and its application to conduction and excitation in nerve*, J. Physiol. 117(1952), 500-544.

[5] Noble, D., *A modification of the Hodgkin-Huxley equations applicable to purkinje fibre action and pacemaker potentials*, J. Physiol. 160(1962), 317-352.

[6] Plant, R. E., *Geometry of the Hodgkin-Huxley model*, Comp. Prog. Biomed. 6(1976a), 85-91.

[7] Plant, R. E., *Periodic solutions of constrained differential equations*,(to appear, 1976b).

[8] Plant, R. E., and M. Kim, *On the mechanism underlying bursting in the Aplysia abdominal ganglion R15 cell*, Math. Biosci. 26(1975), 357-375.

[9] Plant, R. E., and M. Kim, *Mathematical description of a bursting pacemaker neuron by a modification of the Hodgkin-Huxley equations*, Biophys. J. 16(1976), 227-244.

[10] Takens, F., *Constrained Differential Equations, Dynamical Systems-Warwick*, Springer-Verlag Lecture Notes in Math., vol. 468, A. Dond and B. Eckman, eds., 1974, 80-82, New York.

OSCILLATION RESULTS FOR A NONHOMOGENEOUS EQUATION

Samuel M. Rankin, III
Murray State University

Oscillation properties of solutions of the equation $(*)\ y'' + p(t)y = f(t)$ are studied when the equation $(**)\ u'' + p(t)\mu = 0$ is nonoscillatory. The results obtained are found via the transformation $y(t) = u(t)z(t)$ where $u(t)$ is a solution of $(**)$. This substitution transforms $(*)$ into the equation $(u^2(t)z')' = f(t)u(t)$, thus enabling us to characterize the oscillatory behavior of solutions of equation $(*)$ in terms of the forcing function $f(t)$ and the nonoscillatory solutions of equation $(**)$. The need for "explicit" sign conditions on $p(t)$ is eliminated. Typical results are:

<u>THEOREM</u>. If there exists a positive solution $u(t)$ of equation $(**)$ such that for each $T > 0$ and for some $M > 0$

(i) $$\underline{\lim_{t\to\infty}} \int_T^t f(s)u(s)ds = -\infty \qquad \overline{\lim_{t\to\infty}} \int_T^t f(s)u(s)ds = \infty,$$

(ii) $$\left| \int_T^t 1/u^2(s) \int_T^s f(r)u(r)drds \right| \leq M \int_T^t ds/u^2(s)$$

(iii) $$\lim_{t\to\infty} \int_T^t ds/u^2(s) = \infty,$$ then equation $(*)$ is oscillatory.

<u>THEOREM</u>. If there exists a positive solution $u(t)$ of equation $(**)$ such that for T sufficiently large

(i) $$\varliminf_{t\to\infty} \int_T^t 1/u^2(s) \int_T^s f(r)u(r)\,dr\,ds = -\infty$$

$$\varlimsup_{t\to\infty} \int_T^t 1/u^2(s) \int_T^s f(r)u(r)\,dr\,ds = \infty$$

(ii) $\lim_{t\to\infty} \int_T^t ds/u^2(s) < \infty$, then equation (*) is oscillatory.

THE COMPARISON OF A FOUR-COMPARTMENT AND A FIVE-COMPARTMENT MODEL OF ROSE BENGAL TRANSPORT THROUGH THE HEPATIC SYSTEM

Shelley I. Saffer[1], Patricia L. Daniel[1], & Charles E. Mize[2]

Introduction

In many cases, the differentiation among the various types of hepatic disease is a most difficult, if not impossible, task. In the search for techniques that will enable a more conclusive differentiation among physiological abnormalities of the hepatic system, a previously studied four-compartment model of Rose Bengal transport has been expanded to a five-compartmental model. This model is described as a system of coupled difference equations, the solution of which involves non-linear programming techniques. As in the previously studied four-compartmental model, the results are further analyzed as a Markov chain.

Background

It is often very difficult to differentiate the various factors contributing to liver disease. Because the liver performs a multitude of critical tasks involving metabolic functions, vascular functions and secretory functions, an abnormality within the liver could manifest itself as perhaps another physiological dysfunction or vice versa. Also contributing to the difficulty in the differential diagnosis of liver disease is the complicated physical organization of the liver. The liver is a very porous structure consisting of many small lobules which contain sinusoids and tubules. Some of these minute passageways collect arterial blood from the arterial circulation and venous blood from the lower stomach and intestinal region. Other passageways return blood to the venous circulation. Also present within the liver

2 Department of Pediatrics, University of Texas Health Science Center, Dallas, Texas 75235.

1 Department of Medical Computer Science, University of Texas Health Science Center, Dallas, Texas 75235.

lobule are small ducts that collect bile secretion from the liver cells. These small ducts empty into sequentially larger and larger bile ducts which eventually empty into the common bile duct.

Bilirubin Transfer

One of the components of bile is a substance called bilirubin. The pigment biliverdin is chemically reduced to the "insoluble" form of bilirubin which eventually arrives in the liver where it is conjugated into the "soluble" form of bilirubin. Conjugated bilirubin is excreted with the bile or returns to the plasma to be excreted by the kidney. If the conjugated bilirubin is excreted with the bile, it eventually enters the intestine where, through bacterial action, it is transformed into Urobilinogen which exits from the body with the feces giving the feces its usual dark color.

Bilirubin transport is especially important because if the liver is prevented from performing its normal bilirubin conjugation and excretion, an excess of bilirubin may build up in the blood thereby causing the jaundiced, yellow discoloration which is usually associated with liver disease. Such a condition may be indicative of the inability of the liver cells to conjugate bilirubin (a condition more closely associated with hepatitis) or it may be indicative of a blockage within the many passageways in the liver (a condition more closely associated with biliary atresia). Herein lies the principle dilemma in the differential diagnosis of liver disease. Given that liver dysfunction is indicated, it is at times difficult to distinguish between hepatitis or extrahepatic blockage (a blockage within the larger bile ducts in the liver, or in the common bile duct or gall bladder external to the liver) or interhepatic blockage (a blockage in the smaller internal ducts within the liver.) Differential diagnosis is further complicated by the fact that hepatitis way cause swelling which constricts the intrahepatic ducts giving the appearance of atresia.

Rose Bengal

There are a multitude of tests for liver dysfunction, none of which gives a complete and accurate report of liver pathophysiology. Most of these tests are based on either measuring substances in the blood which are related to liver function, or measuring the disappearance rate of injected dyes or radioactive tracers which are removed from the blood by the liver. One of the most popular tracers is radioactive Rose Bengal. This substance acts much like bilirubin in that it is taken up by the liver and excreted in a similar fashion. Upon injection into the blood, Rose Bengal begins to appear in the urine and feces within a few minutes. Thus a measure of radioactivity can be obtained from three sources; blood, urine, and feces.

There are a number of models that have made use of Rose Bengal and similar substances (1,2,3,4). Our previously studied model depicts Rose Bengal transport using a four-compartment scheme (5) which utilizes not only Rose Bengal disappearance in the blood but also Rose Bengal appearance in the urine and feces (Figure 1).

A Four-Compartmental Model (Summary)

The four-compartment model can be described using coupled difference equations:

$$A(t+1) = A(t)*[1 - P12 - P13] + P31*C(t)$$
$$B(t+1) = B(t) + P12*A(t)$$
$$C(t+1) = C(t)*[1 - P31 - P34] + P13*A(t)$$
$$D(t+1) = D(t) + P34*C(t)$$

where $A(0) = 100\%$ $B(0) = C(0) = D(0) = 0\%$
and time (t) goes from 0 to 72 hours in increments of 1 hour.

Amounts in each compartment are stated as a percentage of the whole rather than specific amounts. The Pij's represent a proportion or percentage of the radioactive Rose Bengal that is transferred from compartment i to compartment j during each time interval. The use of percentages for Pij rather than quantitative rates is advantageous when comparing the rates of differ-

FOUR COMPARTMENT MODEL

	P12	P13	P31	P34
Extrahepatic Atresia				
Patient #1	.0147	.2583	.0148	.0002
#2	.0183	.3423	.0581	.0018
Atresia (intermediate)				
Patient #3	.0259	.3703	.0662	.0009
#4	.0251	.3852	.0547	.0010
Hepatitis				
Patient #5	.0425	.5035	.3389	.0411
#6	.0283	.3615	.0690	.0043
#7	.0204	.3422	.0377	.0101
#8	.0523	.5486	.0415	.0135

TABLE 1

FOUR COMPARTMENT MODEL

	Probability in terminating in		Mean time spent in	
	URINE	FECES	BLOOD	LIVER(hours)
Extrahepatic Atresia				
Patient #1	.81	.19	55	949
#2	.64	.36	34	199
Atresia (intermediate)				
Patient #3	.83	.16	32	178
#4	.78	.21	31	216
Hepatitis				
Patient #5	.43	.56	10	13
#6	.56	.43	20	98
#7	.22	.78	10	77
#8	.28	.72	5	23

TABLE 2

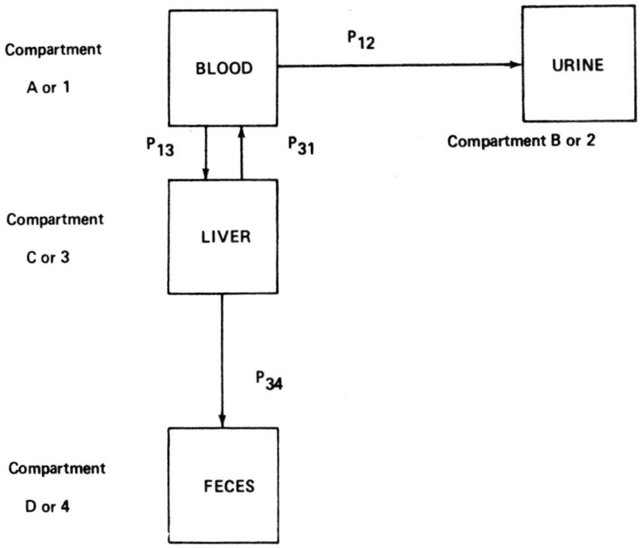

Figure 1. Four compartment model configuration of Rose Bengal transport.

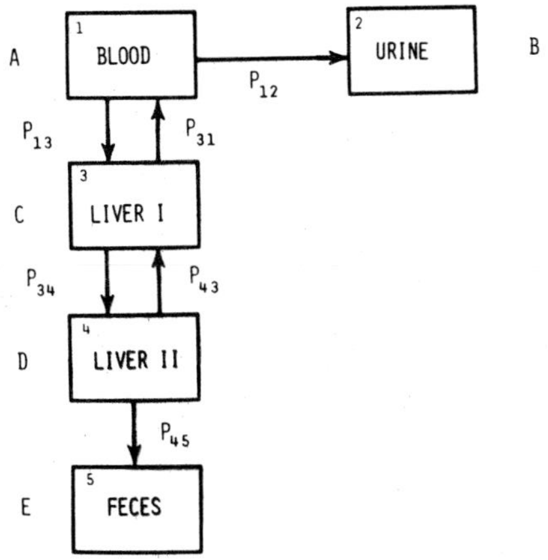

Figure 2. Five compartment model configuration of Rose Bengal transport.

ent patients. It also places a convenient constraint on the P_{ij} parameter:

$$P11 + P12 + P13 = 1 \quad \text{or} \quad 100\%$$
$$P33 + P31 + P34 = 1 \quad \text{or} \quad 100\%$$

Thus there is a 100% probability that a Rose Bengal molecule will remain in the present compartment (state) or travel to a communicating compartment during one time period. These constraints allow the model to be interpreted as a Markov Chain with the states being BLOOD, URINE, LIVER, and FECES and the Pij's being the probability of going from state i to state j in any one time period.

The solution to the above coupled difference equations is accomplished by a "least squares" method which employes non-linear programming techniques (5). A sum of squares objective function is established, each squared term being the difference between an actual patient value and a value generated at the same corresponding time from the above difference equations. Initially, random values are selected for the Pij's (these values must conform to the above constraints). Then amounts in the Blood, Urine, and Fecal compartments are generated and placed with the corresponding patient values of the same time period. The objective function is then evaluated. Next, the non-linear programming algorithms adjust the Pij's seeking to minimize the "sum of squares" objective function (minimizing the difference between the simulated values and actual patient data). A combination of two methods, the Pattern Search method (6) and Gauss Least Squares method (7) is used. Starting with random values for the Pij's, the pattern search method quickly adjusts the Pij's toward a local minimum, and then the Gauss Least-Squares method is utilized to effectively converge to that minimum. This process is repeated several times until it is evident that the global minimum can be distinguished from any local minimum that has appeared. In practice, either the Pij's are adjusted toward the edge of the feasible solution space and become infeasible, or else the same global minimum is reproduced over and over again.

A Five-Compartment Model

One compartmental liver study (1) has suggested that the liver acts as two interacting compartments, an intrahepatic compartment and an extrahepatic compartment. Thus, in an effort to improve the diagnostic capability of the previous 4-compartment model, the liver compartment was expanded into two compartments with a resulting 5-compartment configuration (Figure 2). Because of the generalized nature of the non-linear programming algorithm (which was used to solve the 4-compartment system), the only necessary change was an extension of the objective function to include the system of equations for the 5-compartment model:

$$A(t+1) = A(t)*[1 - P12 - P13] + P31*C(t)$$
$$B(t+1) = B(t) + P12*A(t)$$
$$C(t+1) = C(t)*[1 - P31 - P34] + P13*A(t) + P43*D(t)$$
$$D(t+1) = D(t)*[1 - P43 - P45] + P34*C(t)$$
$$E(t+1) = E(t) + P45*D(t)$$

where $A(\emptyset) = 100\%$ $\quad B(\emptyset) = C(\emptyset) = D(\emptyset) = E(\emptyset) = \emptyset\%$

$$P11 + P12 + P13 = 1$$
$$P33 + P31 + P34 = 1$$
$$P44 + P43 + P45 = 1$$

As with the 4-compartment model, the 5-compartment model can be further analyzed as a Markov Chain. The 5-compartment scheme will yield a 5-state Markov Chain with the states being BLOOD, URINE, LIVER 1, LIVER 2, FECES. Thus the Probability Transition Matrix is constructed as:

	U	F	B	L1	L2
U	1	Ø	Ø	Ø	Ø
F	Ø	1	Ø	Ø	Ø
B	P12	Ø	P11	P13	Ø
L1	Ø	Ø	P31	P33	P34
L2	Ø	P45	Ø	P43	P44

FIVE COMPARTMENT MODEL

	P12	P13	P31	P34	P43	P45
Extrahepatic Atresia						
Patient #1	.0241	.9759	.0876	.0199	.0090	.0006
#2	.0225	.9775	.2600	.0766	.0907	.0043
Atresia (intermediate)						
Patient #3	.0298	.9702	.1636	.0032	.7239	.2761
#4	.0278	.9722	.3249	.1006	.0524	.0016
Hepatitis						
Patient #5	.0405	.6834	.6290	.0556	.00001	.1442
#6	.0391	.9606	.1827	.0266	.8320	.1680
#7	.0288	.9712	.3413	.1390	.0209	.0128
#8	.0817	.9183	.1763	.8247	.2695	.0212

TABLE 3

FIVE COMPARTMENT MODEL

	Probability of terminating in		Mean time spent in		
	URINE	FECES	BLOOD	LIVER(hours)	
				intra	extra
Extrahepatic Atresia					
Patient #1	.62	.36	26	290	602
#2	.63	.36	28	104	84
Atresia (intermediate)					
Patient #3	.85	.14	28	168	1
#4	.75	.24	27	80	150
Hepatitis					
Patient #5	.42	.57	10	10	4
#6	.63	.36	16	82	2
#7	.18	.81	6	15	63
#8	.25	.74	3	12	34

TABLE 4

where $P11 = 1 - P12 - P13$

$P33 = 1 - P31 - P34$

$P44 = 1 - P43 - P45$

Partitioning the Transition Probability Matrix, $[R]$ and $[Q]$ are obtained (8):

$$[R] = \begin{vmatrix} P12 & \emptyset \\ \emptyset & \emptyset \\ \emptyset & P45 \end{vmatrix} \qquad [Q] = \begin{vmatrix} P11 & P13 & \emptyset \\ P31 & P33 & P34 \\ \emptyset & P43 & P44 \end{vmatrix}$$

Also obtained is the Fundamental Matrix $[M]$ (whose elements, Mij, are the mean number of times the process is in transient state j having started from state i);

$$[M] = [I - Q]^{-1}$$

and matrix $[F]$; (where I is the identity matrix)

$$[F] = [M]\cdot[R]$$

whose elements Fij give the probability that the process will finally enter Absorbing State j having started from Transient State i.

Procedure

The same non-linear programming method (a combination of Gauss Least Squares and Pattern Search methods) that was used for the 4-compartment model (5) was employed in the 5-compartment model. This procedure, which works successfully with the 4-compartment model, employs the Pattern Search method to obtain a close proximity to a minimal point, with relatively few iterations, and then utilizes the Gauss Least Squares method to quickly converge to that point. To insure the program worked properly for the 5-compartment scheme, test data were generated from the coupled difference equations which describe the five-compartment system, from time $t = \emptyset$ to 72 using arbitrary values for Pij. In turn, this test data were used as input to the non-linear programming procedure adjusted for the five-compartment model. In all cases, the program converged to the correct values of *Pij* showing that the program was capable of converging to a unique solution (thus producing a global minimum) for the model.

Results Using Patient Data

Next, actual patient data was used as input to the model. In the five-compartment system, the Pattern Search Method did eventually converge (although it was much more time consuming than with the 4-compartment model). However, the Gauss method did not converge because of the wide oscillation in the changes made in the *Pij*'s. Therefore to limit these oscillations, the Levenberg Method, a variation of the Gauss Method was used. The Levenberg Method (9) not only minimizes the sum of squares, but also reduces the change in each parameter such that it ensures the Gauss Method will work. Forming a new expression, which becomes a weighted sum of squares function, the Levenberg method includes the original objective function plus the seuared difference of the value and adjusted value of each parameter *Pij*. The Levenberg Method did converge only when the constraints on the *Pij*'s were relaxed. Unfortunately, one parameter, *P*13, in all cases except one, was infeasible.

It is interesting to note the similarity between the minimum values produced by the Pattern Search Method and those produced by the Levenberg Method. For example, in one case the Pattern Search and Levenberg method yielded respectively: .0241 and .0292 for *P*12; .9759 and 1.57 for *P*13; .0876 and .1096 for *P*31; .0199 and .0119 for *P*34; .0090 and .0063 for *P*43; .0006 and .0008 for *P*45. It appears that the feasible solution space, if interpreted as a hypersurface, slopes toward the infeasible boundary in the direction of *P*13. The direction that the Pattern Search Method takes is toward this infeasible point located by the Levenburg Method. However, the pattern search method stays on the border of the feasible solution space.

Thus, because the Gauss Least Squares method did not converge, and because the Levenberg method produced infeasible results, the Pattern Search method alone was relied upon to produce the Pij values. The "global" minimum was chosen from those values produced by the Pattern Search Method after a number of random starts. We feel less confident, than we did with the 4-compartment model, that the chosen minimum is truly global. However, because most of the minimal points produced by the Pattern Search were close together, and because many of the parameter values were similar to those produced by the Levenberg Method, we feel that the chosen minimum is in close proximity to the global minimum.

Results

The results for the Four and Five Compartment model are given in Tables 1 through 4. Tables 1 and 3 show the Pij values for the Four and Five Compartment models respectively. Tables 2 and 4 show the results of the analysis as a Markov Chain. The mean time in hours that Rose Bengal spends in the Blood and Liver compartments before exiting to an absorbing state is taken from the first row of $[M]$. Table 3 shows the Liver compartment divided into intrahepatic and extrahepatic states. The probability of Rose Bengal terminating in the Fecal or Urine state is taken from $[F]$. This of course is the long-range probability that a Rose Bengal molecule in the system will eventually terminate in the Feces or Urine.

Both the 4-compartment and 5-compartment models appear to be useful in discriminating between Hepatitis and atresia. $P34$, in the 4-compartment model, and $P45$, in the 5-compartment model, are much greater for Hepatitis than for extrahepatic atresia. Hepatitis clearly shows a greater flow from the liver compartment to the fecal compartment than extrahepatic atresia. This is, of course, what one would expect since flow into the extrahepatic passageways would be inhibited in extrahepatic atresia. Further differentiation between Hepatitis and atresia is evidenced in Table 2 and 4. The expected number of time periods that Rose Bengal spends in the liver compartment before exiting to an absorbing state is much greater for atresia. This is a very reasonable result since any blockage would inhibit Rose Bengal from its normal exit to the feces. Hepatitis shows a range from 13 to 98 expected hours for Rose Bengal in the liver (4-compartment model) while the same values for the intermediate and extrahepatic atresia ranges from 178 to 949 hours. Also, the mean time that Rose Bengal spends in the Blood compartment is less in all cases of Hepatitis than in atresia. These ranges, although less dramatic, tend to support the model's ability to reflect a physiological condition and differentiate between the two categories.

The goal of the 5-compartment model, that of gaining insight into the physiological condition within the liver, is not clearly established although it does deserve comment. It is noted from Table 2 and 4 that the distinction between extrahepatic atresia and the intermediate stage is not so obvious as with Hepatitis. The intermediate cases are believed to be originally Hepatitis that progressed to perhaps a degree of cirrhosis and displayed signs compatible with atresia. Thus, these intermediate cases represent the evolution from Hepatitis to atresia and show clinical signs of both conditions. If the intermediate state is in fact a combination of atresia and Hepatitis, then the model might be accurately reflecting the true physiological condition by its inability to produce a clear distinction between the two. Data from the two liver compartments in (Table 4) show a difference in the mean time Rose Bengal was in each compartment. Rose Bengal spent more time in the intrahepatic compartment of patient 2 and 3, but spent more time in the extrahepatic compartment of Patient 1 and 4. Presently, there is no way to accurately compare these results with the true physiological condition. All that can be said is that the model is showing differences in the relationship of these two compartments and more testing will have to be done in order to meaningfully evaluate these relationships and how they apply to each disease condition. However, the information derived still may be clinically useful since clinical steps can be taken at various stages in the disease. It should be noted that the disease progression may not represent a distinct step function but rather a gradual function of time.

Conclusion

The 5-compartment model did confirm the ability of the 4-compartment model to differentiate between Hepatitis and Extrahepatic Atresia. The parameters helpful in making this distinction are $P34$ or $P45$ and the mean time spent in the Blood or Liver compartment.

However, the goal of the 5-compartment model, that of gaining insight into the physiological condition and differentiating between extrahepatic or intrahepatic atresia or the intermediate case, is not clearly established. On one hand, the lack of clinical evidence limits comparison of the model's performance. On the other hand, the lack of a strong convergence of the 5-compartment model (the 4-compartment model did display a strong convergence) leaves doubt as the uniqueness of the Pij values.

Also, because the 5-compartment model always converged on the feasible boundary (except once), one might question the validity of the 5-compartment scheme. $P13$ is driven toward the infeasible boundary because the algorithm, trying to fit the data to the model, must increase $P13$ to beyond its physical limits. This suggests that the compartment configuration may be unable to simulate the true Rose Bengal transport. This further suggests that a 6th compartment, that communicates with the Blood compartment, be added to act as a reservoir and thus allow the Blood compartment data to converge without having to force $P13$ to a maximum. This perhaps will allow the Levenberg method to converge to a unique solution within the feasible region. The 6th compartment would be interpreted as a Body Compartment which takes up and releases Rose Bengal as it is excreted by the kidneys and liver.

A significant point is that such multi-compartmental configurations can conceivably be analyzed with simulation driven by non-linear programming optimization techniques. Also, further analysis as a Markov Chain can yield useful information.

REFERENCES

[1] G. L. Turco, F. Ghemi, G. Molino, and G. Segre, *The kinetics of I-131 Rose Bengal in normal and cirrhotic subjects studied by compartmental analysis and a digital computer*, J. Lab. Clin. Med. 67(1966), p. 983.

[2] A. D. Waxman, P. A. Leins, and J. K. Siemsen, *In vivo dynamic studies of hepatocyte function: A computer method for the interpretation of Rose Bengal kinetics*, Comput. Biomed. Res. 5(1972), p. 1.

[3] K. Winkler, *The kinetics of elimination of Bromosulphalein in man after single injections and continuous infusion*, in The Biliary System (W. Taylor, editor), Philadelphia, Pa.: F. A. Davis Co., p. 551, 1965.

[4] T. G. Richards, *Plasma concentration of Bromasulphalein after single intravenous injections in normal and abnormal human subjects*, The Biliary System (W. Taylor, editor), Philadelphia, Pa. : F. A. Davis Co., p. 567, 1965.

[5] S. Saffer, C. E. Mize, U. N. Bhat, S. A. Szygenda, *Use of non-linear programming and stochastic modeling in the medical evaluation of normal-abnormal liver function*, IEEE Trans. Biomed. Engin. 23(1976), p. 200.

[6] D. J. Wilde, and C. S. Beightler, *Foundations of Optimation*, Englewood Cliffs, N.J.: Prentice Hall, pp. 302-312, 1967.

[7] D. J. Wilde, and S. C. Beightler, *Foundations of Optimation*, Englewood Cliffs, N.J.: Prentice Hall, pp. 298-304, 1967.

[8] U. N. Bhat, *Elements of Applied Stochastic Processes*, New York: John Wiley and Sons, pp. 71-81, 1972.

[9] K. Levenberg, *A method for the solution of certain non-linear problems in Least-Squares*, Quarterly of Applied Math. 2(1944), pp. 164-168.

SOME PROPERTIES OF INCREASING DENSIFYING MAPPINGS

K. L. Singh
Texas A & M University

Introduction

The notion of measure of noncompactness was introduced by C. Kuratowskii (1). It was Darbo (2) who defined the concept of k-set contraction using the notion of measure of noncompactness and proved the following fixed point Theorem (see Theorem 1.1). Darbo's theorem was extended by Furi and Vignoli (3) for densifying mappings. Without being aware of Furi and Vignoli's result Sadovskii (10) also extended the theorem of Darbo, but using different kind of measure of noncompactness (usually called ball measure of noncompactness, see Definition 1.3). Although these two measure of noncompactness share few properties in common (a counter example for the case where they differ may be found in Nussbaum (9), page 127). Nussbaum (9) using the measure of noncompactness that of C. Kuratowskii developed the degree theory for k-set contraction with $k < 1$ and later extended it to densifying mappings. Since the densifying mappings are so general that the generalization of classical fixed point theorems for such kind of mappings is of continuing interest.

In the present paper in section 2, we prove a Theorem for (δ)-increasing densifying mappings, which generalized the Invariance of Domain Theorem. In section 3, we obtained a unique fixed point for uniformly increasing densifying mappings. Finally we have shown that a (ε,∞)-uniformly increasing vectorfield is a bijective mapping.

<u>DEFINITION 1.1</u>. (C. Kuratowskii (1)). Let X be a real Banach space. Let D be a bounded subset of X. The <u>measure of noncompactness</u> of D denoted by $\gamma(D)$ is defined as follows:

$$\gamma(D) = inf\,\{\varepsilon > 0/D \text{ can be covered by a finite number of subsets of diameter} < \varepsilon\}.$$

$\gamma(D)$ has the following properties:

(1) $0 \leq \gamma(D) \leq \delta(D)$, where $\delta(D)$ is the diameter of D.

(2) $\gamma(D) = 0$ if and only if D is precompact (i.e. $\overline{D}$ is compact).

(3) $\gamma(\overline{D}) = 0$ if and only if $\gamma(D) = 0$.

(4) $\gamma(C \cup D) = max\ \{\gamma(C), \gamma(D)\}$.

(5) $C \subset D$ implies $\gamma(C) \leq \gamma(D)$.

(6) $\gamma(C + D) \leq \gamma(C) + \gamma(D)$, where $C + D = \{c + d / c$ in C and d in $D\}$.

(7) $\gamma(S(D,r)) \leq \gamma(D) + 2r$, where $S(D,r) = \{x$ in $X / d\ (x,D) < r\}$.

Closely related to the notion of measure of noncompactness is the concept of k-set contraction introduced by Darbo (2) as follows:

DEFINITION 1.2. Let X be a real Banach space. Let $T : X \to X$ be a continuous mapping. Then T is said to be a k-set contraction if for any bounded but not precompact subset D of X we have

$$\gamma(T(D)) \leq k\gamma(D)$$

for some $k > 0$.

In case $\gamma(T(D)) < \gamma(D)$, for any bounded but not precompact subset D of X such that $\gamma(D) > 0$, then T is called a densifying mapping (Furi and Vignoli (3)).

DEFINITION 1.3. (Sadovskii). Let X be a real Banach space. Let $T : X \to X$ be a continuous mapping. T is said to be condensing if for any bounded but not precompact subset D of X with $\chi(D) > 0$ we have

$$\chi(T(D)) < \chi(D),$$

where $\chi(D)$ denotes the infimum of all real numbers $\varepsilon > 0$ such that D admits a finite ε-net.

DEFINITION 1.4. Let X and Y be two Banach spaces. Let D be a closed and convex subset of X. A mapping $T : D \to Y$ is said to be compact if it is continuous and maps bounded sets into relatively compact sets.

A mapping $T : D \to Y$ is said to be completely continuous if it takes each weakly convergent sequence into strongly convergent sequence.

REMARK 1.1. These two classes of mappings are not comparable. That is neither one is contained in the other. The counter examples demonstrating the difference may be found in Vainberg (21, pp. 14-16). In fact in Vainberg (21) these two mappings have been referred as completely continuous and strongly continuous respectively.

REMARK 1.2. The sum of two k-set contraction is again a k-set contraction. Completely continuous and compact mappings are 0-set contractions. Contraction mapping (a mapping of a closed, bounded and convex subset D of a Banach space into itself satisfying the condition $||T(x) - T(y)|| \leq k||x - y||$, where $k < 1$ and contractive mappings, i.e. $||T(x) - T(y)|| < ||x - y||$) are respectively examples of k-set contraction with $k < 1$ and densifying mappings. The sum of two densifying mappings need not be a densifying mapping (a simple counter example may be found in Singh (12)). However a convex combination of two densifying mappings is always densifying mapping. Moreover the composition of two densifying mappings is always a densifying mapping.

THEOREM 1.1. (Darbo). Let C be a closed, bounded and convex subset of a Banach space X. Let $T : C \to C$ be a k-set contraction with $k < 1$. Then T has a fixed point in C.

DEFINITION 2.1. Let X and Y be two Banach spaces. Let $T : X \to Y$ be a densifying mapping. T is said to be densifying vectorfield on X, provided if T can be expressed in the form

$$T(x) = x - F(x),$$

where $F : X \to Y$ is densifying.

Following Granas (4, pp.32) we have the following definition:

DEFINITION 2.2. Two densifying vectorfields A and B in $L(Y^X)$, (where $L(Y^X)$ is the set of all densifying vectorfields defined on X with values in Y) are said to be homotopic, provided there exists a homotopy $h(x,t) : X \times I \to Y$ between A and B which can be represented by

$$h(x,t) = x - H(x,t),$$

where the mapping $H(x,t) : X \times I \to Y$ is densifying.

DEFINITION 2.3. Two densifying vectorfields T and F mapping X into itself are said to be strongly homotopic if there exists a mapping $H(x,t) : \overline{D} \times I \to X$ such that

(1) $H(x,0) = T(x)$ and $H(x,1) = F(x)$ for every x in $\overline{D}$, where D is bounded open subset of X.

(2) $H(x,t) \neq 0$ for any x in ∂D and t in I, where ∂D denotes the boundary of D.

DEFINITION 2.4. Let X be a real Banach space. Let D be an open, bounded subset of X. A mapping $T : \overline{D} \to X$ is said to be (δ)-increasing at y in D if the following two conditions are satisfied.

(1) $||x|| < \delta$ implies $x + y$ in $\overline{D}$.

(2) $T(y + x) - T(x) = \alpha x$ and $x \neq 0$, then $\alpha > 0$.

THEOREM 2.1. Let X be a real Banach space. Let D be an open, bounded subset of X. A densifying vectorfield $T : D \to X$ is (δ)-increasing at y in D if and only if the mapping $T(y + x) - T(y) = T_y(x)$ is strongly homotopic to the identity mapping I by the homotopy $H(x,t) = (1 - t)x + tT(y)$ on every open subset B of D.

PROOF: Let us assume that T is (δ)-increasing at y. Clearly T_y is a densifying vectorfield. First we note that T_y is continuous, since T is continuous. Now let A be any bounded but not precompact subset of D, then by the definition of T_y we have

$$\{T_y(A) + T(y)\} \subset \{T(y + A)\}.$$

Therefore

$$\gamma\{T_y(A) + T(y)\} \leq \gamma(T_y(A)) + \gamma(y) \leq \gamma(y + A) < \gamma(y) + \gamma(A).$$

(since T is densifying)

Hence

$$(T_y(A)) < \gamma(A) \qquad \text{(since } \gamma(y) = 0).$$

Therefore $I - T_y$ is densifying vectorfield. Clearly for I and T_y the condition (1) of definition 2.3 is satisfied. Indeed,

$$H(x,0) = x = I(x) \quad \text{and} \quad H(x,1) = T_y(x) \quad \text{for every } x \text{ in } D.$$

To finish the proof of the first part of Theorem 2.1, it remains to show that condition (2) of definition 2.3 is also satisfied. Suppose not, that is there exists some bounded set B contained in D such that

$$H(x_0,t_0) = 0 \quad \text{for some } x_0 \text{ in } \partial B \text{ and } t_0 \text{ in } I.$$

Since $x_0 \neq 0$, because it is boundary point of B we must have $t_0 \neq 0$. Indeed, if $t_0 = 0$, then we have

$$H(x_0,t_0) = x_0 = 0$$

a contradiction to the fact that $x_0 \neq 0$. Thus

$$H(x_0,t_0) = (1 - t_0)x_0 + t_0T(x_0) = 0.$$

Hence

$$t_0T_y(x_0) = (t_0 - 1)x_0$$

or

$$T_y(x_0) = \frac{(t_0 - 1)x_0}{t_0} = (1 - (1/t_0))x_0 .$$

Let $\alpha = 1 - (1/t)$, then clearly $\alpha < 0$, a contradiction to the fact that $T_y(x)$ is (δ)-increasing at y.

Conversely let us assume that there exists a bounded, open subset B of D such that the conditions (1) and (2) of definition 2.3 are satisfied. We need to show that T_y is (δ)-increasing. Suppose not, that is T_y is not (δ)-increasing, then we can find a x_0 in $\overline{B}$ such that $T_y(x_0) = \alpha x_0$ and α is not positive. Now

$$T_y(x_0) = \alpha x_0 = (1/t_0)\{H(x_0,t_0) + (t_0 - 1)x_0\}.$$

Let $t_0 = (1/(1 - \alpha))$, then clearly $0 < t_0 < 1$ and for this t_0 we have

$$H(x_0,t_0) = (1 - t_0)x_0 + t_0 T_y(x_0) = (1 - 1/(1 - \alpha))x_0 + (1/(1 - \alpha))\alpha x_0 = (1/(1 - \alpha))\{(1 - \alpha - 1)x_0 + \alpha x_0\} = 0$$

a contradiction to the condition (2) of definition 2.3. Thus the Theorem.

An immediate consequence of Theorem 2.1 we have the following corollary:

COROLLARY 2.1. Let X be a real Banch space. Let D be an open, bounded subset of X. Let $T : \overline{D} \to X$ be a densifying mapping. Furthermore let us assume that T is (δ)-increasing at y in D. Then for any bounded open subset B of D we have

$$Defg(B,T_y,0) = 1.$$

THEOREM 2.2. (Generalized Invariance of Domain Theorem). Let X be a real Banach space. Let D be an open, bounded subset of X. Let $T : \overline{D} \to X$ be (δ)-increasing at every point of an open subset D_1 of D. Moreover let us assume that T is a densifying vectorfield. Then $T(D_1)$ is open.

PROOF: Let us first observe that T is (δ)-increasing at y in D if and only if T_y is (δ)-increasing at zero. Suppose $z = T(x)$, and x in D_1. Since D_1 is open we can select a ball B such that B is contained in D_1 and T is densifying vectorfield on B. We need to show that $T(B)$ contains a neighborhood about z. Since $T_y(x)$ is (δ)-increasing at zero, for any nonzero x_0 in B we have $T_y(x_0) \neq 0$. Indeed, suppose $T_y(x_0) = 0$, $x_0 \neq 0$ in B, then we have $T_y(x_0) = 0.x_0$ which implies that $\alpha = 0$, a contradiction to the fact that $T_y(x)$ is (δ)-increasing at zero. Therefore 0 does not belong to $T_y(\partial B)$.

Now $T_y(\partial B)$ is closed (since T_y according to our assumption is a densifying vectorfield, thus by definition T_y can be written as $T_y = I - G$, where G is densifying moreover $I - G$ is closed by Lemma 1, pp. 80 (9), therefore T_y is closed), therefore there exists C contained in B such that $C \cap T_y(\partial B) \neq \phi$. Let p be any element in $z + C$. Consider the densifying vectorfield

$$g(x) = T(y + x) - p \quad \text{on} \quad B.$$

Clearly $g(x) \neq 0$ for all x in ∂B, since

$$T_y(x) - g(x) = T(y + x) - T(y) - T(y + x) + p = p - T(y)$$
$$= p - z \quad \text{in} \quad C.$$

Define the homotopy $H(x,t) : \overline{B} \times I \to X$ as follows

$$H(x,t) = tg(x) + (1 - t)T_y(x).$$

Clearly $H(x,t)$ being the convex combination of two densifying vectorfields is again a densifying vectorfield. We claim that $H(x,t) \neq 0$ for all x in ∂B and t in I. We have following three cases:

Case 1. If $t = 0$, then $H(x,0) = t_y(x) \neq 0$.

Case 2. If $t = 1$, then $H(x,1) = g(x) \neq 0$.

Case 3. If $0 < t < 1$, then we have

$$||H(x,t)|| = ||t(g(x) - T_y(x)) + T_y(x)|| > t\{||g(x)||\}$$
$$- ||T_y(x)||\} + ||T_y(x)||$$
$$= t||g(x)|| + (1 - t)||T_y(x)|| > 0.$$

Thus by homotopy theorem the degree is constant in t for t in I. Hence

$$Deg\ (H(x,0),B,0) = Deg\ (H(x,1),B,0)$$

or $$Deg\ (g,B,0) = Deg\ (T_y,B,0) = 1,$$ by corollary 2.1.

Therefore there exists x in B such that $g(x) = 0$, i.e. $p = T(y + x)$. Since $y + x$ is contained in $y + B \subset D_1$. Thus the theorem.

DEFINITION 2.5. Let X be a real Banach space. Let D be an open, bounded subset of X. A mapping $T : \overline{D} \to X$ is said to be

<u>uniformly increasing</u> at y in D if there exists an open set B and $\varepsilon > 0$ such that the following conditions are satisfied:

(1) $y + B$ is contained in D.

(2) If there exists a nonzero x in $\overline{B}$ such that $||T_y(x) - \alpha x|| < \varepsilon||x||$, then $\alpha > 0$.

Using Corollary 2.1 now we present a generalization of contraction mapping namely for uniformly increasing mapping T with some other assumptions.

<u>THEOREM 2.3.</u> Let us assume that

(1) The mapping $T : X \to X$ is uniformly increasing at every point y of X and $B = X$ for all y in X.

(2) There exists an element y in X such that $||T(y)|| \leq \varepsilon||y||$.

(3) The mapping $F(x) = x - T(x)$ is densifying on $S(y,r) = \{x$ in $X/||y - x|| \leq r\}$. Then the equation $F(x) = x$ has a unique solution in $S(y,r)$.

<u>Proof</u>: First let us note that if T is uniformly increasing at y, then T is increasing at y. The condition $B = X$ clearly implies the uniqueness of the solution of F. Indeed, let $F(x) = x$ and $F(y) = y$. Then

$$0 = T(x) - T(y) = x - F(x) - y + F(y) = T_y(x - y) = 0.(x - y)$$

a contradiction that T is increasing at y, unless $x = y$.

It remains to show the existence. For the y as in condition (2) of Theorem 2.3 choose the set $S(y,r)$ such that 0 is an interior point of $S(y,r)$. Define the mapping G as follows:

$$G(x) = T(x + y).$$

Then G is densifying mapping on $S(r)$. Indeed, the continuity of G follows from the continuity of T. Let C be any bounded but not precompact subset of $S(x,r)$. Then

$$G(C) = T(C + y).$$

Hence

$$\gamma(G(C)) = \gamma(T(C + y)) < \gamma(T(C)) + \gamma(y) < \gamma(C) + \gamma(y)$$

$$= \gamma(C), \quad \text{since} \quad \gamma(y) = 0.$$

Now define the homopoty $H(x,t) : \overline{S(r)} \times I \to X$ as follows

$$H(x,t) = (1 - t)x + tG(x).$$

Then clearly $H(x,t)$ being convex combination of densifying mappings is densifying. Moreover it can be easily seen that $H(x,t)$ is uniformly continuous in t for t in I. Thus $H(x,t)$ is well defined homotopy.

We claim that $G(x)$ is strongly homotopic to the identity mapping. Indeed, $H(x,0) = x = I(x)$, $H(x,1) = G(x)$. Thus condition (1) of definition 2.3 is satisfied. To show that condition (2) of definition 2.3 is also satisfied we proceed as follows. Suppose not, i.e. $H(x_0,t_0) = 0$ for some x_0 in $\partial S(r)$ and t_0 in I. Now $H(x_0,t_0) = 0$ implies $(1 - t_0)x_0 + t_0G(x_0) = 0$. (A).

From (A) we conclude that $t_0 \neq 0$, otherwise we get $x_0 = 0$, a contradiction that x is a boundary point. Also from (A) we have

$$G(x_0) = ((t_0 - 1)/t_0)x_0 = (1 - (1/t_0)x_0.$$

Finally consider

$$||T_y(x_0) - (1 - (1/t_0))x_0|| = ||T(y + x_0) - (1 - (1/t_0))x_0 - T(y)||$$

$$= ||(1/t_0)\{t_0G(x_0) - (t_0 - 1)x_0\} - T(y)||$$

$$= ||(1/t_0)\{t_0G(x_0) + (1 - t_0)x_0\} - T(y)||$$

$$= ||T(y)|| \leq \varepsilon||y|| < \varepsilon.r = \varepsilon||x||.$$

This implies that $1 - (1/t_0) > 0$, which is impossible. Therefore by Corollary 2.1.

$$Deg\ (T_y,S(r),0) = 1 \neq 0.$$

Thus there exists an element x such that $T_y(x_0) = 0$ or $T(y + x_0) = x_0 + y - F(x_0 + y) = 0$, hence $F(x_0 + y) = x_0 + y$.

Taking the counterpart of definition 2.4 we have

DEFINITION 3.1. A mapping $T : \overline{D} \to X$ is said to be (δ)-increasing at y in D if T satisfies the following two conditions:

(1) $||x|| < \delta$ implies $y + x$ in D;

(2) $T(y + x) - T(y) \neq \alpha x$ if $\alpha < 0$ and $0 < ||x|| < \delta$.

Before we state and prove our Lemma 3.1 and Lemma 3.2, let us make the following remarks.

REMARK 3.1. A mapping T may be (∞)-increasing without being (ε,δ)-uniformly increasing for any $\varepsilon > 0$ and $\delta > 0$. For example, let X be a real Hilbert space with a complete normalized orthogonal sequence $\{e_i\}$. Then the linear mapping

$$T(x) = \sum_{i=1}^{\infty} \alpha_i(x,e_i), \quad \text{where } \alpha_i > 0 \text{ and}$$

$$\lim_{i\to\infty} \alpha_i = 0 \quad \text{is } (\infty)\text{-increasing at } 0,$$

but is not (ε,δ)-uniformly increasing for any $\varepsilon > 0$ and $\delta > 0$.

REMARK 3.2. A mapping T may be (δ)-increasing at every point of an open set, but it is not necessarily one-to-one. However we have the following Lemma:

LEMMA 3.1. If a mapping $T : X \to X$ is (∞)-increasing at every point of X, then T is one-to-one.

Proof: Let x_1 and x_2 be two points in X such that $x_1 \neq x_2$. Since we are in a Banach space we can write $x_1 = x_2 + a$, where $a \neq 0$. We need to show that $T(x_1) \neq T(x_2)$. It is enough to show that $||T(x_1) - T(x_2)|| \neq 0$. Now

$$||T(x_1) - T(x_2)|| = ||T(x_2 + a) - T(x_2)|| \neq |\alpha|\,||a|| \neq 0.$$

Thus $T(x_1) \neq T(x_2)$.

DEFINITION 3.2. A mapping $T : \overline{D} \to X$ is said to be (ε,δ)-uniformly increasing at y in D if T satisfies the following conditions:

(1) $||x|| < \delta$ implies that $x + y$ in D;

(2) $||T(y + x) - T(y) - \alpha x|| > \varepsilon||x||$ if $\alpha < 0$ and

$0 < ||x|| < \delta$.

REMARK 3.3. It is evident that if a mapping $T : \overline{D} \to X$ is (ε,δ)-uniformly increasing at y, then T is (δ)-increasing at y.

LEMMA 3.2. If a continuous mapping $T : X \to X$ is (ε,∞)-uniformly increasing at every point of X, then for any non-positive number α, the range of $T(x) - \alpha x$ is closed.

Proof: Let $H = \{y \text{ in } X/y = T(x) - \alpha x\}$. Let $\{y_n\}$ be a sequence in H such that $y_n \to y$. We need to show that y belongs to H. Now by the definition $y_n = T(x_n) - \alpha x_n$. Taking $y = x_m$ and $x = x_n - x_m$ in condition (2) of definition 3.2 we have

$$||T(x_n) - T(x_m) - \alpha(x_n - x_m)|| > \varepsilon||x_n - x_m||.$$

Hence $$||T(x_n) - T(x_m)|| + |\alpha|||x_n - x_m|| > \varepsilon||x_n - x_m|| \quad (1).$$

Taking $\alpha = -\varepsilon/2$, we clearly have $\alpha < 0$. Thus from (1) we have

$$||y_n - y_m|| > \varepsilon/2||x_n - x_m||.$$

Therefore we conclude that $\{x_n\}$ is a Cauchy sequence. Applying the continuity of T we have $y = \lim_{n\to\infty} y_n = \lim_{n\to\infty} \{T(x_n) - \alpha(x_n)\} = T(x) - \alpha(x)$. Thus y belongs to H.

THEOREM 3.2. Let $T : X \to X$ be a mapping satisfying the following conditions:

(1) T is (ε,∞)-uniformly increasing at every point of X;

(2) T is a densifying vectorfield.

Then T is a homeomorphism.

Proof: Since T is (ε,∞)-uniformly increasing at every point of X, therefore by remark 3.3 we conclude that T is (∞)-increasing at every point of X. Thus from Lemma 3.1 it follows that is one-to-one. Since T is densifying vectorfield, therefore T being the difference of two continuous mappings is a continuous mapping, thus from Lemma 3.2 it follows that $T(X)$ is closed. We only need to show that $T(X)$ is open.

Let y_0 be in $T(X)$, then $y_0 = T(x_0)$ for some x_0 in X.

Therefore there exists an open ball $B(r) = \{x \text{ in } X / ||x|| < r\}$ such that x_0 is contained in $B(r)$. By condition (2), T is densifying vectorfield on the closed ball $\overline{B(r)}$. Thus from Theorem 2.2 we conclude that $T(B(r))$ is an open set in X and evidently y_0 belongs to $T(B(r))$ which is contained in $T(X)$. Thus $T(X)$ is open, hence T must be onto. The continuity of the inverse mapping follows from condition (2) of definition 3.2 by taking $\alpha = 0$. Indeed, taking $\alpha = 0$ the condition (2) of definition 3.2 reduces to

$$||T(x + y) - T(y)|| > \varepsilon ||x|| \quad \text{for} \quad 0 < ||x|| < \infty \qquad (2).$$

Since from above T^{-1} exists, setting $T(x + y) = u$ and $T(y) = v$, we have

$$x + y = T^{-1}(u) \qquad (3).$$

and

$$y = T^{-1}(v) \qquad (4).$$

From (3) and (4) we have $x = T^{-1}(u) - T^{-1}(v)$. Thus from (2) we get

$$||u - v|| > \varepsilon ||T^{-1}(u) - T^{-1}(v)||$$

or

$$||T^{-1}(u) - T^{-1}(v)|| < (1/\varepsilon) ||u - v||.$$

Thus T^{-1} satisfies the Lipschitz condition, hence is continuous.

DEFINITION 4.1. Let $T : X \to Y$ be a mapping of a metric space X onto a metric space Y and let $y_0 \in T(X)$. We say that $T : X \to Y$ is open at the point y_0 if there exists a spherical region $S(y_0, r_0)$ in Y such that $S(y_0, r_0)$ is contained in $T(X)$.

Now let $T : X \to X$ be a mapping of a Banach space X into itself and denote by

$$M(x_0, r) = \sup_{||x - x_0|| \leq r} ||T(x) - T(x_0)||.$$

THEOREM 4.1. Let $T : X \to X$ be a densifying mapping of a Banach space X into itself and let $F(x) = x - T(x)$. If for the point $y_0 \in T(X)$, there exists a point $x_0 \in F^{-1}(y_0)$ such that

$$(1) \qquad \inf_{0<r<\infty} \frac{M(x_0,r)}{r} < 1 ,$$

then $F : X \to X$ is open at the point $y_0 \in F(X)$.

Proof: Let $H(x) = x - (F(x) - y)$. We have

$$
\begin{aligned}
(2) \quad ||H(x) - x_0|| &= ||x - F(x) - x_0 + y|| = ||T(x) - x_0 + y|| \\
&= ||T(x) - T(x_0) + T(x_0) - x_0 + y|| \\
&\leq ||T(x) - T(x_0)|| + ||T(x_0) + y - x_0|| \\
&= ||T(x) - T(x_0)|| + ||y_0 - y||
\end{aligned}
$$

where $y_0 = F(x_0)$. By (1) we can choose r such that for x satisfying $||x - x_0|| \leq r$ there is $||T(x) - T(x_0)|| \leq r_1 < r$. Thus by choosing r_0 such that $0 < r_0 < r - r_1$, we obtain for any y in $S(y_0,r_0)$ that $||H(x) - x_0|| \leq r_1 + (r - r_1) = r$. Thus $H(x)$ maps the closed spherical region $\overline{S}(x_0,r)$ into itself. Since T is densifying, $H(x) = T(x) + y$ is also densifying. Hence by Furi and Vignoli's Theorem (3) we conclude that there exists a point x^* such that $x^* = H(x^*) = x^* - F(x^*) + y$, i.e. $F(x^*) = y$. Hence every point y of $S(x_0,r_0)$ belongs to the image $F(X)$ of X and therefore $F : X \to X$ is open at the point $y_0 \in F(X)$.

REMARK 4.1. If X is finite dimensional Euclidean space and $T : X \to X$ is continuous mapping, then T is completely continuous mapping, and hence a 0-set contraction. Thus as a corollary of Theorem 4.1 we have

COROLLARY 4.1. Let $T : X \to X$ be a continuous mapping of the n-dimensional Euclidean space X into itself and let $F(x) = x - T(x)$. If y_0 in $F(X)$ and if there exists a point x_0 in $F^{-1}(y_0)$ such that $\inf_{0<r<\infty} M(x_0,r)/ r < 1$, then $F : X \to X$ is open at the point y_0 in $F(X)$.

Also as a corollary of Theorem 4.1 we have the following result of M. Reichbach (23).

COROLLARY 4.2. Let $F : X \to X$ be a completely continuous mapping of a Banach space X into itself and let $A(x) = x - F(x)$. If, for some point y_0 in $A(X)$, there exists a point x in $A^{-1}(y_0)$ such that $\inf_{0<r<\infty} M(x,r)/ r < 1$, then $A : X \to X$ is open at the point y_0 in $A(X)$.

If $T : X \to X$ is a locally contractive mapping, i.e. a mapping such that for every point x_0 there exists a spherical region $S(x_0,r_0)$ and a number $\alpha = \alpha(x_0)$ $0 < \alpha < 1$, such that for every point x of $S(x_0,r_0)$ there is $||T(x) - T(x_0)|| \le \alpha||x - x_0||$, then clearly $\inf_{0<r<\infty} M(x_0,r)/ r \le \alpha < 1$ for every x_0, moreover T is densifying. Thus by Theorem 4.1 we conclude:

COROLLARY 4.3. Let $T : X \to X$ be locally contractive mapping of a Banach space X into itself, then $F(x) = x - T(x)$ is open at every point y_0 in $F(X)$ and therefore $F(X)$ is open in X.

REMARK 4.2. A Theorem similar to that of Theorem 4.1 along with the structure of fixed point set for ψ-set contraction with respect to convexity (a concept introduced by J. Eisenfeld and V. Lakshmikantham (24)) will appear somewhere else.

REFERENCES

[1] C. Kuratowskii, *Topology, Volume 1, (Section 34)*, Academic Press, New York, 1964.

[2] G. Darbo, *Punti Unit in Transformazioni a Codomino Noncompactto*, Rend. Sem. Mat. Padova, 24, pp. 84-92 (1955).

[3] M. Furi and A. Vignoli, *On α-nonexpansive Mappings and Fixed Points*, Acad. Nazionale Dei Lincei, Series 8, Volume XLVIII, pp. 131-34 (1970).

[4] Andrezej Granas, *Introduction to the Topology of Functional Spaces*, Lecture Note, University of Chicago, 1961 (Spring).

[5] K. Borsuk, *Theory of Retracts*, Warsaw (1967).

[6] F. E. Browder, *The Solvability of Nonlinear Functional Equations*, Duke Math. Journal, 30, pp. 557-566 (1963).

[7] M. A. Krasnoselskii, *Topological Methods for the Study of Nonlinear Integral Equations*, Pergamon Press, New York, (1964).

[8] W. V. Petryshyn, *Fixed Point Theorems for Various Class of 1-set Contractive and 1-ball Contractive Mappings in Banach*

Spaces, Trans. Amer. Math. Soc. 182, pp. 323-352 (1973).

[9] R. D. Nussbaum, *The Fixed Point Index and Fixed Point Theorems for k-set Contractions*, Ph.D. dissertation, University of Chicago (1969).

[10] B. N. Sadovskii, *A Fixed Point Principle*, Journal of Functional Analysis and Applications, 1, pp. 74-76 (1967).

[11] B. N. Sadovskii, *Limit Compact and Condensing Operators*, Russian Math. Survey, 27, pp. 85-155 (1972).

[12] K. L. Singh, *Fixed Point Theorems for Densifying Mappings I*, The Mathematics Students, Volume XL, No. 3, pp. 283-288 (1972).

[13] K. L. Singh, *Some Applications of Darbo's Theorem*, Acad. Nazionale Dei Lincei, Vol. LVIII, No. 6, June (1975), pp. 880-886.

[14] K. L. Singh, *Eigenvalues of Densifying Mappings*, Acad. Nazionale Dei Lincei, Vol. LVIII, No. 6, June (1975), pp. 34-39.

[15] K. L. Singh, *A Note on The Spectra of Uniformly Increasing Densifying Mappings*, Bull. De 1 Ac. Pol. Des Sc. (Accepted).

[16] K. L. Singh, *Two Theorems Characterizing Increasing k-set Contraction Mappings*, Acad. Nazionale Dei, Lincei (Accepted).

[17] K. L. Singh and S. Srivastava, *Fixed Point Theorems for Densifying Mappings II*, Rivista Di Mat. Univ. Parma (Accepted).

[18] K. L. Singh, *Construction of Fixed Point Theorems for Densifying Mappings*, Rivista Di Mat. Univ. Parma (Accepted).

[19] Sadayuki Yamamuro, *A Characterization of the Increasing Mappings*, Yokohama Math. Journal, Volume XIII, No. 1, pp. 1-5, (1965).

[20] Sadayuki Yamamuro, *On the Derivative of Increasing Mappings*, Yokohama Math. Journal, Vol. XIII, No. 1, pp. 7-9 (1965).

[21] M. M. Vainberg, *Variational Methods for the Study of Nonlinear Operators*, Holden-Day Publishing Company, San Francisco (1964).

[22] M. M. Vainberg, *Variational Method and Method of Monotone Operators*, Halsted Press, New York (1973).

[23] M. Reichbach, *Fixed-points and Openness*, Proc. Amer. Math. Soc. 12(1961), pp. 734-36.

[24] J. Eisenfeld and V. Lakshmikantham, *Remarks on Nonlinear Contraction & Comparison Principle in Abstract Cones*, Technical Report No. 25, June 1975, Univ. of Texas at Arlington.

[25] K. L. Singh, *Some Applications of Kuratowskii's Measure of Noncompactness*, Atti. Accad. Naz. Dei Lincei, V. IX, 1976.

OSCILLATION IN A NON-LINEAR PARABOLIC MODEL OF SEPARATED, COOPERATIVELY COUPLED ENZYMES

H. D. Thames, Jr.*, and D. G. Aronson**

1. *Introduction*

The control of gene expression at the molecular level has been most extensively studied in bacteria, especially with regard to enzyme synthesis and activity. It has been found that bacteria are capable of growing on any of a number of nutrients, since they can synthesize the enzyme appropriate to the catabolism of a carbon source in response to its presence. This phenomenon has been called inductive synthesis. In the Jacob-Monod operon model (1961) an enzyme is induced when the substrate (or one of its analogues) binds and inactivates repressor molecules thus allowing the transcription of the mRNA which codes for the enzyme. The transcription, which is localized at the cell nucleus, is followed by translation, which occurs principally in the vicinity of the cell membrane. The translation of the mRNA into the enzyme consumes the substrate which was initially responsible for the synthesis of the mRNA. Thus induction provides an example of a feedback circuit in which the interacting components are spatially localized.

A similar instance is provided by a metabolic pathway in which the synthesis of an unstable enzyme within the pathway is controlled by one of its ultimate products. In this case the product ("effector" molecule) inhibits synthesis of mRNA coding for the unstable enzyme. To the extent that the latter is translated from available messenger, it "activates" synthesis of the effector molecule. Once again we have a spatially localized feedback circuit, but one in which the negative response occurs at the nucleus, and the positive response in the cytosol.

* Department of Biomathematics, Univ. of Texas System Cancer Center M.D. Anderson Hospital & Tumor Institute, Houston, Texas.

** School of Mathematics, University of Minnesota, Minneapolis, Minnesota.

Since the responses are of positive-negative character, it is natural to inquire whether oscillations in the levels of the components can occur. Periodic enzyme synthesis is observed in proliferating cells. While total cell mass increases roughly linearly in time, synthesis of particular enzymes occurs at discrete times in the cell cycle. It is of some interest to know whether these periodic syntheses result from an endogenous mechanism involving the species in question. In the following we present a model of a localized, two-component feedback circuit which possesses stable periodic solutions. Its application to cellular circadian rhythm has been discussed by Thames and Elster (1976).

II. *Model*

In Figure 1 the active sites of enzymes E_1 and E_2 occur at the surface of the membranes $M_1(x = 0)$ and $M_2(x = \ell)$, respectively, where enzyme E_i is immobilized in membrane M_i. The membranes are open to passage of the substrates of E_1 and E_2, but impermeable to their respective products, U and V. These diffuse between the membranes on the interval $(0,\ell)$, and experience first-order decay there from processes not involving E_1 and E_2. Finally, the rate of production of U by E_1 is modulated by V, and that of V by E_2 is affected by U. After normalization to dimensionless variables the model takes the form

$$\left.\begin{aligned} u_t &= u_{xx} - q^2 u \\ v_t &= v_{xx} - q^2 v \end{aligned}\right\} \quad \text{in} \quad (0,1) \times R^+;$$

$$\left.\begin{aligned} &u_x(0,t) = -pqf(v(0,t)); \; v_x(1,t) = pq(1 - f(u(1,t))) \\ &u_x(1,t) = v_x(0,t) = 0 \end{aligned}\right\} \text{in } R^+. \tag{1}$$

where u and v are the normalized concentrations of U and V, respectively. For simplicity we have taken the decay rate $q > 0$ and the parameter $p > 0$ the same for both products. The function f is assumed to have the form

$$f(w) = w^2/(1 + w^2). \tag{2}$$

Sigmoid response functions of the form (2) have been discussed in connection with other biochemical feedback control circuits by Othmer (1976).

As a model of controlled enzyme synthesis, the model (1) may be interpreted as follows. V plays the rôle of mRNA which translates for the enzyme U. An intermediate or end-product (not explicitly included in the model) of the pathway of which U is a member activates repression of the operon controlling transcription of the U-gene, inhibiting production of V. The "enzyme" E_2 represents the mRNA-synthetase, which transcribes the U-gene and which is localized in the nucleus (represented by M_2 at $x = \ell$). The "enzyme" E_1 represents the protein synthesis machinery localized at the ribosomes (represented by M_1 at $x = 0$).

III. *Stability of Equilibrium State. Periodic Solutions*

Equilibrium solutions of the system (1) have the form

$$\begin{aligned} U(x) &= a \cosh q(1 - x) \\ V(x) &= b \cosh qx \end{aligned} \tag{3}$$

where $a = U(1)$ and $b = V(0)$, and the boundary conditions require

$$\begin{aligned} a &= \lambda f(b) \\ b &= \lambda(1 - f(a)), \end{aligned} \tag{4}$$

where

$$\lambda = p/\sinh q.$$

The function $h(a) = a - \lambda f\{\lambda(1 - f(a))\}$ has the properties: $h(0) < 0$, $h'(a) > 0$, and $h(a) \underset{-}{\sim} a$ as $a \uparrow \infty$. Thus for each $\lambda > 0$ the system (4) has a unique solution $(a(\lambda), b(\lambda))$ which determines the equilibrium solution (3).

Linearization of the system (1) about the equilibrium state equation (3) leads to the linear boundary value problem

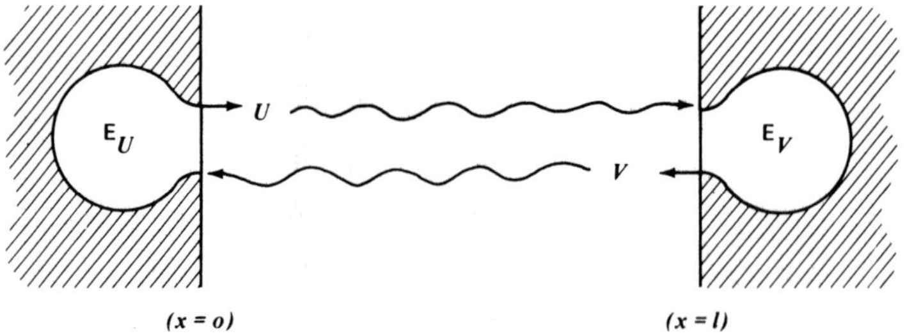

Figure 1. The enzyme E_U (E_1 in text) is bound to the membrane M_1 at $x = 0$. Its product U affects the activity of E_V (E_2 in text), bound to membrane M_2 at $x = \ell$, whose product V affects in turn the activity of E_U.

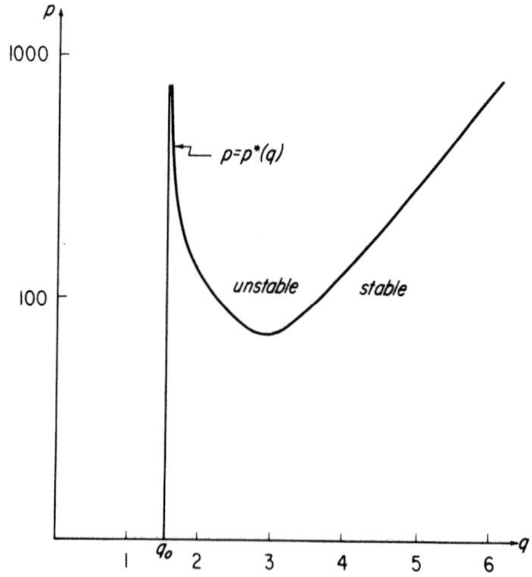

Figure 2. Locus of marginal linear stability, $p^*(q)$. The threshold $q_0 = 1.55973...$

$$\left.\begin{aligned} \xi_t &= \xi_{xx} - q^2\xi \\ \zeta_t &= \zeta_{xx} - q^2\zeta \end{aligned}\right\} \text{ in } (0,1) \times R^+,$$

$$\left.\begin{aligned} \xi_x(0,t) &= -pqf'(b)\zeta(0,t) \\ \zeta_x(1,t) &= -pqf'(a)\xi(1,t) \\ \xi_x(1,t) &= \zeta_x(0,t) = 0 \end{aligned}\right\} \text{ in } R^+, \tag{5}$$

where $\xi = u-U, \zeta = v - V$. Problem (5) admits solutions of the form $\xi = \phi(x)exp(-m^2t)$, $\zeta = \psi(x)exp(-m^2t)$, where $\phi(x) = \alpha\, cos\, \mu(1 - x)$, $\psi(x) = \beta\, cox\, \mu x$, and $\mu^2 = m^2 - q^2$. The numbers α,β are determined from the boundary conditions by means of the homogeneous linear system

$$\begin{aligned} \alpha\, \mu\, sin\, \mu + \beta\, p\, q\, f'\,(b) &= 0\ , \\ \alpha\, p\, q\, f'\,(a) - \beta\, \mu\, sin\, \mu &= 0\ . \end{aligned} \tag{6}$$

This system has a nontrivial solution if and only if

$$\mu\, sin\, \mu = \pm\, i\, \kappa, \tag{7}$$

where

$$\kappa \equiv p\, q\, \sqrt{f'(a)f'(b)}.$$

For the sake of brevity we present only the results of the analysis of the solutions of (7). The details can be found elsewhere (Aronson and Thames, 1976).

Let $\mu = \rho + i\sigma$. For each integer $\nu \geq 0$ equation (7) has a unique conjugate pair of solutions $\rho_\nu \pm i\sigma_\nu$ with $\nu\pi < \rho_\nu < (2\nu + 1)\pi/2$ and $\sigma_\nu > 0$. Moreover, the σ_ν decrease as ν increases. The equilibrium state (3) is said to be (linearly) stable if $Re\ m^2 > 0$, that is, if

$$\sigma_0^2 < \rho_0^2 + q^2. \tag{8}$$

It is said to be unstable if

$$\sigma_0^2 > \rho_0^2 + q^2. \tag{9}$$

There exists a $q_0 \in R^+$ such that (8) holds for each $q \in (0, q_0)$ and all $p \in R^+$. On the other hand, if $q > q_0$ then there exists a $p^*(q) \in R^+$ such that (8) holds for all $p < p^*(q)$ and (9) holds for all $p > p^*(q)$. The graph of $p = p^*(q)$ is shown in figure 2. For each $q > q_0$ the equilibrium state (3) loses its stability as p passes through the critical value $p^*(q)$. This loss of stability occurs because a complex conjugate pair of eigenvalues of problem (5) crosses the imaginary axis and, in fact, they do so with speed $\frac{\partial}{\partial p} \kappa\ (p^*(q), q) \neq 0$. Thus the Hopf bifurcation theorem can be applied to prove the existence of a one-parameter family of periodic solutions (of the nonlinear problem (1)) parameterized by p, at least for $p > p^*(q)$ sufficiently small. Numerical studies indicate that these periodic solutions are stable and that their amplitudes are proportional to $\sqrt{p - p^*(q)}$.

IV. *Discussion*

An interesting question is whether, for realistic values of diffusivities, decay rates, and enzyme maximum turnover rates, the values of p and q that may "naturally" occur can lead to oscillation. The parameter q has the form

$$q = \ell\sqrt{\gamma/D} \tag{10}$$

where ℓ = separation between membranes M_1 and M_2 (Figure 1), γ = first-order decay rate of U and V, and D = diffusion coefficient of U and V. The threshold $q_0 \simeq 1.56$.

First, it is necessary that the species V and U exhibit rapid turnover, with half-lives in the range of a few hours. In this case γ is in the range of a few tenths (sec^{-1}). Since $\ell^2 \sim 10^{-3} cm$, and $q^2 > q_0^2$, for periodic solution we require

$$D \lesssim 10^{-7} cm^2 sec^{-1}. \tag{11}$$

The period (dimensionless) of solutions of the system (1) is roughly independent of p, and has the value $\underset{\sim}{\sim}1$ when $q = 2$. Hence, at the upper limit of (11), $D = 10^{-7} cm^2 sec^{-1}$, the expected period of the solutions (ℓ^2/D) is in the range of a few seconds.

The biological examples of localized feedback circuits discussed at the outset, as well as cellular circadian oscillations (Thames and Elster, 1976), are characterized by periods of tens of hours. These would require values of D four to five orders of magnitude smaller than the upper limit of inequality (11). Physically, what is indicated by such lengthy periods is a substantial delay in transport between M_1 and M_2, delay of a much greater magnitude than can be accounted for by diffusion.

In their processing by and traversal of the nuclear membrane, certain RNA species experience delays of the order of 1/2 hour (Thompson, 1972). If diffusion is made slow enough to mimic such a long delay, the period of the solutions of system (1) is greatly increased. This suggests that the delays imposed by the nuclear membrane on the components of a feedback repression control circuit may be responsible for the long (circadian) periods observed in some rhythms in eukaryotic cells (those possessing a nuclear membrane). Such long periods have never been observed in prokaryotes. The most promising direction for future development of the model presented here is the inclusion of a third component, whose kinetics are localized at the "nucleus" and incorporate the observed time lags in RNA processing more realistically than can be done by lumping them into a pseudo "diffusion" coefficient.

REFERENCES

[1] Aronson, D. G., and H. D. Thames, Jr., *J. Theoret. Biol.*, 1976, submitted.

[2] Jacob, F., and J. Monod, *J. Mol. Biol.* 3(1961), 318-56.

[3] Othmer, H. G., *J. Math. Biol.* 3(1976), 53-78.

[4] Thames, H. D., and A. D. Elster, *J. Theoret. Biol.* 59(1976), 415-27.

[5] Thompson, T. B., *In Temporal Aspects of Therapeutics*, J. Urquhart and F. E. Yates, editors), New York: Plenum Press, 1972.

ITERATIVE TECHNIQUES FOR INVERSION OF THE NONLINEAR PERRIN EQUATIONS FOR DIFFUSION OF SPHEROIDS

A. Kent Wright and John E. Baxter
University of Tennessee Center for the Health Sciences

Experimental studies on the size and shape of macromolecules in solution are often based on hydrodynamic procedures which yield information on the translational and/or rotational diffusion tensor. The methods associated with observation of the translation diffusion tensor are most often expressed as the trace of the tensor which corresponds to the arithmetic mean of three principal translational diffusion coefficients, denoted here as D. On the other hand, methods associated with observation of the rotational diffusion tensor yield estimates of the individual principal rotational diffusion coefficients, which are denoted here as R_1, R_2, and R_3. Since each of the diffusion coefficients is a function of the size and shape of the macromolecule in question, any combination of these four coefficients taken three at a time may in principle be used to solve for three independent shape and size parameters. Usually, these correspond to the semiaxes of an equivalent asymmetric ellipsoid.

A large number of macromolecules yield experimental values of the diffusion coefficients such that within experimental error, two of the coefficients are indistinguishable. As such, we have $R_1 = R_2 \gtrless R_3$. In this case the macromolecule is describable in terms of an ellipsoid of revolution; oblate for the upper inequality and prolate for the lower one. By convention, R_3 corresponds to rotation of the ellipsoid of revolution about the symmetry axis. In this paper we will be solely concerned with ellipsoids of revolution, hereafter referred to as spheroids, and with inversion procedures for two pairs of coefficients: R_1 and R_3, and R_1 and D.

Inversion Procedure for R_1 and R_3

Perrin [1] has related the rotational diffusion coefficients for prolate spheroids in terms of the semiaxes by means of the equations

$$R_1' = F(\varepsilon)/a_3^3$$
$$R_3' = g(\varepsilon)/a_3^3$$

where $$f(\varepsilon) = 2/(2 - \varepsilon^2)(1 - \varepsilon^2) - [(1 + \varepsilon^2)/(2 - \varepsilon^2)]g(\varepsilon)$$
$$g(\varepsilon) = (1/2\varepsilon^3)\ [2\varepsilon/(1 - \varepsilon^2) - ln((1 + \varepsilon)/(1 - \varepsilon))]$$

and the argument is the eccentricity $\varepsilon = (1 - a_1^2/a_3^2)^{1/2}$. The prime indicates that the experimental conditions of temperature and solvent viscosity have been removed by means of the equation $R' = (16\pi\eta/3kT)R$.

Similarly, the Perrin equations for the reduced rotary diffusion coefficients in the oblate case are

$$R_1' = k(\varepsilon)/a_1^3$$
$$R_3' = h(\varepsilon)/a_1^3$$

where $$k(\varepsilon) = \frac{1}{2 - \varepsilon^2}\left[2(1 - \varepsilon^2)^{1/2} - (1 - 2\varepsilon^2)h(\varepsilon)\right] \qquad (2)$$
$$h(\varepsilon) = \frac{1}{\varepsilon^3}\left[Sin^{-1}\varepsilon - \varepsilon(1 - \varepsilon^2)^{1/2}\right]$$

and the argument is the eccentricity $\varepsilon = (1 - a_3^2/a_1^2)^{1/2}$. In the prolate case, we note that $(R_3/R_1) > 1$ is a function only of the eccentricity and in the oblate case $(R_1/R_3) > 1$ is a function only of the eccentricity. Therefore, the ratio of the experimental values determines whether the particle is prolate or oblate.

In both cases, R_1' and R_3' are families of intersecting curves in a_1, a_3 space. Intersection of a given R_1' and R_3' occurs at a unique eccentricity in the prolate case, since

R'_3/R'_1 is a monotonic increasing function. Determination of this eccentricity by a numerical iterative technique in addition with the experimental reduced diffusion coefficients allows evaluation of the semiaxial lengths of the spheroid.

In the oblate case, R'_1/R'_3 shows a maximum 1.256 at $\varepsilon =$ 0.91. For ratio values greater than the maximum, the two diffusion coefficients therefore imply polydispersity, and we consider the case no further. For values less than the maximum, R'_1 and R'_3 intersect at two distinct values, one of which corresponds to the actual case. In order to distinguish which case represents the true dimensions of the spheroid, additional experimental information, such as D, is necessary.

Inversion Procedure for R_1 and D

The Perrin [1] equations for the translational diffusion coefficient for spheroids are

$$D' = d(\varepsilon)/a_3 \tag{3}$$

where $$d(\varepsilon) = ln[(1 + \varepsilon)/(1 - \varepsilon^2)^{1/2}]/\varepsilon$$

and where the argument denotes the eccentricity $\varepsilon = (1-a_1^2/a_3^2)^{1/2}$ in the prolate case, and in the oblate case

$$D' = t(\varepsilon)/a_1 \tag{4}$$

where $$t(\varepsilon) = \frac{Tan^{-1}\left[\frac{\varepsilon^2}{1 - \varepsilon^2}\right]^{1/2}}{\varepsilon}$$

and the eccentricity in this case is $(1 - a_3^2/a_1^2)^{1/2}$. The reduced diffusion coefficient is given by $D' = (6\pi\eta/kT)D$ in both cases. The equations for R'_1 in both the prolate and oblate cases are given by Eq. 1 and 2. In the prolate case D'^3/R'_1 increases from 1.5 to 51 as the eccentricity increases from 0 to 1. In the oblate case D'^3/R'_1 decreases from 1.5 to 1.453 and then increases to 2.467 as the eccentricity increases from 0 to 1.

The numerical method of solution for both cases is identical to that presented earlier for $R1$ and $R3$. Therefore, for ratios less than 1.5, two solutions exist for the oblate case, one of which is the true solution. For ratios greater than 1.5 and less than 2.467 two solutions exist, one oblate and one prolate, again with only one being the true solution. For ratios greater than 2.467 and less than 51 a unique solution for the prolate case exists. Therefore, for the first two stated limits, additional experimental data is required for the determination of the true solution.

Discussion

In the study of globular proteins, many of the interesting cases correspond to spheroids of moderate axial ratio. Therefore, observations of the R_1 and R_3 type are more likely to yield a unique solution since the prolate case is uniquely determined by the data, and the oblate case yields two solutions, one of axial ratio less than 0.91 and the other greater than this value. However, since we are mainly concerned with globular proteins of moderate axial ratio, that solution with the smaller axial ratio is more likely to be the true solution.

Specific iterative techniques applied to esperimental data for R_1 and R_3, and R_1 and D are separately discussed by Wright et al. [2], and by Wright and Baxter [3].

For rigid structural models composed of spherical subunits hydrodynamic techniques may be used to test for alternative structures. Bloomfield et al. [4] have presented a method of calculating D in terms of the hydrodynamic forces acting on individual subunits. His method is based on the total force $\vec{F}$ acting on the array of subunits being proportional to the relative velocity of the particle moving through the solvent. The proportionality tensor is defined as the translational frictional tensor, and the trace of the reciprocal tensor is proportional to the diffusion coefficient D. Since $\vec{F} = \sum_{\ell=1}^{n} \vec{F}_\ell$, the total torque N may be calculated by means of the equation $\vec{N} = \sum_{\ell=1}^{n} \vec{N}_\ell$, where $\vec{N}_\ell = \vec{r}_\ell \times \vec{F}_\ell$ & $\vec{r}_\ell$ denotes the vector from the center of diffusion [5] to the ℓth subunit. Assuming the particle to possess only rotational motion, we may substitute for the translational velocities appearing in Bloomfield et al. the values $\vec{u}_\ell = \vec{\omega} \times \vec{r}_\ell$, and $\vec{v}_\ell^{\,0} = 0$, where $\vec{u}_\ell$ denotes the velocity of the ℓth subunit and $\vec{v}_\ell^{\,0}$ denotes the velocity of the unperturbed solvent at the position of the ℓth subunit. By means of the above equation we may obtain the rotational frictional tensor, ζ, defined by the equation $\vec{N} = \zeta \cdot \vec{\omega}$. By symmetry, this tensor is symmetric and as such may be diagonalized, yielding the principal rotational frictional coefficients. The principal rotational diffusion coefficients are proportional to the reciprocal values of these frictional coefficients [6], [7], [8].

REFERENCES

[1] F. Perrin, *Mouvement Brownien d'un Ellipsoide (I). Dispersion Dielectrique Pour des Molecules Ellipsoidales*, J. Phys. Radium 5(1934), 497-511.

[2] A. K. Wright, R. C. Duncan, and K. A. Beekman, *A numerical inversion of the Perrin equations for rotational diffusion constants for ellipsoids of revolution by iterative techniques*, Biophys. J. 13(1973), 795-803.

[3] A. K. Wright, and J. E. Baxter, *A numerical inversion of the Perrin equations for rotational and translational diffusion constants by iterative techniques*, Biophys. J. 16(1976), 931-938.

[4] V. Bloomfield, W. O. Dalton, and K. E. Van Holde, *Frictional coefficients of multisubunit structures, I, Theory*, Biopolymers 5(1967), 135-148.

[5] H. Brenner, *Coupling between the translational and rotational Brownian motions of rigid particles of arbitrary shape, II, General Theory*, J. Colloid Interface Sci. 23(1967), 407-436.

[6] A. K. Wright, and J. E. Baxter, *Theoretical calculation of the rotational frictional drag tensor of multisubunit structures*, Fed. Proc. 35(1976), 1455.

[7] A. K. Wright, and J. E. Baxter, *A comparison of the structure of Bovine, CuZn superoxide dismutase in solution with the crystal structure*, to be published.

[6] J. A. McCammon, and J. M. Deutch, *Frictional properties of nonspherical multisubunit structures, Application to tubules and cylinders*, Biopolymers 15(1976), 1397-1408.

A
B 7
C 8
D 9
E 0
F 1
G 2
H 3
I 4
J 5